High-Dimensional Regression Analysis and Artificial Intelligence

Asaf Hajiyev · Jiuping Xu

High-Dimensional Regression Analysis and Artificial Intelligence

Theory, Methods and Applications

Springer

Asaf Hajiyev
Institute of Control Systems
Ministry of Science and Education
Azerbaijan Republic
Baku, Azerbaijan

Jiuping Xu
Sichuan University
Business School
Chengdu, Sichuan, China

ISBN 978-981-95-2513-3 ISBN 978-981-95-2514-0 (eBook)
https://doi.org/10.1007/978-981-95-2514-0

The original submitted manuscript has been translated into English. The translation was done using artificial intelligence. A subsequent revision was performed by the author(s) to further refine the work and to ensure that the translation is appropriate concerning content and scientific correctness. It may, however, read stylistically different from a conventional translation.

This Springer imprint is published by the registered company Springer Nature Singapore Pte Ltd.
The registered company address is: 152 Beach Road, #21-01/04 Gateway East, Singapore 189721, Singapore

The book is dedicated to the bright memory of the National Leader of Azerbaijan, Heydar Aliyev

Preface

Regression analysis has long been a foundational approach for quantifying relationships among variables, offering both predictive capabilities and interpretable parameter estimates. By explicitly accounting for stochastic variations, it enables researchers to make reliable predictions while drawing meaningful inferences about underlying data-generating processes. However, modern datasets are often characterized by nonlinearity, heterogeneity, and high dimensionality, posing significant challenges to classical regression methods. In response, advanced techniques have been developed to balance model flexibility and parameter interpretability an aspect that remains essential across a wide range of scientific disciplines, including physics, engineering, biology, economics, and medicine.

These challenges become even more critical in the context of artificial intelligence (AI), where data-driven models have grown rapidly in scale and complexity. While AI systems have demonstrated remarkable predictive performance, this success has often come at the expense of interpretability. Regression principles, however, remain highly relevant to modern AI, as they underpin the parameterized modeling of input output relationships a core characteristic of contemporary AI systems. This foundational connection makes regression particularly valuable in efforts to recover parameter-level interpretability, which is increasingly important as AI models become more intricate and opaque.

The decline of parameter interpretability has led directly to the well-recognized black-box problem in AI. Unlike classical regression models, where parameter estimates transparently quantify the contributions of inputs, modern AI models often obscure this information due to parametric opacity and structural complexity. Such opacity can mask hidden biases, induce feedback loops that exacerbate errors, and impede effective oversight and risk management even when model parameters and code are fully available. More broadly, this erosion of transparency undermines accountability and public trust, as emphasized by international organizations. Enhancing parameter interpretability is therefore essential to improving the reliability and trustworthiness of AI systems.

This book is the product of a long collaboration between two academic teams: Prof. Asaf Hajiyev's team at the Institute of Control Systems, Ministry of Science and

Education, the Azerbaijan Republic, and Prof. Jiuping Xu's team at Sichuan University, a key university directly under the Ministry of Education of China. Our partnership, initiated in 2010, brings together complementary expertise. Professor Asaf Hajiyev has made significant contributions to probability theory and mathematical statistics, while Prof. Jiuping Xu has conducted pioneering research in foundational modeling for computer science and management science. Through the integration of these disciplines, we have developed innovative methodologies for evaluating the stability of AI interpretability using regression-based confidence regions, including confidence ellipsoids and confidence bands. This book represents the culmination of years of collaboration, advancing fundamental mathematical modeling to improve parameter interpretability in artificial intelligence and supporting real-world tasks that demand robust and explainable regression frameworks.

In this book, a novel high-dimensional nonlinear regression model is introduced to address, in part, the challenges of evaluating the stability and confidence of large-scale models' interpretability. The book begins by reviewing foundational concepts in regression analysis, and discussing the current state and challenges of AI interpretability. Through an in-depth exploration of linear and nonlinear regression models, the core principles of data-driven linear regression are explained. To enhance the explanatory power of regression models, variable-parameter regression models are further investigated and extended to variable-parameter nonlinear regression models. To handle complex relationships, the Gauss-Newton iterative method is incorporated, ensuring the stability of high-dimensional nonlinear regression. The Confidence Interval-based Credibility Evaluation (CICE) framework is introduced to combine statistical indicators like interval width, center deviation, and accuracy into a single score to assess explanation stability and reliability. Validated through case studies in engineering, finance, and time series prediction, CICE improves model transparency and identifies unstable interpretations. Overall, the book presents a coherent framework for interpretable AI, integrating regression modeling, confidence region construction, and credibility evaluation to enhance interpretability and statistical accountability, fostering more trustworthy AI systems.

This book provides a systematic treatment of regression analysis with a focus on improving parameter interpretability in large-scale models. Chapter 1 introduces the fundamental concepts and theoretical developments of both regression analysis and AI explainability, highlighting their interconnections. Chapter 2 reviews essential probability theory and mathematical statistics, covering random variables, measure spaces, probability distributions, parameter estimation (including least squares and maximum likelihood methods), and asymptotic theory, which serve as the foundation for analyzing model consistency and convergence. Chapter 3 focuses on the effects of correlated errors in linear regression, establishing parameter convergence conditions to ensure the consistency and asymptotic normality of covariance estimators. These results are further extended to nonlinear regression models, where asymptotic properties and confidence interval construction are explored. Chapter 4 introduces variable-parameter regression models and systematically studies M-estimators and generalized regression models within the framework of robust statistics. By addressing non-normal errors and outliers, these methods improve model adaptability. The chapter also establishes the robustness of the generalized regression model through theoretical analysis of covariance estimation. Chapter 5 introduces the Confidence Interval-based Credibility Evaluation (CICE) framework, which integrates multiple statistical indicators into a unified score to assess the stability and reliability of model explanations. Through real-world case studies in engineering, finance, and time series prediction, the effectiveness of CICE in detecting unstable interpretations and enhancing model transparency is demonstrated.

We would like to express our sincere gratitude to all those who contributed to the realization of this book. First and foremost, we extend our deepest thanks to Professor Yuri Belyaev (Umea University, Sweden), Dr. Yanbing Hou, Dr. Jianglin Ma, and Dr. Yuhan Li from Sichuan University, whose dedicated efforts in writing and developing this work were indispensable. Their expertise and commitment have been the foundation of this project. We are also profoundly grateful to Wayne Hu, our editor, whose invaluable guidance and meticulous attention to detail were essential in navigating the publishing process. Wayne facilitated key connections with publishers and provided insightful feedback, ensuring the highest standards of quality for this book. Special thanks are due to Dr. Narmina Abdullayeva, Dr. Turan Mammadov, for their thoughtful advice and significant support, and Dr. Nabathanum Hajiyeva, who created the necessary conditions for the effective preparation of this book. This book is a testament to the collective efforts of many talented individuals, and we are deeply grateful to each of them.

Baku, Azerbaijan
Chengdu, China
May 2025

Asaf Hajiyev
Jiuping Xu

Contents

Acronyms

AI	Artificial Intelligence
ANOVA	Analysis of Variance
AR	Autoregressive Model
AUC	Area Under the Curve
CICE	Confidence Interval-based Credibility Evaluation
Clsprc	Closing price
CSMAR	China Stock Market & Accounting Research
GLMs	Generalized Linear Models
Hiprc	Highest price
ICE	Individual Conditional Expectation
IRLS	Iterative Reweighted Least Squares
LASSO	Least Absolute Shrinkage and Selection Operator
L-BFGS	Limited-memory Broyden-Fletcher-Goldfarb-Shanno
LIME	Local Interpretable Model-agnostic Explanations
Loprc	Lowest price
LSE	Least Squares Estimator
LSTM	Long Short-Term Memory
MAE	Mean Absolute Error
MAPE	Mean Absolute Percentage Error
ML	Machine Learning
MLE	Maximum Likelihood Estimation
MSE	Mean Squared Error
OLS	Ordinary Least Square
Opnprc	Opening Price
PDP	Partial Dependence Plots
RMSE	Root Mean Square Error
ROC	Receiver Operating Characteristic
RSS	Residual Sum of Squares
SHAP	Shapley Additive Explanations
SLR	Simplest Linear Regression
SVM	Support Vector Machines

SVR	Support Vector Regression
TBL-TE	Noise Turbulent Boundary Layer-Trailing Edge Noise
XAI	Explainable AI
XGBoost	eXtreme Gradient Boosting

List of Figures

List of Tables

Chapter 1
Introduction

I would rather discover a single causal relationship than be king of Persia.

— Democritus (B.C.460–B.C.370)

Regression analysis has long served as a cornerstone of scientific inference. Initially developed to extract meaningful patterns from short-term observations and describe variable relationships, it has since evolved into a fundamental tool for modeling uncertainty. However, the classical assumptions underpinning traditional regression models—such as homoscedasticity and repeatable observations—often break down in modern, high dimensional, and complex data environments. This breakdown necessitates the development of more robust and broadly applicable regression frameworks. At the same time, the rise of artificial intelligence has introduced powerful predictive models that often lack interpretability, raising critical concerns about model transparency and reliability. Although local surrogate methods such as Local Interpretable Model-agnostic Explanations (LIME) have become mainstream tools for explaining black-box models, they typically offer limited guarantees in terms of statistical stability. Against this backdrop, this book proposes a novel framework that integrates confidence ellipsoids and confidence bands to bridge statistical inference with AI interpretability. By enabling quantitatively rigorous and comparable explanations, this approach aims to enhance the trustworthiness of model interpretations in both classical and modern contexts.

1.1 Statistical Foundations and Stability Challenges

As one of the earliest tools in statistical inference, regression analysis emerged from the scientific need to extract patterns from limited observational data. Originating from the method of least squares proposed by Gauss and Legendre, classical regression models rely on a set of idealized assumptions—including error independence, homoscedasticity, and repeatable observations—to ensure the stability of estimates

A. Hajiyev and J. Xu, *High-Dimensional Regression Analysis and Artificial Intelligence*,
https://doi.org/10.1007/978-981-95-2514-0_1

and the validity of inference. However, as data environments grow increasingly complex, these assumptions are often violated in real-world applications. Challenges such as single-point observations, heteroscedastic errors, and high-dimensional feature structures expose the limitations of traditional point estimation, leading to substantial uncertainty and instability. In response, this section begins by tracing the historical development of regression, then examines the challenges posed by modern data, and finally introduces a stability-oriented framework based on confidence ellipsoids and confidence bands to enable reliable inference in contemporary settings.

1.1.1 Least Squares and Origins of Regression

The origins of regression analysis can be traced to the scientific pursuit of extracting reliable patterns from limited observational data. In early applications such as astronomy and geodesy, both Gauss and Legendre independently proposed the method of least squares to address the instability inherent in short-term or noisy measurements. This technique laid the foundation for regression modeling and ushered in a new era of statistical inference. This section begins by revisiting the scientific context in which least squares was developed, highlighting the idealized assumptions underpinning the classical framework, and setting the stage for subsequent discussions on the challenges posed by real-world data and the need to reconstruct inference stability.

1.1.1.1 Least Squares Foundations for Regression

The historical development of regression analysis can be traced back to one of the earliest problems in scientific inference: how to extract reliable quantitative relationships from limited or noisy observations. This challenge was especially prominent in fields such as astronomy, where precise measurements were difficult and often inconsistent. In response, statisticians developed the method of least squares—a foundational technique that laid the groundwork for modern regression analysis.

The method was formally introduced by Adrien-Marie Legendre in 1805 and later expanded upon by Carl Friedrich Gauss in 1809. Gauss claimed to have applied the method as early as 1795, but he did not publish his results until years later. While Legendre proposed least squares as a general approach to minimizing observational errors, Gauss applied it specifically to determine the orbit of the asteroid Ceres, significantly advancing its theoretical justification [26, 39, 84].

This early development of the least squares method not only addressed the pressing need to extract reliable estimates from noisy or limited astronomical data, but also laid the foundation for what would become a general and widely applicable statistical technique: regression analysis.

Regression analysis is a statistical technique used to analyze, model, and quantify the relationship between a dependent variable and one or more independent variables. The primary goal is to estimate and interpret this relationship to enable predictions or inferences about the dependent variable based on the values of the independent variables [23]. At its core, regression analysis involves fitting a model to data and estimating the parameters that best represent the relationship between the variables. The simplest form is simple linear regression, which models a linear relationship between two variables using a straight line. Beyond simple linear regression, more advanced regression models can capture complex relationships, including those involving multiple predictors (as in multiple regression) or nonlinear patterns. Many regression techniques, especially linear regression models, can be formulated using matrices and vectors from linear algebra.

This foundational structure allows regression to serve not only as a method for error reduction, but also as a general approach to identifying and quantifying relationships between variables—a direction that would define its central role in modern scientific inference.

1.1.1.2 From Error Minimization to Relationship Modeling

The term "regression" was introduced much later, in the late nineteenth century, by Francis Galton, a British polymath who studied the relationship between the heights of parents and their children [14, 22, 25]. Galton observed that while tall parents tend to have tall children, their children's heights would often regress toward the mean height of the population. This phenomenon, which he termed "regression toward the mean", sparked further developments in the statistical field. Galton's work on regression toward the mean provided initial insights, but it was Karl Pearson and Udny Yule who extended these ideas into a formal statistical framework, developing key concepts such as the correlation coefficient and refining regression analysis methods. Pearson's mathematical formalization of correlation and regression was instrumental in the evolution of modern statistical theory.

Building on Galton's and Yule's work, Karl Pearson, a prominent British statistician, played a crucial role in the development of regression analysis as a formal statistical method. Pearson mathematically defined the correlation coefficient, which quantifies the strength and direction of a linear relationship between two variables. His work laid the foundation for the modern statistical treatment of regression analysis, particularly in its theoretical structure [51]. In addition, Pearson contributed significantly to statistical methodology by advancing techniques such as the chi-square test, principal component analysis, and the development of a broader family of probability distributions [50]. His research provided a more rigorous mathematical framework for understanding variable relationships, shaping many modern regression analysis techniques [48, 49, 67].

As regression analysis developed, its terminology and conceptual framework evolved. The term "regression" was originally coined by Galton to describe the phenomenon of "regression toward the mean", but over time, it came to represent the broader statistical method of modeling relationships between variables. This shift reflected a deeper understanding that variables often exhibit structured, quantifiable relationships rather than being purely independent. Pearson's formalization of correlation and regression further reinforced this perspective, providing a rigorous statistical foundation for analyzing such dependencies [51].

Following Pearson's contributions, regression analysis continued to evolve, becoming a central tool in statistical modeling by the twentieth century. The advent of multiple regression, in particular, enabled the simultaneous analysis of multiple predictors, allowing for more accurate and nuanced models of complex relationships in fields such as economics, social science, and medicine [1, 7, 11, 16, 17, 19, 21, 44, 47]. Today, regression analysis remains a core tool in data analysis and statistical modeling [4, 5, 10, 14, 15, 18, 24, 30, 32, 37, 38, 40–43]. Regression models can be broadly categorized into linear [12, 31, 52, 60, 68, 70] and nonlinear types [6, 8, 31, 56, 58, 59, 63, 72]. Additionally, there exist various specialized regression models, including nonparametric, hybrid, mixed effects, logistic, and spline regression, among others [1, 2, 13, 27, 29, 45, 54, 66, 73, 90, 91].

As regression analysis became more sophisticated, a need arose for rigorous methods to assess model validity and parameter estimation. One of the most pivotal figures in this advancement was Ronald A. Fisher, a British statistician whose contributions revolutionized the field. Fisher's groundbreaking work in the early twentieth century helped solidify the statistical underpinnings of regression analysis. Fisher formally introduced analysis of variance (ANOVA), building upon earlier work on statistical variability. ANOVA became crucial for understanding the variability in regression models and testing the significance of explanatory variables [20]. Fisher's book, Statistical Methods for Research Workers, laid the foundation for modern statistical inference in regression analysis. He developed the maximum likelihood estimation (MLE) method and extended the theory of linear models, particularly in experimental design and statistical inference [3, 20, 36]. Fisher's work established the framework for testing hypotheses in regression models and provided the tools necessary for handling complex data in scientific research. His influence on regression analysis and statistical methods has been immense.

By the mid-twentieth century, regression analysis had expanded beyond simple linear models to accommodate more complex relationships. A major development in this evolution was the introduction of generalized linear models (GLMs), which allowed regression analysis to be applied to a wider range of data types, such as binary and count data, where traditional linear regression models would be inappropriate. John Nelder and Robert Wedderburn formalized GLMs, integrating the concept of a link function into a unified framework [46]. The advent of GLMs enabled statisticians to model data that did not meet the assumptions of linear regression, significantly broadening the applicability of regression analysis in fields such as medicine, economics, and social sciences.

The development of regression analysis has been shaped by significant contributions from pioneering statisticians [90]. From Galton's early observations of biological phenomena to Pearson's mathematical formalization of regression and Fisher's statistical advancements, the field has evolved substantially. Additionally, scholars like U. Yule [71, 72], C.R. Rao [53, 55, 57], Seber [61–65], Huber [30–35], Kendall [74], and other scholars [89] from different continents and countries have played crucial roles in furthering the field. The introduction of generalized linear models, robust regression techniques, and computational methods continues to broaden the scope and applicability of regression analysis. This rich history, combined with the contributions of key figures, has made regression analysis an indispensable tool in studying relationships between variables across numerous disciplines [28, 40, 44, 47, 63, 69, 83].

As regression analysis evolved from its original purpose of minimizing observational errors, it gradually became a foundational tool for identifying and quantifying structured relationships among variables. Today, regression is not merely a computational technique—it represents a general statistical language for formulating hypotheses, testing theories, and guiding empirical research. This transformation marks its establishment as one of the cornerstones of modern scientific inference.

1.1.1.3 Classical Regression Assumptions for Interpretability

Classical regression models rely on a set of well-established statistical assumptions that provide a foundation for inference. Chief among these are: (1) the independence of error terms, (2) homoscedasticity—that is, constant error variance across observations. These assumptions enable analytical derivation of key properties such as unbiasedness, consistency, and minimum variance of estimators, particularly in the context of least squares estimation.

Under these conditions, the variance of the random errors is either assumed to be known or treated as constant but unknown. This facilitates estimating the error variance, often via the mean of squared residuals, and deriving elements of the covariance matrix of the estimated parameters. These calculations are essential for constructing confidence intervals, ellipsoids, and bands, which in turn support inference about the unknown function or model behavior.

Additionally, the assumption of repeated observations at fixed design points is often implicit in the justification of these techniques. It allows for empirical estimation of variability and reduces the reliance on strong parametric assumptions. The classical framework further assumes that error distributions are well behaved—often Gaussian or sub-Gaussian—which ensures desirable asymptotic properties and simplifies the mathematical treatment of estimation and inference.

While these assumptions are foundational to the theoretical success of regression analysis, they are idealized and often difficult to satisfy in complex, real-world datasets. The next section discusses how modern data environments—particularly those involving non-replicable observations, heteroscedasticity, and structured dependence—challenge the validity of classical regression assumptions and highlight the need for more robust inferential approaches.

1.1.2 Breakdown of Classical Assumptions in Real Data

While classical regression models rely on well-defined assumptions to support inference, real-world data environments often present a starkly different landscape. In many practical domains—such as personalized medicine, finance, and engineering systems—measurements at each observation point are typically obtained only once. The assumption of repeatable experiments, central to many traditional inference methods, becomes infeasible. Repeating an experiment may require considerable cost, time, or may simply be impossible due to ethical or physical constraints. As a result, a common scenario in applied regression is that each design point yields no more than one observation, making it extremely difficult to estimate error variability empirically. Some scholars are trying to estimate the variance of random errors [56, 75–79], but it needs additional conditions for variances, and sometimes it costs a lot of money or even is impossible. Hence, it is necessary to create new approaches and methods for investigating regression models where the variances of errors are unknown and different. Regression models where the variances of random errors are different and unknown, and at each point of observation, there is no more than one observation have been investigated in [70, 80–82], where the new universal approaches for calculating the l.s.e. and construction of a confidence band for unknown functions have been provided.

This lack of repeated measurements interacts with another major issue: the structural complexity of error terms. In many modern applications, error variances are not only unknown but also potentially different across observations—introducing heteroscedasticity. Furthermore, the independence assumption of error terms is often violated due to temporal, spatial, or structural dependence among samples. These issues are compounded when working with high-dimensional models, where the number of unknown parameters increases and the underlying stochastic structure becomes more fragile. In such settings, conventional error modeling strategies, such as assuming constant variance or independent noise, often break down.

The combined effect of these realities is that point estimates produced by classical regression methods can become unstable and unreliable. Without adequate control over variance estimation or the ability to empirically assess uncertainty, model-based inference loses its robustness. More critically, the absence of repeated observations and homoscedastic structure undermines the construction of reliable confidence regions—both for parameters and for predicted functions—calling into question the credibility of inferential conclusions.

These challenges underscore the need for more flexible, data-aware modeling frameworks that can accommodate observational asymmetry, error heterogeneity, and structural dependence. The limitations highlighted in this section motivate the development of enhanced tools for inference, particularly those capable of incorporating uncertainty and ensuring stability in the face of imperfect data conditions. This is precisely the focus of the subsequent chapters.

1.1.3 Confidence Regions for Unstable Inference

Classical regression analysis centers around point estimation—inferring a single best-fitting parameter or function from the data. While such estimates are mathematically convenient and often interpretable, they fail to express the uncertainty inherent in statistical modeling. In complex data environments, where measurement noise, model misspecification, or limited observations are present, relying solely on point estimates can lead to unstable or misleading conclusions. Thus, a more rigorous approach to inference must extend beyond point values and include the structure of uncertainty itself.

Confidence structures such as confidence intervals, confidence ellipsoids, and confidence bands provide essential tools for expressing this uncertainty. A confidence ellipsoid describes a region in the parameter space that likely contains the true parameter vector, incorporating both the variance and correlation structure of the estimators. Similarly, a confidence band offers a visual and quantitative depiction of the uncertainty associated with the function predictions across a range of input values. These structures not only reflect the variability in the estimates but also serve as a foundation for quantifying the stability and reliability of inference.

Constructing these confidence regions, however, is far from straightforward in real-world scenarios. In many practical settings, the variances of random errors are unknown and possibly heterogeneous. More critically, when only a single observation is available at each input point—a common situation in domains such as medicine, economics, or remote sensing—the empirical estimation of error variance becomes impossible. Traditional inference procedures often assume known or constant variance, but such assumptions break down under these conditions. Consequently, the construction of meaningful confidence regions requires new strategies that can operate under incomplete, noisy, and non-replicable data structures.

Addressing these limitations demands a shift from classical estimators toward a broader framework that integrates estimation with uncertainty quantification. Motivated by this need, recent developments have focused on establishing the conditions for the existence and convergence of statistical estimators to the true values of unknown parameters, leveraging modern computational tools to deepen the study of regression analysis and offer effective recommendations for practical problems. Particular emphasis is placed on constructing confidence regions for unknown functions, providing rich statistical information, especially regarding the variances of random observation errors. Methods are introduced for constructing confidence ellipsoids and confidence bands in regression models with increasing dimensionality and non-standard error conditions, without requiring additional assumptions on error variances [85–88]. These advances aim not only to restore inferential stability lost in classical settings but also to enable reliable interpretation in modern, data-constrained environments.

The preceding sections have traced the origins of regression analysis, clarified its classical assumptions, and critically examined the mismatch between theoretical ideals and the practical realities of modern data. As demonstrated, point estimates alone are insufficient to support reliable inference, especially when observations are sparse, error structures are complex, and replicability is limited. These limitations have renewed interest in confidence structures—such as ellipsoids and bands—not as novel tools, but as essential components for expressing statistical uncertainty and reinforcing model stability.

However, as regression analysis is increasingly embedded within modern machine learning systems, its role extends beyond estimation and prediction. In these contexts, understanding why a model behaves a certain way becomes as important as understanding what it predicts. This growing demand for interpretability raises new challenges, especially when data conditions are imperfect and model complexity is high. The next section explores the evolving role of regression in artificial intelligence—both as a predictive engine and as a statistical surrogate for explaining black-box decisions—and argues for a stability-aware framework for trustworthy interpretation.

1.2 Regression Models for AI Interpretability

As regression analysis continues to evolve, its role has expanded beyond traditional statistical modeling in the domain of artificial intelligence. In modern AI systems, regression models are used not only for prediction, but also as interpretable surrogates for complex black-box algorithms. Techniques such as LIME and SHAP rely on local regression approximations to explain the output of models like XGBoost, support vector machines (SVM), and deep neural networks. However, this shift from prediction to interpretation introduces a new set of challenges. While predictive accuracy remains important, interpretability now demands its own standards—particularly around the reliability and stability of explanations. As models grow more complex and high-dimensional, local approximations become sensitive to perturbations and data sparsity, raising questions about the consistency and statistical credibility of explanation outcomes. This section explores how regression models are being repurposed as explanatory tools in AI, and why a statistical perspective—rooted in uncertainty quantification—is essential for building trustworthy interpretability frameworks.

1.2.1 Regression in Interpretable Machine Learning

In this section, first, we examine the mathematical parallels between regression and other machine learning models, highlighting shared foundations in function approximation and optimization. Then, we analyze how regression has been adapted to serve interpretive purposes in tools like LIME, which rely on local linear surrogates. Finally, we identify critical structural mismatches that arise when such surrogates are used in high-dimensional, nonlinear contexts—pointing to deeper limitations in their capacity to deliver reliable explanations.

1.2.1.1 Structural Links Between Regression and AI Models

The foundational architecture of many machine learning models shares a deep theoretical connection with classical regression analysis. The neuron model proposed by McCulloch and Pitts in 1943, maps inputs to outputs through weighted summation and a threshold function, resembling the mathematical form of generalized linear models. Regression analysis and machine learning models share a common mathematical foundation. The following discussion deeply explores their connection from three key perspectives: model representation, model optimization, and generalization ability.

(1) Model representation. Both regression analysis and machine learning models adhere to the fundamental framework of function approximation, reflecting a shared mathematical modeling paradigm. At their core, both approaches construct input–output mappings using parametric or piecewise-parametric models: regression analysis relies on predefined linear or nonlinear functions (e.g., polynomial terms, interaction terms) to explicitly formulate mathematical models, with parameters estimated through error minimization. In contrast, many machine learning models—such as tree ensembles, kernel-based methods, or neural networks—construct nonlinear decision functions through structured combinations of simpler components, optimized to best capture the underlying data distribution. Despite differences in structure, both classes of models ultimately aim to approximate the true data-generating process via "input features—function transformation—predicted output".

In terms of feature representation, regression analysis enhances model expressiveness through manually designed feature expansions (e.g., polynomial regression), whereas machine learning models often perform automatic feature transformations. For example, gradient-boosted decision trees can capture feature interactions implicitly, while neural networks extract hierarchical patterns through layer-wise nonlinear transformations. These represent two distinct paths in feature engineering: the former explicitly defines nonlinear relationships, while the latter implicitly learns nonlinear representations from data. Traditional regression also extends its nonlinear modeling capabilities through basis function expansions (e.g., ridge regression with regularization), while machine learning models overcome linear constraints by increasing model depth, allowing nonlinear splits, or using kernel tricks. Both can thus be viewed as progressive extensions from linear to nonlinear spaces. This evolution from simple linear structures to complex model architectures underscores their mathematical commonality: both construct abstract data representations through function composition and parameter optimization, rooted in the same function approximation theory.

(2) Model optimization. Regression analysis and machine learning models share the same mathematical framework in model optimization, both aiming to minimize a loss function and relying on parameter updates driven by error gradients or their surrogates. From an optimization perspective, both approaches define a differentiable (or sub-differentiable) loss function (e.g., mean squared error, MSE) to quantify prediction error and adjust model parameters accordingly. Unregularized linear

regression allows for an analytical solution via the least squares method, while regularized or more complex regression models—like penalized regression or robust loss functions—typically require numerical optimization methods such as gradient descent. Many machine learning models, including XGBoost or deep learning architectures, rely on iterative optimization techniques—gradient boosting, backpropagation, or variants of stochastic gradient descent—to adjust weights or split thresholds. While regression analysis can yield closed-form solutions under certain conditions, machine learning models often employ adaptive optimization algorithms (e.g., Adam, AdaBoost, or Newton boosting) to enhance convergence efficiency and navigate complex loss.

Moreover, optimization processes in both paradigms are rooted in shared principles of gradient-based learning. In regression analysis, polynomial regression employs explicit feature transformations, where gradient computation involves only a single-layer parameter update. In contrast, modern ML models recursively optimize layered or ensemble structures, either through backpropagation in deep architectures or residual updates in boosting frameworks. Despite these architectural differences, both methods follow the same principle: optimizing a prediction function by minimizing error through iterative parameter adjustment.

(3) Generalization ability. Regression analysis and machine learning models adopt similar strategies to mitigate the risk of overfitting, which is akin to "memorization" in machine learning. In regression models, regularization techniques such as ridge regression (L2 regularization) and LASSO (L1 regularization) constrain parameter values to prevent overfitting. Ridge regression shrinks all coefficients proportionally, while LASSO encourages sparsity by driving some coefficients to exactly zero. Machine learning models employ techniques like early stopping, dropout (in deep learning), or pruning (in decision trees), which control complexity by limiting model capacity or penalizing instability. Fundamentally, these approaches promote a balance between flexibility and generalizability, preventing models from capturing spurious correlations in the training data that may hinder their performance on unseen instances.

Despite differences in methodology, both approaches share the same objective of assessing model reliability. Regression analysis often uses statistical tests, such as p-values, to assess the significance of individual parameters, ensuring that only robust relationships are retained in the model. In contrast, machine learning models are typically evaluated through cross-validation, hold-out testing, and theoretical bounds from learning theory, which focus on understanding how well the model generalizes to unseen data. This distinction is analogous to two approaches in education: one focuses on evaluating specific problem-solving steps, while the other assesses the ability to apply learned concepts to new and unfamiliar problems. Although these methods differ in implementation, they ultimately serve the same purpose, ensuring that learned representations extend beyond specific training examples to new, unseen data.

1.2.1.2 Regression for Prediction and Interpretation in AI

In modern machine learning systems, regression remains a fundamental task in supervised learning, especially when the goal is to predict continuous outcomes or perform ranking and scoring. A wide variety of models—including support vector machines (in their regression form), gradient-boosted trees, and deep learning architectures like LSTM—are routinely adapted to perform regression tasks. While these models differ significantly in architecture, they share the core objective of minimizing prediction error over continuous targets, often using loss functions such as mean squared error. In this sense, regression serves as a key functional objective embedded within diverse machine learning pipelines.

Simultaneously, regression models have been repurposed as explanatory tools to approximate the local behavior of more complex, black-box systems. In particular, the LIME (Local Interpretable Model-agnostic Explanations) framework constructs local linear regressors to mimic the decision boundary of models such as deep neural networks, support vector machines, or tree ensembles. These surrogate models are trained on perturbed samples around a target prediction and are expected to provide an interpretable approximations of the underlying model's behavior in a local neighborhood. By design, LIME treats the original model as a black box, making no assumption about its internal structure, and relies solely on input–output queries to build the local regression.

This dual usage of regression—prediction versus explanation—gives rise to fundamentally different evaluation concerns. In predictive modeling, performance is typically assessed by aggregate metrics such as MSE, R^2, or out-of-sample accuracy. The goal is global generalization. In contrast, surrogate regression models for interpretation are judged not by their overall fit to the data, but by their local fidelity and explanatory coherence: how well does the local regressor mimic the original model's behavior in the immediate vicinity of a target instance? Additionally, explanation models should ideally exhibit stability, i.e., minimal sensitivity to small perturbations in input data or sampling noise.

However, using simple linear models as surrogates in high-dimensional, nonlinear environments present serious limitations. The assumption that local linearity holds near the decision boundary of a complex model can be overly optimistic, particularly when interactions and nonlinearities are prevalent. As we will explore in the next section, this tension between model complexity and interpretability raises important questions about the reliability, robustness, and statistical credibility of local explanations.

1.2.1.3 Limits of Linear Surrogates in High-Dimensional Settings

While linear surrogate models such as those used in LIME offer a straightforward and interpretable approximation of black-box decision boundaries, their applicability in high-dimensional and nonlinear settings is inherently limited. LIME operates under the assumption that the complex decision surface of a model can be locally approximated by a linear function within a small neighborhood of the input instance. However, when the underlying model is highly nonlinear—as is often the case with deep networks, ensemble methods, or kernel-based learners—this assumption becomes problematic.

In high-dimensional spaces, local linearity is a weak guarantee. The data is often sparse and fragmented, and small perturbations around a target point may not preserve meaningful neighborhood structure. As a result, the local samples generated for surrogate training may lie on a distorted or disjointed surface, undermining the ability of a linear model to faithfully capture the original model's behavior. This mismatch between the linear surrogate and the true nonlinear function leads to poor local fidelity, even when the surrogate achieves a low training error on the sampled neighborhood.

These structural limitations of linear surrogate models point to a tension between simplicity and representational adequacy. While interpretability favors linear forms, complex machine learning models often make decisions based on nonlinear interactions, threshold effects, and high-order feature compositions. Relying on a linear approximation in such cases risks sacrificing explanatory accuracy for the sake of visual or computational simplicity.

In sum, the use of linear surrogates in high-dimensional, nonlinear settings introduces a structural gap that undermines the validity of local explanations. This motivates the need to go beyond visual interpretability and introduce statistical tools that can assess explanation quality, quantify uncertainty, and stabilize local interpretations. These concerns are the focus of the next section.

The analysis in this section reveals a fundamental tension: while regression offers a computationally efficient and conceptually simple means for local interpretation, its use as a surrogate model introduces significant risks in complex, high-dimensional settings. Linear approximations may fail to capture key decision dynamics, especially in cases involving nonlinear interactions, sparse neighborhoods, or sensitive thresholds.

These structural issues are not just theoretical—they have practical consequences for the stability, reliability, and repeatability of explanations. As we will discuss in the next section, the lack of statistical assessment in current interpretability frameworks makes it difficult to evaluate how much trust should be placed in a given explanation. Understanding and quantifying this uncertainty is essential for building truly trustworthy interpretable AI.

1.2.2 Statistical Credibility of AI Explanations

As interpretable machine learning methods become increasingly central to responsible AI, the ability to assess the statistical reliability of explanations has become a pressing concern. While existing tools like LIME and SHAP offer intuitive and accessible views into model behavior, their interpretive outputs are often presented without any formal measure of uncertainty or robustness. This is particularly problematic in high-dimensional or complex settings, where explanations may vary substantially depending on local sampling, noise, or perturbation design.

This section examines the critical issue of statistical credibility in AI interpretability. We first discuss how current local surrogate methods fail to provide uncertainty quantification, and why this omission undermines explanation stability. We then turn to a bibliometric analysis of the literature, which highlights the disconnect between current methodological trends and the statistical foundations necessary for building trustworthy, explainable AI systems.

1.2.2.1 Lack of Uncertainty Quantification in Local Explanation Models

Despite their widespread adoption, local surrogate-based interpretation methods such as LIME and SHAP lack formal statistical mechanisms for quantifying uncertainty. These methods typically construct an interpretable proxy—often a linear model—to approximate the behavior of a complex black-box model within a localized neighborhood of the input space. However, the resulting explanations are treated as deterministic outputs, with no accompanying assessment of their reliability, variability, or confidence.

This omission is not merely a technical detail—it has significant implications for the trustworthiness of interpretability systems. In practical settings, explanations derived from surrogate models can exhibit considerable instability. When perturbing the same input point multiple times, LIME-like models may yield drastically different feature attributions due to randomness in neighborhood sampling or sensitivity to high-dimensional boundary structures. Similarly, SHAP values, while grounded in cooperative game theory, rely on strong assumptions about feature independence or a specific background distributions that are rarely met in real-world data. Without a way to assess the robustness of these explanations across perturbations, users are left with interpretability results that may appear precise but are in fact fragile and potentially misleading.

Another core limitation is the absence of confidence structures—such as intervals, ellipsoids, or uncertainty bands—that can express the range or variability of an explanation. In regression, such structures are essential for inference, enabling users to distinguish between statistically meaningful and noise-driven patterns. Their absence in explanation models implies that interpretability is treated more as a visualization exercise than a formal inferential process. As a result, model explanations become difficult to compare, validate, or systematically improve.

These issues are particularly acute in high-dimensional, sparse, or nonlinear settings, where the approximation error of local surrogates tends to be large and explanations are highly sensitive to sampling design. In such contexts, interpretability systems must not only produce explanations but also qualify them with appropriate statistical indicators of uncertainty and stability. Without these, explanations cannot support rigorous downstream tasks such as scientific reasoning, clinical decision-making, or policy evaluation.

In summary, the current generation of surrogate-based interpretability methods lacks the statistical indicators needed to evaluate the credibility of their outputs. Addressing this gap requires rethinking interpretability not only as a communicative task, but also as a form of local statistical inference. The next section supports this argument through a bibliometric analysis of the explainable AI literature, which highlights the disconnect between current methodological trends and the statistical rigor necessary for robust explanation.

1.2.2.2 Statistical Trends in Explainability Literature

We conducted a bibliometric analysis of 1038 articles on AI model explainability from the Web of Science Core Collection (2019–2025), using CiteSpace for keyword co-occurrence, clustering, and burst detection. The results reveal a rapidly growing body of research focused on visual explanation, domain applications, and general-purpose interpretability frameworks.

However, these visualizations also expose a notable gap: very few studies explicitly incorporate statistical foundations such as confidence intervals, uncertainty quantification, or robustness assessments into their interpretability methods. As shown in Fig. 1.1 while keywords like "Explainable AI", "Deep Learning", and "Prediction" dominate, terms associated with statistical validation—e.g., "confidence", "variance", or "interval"—are largely absent.

Clustering analysis further reinforces this finding. As seen in Fig. 1.2 mainstream research topics center on "Predictive Models", "Deep Learning", and "Visual Explainability", with heavy emphasis on healthcare, cybersecurity, and human-AI interaction. Yet none of the prominent clusters systematically addresses the mathematical rigor of explanation methods. This reflects an over-reliance on visual clarity and domain adaptation, rather than on statistical inference or interpretive consistency.

These results support the motivation for this book: while explainability is gaining momentum across disciplines, its current trajectory lacks the statistical tools needed to ensure trustworthy interpretation. In particular, confidence-based evaluation, uncertainty-aware explanations, and statistically grounded surrogate models remain underexplored. This deficiency calls for a new class of explanation methods—ones that are not only interpretable but also statistically credible.

1.2.2.3 Common Local Explanation Techniques and Limits

In summary, AI interpretability is generally categorized into intrinsic and post-hoc interpretability. Intrinsic interpretability refers to models that are inherently designed to be interpretable, such as linear regression. However, simple models may fail to capture complex data patterns and often perform worse than more sophisticated models like deep learning. Given the complexity and high dimensionality of AI models and

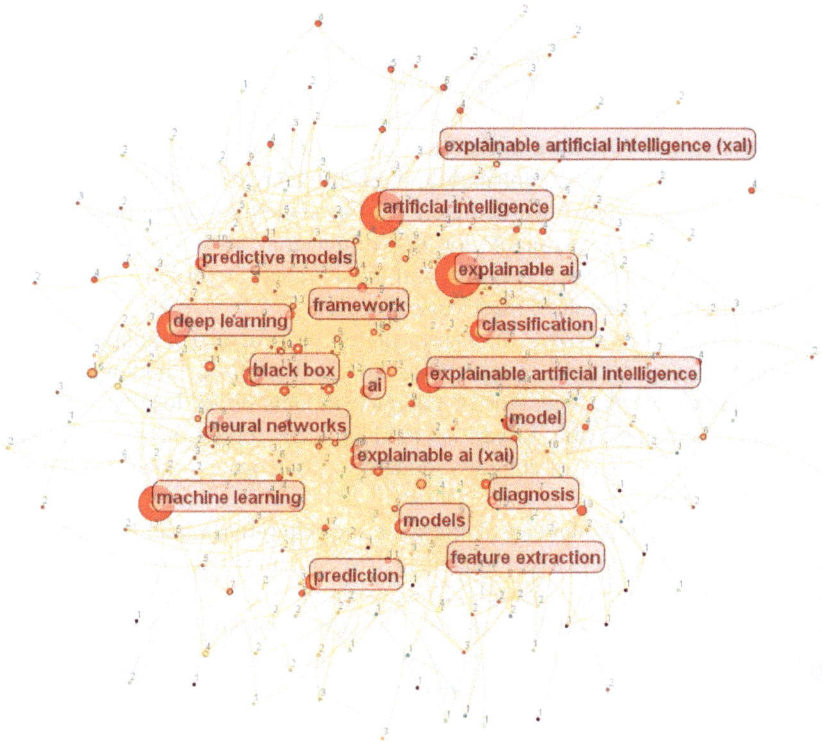

Fig. 1.1 Keyword burst analysis on AI model explainability in Web of Science

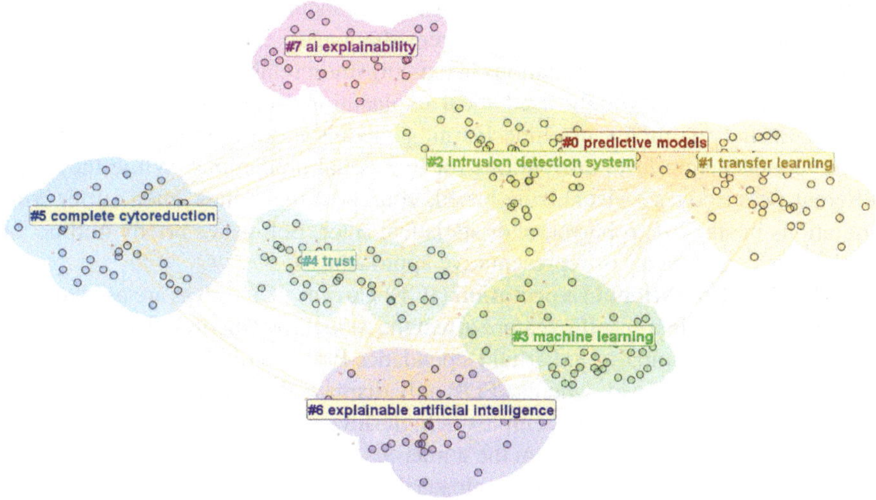

Fig. 1.2 Keyword cluster analysis on AI model explainability in Web of Science

data, post-hoc interpretability is commonly used. Post-hoc interpretability refers to techniques applied after model training to explain complex models. Existing post-hoc methods can be broadly classified into three categories, each with its own strengths and limitations.

The first category is feature importance or weight-based methods, which interpret models by analyzing the contribution of each feature to the prediction, such as through importance scores or learned weights. These methods are intuitive and easy to implement, making them suitable for high-dimensional data. However, they may overlook interactions between features, and the assigned importance scores do not always fully reflect the model's true decision logic.

The second category involves surrogate models or local approximation methods. These approaches use simple, interpretable models (e.g., linear models or decision trees) to approximate the behavior of complex models either locally or globally. While flexible and applicable to various models, local approximations may fail to capture the global behavior of complex models, and surrogate models can introduce additional bias or errors.

The third category includes causal inference and generative model-based methods, which aim to uncover deeper relationships between features and outcomes through causal reasoning or generative models. These methods offer more fundamental explanations, making them particularly useful for causal analysis. However, they often require strong assumptions and additional data support. Moreover, they tend to have high computational costs, and generative models themselves can be complex and difficult to train.

Overall, these methods seek to balance model complexity and interpretability, but each has its own scope and limitations. Current techniques primarily provide descriptive explanations, and human intuition often serves as a bridge between models and reality, potentially introducing harmful biases. Advancing more efficient and precise interpretability techniques remains an urgent challenge.

To better understand the limitations of existing interpretability systems, it is essential to review several widely used local explanation techniques. These methods attempt to make complex machine learning models more interpretable, typically by providing feature-level attributions for individual predictions. While they offer different perspectives—visual, rule-based, counterfactual—they often suffer from limitations in stability, robustness, or statistical rigor. Below, we briefly outline the mechanisms and caveats of representative approaches.

Individual Conditional Expectation (ICE) Curves. ICE curves are used to visualize how changes in a specific feature affect the model's prediction. Each curve represents a sample. Compared to Partial Dependence Plots (PDP), ICE retains information at the individual level, whereas PDP only displays the average trend. However, ICE curves can become overly complex in high-dimensional data, making it difficult to extract meaningful insights from the numerous curves. Therefore, ICE is typically analyzed in conjunction with PDP or clustering methods to facilitate interpretation.

LIME (Local Interpretable Model-agnostic Explanations). LIME generates perturbation samples within the local neighborhood of the target instance and trains an interpretable surrogate model (such as a linear model or decision tree) based on these samples to explain the predictions of the original model. LIME offers strong interpretability, but its explanations can be influenced by the choice of neighborhood and perturbation strategy, potentially leading to instability. Additionally, LIME focuses only on the local region and does not directly contribute to understanding global feature importance.

Anchors. Anchors generate rule-based explanations that satisfy specific precision and coverage constraints. These rules apply only when certain conditions are met, offering more constrained and precise explanations than LIME. However, Anchors requires maintaining high precision while ensuring sufficient coverage, which can result in overly complex or overly conservative rules in high-dimensional data, limiting their scalability.

Counterfactual Explanations. Counterfactual explanations aim to answer the question, "How can the input be modified to cause a specific change in the model's prediction?" By adjusting input features, counterfactuals identify the minimal changes needed to alter the model's output. However, the generated counterfactual samples may violate real-world constraints (e.g., reducing age by 10 years) or fall outside the support of the training data, reducing their feasibility. Thus, additional constraints are needed to ensure the realism and actionability of the explanations.

Influence Functions. Influence functions measure the contribution of training samples to a model's prediction. By calculating the second derivative (Hessian matrix) of the loss function with respect to training samples, they assess the sensitivity of predictions to training data points. Due to the high computational cost of Hessian-based methods, influence functions are difficult to scale to deep models and often require approximate variants, such as stochastic estimation, to be practical.

SHAP (Shapley Additive Explanations). SHAP is based on cooperative game theory and provides a theoretically grounded method for distributing feature attributions. It supports both local explanations (for individual predictions) and global interpretations (for feature importance across the dataset). In local use, SHAP computes the marginal contribution of each feature to a prediction. In global use, it aggregates Shapley values across samples. SHAP also supports multiple visualization tools (e.g., summary plots, dependence plots). However, SHAP is computationally expensive, especially for large models, and requires approximations (e.g., Kernel SHAP, Tree SHAP). Moreover, its assumption of feature independence can bias results when strong feature correlations are present.

Summary. While the above methods offer interpretive insights into complex models, they are not designed to assess the credibility or consistency of their own outputs. Most lack mechanisms for uncertainty quantification, robustness guarantees, or statistical validation.

Together, these theoretical and empirical insights underscore a gap in modern explainability research: the absence of a statistical framework for evaluating the credibility, consistency, and stability of explanations. Without tools to quantify interpretive uncertainty or validate explanation robustness, users are left to trust visual outputs that may be inherently fragile or misleading. In the next section, we build upon these findings by introducing the concepts of stability and consistency in explanation evaluation. These concepts provide the foundation for our proposed statistical framework.

1.2.3 Evaluating Explanation Stability and Trustworthiness

Beyond interpretability itself, a crucial and often overlooked dimension of explainable AI (XAI) is the evaluation of explanations—not only in terms of human readability, but in terms of stability, consistency, and trustworthiness. As model explanations increasingly influence high-stakes decisions in medicine, finance, and public policy, the need for reliable and repeatable interpretive outputs becomes paramount. Yet, most current explanation methods lack the tools needed to quantify how robust an explanation is, how consistent it remains across perturbations, or how much confidence a user can place in its output.

We define stability as the degree to which an explanation remains consistent when the underlying data or sampling conditions are slightly perturbed. A stable explanation should yield similar insights even when local neighborhoods vary, noise is introduced, or redundant features are altered. Meanwhile, trustworthy explanations require more than stability—they demand formal structures to express uncertainty.

Unfortunately, existing methods such as LIME and SHAP provide no standardized mechanism to express or compare these properties. Two explanations generated for the same model and instance may look dramatically different due to randomness in sampling or feature interaction effects, yet no metric exists to assess whether one is more trustworthy than the other. The lack of statistical evaluation metrics make it impossible to validate or optimize the explanation outputs in a principled manner. This stands in sharp contrast to predictive modeling, where notions of generalization error, bias-variance tradeoffs and confidence intervals are standard.

This gap is not merely conceptual—it is foundational. Without quantifiable criteria, explanations cannot be meaningfully evaluated, compared, or improved. More importantly, they cannot be integrated into decision-making pipelines where uncertainty and accountability are essential. As the demands on interpretability grow, so too must the standards for evaluating it.

This book begins from that premise. We argue that explanation systems must not only produce visually interpretable results, but must also support statistically credible inference. We propose a framework that brings the rigor of confidence estimation, uncertainty bands, and stability analysis into the domain of interpretability, offering a pathway toward more trustworthy and quantifiable AI explanations.

1.3 Designing Statistically Credible Explanations

Most existing AI Explainability methods focus on visual clarity or heuristic reasoning, often lacking the statistical rigor necessary to support reliable decision-making. This book proposes a methodological framework that integrates regression-based surrogate modeling with the construction of statistical confidence regions to enable interpretable yet statistically grounded explanations. Our focus is on building local explanation models that not only describe model behavior but also quantify uncertainty and provide formal inferential guarantees.

Rather than relying on black-box explainer systems with opaque internals or unverifiable heuristics, we adopt a transparent approach: approximating complex model predictions using interpretable regression models, and then constructing confidence ellipsoids and bands to represent parameter- and function-level uncertainty. This approach is local by design, but offers mathematically verifiable structures that can support downstream evaluation and comparison.

This section outlines the key goals and methodological contributions of the book. We describe the motivation for using local regression models in high-dimensional and nonlinear settings, explain the interpretive advantages of confidence regions, and summarize the scope, significance, and known limitations of our approach.

1.3.1 Regression Based Local Interpretation

In high-dimensional and nonlinear modeling contexts, constructing stable and interpretable surrogate models remains a core challenge for explainable AI. This book adopts a statistically grounded approach to local interpretability by using regression-based surrogate models to approximate complex predictive behaviors and then constructing confidence regions—such as confidence ellipsoids and confidence bands—around these local approximations. The goal is to offer not only descriptive explanations but also formal statistical structures that allow for uncertainty-aware interpretation.

High-dimensional nonlinear regression models serve as the mathematical backbone of this framework. They offer a powerful mechanism for locally mimicking the behavior of black-box models through explicitly parameterized surrogate models. By capturing the input–output mapping of a target model in a localized region, regression-based surrogates allow us to extract interpretable patterns from complex models that would otherwise be opaque. These patterns, embedded in the parameter structure of the regression model, support both interpretive clarity and statistical testing.

The strength of this approach lies in its mathematical transparency and extensibility. Compared to traditional interpretation techniques based on local approximation heuristics or perturbation-based visualizations, regression models simultaneously capture the independent influence of each input feature and its interactions with other variables. For example, when applied to image classification models, regression coefficients may reflect the relative importance of pixel-level features, while interaction terms can reveal how different visual elements contribute jointly to prediction outcomes. These structural insights enable the modeling of behavioral regularities rather than relying on isolated case-by-case explanations.

In practical applications, this framework offers a unified pathway from interpretability to model optimization. For instance, in feature importance analysis, parameter significance levels can be translated into actionable metrics for feature selection or engineering. In prediction tasks involving complex interactions, higher-order terms in the regression surrogate can uncover implicit decision logic, such as identifying nonlinear association rules in financial risk assessment. Additionally, redundant or low-significance interaction terms can be pruned, providing insights for simplifying model architectures without sacrificing performance. In this sense, regression-based interpretation serves not only as a diagnostic tool but also as a core component in model refinement and iteration.

Overall, high-dimensional regression, when paired with confidence region construction, offers a theoretically grounded and practically effective pathway for interpretable machine learning—particularly in settings where statistical credibility is essential for downstream decision-making.

1.3.2 Practical Significance for AI Explainability

While many interpretability methods provide descriptive insights into model behavior, few offer mechanisms to evaluate the credibility or stability of those insights. The practical significance of this book's approach lies in its ability to transform interpretive outputs into statistically grounded objects—namely, confidence regions—that enable rigorous, quantitative evaluation of explanations.

By constructing confidence ellipsoids (for parameters) and confidence bands (for predicted functions), our framework goes beyond attributing importance: it quantifies the uncertainty surrounding each explanation. This is essential for high-stakes decision-making environments where explanations must be verifiable, comparable, and robust. For instance, in medical diagnostics or financial forecasting, it is not enough to know that a feature is influential—we must also know how confident we are in that assessment, and whether small changes in inputs or training data would lead to radically different conclusions.

These confidence regions provide a bridge between statistical inference and interpretability. They enable hypothesis testing over interpretive claims (e.g., "is this feature significantly contributing to the output?"), allow comparison between different surrogate models or explanation techniques, and support the development of standardized interpretability benchmarks. In settings with multiple competing models or explanation methods, this framework helps identify which explanations are not only informative, but also stable under perturbation and generalizable to nearby data instances.

Moreover, the construction of such regions makes interpretability usable in real-world workflows—not as static visualizations, but as dynamic, testable structures that can be integrated into safety validation pipelines, bias auditing procedures, or regulatory documentation. They enable AI systems to satisfy a growing demand: not only to predict accurately, but to explain transparently, and with measurable confidence.

In sum, confidence-based interpretability marks a paradigm shift—from explaining what a model does, to explaining it in ways that are "statistically accountable, quantitatively comparable, and operationally trustworthy".

1.3.3 Scope and Limits of the Proposed Approach

While the proposed framework introduces statistical rigor to local interpretability, it is not without limitations. Most notably, the methodology remains fundamentally "local" in nature. That is, confidence regions are constructed around specific data instances or small neighborhoods, and the resulting explanations characterize model behavior only within those localized regions. Although this allows for tractable inference and focused explanation, it also implies that the framework cannot guarantee global fidelity or completeness in characterizing the black-box model.

This local restriction is not merely a design choice, but a reflection of current mathematical and computational constraints. Performing global regression over high-dimensional nonlinear models—particularly those with complex decision boundaries—requires either unrealistic assumptions (e.g., full functional form knowledge) or prohibitive computational resources. To date, the field lacks general-purpose methods for statistically valid, scalable, and model-agnostic global explanation. As a result, local approaches like ours represent a pragmatic compromise between interpretability, fidelity, and feasibility.

In addition, the accuracy and reliability of the constructed confidence regions depend heavily on the quality of local sampling and model fit. In high-dimensional settings, generating meaningful perturbation samples near a target instance becomes increasingly difficult due to the "curse of dimensionality". Poor sampling can lead to unstable surrogate models, which in turn affect the accuracy and shape of the resulting confidence ellipsoids or bands. Moreover, the regression model itself assumes certain functional smoothness and structure that may not always be valid in complex real-world black-box models. In such cases, the statistical validity of the explanation may be compromised.

In short, this framework lays the groundwork for statistically credible interpretability but must be understood within its natural scope: "local, instance-based explanation under structural and sampling assumptions". These limitations are not shortcomings of the framework alone, but reflect the broader challenges faced by the explainable AI community in balancing interpretability, scalability, and statistical soundness.

In this book, a novel high-dimensional nonlinear regression model is introduced to address, in part, the interpretability challenges of large-scale models. The book begins by reviewing foundational concepts in AI and discussing the current state and challenges of AI interpretability. Through an in-depth exploration of linear and nonlinear regression models, the core principles of data-driven linear regression are explained. To enhance the explanatory power of regression models, variable-parameter regression models are further investigated and extended to variable-parameter nonlinear regression models. To handle complex relationships, the Gauss–Newton iterative method is incorporated, ensuring the stability of high-dimensional nonlinear regression.

To assess the credibility of explanations derived from these models, the final chapter introduces a unified evaluation framework known as the Confidence Interval-based Credibility Evaluation (CICE). This framework aggregates multiple statistical indicators—such as interval width, center deviation, and accuracy—into a single score to evaluate the stability and reliability of explanations across different settings. The CICE framework is then validated through real-world case studies in engineering, finance, and time series prediction. These applications demonstrate the framework's effectiveness in identifying unstable interpretations and improving model transparency.

Taken together, the methods presented in this book form a coherent framework for interpretable AI grounded in statistical inference. By integrating regression modeling, confidence region construction, and credibility evaluation, the book aims to bridge the gap between interpretability and statistical accountability—paving the way toward more trustworthy, explainable, and auditable AI systems.

References

1. Mithun Kumar Acharjee and Kumer Pial Das. Polynomial spline regression: Theory and application. *arXiv preprint* arXiv 2212.14777, 2022.
2. Abdelmonem A Afifi and Stanley P Azen. *Statistical analysis: a computer oriented approach*. Academic press, 2014.
3. M Rauf Ahmad and S Ejaz Ahmed. A robust approach to statistical quality control for high-dimensional non-normal data. In *Advances and Innovations in Statistics and Data Science*, pages 121–134. Springer, 2022.
4. Theodore Wilbur Anderson and John B Taylor. Strong consistency of least squares estimates in normal linear regression. *The Annals of Statistics*, pages 788–790, 1976.
5. Sam Ansari and Ali Bou Nassif. A comprehensive study of regression analysis and the existing techniques. In *2022 Advances in Science and Engineering Technology International Conferences (ASET)*, pages 1–10. IEEE, 2022.
6. Robert Bartels. Re-interpreting r2, regression through the origin, and weighted least squares. *Univ. Sydney Bus. Sch*, 2015.
7. Alexandre Belloni and Victor Chernozhukov. ℓ_1 - penalized quantile regression in high-dimensional sparse models. 2011.
8. Michael T Braun and Frederick L Oswald. Exploratory regression analysis: A tool for selecting models and determining predictor importance. *Behavior research methods*, 43:331–339, 2011.
9. Matias D Cattaneo, Rajita Chandak, and Jason M Klusowski. Convergence rates of oblique regression trees for flexible function libraries. *The Annals of Statistics*, 52(2):466–490, 2024.

10. Matias D Cattaneo, Nicolás Idrobo, and Rocio Titiunik. A practical introduction to regression discontinuity designs: Extensions. *arXiv preprint* arXiv:2301.08958, 2023.
11. Samprit Chatterjee and Ali S Hadi. *Regression analysis by example*. John Wiley & Sons, 2015.
12. Harold Cramer. Mathematical methods of statistics, princeton univ. *Press, Princeton, NJ*, 27, 1946.
13. Patrick L Davies. Aspects of robust linear regression. *The Annals of statistics*, pages 1843–1899, 1993.
14. P Diaconis and DA Freedman. On inconsistent m-estimators. *Ann. Statist*, 10:454–461, 1982.
15. Timo Dimitriadis, Tobias Fissler, and Johanna Ziegel. Characterizing m-estimators. *Biometrika*, 111(1):339–346, 2024.
16. Hilmar Drygas. *BLUE in linear regression models*, volume 2. Probab. Mathem. Statist. Fasc., 1990.
17. RL Eubank. Approximate regression models and splines. *Communications in Statistics-Theory and Methods*, 13(4):433–484, 1984.
18. Jianqing Fan, Quefeng Li, and Yuyan Wang. Estimation of high dimensional mean regression in the absence of symmetry and light tail assumptions. *Journal of the Royal Statistical Society Series B: Statistical Methodology*, 79(1):247–265, 2017.
19. Mark Finster. Estimation in the general linear model when the accuracy is specified before data collection. *The Annals of Statistics*, 13(2):663–675, 1985.
20. Ronald Aylmer Fisher. Statistical methods for research workers. In *Breakthroughs in statistics: Methodology and distribution*, pages 66–70. Springer, 1970.
21. Candace Flatt and Ronald L Jacobs. Principle assumptions of regression analysis: Testing, techniques, and statistical reporting of imperfect data sets. *Advances in Developing Human Resources*, 21(4):484–502, 2019.
22. David A Freedman. Bootstrapping regression models. *The annals of statistics*, 9(6):1218–1228, 1981.
23. Rudolf J Freund, William J Wilson, and Ping Sa. *Regression analysis*. Elsevier, 2006.
24. Jerome Friedman, Trevor Hastie, and Robert Tibshirani. Additive logistic regression: a statistical view of boosting (with discussion and a rejoinder by the authors). *The annals of statistics*, 28(2):337–407, 2000.
25. Francis Galton. Regression towards mediocrity in hereditary stature. *The Journal of the Anthropological Institute of Great Britain and Ireland*, 15:246–263, 1886.
26. Carl Friedrich Gauss. *Theoria motus corporum coelestium in sectionibus conicis solem ambientium*, volume 7. FA Perthes, 1877.
27. Vi͡acheslav Leonidovich Girko. Introduction in general statistical analysis. *Theory of Probab. Appl*, 2:252–265. (in Russian).
28. Asaf H Hajiyev. Regression models with increasing numbers of unknown parameters. *International Encyclopedia. Statist. Sciences, Springer*, pages 1208–1211, 2011.
29. Wolfgang Härdle. *Applied nonparametric regression*. Number 19. Cambridge university press, 1990.
30. Ronald R Hocking. A biometrics invited paper. the analysis and selection of variables in linear regression. *Biometrics*, pages 1–49, 1976.
31. Arthur E Hoerl and Robert W Kennard. Ridge regression: applications to nonorthogonal problems. *Technometrics*, 12(1):69–82, 1970.
32. Arthur E Hoerl and Robert W Kennard. Ridge regression: Biased estimation for nonorthogonal problems. *Technometrics*, 12(1):55–67, 1970.
33. Hsin-Hsiung Huang and Qing He. Nonlinear regression analysis. *arXiv preprint* arXiv:2402.05342, 2024.
34. Mengyu Huang. Theory and implementation of linear regression. In *2020 International conference on computer vision, image and deep learning (CVIDL)*, pages 210–217. IEEE, 2020.
35. Peter J Huber. Robust estimation of a location parameter. In *Breakthroughs in statistics: Methodology and distribution*, pages 492–518. Springer, 1992.

36. Sylvie Huet, Annie Bouvier, Marie-Anne Poursat, Emmanuel Jolivet, and AM Bouvier. *Statistical tools for nonlinear regression: a practical guide with S-PLUS and R examples*. Springer, 2004.
37. Ingrid Kreuzmair and Helge Toutenburg. Sensitivity of the restricted least squares estimator under mis specified restrictions and multicollinearity. *Advances and Applications in Statistics*, 8(2):153–175, 2008.
38. David C Lay. *Linear algebra and its applications*. Pearson Education India, 2003.
39. Adrien Marie Legendre. *Nouvelles méthodes pour la détermination des orbites des comètes: avec un supplément contenant divers perfectionnemens de ces méthodes et leur application aux deux comètes de 1805*. Courcier, 1806.
40. Weihao Lu, Haobo Zhang, Yicheng Li, Manyun Xu, and Qian Lin. Optimal rate of kernel regression in large dimensions. *arXiv preprint* arXiv:2309.04268, 2023.
41. Adewale F Lukman, BM Golam Kibria, Cosmas K Nziku, Muhammad Amin, Emmanuel T Adewuyi, and Rasha Farghali. Kl estimator: Dealing with multicollinearity in the logistic regression model. *Mathematics*, 11(2):340, 2023.
42. Edmond Malinvaud. The consistency of nonlinear regressions. *The Annals of Mathematical Statistics*, 41(3):956–969, 1970.
43. M.B Malyutov. Lower boarders for average duration sequentially design of experiment. *Izvestiya vuzo*, 11:19–41, 1983. (in Russian).
44. CA McGilchrist. Regression analysis of dependent error models. *Bulletin of the Australian Mathematical Society*, 34(2):199–209, 1986.
45. Douglas C Montgomery, Elizabeth A Peck, and G Geoffrey Vining. *Introduction to linear regression analysis*. John Wiley & Sons, 2021.
46. John Ashworth Nelder and Robert WM Wedderburn. Generalized linear models. *Journal of the Royal Statistical Society Series A: Statistics in Society*, 135(3):370–384, 1972.
47. Aydin Öztürk. Weighted least squares estimation of location and scale parameters. *American Journal of Mathematical and Management Sciences*, 7(1-2):113–129, 1987.
48. Michael Panik. *Regression modeling: Methods, theory, and computation with SAS*. Chapman and Hall/CRC, 2009.
49. Iain Pardoe. *Applied regression modeling*. John Wiley & Sons, 2020.
50. Karl Pearson. Vii. mathematical contributions to the theory of evolution.-iii. regression, heredity, and panmixia. *Philosophical Transactions of the Royal Society of London. Series A, containing papers of a mathematical or physical character*, (187):253–318, 1896.
51. Karl Pearson. Notes on the history of correlation. *Biometrika*, 13(1):25–45, 1920.
52. Stephen Portnoy. Asymptotic behavior of m-estimators of p regression parameters when p 2/n is large. i. consistency. *The Annals of Statistics*, pages 1298–1309, 1984.
53. Yurii Vasil'evich Prohorov and Yurii Anatol'evich Rozanov. *Probability theory: basic concepts, limit theorems, random processes*, volume 18. Springer, 1969.
54. Tibshriani R. *High-Dimensional Regression: Ridge Advanced Topics in Statistical Learning*. 2023.
55. Enayetur Raheem, S Ejaz Ahmed, and Shuangzhe Liu. Stein-rule m-estimation in sparse partially linear models. *Japanese Journal of Statistics and Data Science*, 7(1):507–535, 2024.
56. C Radhakrishna Rao. Estimation of heteroscedastic variances in linear models. *Journal of the American Statistical Association*, 65(329):161–172, 1970.
57. Calyampudi Radhakrishna Rao. *Linear statistical inference and is applications*. John Wiley & Sons, Incorporated, 1965.
58. DA Ratkowsky. Handbook of nonlinear regression models, 1990.
59. Hossein Riazoshams, Habshah Midi, and Gebrenegus Ghilagaber. *Robust nonlinear regression: with applications using R*. John Wiley & Sons, 2018.
60. Christian Ritz and Jens Carl Streibig. *Nonlinear regression with R*. Springer, 2008.
61. G.F. Seber. Linear hypotheses and induced tests. *Biometrika*, 51(1/2):41–47, 1964.
62. G.F. Seber. Orthogonality in analysis of variance. *The Annals of Mathematical Statistics*, 35(2):705–710, 1964.

63. G.F. Seber and AJ Lee. Linear regression analysis, wiley-interscience. *Google Scholar*, pages 97–118, 2003.
64. G.F Seber. *Logistic Regression. Wiley Series in Probability and Statistics*. 2008.
65. G.F Seber and C.J Wild. *Nonlinear regression*. NY, Wiley, 2003.
66. Stephen M Stigler. Gauss and the invention of least squares. *the Annals of Statistics*, pages 465–474, 1981.
67. Dana L Thomas and David R Thomas. Confidence bands for percentiles in the linear regression model. *Journal of the American Statistical Association*, 81(395):705–708, 1986.
68. JOHNW TUKEY. Exploratory data analysis: past, present and future. 1993.
69. Chien-Fu Wu. Asymptotic theory of nonlinear least squares estimation. *The Annals of statistics*, 9(3):501–513, 1981.
70. Chien-Fu Jeff Wu. Jackknife, bootstrap and other resampling methods in regression analysis. *the Annals of Statistics*, 14(4):1261–1295, 1986.
71. George Udny Yule. On the significance of bravais' formulae for regression, &c., in the case of skew correlation. *Proceedings of the Royal Society of London*, 60(359-367):477–489, 1897.
72. George Udny Yule. Iii. on the theory of consistence of logical class-frequencies, and its geometrical representation. *Philosophical Transactions of the Royal Society of London. Series A, Containing Papers of a Mathematical or Physical Character*, 197(287-299):91–133, 1901.
73. Le Zhang, Zenglin Shi, Ming-Ming Cheng, Yun Liu, Jia-Wang Bian, Joey Tianyi Zhou, Guoyan Zheng, and Zeng Zeng. Nonlinear regression via deep negative correlation learning. *IEEE transactions on pattern analysis and machine intelligence*, 43(3):982–998, 2019.
74. Eckhard Liebscher. Kendall regression coefficient. *Computational Statistics & Data Analysis*, 157:107140, 2021.
75. Wayne A. Fuller and J. N. K. Rao. Estimation for a linear regression model with unknown diagonal covariance matrix. *The Annals of Statistics*, pages 1149–1158, 1978.
76. Andrej Pázman. Distribution of the weighted LS estimates in nonlinear models with symmetrical errors. *Kybernetika*, 24(6):413–427, 1988.
77. Lih-Yuan Deng and C. F. J. Wu. Estimation of variance of the regression estimator. *Journal of the American Statistical Association*, 82(398):568–576, 1987.
78. C. Radhakrishna Rao. Estimation of variance and covariance components in linear models. *Journal of the American Statistical Association*, 67(337):112–115, 1972.
79. V. I. Pagurova. On confident estimation in general linear in case of nonhomogeneous variances. *Theory of Probability and Its Applications*, 27(1):384–388, 1982. (in Russian)
80. Asaf H. Hajiyev. Statistical estimation in linear models. *Theory of Probability and Its Applications*, 34:764–768, (in Russian).
81. Yury K. Belyaev and Asaf H. Hajiyev. Statistical data analysis. Izv. AN USSR, Technical Cybernetics, 1979. (in Russian)
82. Asaf H. Hajiyev. Linear regression models with increasing numbers of unknown parameters. *Russian Academy of Sciences, Doklady Mathematics*, 399(3):1–5, 2004.
83. Frederick Mosteller and John W. Tukey. Data analysis and regression. A second course in statistics. *Addison-Wesley series in behavioral science: quantitative methods*, 1977.
84. Oscar Sheynin. Gauss and the Method of the Least Squares/Gauß und die Methode der kleinsten Quadrate. *Jahrbücher für Nationalökonomie und Statistik*, 219(3-4):458–467, 1999.
85. A. H. Hajiyev. Construction of a confident region for linear function. Preprint, Chalmers University, Gothenburg, Sweden, 1984, 15 pages.
86. A. H. Hajiyev. On estimation in regression models. *Bulletin of Moscow University, Series Mathematics and Mechanics*, (3):8–13, 1989.
87. A. H. Hajiyev. Estimation of unknown parameters and construction of confidence region. *Mathematical Sciences*, 75(6):42–50, 1995.
88. A. H. Hajiyev. Mathematical theory of regression models with an increasing number of unknown parameters. Nonlinear models. Research Report, Department of Computer Science, Lincoln University, No. 1, p. 24, 2004.
89. Jeffrey M. Stanton. Galton, Pearson, and the peas: A brief history of linear regression for statistics instructors. *Journal of Statistics Education*, 9(3), 2001.

90. Frederick Mosteller and John W. Tukey. Data analysis and regression. A second course in statistics. *Addison-Wesley series in behavioral science: quantitative methods*, 1977.
91. Ludwig Fahrmeir, Thomas Kneib, Stefan Lang, and Brian Marx. Regression: Models, Methods and Applications. Springer, 2021.

Chapter 2
Elements

La filosofia è scritta in questo grandissimo libro che continuamente ci sta aperto innanzi a gli occhi (io dico l'universo), ma non si può intendere se prima non s'impara a intender la lingua, e conoscer i caratteri, ne' quali è scritto. Egli è scritto in lingua matematica...

— Galileo Galilei (1564–1642)

This chapter establishes the mathematical foundation for regression models with an increasing number of unknown parameters, drawing on linear algebra, probability theory, and mathematical statistics. These disciplines provide essential tools for analyzing high-dimensional data and complex models. The linear algebra section introduces fundamental concepts, including vectors, matrices, and their operations, such as addition, multiplication, transposition, inversion, determinants, traces, eigenvalues, and eigenvectors. These elements form the theoretical basis for subsequent proofs and model derivations. Before formally introducing probability theory and mathematical statistics, essential concepts from mathematical analysis are presented. Topics such as limits, continuity, derivatives, and differentials are discussed to facilitate the study of the asymptotic properties of parameter estimators. These mathematical tools are indispensable for understanding the behavior of estimators as the sample size increases and plays a fundamental role in the theoretical development of regression models. The probability theory and mathematical statistics section begins with random variables, their distributions, expectations, variances, and limit theorems, which form the foundation of statistical inference. The mathematical characterization of random variables and probability distributions is essential for analyzing stochastic errors in regression models, while limit theorems provide theoretical support for large-sample inference. Building on this, measure theory and probability spaces are introduced, including probability measures and measurable functions. Next, parameter estimation methods, including the least squares estimator (LSE) and maximum

A. Hajiyev and J. Xu, *High-Dimensional Regression Analysis and Artificial Intelligence*,
https://doi.org/10.1007/978-981-95-2514-0_2

likelihood estimator (MLE), are presented, with an analysis of their consistency and asymptotic normality. Finally, hypothesis testing and confidence intervals are discussed as key tools for model evaluation and uncertainty quantification in regression analysis.

2.1 Linear Algebra Basics

Linear algebra is the branch of mathematics that focuses on the study of vector spaces and their mappings, with vectors and matrices serving as its foundational tools. A vector is a quantity characterized by both magnitude and direction, while a matrix is a fundamental tool for describing linear transformations. Key concepts in linear algebra, such as matrix operations, properties of vectors, and the relationships between matrices and vectors, provide a solid theoretical foundation for understanding and solving regression models. This monograph covers essential linear algebra topics, including matrix operations (addition, multiplication, inversion, transposition, and trace), linear independence and basis concepts, eigenvalues and eigenvectors, and the matrix formulation and derivation of the least squares method. These concepts are central to the understanding and application of both linear and nonlinear regression models. Mastery of these topics equips readers to efficiently analyze parameter estimation in regression analysis.

2.1.1 Vector and Vector Space

The concepts of the field, vector, vector space, linear combination, linear independence, subspace, bases, dimension, sum, direct sum and norm are introduced.

Vectors and Linear Combinations

A set is a collection of objects, and an element of a set is called a member. Symbols are used to represent specific sets, such as $\mathbb{R}$ for the set of real numbers and $\mathbb{C}$ for the set of complex numbers. The phrase "x is a real number" is equivalent to "$x \in \mathbb{R}$", and "x is an n-tuple of real numbers" is equivalent to "$x \in \mathbb{R}^n$".

We write $u \in S$ to mean "u is an element of set S". If every element of set S' is also an element of S, then S' is a subset of S, denoted $S' \subseteq S$. If S' is a proper subset of S, we denote it as such. The intersection of sets S_1 and S_2, denoted $S_1 \cap S_2$, consists of elements common to both sets. The union of S_1 and S_2, denoted $S_1 \cup S_2$, consists of elements that are in either or both sets.

Definition 2.1 (*Lang* [6]) Let K be a subset of the complex numbers $\mathbb{C}$. We shall say that K is a field if it satisfies the following conditions:

(1) If x, y are elements of K, then $x + y$ and xy are also elements of K.

(2) If $x \in K$, then $-x$ is also an element of K. If furthermore $x \neq 0$, then x^{-1} is an element of K.

(3) The elements 0 and 1 are elements of K.

It can be observed that both $\mathbb{R}$ and $\mathbb{C}$ are fields.Let $\mathbb{Q}$ denote the set of rational numbers, i.e., the set of all fractions $\frac{m}{n}$, where m, n are integers and $n \neq 0$. It is easily verified that $\mathbb{Q}$ is a field. Let $\mathbb{Z}$ denote the set of all integers. Then $\mathbb{Z}$ is not a field because condition (2) is not satisfied.

The essential property of a field is that it is a set of elements that can be added and multiplied in such a way that the ordinary rules of arithmetic are satisfied, and in such a way that division by non-zero elements is possible. Let K and L be fields, and assume that K is a subset of L. In this case, we define K as a subfield of L. Consequently, all the fields under consideration are subfields of the complex numbers. Specifically, we can assert that $\mathbb{R}$ is a subfield of $\mathbb{C}$, and $\mathbb{Q}$ is a subfield of $\mathbb{R}$.

A vector of dimension n over a number field K is an n-tuple over K:

$$\mathbf{a} = (a_1, a_2, \ldots, a_n),$$

where $a_i \in K$ is called the i-th component for $1 \leq i \leq n$. The set V of all vectors of dimension m over K is called the vector space of dimension n over K. V is also denoted by K^n, the Cartesian n-space. The vector $(0, \ldots, 0)$, whose components are all 0, is called the zero vector and is denoted by $\boldsymbol{0}$ or simply by 0.

Sometimes a vector can be written as

$$\mathbf{a}^T = \begin{pmatrix} a_1 \\ \vdots \\ a_m \end{pmatrix}$$

this is just a different way of writing. This form of vector is usually called a **column vector**.

The **magnitude** (or **length**) of a vector is a non-negative real number that measures the size of the vector. In Euclidean space, for a vector $\mathbf{x} = (x_1, x_2, \ldots, x_n)$, its magnitude is defined as:

$$\|\mathbf{x}\| = \sqrt{x_1^2 + x_2^2 + \cdots + x_n^2}$$

this represents the distance from the origin to the point $(x_1, x_2, \ldots, x_n)$, commonly referred to as the Euclidean length of the vector.

the magnitude satisfies the following properties:

Non-negativity: $\|\mathbf{x}\| \geq 0$, and $\|\mathbf{x}\| = 0$ if and only if $\mathbf{x} = \mathbf{0}$.

Homogeneity: For any scalar α and vector $\mathbf{x}$, $\|\alpha\mathbf{x}\| = |\alpha|\|\mathbf{x}\|$.

Triangle inequality: For any vectors $\mathbf{x}, \mathbf{y}$, $\|\mathbf{x} + \mathbf{y}\| \leq \|\mathbf{x}\| + \|\mathbf{y}\|$.

Vectors can undergo a dot product operation. Given two vectors $\mathbf{x} = (x_1, x_2, \ldots, x_n)$ and $\mathbf{y} = (y_1, y_2, \ldots, y_n)$ in $\mathbb{R}^n$, their **dot product** is defined as:

$$\mathbf{x} \cdot \mathbf{y} = \sum_{i=1}^{n} x_i y_i$$

The dot product satisfies the following properties:

Commutativity: $\mathbf{x} \cdot \mathbf{y} = \mathbf{y} \cdot \mathbf{x}$.

Linearity: For any vectors $\mathbf{x}$, $\mathbf{y}$, $\mathbf{z}$ and scalar α, it follows that:

$$\mathbf{x} \cdot (\alpha \mathbf{y} + \mathbf{z}) = \alpha(\mathbf{x} \cdot \mathbf{y}) + \mathbf{x} \cdot \mathbf{z}$$

Positive definiteness: For any vector $\mathbf{x}$, $\mathbf{x} \cdot \mathbf{x} = \|\mathbf{x}\|^2 \geq 0$, and $\mathbf{x} \cdot \mathbf{x} = 0$ if and only if $\mathbf{x} = \mathbf{0}$.

Let V be a vector space over the field $\mathbb{F} \in \{\mathbb{R}, \mathbb{C}\}$. A mapping $\langle \cdot, \cdot \rangle : V \times V \to \mathbb{F}$ is called an *inner product* if, for all $\mathbf{x}, \mathbf{y}, \mathbf{z} \in V$ and all $\alpha \in \mathbb{F}$,

Conjugate symmetry: $\langle \mathbf{x}, \mathbf{y} \rangle = \overline{\langle \mathbf{y}, \mathbf{x} \rangle}$.

Linearity in the first argument: $\langle \alpha \mathbf{x} + \mathbf{y}, \mathbf{z} \rangle = \alpha \langle \mathbf{x}, \mathbf{z} \rangle + \langle \mathbf{y}, \mathbf{z} \rangle$.

Positive definiteness: $\langle \mathbf{x}, \mathbf{x} \rangle > 0$ for $\mathbf{x} \neq \mathbf{0}$, and $\langle \mathbf{x}, \mathbf{x} \rangle = 0 \iff \mathbf{x} = \mathbf{0}$.

For $V = \mathbb{F}^n$ with the standard basis, this inner product can be written in transpose form as

$$\langle \mathbf{x}, \mathbf{y} \rangle = \mathbf{x}^T \mathbf{y} = \sum_{i=1}^{n} x_i y_i$$

which, when $\mathbb{F} = \mathbb{R}$, reduces to the dot product above.

A vector space V over a field K consists of a set of elements that can be added together and multiplied by elements of K in such a way that the sum of any two elements in V is still an element of V, the product of an element of V by an element of K is an element of V. Let $\alpha = (a_1, \ldots, a_n)$, $\beta = (b_1, \ldots, b_n)$ be two vectors in V, the **Addition** is defined as:

$$\alpha + \beta = (a_1 + b_1, \ldots, a_n + b_n).$$

Let $c \in F$ and $\alpha = (a_1, \ldots, a_n)$, we define **scalar multiplication** of α by c as

$$c\alpha = (ca_1, \ldots, ca_m).$$

We can also describe scalar multiplication by a number c as multiplying each component of α by c. The following properties can be easily proven to hold:

Commutativity: Given α, β of V, we have $\alpha + \beta = \beta + \alpha$,

Associativity: Given α, β, γ of V, we have $(\alpha + \beta) + \gamma = \alpha + (\beta + \gamma)$,

Neutral Element : There is an element of V, denoted as $\mathbf{0}$, for all elements α of V,such that,

$$\boldsymbol{0} + \alpha = \alpha + \boldsymbol{0} = \alpha,$$

Additive Inverse: Given an element α of V, there exists an element $-\alpha$, such that

$$\alpha + (-\alpha) = \boldsymbol{0},$$

Meanwhile, $\alpha - \beta$ is defined to be $\alpha + \beta$
Distributivity over vectors: If c is a number, then

$$c(\alpha + \beta) = c\alpha + c\beta,$$

Distributivity over numbers: If c_1, c_2 are two numbers,

$$(c_1 + c_2)\alpha = c_1\alpha + c_2\alpha,$$

Compatibility of Multiplication: If a, b are two numbers, then

$$(ab)\beta = a(\beta),$$

Multiplicative Identity: For all elements u of V, we have

$$1 \cdot u = u \quad \text{(1 here is the number one).}$$

Definition 2.2 (*Lang* [6]) Let V be a vector space, and let W be a subset of V. We define W to be a **subspace** if W satisfies the following conditions:
(1) If v, w are elements of W, their sum $v + w$ is also an element of W.
(2) If v is an element of W and c is a number, then cv is an element of W.
(3) The element 0 of V is also an element of W.

W is also a vector space, so it satisfies all the properties of a vector space.

Definition 2.3 (*Lang* [6]) Let V be an arbitrary vector space, and let $\alpha_1, \alpha_2, \ldots, \alpha_n$ be elements of V. Let $k_1, k_2, \ldots, k_n$ be numbers. An expression of the type

$$\beta = k_1 v_1 + k_2 v_2 + \cdots + k_n v_n \tag{2.1}$$

is called **a linear combination** of $\alpha_1, \alpha_2, \ldots, \alpha_n$.

Let W be the set of all linear combinations of $v_1, v_2, \ldots, v_n$. Then W is a subspace of V. The subspace W is called the subspace generated by $v_1, v_2, \ldots, v_n$.

Definition 2.4 (*Axler* [1]) The set of all linear combinations of a list of vectors $v_1, \ldots, v_m$ in V is called the span of $v_1, \ldots, v_m$, denoted by $\text{span}(v_1, \ldots, v_m)$. In other words,

$$\operatorname{span}(v_1, \ldots, v_m) = \{a_1v_1 + \cdots + a_mv_m \mid a_1, \ldots, a_m \in F\}. \tag{2.2}$$

The span of the empty list () is defined to be {0}.

Bases and Dimension of a Vector Space

Let V be a vector space over the field K, and let $\alpha_1, \ldots, \alpha_n$ be elements of V. We shall say that $v_1, \ldots, \alpha_n$ are **linearly dependent** over K if there exist elements $k_1, \ldots, k_n$ in K, not all equal to 0, such that

$$k_1v_1 + k_2v_2 + \cdots + k_nv_n = 0. \tag{2.3}$$

If there do not exist such numbers, then we say that $v_1, \ldots, v_n$ are linearly independent. In other words, vectors $v_1, \ldots, v_n$ are linearly independent if and only if the following condition is satisfied:

Whenever $k_1, \ldots, k_n$ are numbers in K such that $k_1v_1 + k_2v_2 + \cdots + k_nv_n = 0$, then $k_i = 0$ for all $i = 1, \ldots, n$.

Lemma 2.1 (Axler [1]) *Suppose $v_1, \ldots, v_m$ is a linearly dependent list in V. Then there exists $k \in \{1, 2, \ldots, m\}$ such that*

$$v_k \in span(v_1, \ldots, v_{k-1}).$$

furthermore, if k satisfies the condition above and the k-th term is removed from $v_1, \ldots, v_m$, then the span of the remaining list equals $span(v_1, \ldots, v_{k-1}, v_{k+1}, \ldots, v_m)$.

For a linearly dependent set of vectors $v_1, \ldots, v_m$, there exists an index k such that $v_k \in \operatorname{span}(v_1, \ldots, v_{k-1})$. Removing v_k from the list does not affect the span of the remaining vectors. This lemma highlights the issue of multicollinearity in regression, where one feature can be expressed as a combination of others. Removing redundant features does not change the model's explanatory power, helping simplify the model and prevent overfitting.

If elements $v_1, \ldots, v_n$ of V generate V and in addition are linearly independent, then $\{v_1, \ldots, v_n\}$ is called a basis of V. In other words, A basis of V is a list of vectors in V that is linearly independent and spans V.

Theorem 2.1 (Lang [6]) *Let V be a vector space. Let $v_1, \ldots, v_n$ be linearly independent elements of V. Let $x_1, \ldots, x_n$ and $y_1, \ldots, y_n$ be numbers. Suppose that we have*

$$\sum_{i=1}^{n} x_iv_i = \sum_{i=1}^{n} y_iv_i. \tag{2.4}$$

Then $x_i = y_i$ for $i = 1, \ldots, n$.

Theorem 2.1 states that for a set of linearly independent vectors, the coefficients in any linear combination of those vectors are uniquely determined. This ensures that in problems involving such vectors, such as linear regression, the solution is unique.

Theorem 2.2 (Lang [6]) *Let* $\{v_1, \ldots, v_n\}$ *be a set of generators of a vector space* V. *Let* $\{v_1, \ldots, v_r\}$ *be a maximal subset of linearly independent elements. Then* $\{v_1, \ldots, v_r\}$ *is a basis of* V.

Theorem 2.2 asserts that the largest set of linearly independent vectors within a generating set forms a basis for the vector space. In linear regression, this concept is crucial as it ensures that the chosen features (variables) are independent and do not suffer from multicollinearity, leading to a well-defined and interpretable model.

Theorem 2.3 (Lang [6]) *Let* V *be a vector space over the field* K. *Let* $\{v_1, \ldots, v_m\}$ *be a basis of* V *over* K. *Let* $w_1, \ldots, w_n$ *be elements of* V, *and assume that* $n > m$. *Then* $w_1, \ldots, w_n$ *are linearly dependent.*

Theorem 2.3 above highlights an important property of vector spaces: if the number of vectors exceeds the dimension of the space (i.e., the size of a basis), then those vectors must be linearly dependent. This provides a crucial insight into the relationship between the number of vectors and the structure of the vector space. It directly leads us to the concept of dimension, which captures the number of independent directions (or basis vectors) required to span the entire space. The following definition formalizes this idea,

Definition 2.5 (*Axler* [1]) The dimension of a finite-dimensional vector space is the length of any basis of the vector space and is denoted by $\dim V$.

Sums and Direct Sums

In order to effectively combine and decompose vector spaces or their subspaces, it is necessary to introduce the concepts of **sum** and **direct sum**. This allows us to determine how elements from different subspaces combine to form a larger vector space, particularly when the subspaces do not overlap.

Definition 2.6 (*Axler* [1]) Suppose $V_1, \ldots, V_m$ are subspaces of V. The sum of $V_1, \ldots, V_m$, denoted by

$$V_1 + \cdots + V_m = \{v_1 + \cdots + v_m : v_1 \in V_1, \ldots, v_m \in V_m\}. \tag{2.5}$$

is the set of all possible sums of elements of $V_1, \ldots, V_m$.

The **sum** is the set of all vectors obtained by selecting one element from each subspace and adding them together. This construction allows us to combine elements from multiple subspaces to form a larger space. A special situation is called **direct sum** defined as:

Definition 2.7 (*Axler* [1]) Suppose $V_1, \ldots, V_m$ are subspaces of V.

(1) The sum $V_1 + \cdots + V_m$ is called a direct sum if each element of $V_1 + \cdots + V_m$ can be written in only one way as a sum $v_1 + \cdots + v_m$, where each $v_k \in V_k$.

(2) If $V_1 + \cdots + V_m$ is a direct sum, then $V_1 \oplus \cdots \oplus V_m$ denotes $V_1 + \cdots + V_m$, with the $\oplus$ notation serving as an indication that this is a direct sum.

Direct Sum is the sum of multiple subspaces, which is called a direct sum if each element can be uniquely written as the sum of elements from each subspace.More specifically, let $V_1, V_2, \ldots, V_m$ be subspaces of a vector space V. If each element of $V_1 + \cdots + V_m$ can be written in only one way as $v_1 + \cdots + v_m$, where $v_k \in V_k$, then $V_1 + \cdots + V_m$ is called a direct sum. This means that the elements of the direct sum do not have overlapping parts from different subspaces, and each element in the direct sum has a unique representation.

In short, the direct sum provides a way to combine subspaces into a new vector space, where each vector in the sum has a unique decomposition. More precisely, given a vector space V and subspaces $V_1, V_2, \ldots, V_m$, if the intersection of these subspaces contains only the zero vector, that is,

$$V_1 \cap V_2 \cap \cdots \cap V_m = \{0\},$$

then their direct sum (denoted by $V_1 \oplus V_2 \oplus \cdots \oplus V_m$) is the sum of these subspaces.

Theorem 2.4 (Axler [1]) *If V_1 and V_2 are subspaces of a finite-dimensional vector space, then*

$$\dim(V_1 + V_2) = \dim V_1 + \dim V_2 - \dim(V_1 \cap V_2). \tag{2.6}$$

Specially, if $V_1 + V_2 = V_1 \oplus V_2$, then

$$\dim(V_1 + V_2) = \dim V_1 + \dim V_2 \tag{2.7}$$

The Theorem 2.4 further explains that the dimension of the sum of two subspaces V_1 and V_2 is the sum of their dimensions minus the dimension of their intersection; if $V_1 + V_2$ is a direct sum, its dimension is the sum of the individual dimensions.

It should also be noted that, for any finite-dimensional vector space V and its subspace W, there exists a subspace U such that V can be written as the direct sum of W and U.

Norm

In mathematics, a **norm** is a function that assigns a non-negative real number to each vector in a vector space, which measures the "size" or "length" of the vector. For a vector $\mathbf{x} \in V$ in a vector space V, a norm is a function $\|\mathbf{x}\| : V \to \mathbb{R}$ that satisfies the following properties:

Non-negativity: For all $\mathbf{x} \in V$, $\|\mathbf{x}\| \geq 0$, and $\|\mathbf{x}\| = 0$ if and only if $\mathbf{x} = \mathbf{0}$.
Homogeneity: For any scalar α and vector $\mathbf{x} \in V$, $\|\alpha\mathbf{x}\| = |\alpha|\|\mathbf{x}\|$.
Triangle inequality: For any vectors $\mathbf{x}, \mathbf{y} \in V$, $\|\mathbf{x} + \mathbf{y}\| \leq \|\mathbf{x}\| + \|\mathbf{y}\|$.

Example 2.1 Euclidean norm (L^2 norm): The Euclidean norm of a vector $\mathbf{x} = (x_1, x_2, \ldots, x_n)$ is defined as:

$$\|\mathbf{x}\|_2 = \sqrt{x_1^2 + x_2^2 + \cdots + x_n^2} \tag{2.8}$$

Example 2.2 **Manhattan norm (L^1 norm)**: The Manhattan norm of a vector $\mathbf{x} = (x_1, x_2, \ldots, x_n)$ is defined as:

$$\|\mathbf{x}\|_1 = |x_1| + |x_2| + \cdots + |x_n| \tag{2.9}$$

Example 2.3 **The L^p norm** (also known as the p-norm) is a family of norms used to measure the size of vectors in a vector space. For a vector $\mathbf{x} = (x_1, x_2, \ldots, x_n) \in \mathbb{R}^n$, the L_p norm is defined as:

$$\|\mathbf{x}\|_p = \left(\sum_{i=1}^{n} |x_i|^p\right)^{\frac{1}{p}} \tag{2.10}$$

Example 2.4 **Frobenius norm**: The Frobenius norm of a matrix $A = [a_{ij}] \in \mathbb{R}^{m\times n}$ is defined as:

$$\|A\|_F = \sqrt{\sum_{i=1}^{m}\sum_{j=1}^{n} |a_{ij}|^2} \tag{2.11}$$

2.1.2 Matrices

The concept of the definition of matrices, rank, matrix addition, matrix scalar multiplication, the transpose of matrix, the multiplication of matrices, inverse matrix and invertibility.

The Definition of Matrices

Let us introduce a new concept, matrices. Let K be a field, and let n and m be two integers greater than or equal to 1. A matrix is defined as a rectangular array of numbers from K, typically written as:

$$\begin{pmatrix} a_{11} & a_{12} & a_{13} & \ldots & a_{1n} \\ a_{21} & a_{22} & a_{23} & \ldots & a_{2n} \\ \vdots & \vdots & \vdots & \ddots & \vdots \\ a_{m1} & a_{m2} & a_{m3} & \ldots & a_{mn} \end{pmatrix}$$

This object is called a matrix over K. The elements a_{ij} of the matrix are indexed by row $i = 1, \ldots, m$ and column $j = 1, \ldots, n$. Each individual element a_{ij} is referred to as the ij-entry or the ij-component of the matrix. If $m = n$, we say that it is a **square matrix**, if $a_{ij} = 0$ for all i, j, we say that it is a **zero matrix**.

An **upper triangular matrix** is a matrix in which all the elements below the main diagonal are zero. A **lower triangular matrix** is a matrix in which all the elements above the main diagonal are zero.

In linear algebra, the **rank** of a matrix is an important quantity that describes the linear independence of the matrix's rows or columns. The rank of a matrix can be defined as follows:

(1) **Row rank**: The row rank of a matrix A is the maximum number of linearly independent row vectors of the matrix. In other words, the row rank is the dimension of the row space of the matrix, i.e., the number of linearly independent row vectors that span the row space of A.

(2) **Column rank**: The column rank of a matrix A is the maximum number of linearly independent column vectors of the matrix. The column rank is the dimension of the column space of the matrix, i.e., the number of linearly independent column vectors that span the column space of A.

The **Rank Theorem** states that the **row rank** and **column rank** of a matrix are always equal, that is,

Theorem 2.5 (Axler [1]) *Suppose A is an m-by-n matrix, then the column rank of A is equal to the row rank of A.*

This result is a fundamental theorem in linear algebra, which shows that the rank of a matrix is the same regardless of whether we consider the row vectors or the column vectors. Therefore, we usually refer to the rank of a matrix simply as "rank" without distinguishing between row rank and column rank.

Basic Operations of the Matrix

Matrix Addition:

Let $A = (a_{ij})_{m\times n}$ and $B = (b_{ij})_{m\times n}$ be two matrices of the same size $(m \times n)$. The sum of A and B, denoted by $A + B$, is the matrix C where each element c_{ij} is the sum of the corresponding elements of A and B:

$$A + B = C = (c_{ij})_{m\times n}, \quad \text{where} \quad c_{ij} = a_{ij} + b_{ij}.$$

Matrix Scalar Multiplication:

Let $A = (a_{ij})_{m\times n}$ be a matrix and α be a scalar. The scalar multiplication of A by α, denoted by αA, is the matrix B where each element b_{ij} is the product of α and a_{ij}:

$$B = \alpha A = \alpha(a_{ij})_{m\times n}.$$

The Transpose of Matrix

The **transpose** of a matrix A, denoted by A^T, is a new matrix obtained by swapping the rows and columns of A. Specifically, if $A = [a_{ij}]$ is an $m \times n$ matrix, then the transpose of A is an $n \times m$ matrix $A^T = [b_{ji}]$, where $b_{ji} = a_{ij}$, the element in the i-th row and j-th column of A is placed in the j-th row and i-th column of A^T:

$$A = \begin{pmatrix} a_{11} & a_{21} & \dots & a_{m1} \\ a_{12} & a_{22} & \dots & a_{m2} \\ \vdots & \vdots & \ddots & \vdots \\ a_{1n} & a_{2n} & \dots & a_{mn} \end{pmatrix} \longrightarrow A^T = \begin{pmatrix} a_{11} & a_{12} & \dots & a_{1n} \\ a_{21} & a_{22} & \dots & a_{2n} \\ \vdots & \vdots & \ddots & \vdots \\ a_{m1} & a_{m2} & \dots & a_{mn} \end{pmatrix}$$

A **symmetric matrix** is a square matrix that is equal to its transpose. Specifically, a matrix $A = [a_{ij}]$ is symmetric if it satisfies the condition: $A = A^T$.

A **diagonal matrix** is a square matrix in which all the off-diagonal elements are zero. Specifically, a matrix $A = [a_{ij}]$ is diagonal if $a_{ij} = 0$ for all $i \neq j$. That is:

$$D = \begin{pmatrix} a_{11} & 0 & \dots & 0 \\ 0 & a_{22} & \dots & 0 \\ \vdots & \vdots & \ddots & \vdots \\ 0 & 0 & \dots & a_{nn} \end{pmatrix}$$

The elements $a_{11}, a_{22}, \dots, a_{nn}$ on the main diagonal can be any numbers, but the rest of the elements are all zero. An **identity matrix** is a special type of diagonal matrix where all the diagonal elements are equal to 1. The identity matrix acts as the multiplicative identity for matrices, meaning that for any matrix A, we have $A \cdot I = I \cdot A = A$.

Multiplication of Matrices

Definition 2.8 (*Axler* [1]) Suppose A is an m-by-n matrix and B is an n-by-p matrix. Then AB is defined to be the m-by-p matrix whose entry in row j, column k, is given by the equation

$$(AB)_{jk} = \sum_{r=1}^{n} A_{jr} B_{rk}. \tag{2.12}$$

Thus, the entry in row j, column k of AB is computed by taking row j of A and column k of B, multiplying together corresponding entries, and then summing.

Example 2.5 Let $A = [a_{ij}]$ be an $m \times n$ matrix and $B = [b_{ij}]$ be an $n \times p$ matrix. The product AB is an $m \times p$ matrix, where the entry in the j-th row and k-th column of the product matrix is given by:

$$(AB)_{jk} = \sum_{r=1}^{n} a_{jr} b_{rk}. \tag{2.13}$$

Thus, the matrix multiplication AB is given by:

$$AB = \begin{pmatrix} \sum_{r=1}^{n} a_{1r}b_{r1} & \sum_{r=1}^{n} a_{1r}b_{r2} & \dots & \sum_{r=1}^{n} a_{1r}b_{rp} \\ \sum_{r=1}^{n} a_{2r}b_{r1} & \sum_{r=1}^{n} a_{2r}b_{r2} & \dots & \sum_{r=1}^{n} a_{2r}b_{rp} \\ \vdots & \vdots & \ddots & \vdots \\ \sum_{r=1}^{n} a_{mr}b_{r1} & \sum_{r=1}^{n} a_{mr}b_{r2} & \dots & \sum_{r=1}^{n} a_{mr}b_{rp} \end{pmatrix}$$

Example 2.6 Let $A = [a_{ij}]$ be an $m \times n$ matrix, let B be an $n \times 1$ column vector, represented as: $B = \begin{pmatrix} b_1 & b_2 & \cdots & b_n \end{pmatrix}^T$.

The multiplication of A and B, denoted as AB, is an $m \times 1$ column vector: $AB = \begin{pmatrix} y_1 & y_2 & \cdots & y_m \end{pmatrix}^T$, where each component y_i is calculated as:

$$y_i = \sum_{j=1}^{n} a_{ij}b_j = a_{i1}b_1 + a_{i2}b_2 + \cdots + a_{in}b_n, \quad i = 1, 2, \dots, m. \tag{2.14}$$

In other words, each element y_i of the result vector AB is obtained by taking the dot product of the i-th row of matrix A with the column vector B.

Example 2.7 Let A be a $1 \times m$ row vector, represented as: $A = \begin{pmatrix} a_1 & a_2 & \cdots & a_m \end{pmatrix}$, let $B = [b_{ij}]$ be an $m \times n$ matrix, the product of A and B, denoted as AB, is a $1 \times n$ row vector: $AB = \begin{pmatrix} c_1 & c_2 & \cdots & c_n \end{pmatrix}$, where each component c_j is calculated as:

$$c_j = \sum_{i=1}^{m} a_i b_{ij} = a_1 b_{1j} + a_2 b_{2j} + \cdots + a_m b_{mj}, \quad j = 1, 2, \dots, n. \tag{2.15}$$

In other words, each element c_j of the result vector AB is obtained by taking the dot product of the row vector A with the j-th column of matrix B.

Example 2.8 Linear system of equations is a collection of one or more linear equations involving the same set of variables. A general form of a linear system with m equations and n unknowns is as follows:

$$\begin{cases} a_{11}x_1 + a_{12}x_2 + \cdots + a_{1n}x_n = b_1 \\ a_{21}x_1 + a_{22}x_2 + \cdots + a_{2n}x_n = b_2 \\ \vdots \\ a_{m1}x_1 + a_{m2}x_2 + \cdots + a_{mn}x_n = b_m \end{cases}$$

Thus, the system can be written as:

$$AX = B \tag{2.16}$$

where A is the $m \times n$ matrix of coefficients a_{ij}; X is the column vector of unknowns $x_1, x_2, \dots, x_n$; B is the column vector of constants $b_1, b_2, \dots, b_m$.

Example 2.9 Quadratic forms $q(x_1, x_2, \ldots, x_n)$ are homogeneous quadratic polynomials of n variables. In the cases of one, two, and three variables they are called unary, binary, and ternary quadratic forms and have the following explicit form:

$$
\begin{aligned}
q(x_1) &= a_1 x_1^2, (unary) \\
q(x_1, x_2) &= a_1 x_1^2 + a_2 x_1 x_2 + a_3 x_2^2, (binary) \\
q(x_1, x_2, x_3) &= a_1 x_1^2 + a_2 x_1 x_2 + a_3 x_2^2 + a_4 x_1 x_3 + a_5 x_2 x_3 + a_6 x_3^2, (ternary)
\end{aligned}
\tag{2.17}
$$

where $a_1, a_2, \ldots, a_6$ are the coefficients.

Any $n \times n$ matrix A determines a quadratic form q_A of n variables by

$$q_A(x_1, x_2, \ldots, x_n) = \sum_{i=1}^{n} \sum_{j=1}^{n} a_{ij} x_i x_j \tag{2.18}$$

Matrix multiplication has many important properties, such as the associative property, distributive property, and properties of the transpose operation, which are outlined as follows:

Theorem 2.6 (Lang [6]) *Let A, B, C be matrices. Assume that A, B can be multiplied, and A, C can be multiplied, and B, C can be added. Then A, $B + C$ can be multiplied, and we have*

$$A(B + C) = AB + AC. \tag{2.19}$$

If x is a number, then

$$A(xB) = x(AB). \tag{2.20}$$

Theorem 2.7 (Lang [6]) *Let A, B, C be matrices such that A, B can be multiplied and B, C can be multiplied. Then A, BC can be multiplied. So can AB, C, and we have*

$$(AB)C = A(BC). \tag{2.21}$$

Theorem 2.8 (Lang [6]) *Let A, B be matrices which can be multiplied. Then $(AB)^T = B^T A^T$.*

Matrix multiplication generally does not satisfy the commutative property. That is, for any matrices A and B, we typically have:

$$A \cdot B \neq B \cdot A$$

Matrix multiplication can only satisfy the commutative property under specific conditions, such as when A and B are diagonal matrices or have certain special structures.

Inverse Matrix and Invertibility

Let $A \in \mathbb{R}^{n \times n}$ be a square matrix. A matrix $A^{-1} \in \mathbb{R}^{n \times n}$ is called the inverse of A if and only if the following condition holds:

$$A \cdot A^{-1} = A^{-1} \cdot A = I_n \tag{2.22}$$

where I_n is the identity matrix of size $n \times n$. The matrix A^{-1} is unique if it exists, and it satisfies the property that multiplying it by A (in either order) results in the identity matrix. If such a matrix A^{-1} exists, we say that A is invertible; otherwise, A is singular.

A square matrix $A \in \mathbb{R}^{n \times n}$ is invertible if and only if the following conditions hold:

(1) **Full Rank**: The matrix A has full rank, i.e., $\text{rank}(A) = n$. This implies that the rows (and columns) of A are linearly independent.

(2) **Non-zero Determinant**: The determinant of A is non-zero, i.e., $\det(A) \neq 0$. This condition ensures that A is non-singular and thus invertible (The definition of the determinant will be introduced in the next subsection).

(3) **Existence of a Unique Solution**: The linear system $A\mathbf{x} = \mathbf{b}$ has a unique solution for every vector $\mathbf{b} \in \mathbb{R}^n$. This condition guarantees that A is invertible.

Positive Definite Matrices

Definition 2.9 Matrix A is positive definite if, for any non-zero vector x, the following ratio is held $x^T Ax > 0$.

There are some criteria (necessary and sufficient conditions) for defining when the matrix can be positively definite are given.

A matrix is positively definite if and only if all its eigenvalues are positive.

A square matrix A is positive definite if it is symmetric and its eigenvalues λ take positive values, that is $\lambda > 0$.

If A is a positive definite matrix, then it is an invertible matrix (i.e., there exists matrix B for which $BA = AB = I$) and $\det A > 0$.

Matrix is positively definite if and only if there exists such *nonsingular* matrix B that $A = B^T B$.

Symmetric matrix A is positively definite if and only if all corner minors are positive.

2.1.3 *Determinants*

The concepts of the definition of determinants, the computation methods of determinants(by using cofactor), the properties of determinants are introduced.

The Definition of Determinants

The determinant of a square matrix is a scalar value that is computed from the elements of the matrix. It plays a crucial role in various areas of linear algebra, such as solving linear systems, matrix inversion, and eigenvalue analysis. For a 2×2 matrix $A = \begin{pmatrix} a & b \\ c & d \end{pmatrix}$, the determinant is defined as:

$$\det(A) = \begin{vmatrix} a & b \\ c & d \end{vmatrix} = ad - bc \tag{2.23}$$

For larger matrices, the determinant is typically defined recursively using the cofactor expansion method.

Definition 2.10 The cofactor of an element a_{ij} in an $n \times n$ matrix $A = [a_{ij}]$ is defined as the product of two components:

(1) **Minor:** This is the determinant of the $(n-1) \times (n-1)$ submatrix obtained by deleting the i-th row and the j-th column from matrix A, denoted as $\det(A_{ij})$, where A_{ij} is the submatrix formed by removing the i-th row and the j-th column.

(2) **Sign factor:** The cofactor is multiplied by a sign factor $(-1)^{i+j}$, which alternates according to the checkerboard pattern. The sign is positive when $i + j$ is even, and negative when $i + j$ is odd.

Thus, the cofactor C_{ij} of element a_{ij} is given by:

$$C_{ij} = (-1)^{i+j} \det(A_{ij}) \tag{2.24}$$

where A_{ij} is the submatrix formed by deleting the i-th row and the j-th column, and $(-1)^{i+j}$ is the sign factor.

The determinant can be expanded using any row i or column j. The general formulas are:

Expanding along row i:

$$\det(A) = \sum_{j=1}^{n} a_{ij} C_{ij} = \sum_{j=1}^{n} a_{ij} (-1)^{i+j} \det(A_{ij}) \tag{2.25}$$

Expanding along column j:

$$\det(A) = \sum_{i=1}^{n} a_{ij} C_{ij} = \sum_{i=1}^{n} a_{ij} (-1)^{i+j} \det(A_{ij}) \tag{2.26}$$

Specifically, for expansion along the first row, the formula is:

$$\det(A) = \sum_{j=1}^{n} (-1)^{1+j} a_{1j} \det(A_{1j}) \tag{2.27}$$

Example 2.10 For a 3 × 3 matrix

$$A = \begin{pmatrix} a_{11} & a_{12} & a_{13} \\ a_{21} & a_{22} & a_{23} \\ a_{31} & a_{32} & a_{33} \end{pmatrix}$$

The determinant of A is denoted as:

$$\text{Det}(A) = \begin{vmatrix} a_{11} & a_{12} & a_{13} \\ a_{21} & a_{22} & a_{23} \\ a_{31} & a_{32} & a_{33} \end{vmatrix}$$

then, the determinant can be expanded along the first row as:

$$\text{Det}(A) = a_{11} \begin{vmatrix} a_{22} & a_{23} \\ a_{32} & a_{33} \end{vmatrix} - a_{12} \begin{vmatrix} a_{21} & a_{23} \\ a_{31} & a_{33} \end{vmatrix} + a_{13} \begin{vmatrix} a_{21} & a_{22} \\ a_{31} & a_{32} \end{vmatrix}$$

Definition 2.11 (*Axler* [1]) Let $A \in \mathbb{R}^{n\times n}$ be a square matrix. The *corner minors* (also called *leading principal minors*) of A are defined as the determinants of its upper-left $k \times k$ submatrices:

$$\Delta_k = \det(A_k)$$

where A_k is the $k \times k$ submatrix obtained by taking the first k rows and the first k columns of A.

Properties of Determinants

Theorem 2.9 (Lang [6]) *The determinant satisfies the following properties:*

(1) As a function of each column vector, the determinant is linear. That is, if the j-th column A^j is equal to a sum of two column vectors, say $A^j = C + C'$, then

$$D(A^1, \ldots, C + C', \ldots, A^n) = D(A^1, \ldots, C, \ldots, A^n) + D(A^1, \ldots, C', \ldots, A^n).$$

(2) If t is a number, then

$$D(A^1, \ldots, tA^j, \ldots, A^n) = tD(A^1, \ldots, A^j, \ldots, A^n).$$

(3) If two adjacent columns are equal, i.e., if $A^j = A^{j+1}$ for some $j = 1, \ldots, n-1$, then the determinant $D(A)$ is equal to 0.

(4) If I is the unit matrix, then $D(I) = 1$.

Theorem 2.10 (Lang [6]) *The determinant satisfies the rule for expansion according to rows and columns, and Det(A) = Det(A^T). In other words, the determinant of a matrix is equal to the determinant of its transpose.*

Besides the properties mentioned above, the determinant also satisfies the following properties:

(1) **Swapping Rows or Columns:** If two rows or columns of a matrix are swapped, the determinant changes its sign.

(2) **Scaling a Row or Column:** If all the elements of a row or column are multiplied by a scalar k, the determinant is also multiplied by k:

$$\det(kA) = k \cdot \det(A)$$

(3) **Upper or Lower Triangular Matrices:** For upper or lower triangular matrices, the determinant is the product of the diagonal elements:

$$\det(A) = \prod_{i=1}^{n} a_{ii}$$

(4) **Product of Matrices:** The determinant of the product of two matrices equals the product of their determinants:

$$\det(AB) = \det(A) \cdot \det(B)$$

(5) **Invertibility of Matrices:** A matrix is invertible if and only if its determinant is non-zero:

$$\det(A) \neq 0 \quad \text{if and only if} \quad A \text{ is invertible}$$

(6) **Linear Dependence:** If any row or column of a matrix is a linear combination of the other rows or columns, the determinant is zero.

2.1.4 Eigenvalues and Eigenvectors

The study of eigenvalues and eigenvectors forms a cornerstone in understanding linear transformations and their structural properties. We will then define eigenvalues, explore their properties, and present important related theorems.

Linear map

Definition 2.12 (*Axler* [1]) Let V and W be vector spaces over a field K. A function $T : V \to W$ is called a linear map (or linear transformation) if it satisfies the following two properties for all vectors $\mathbf{u}, \mathbf{v} \in V$ and scalars $\alpha \in K$:

(1) **Additivity**:

$$T(\mathbf{u} + \mathbf{v}) = T(\mathbf{u}) + T(\mathbf{v}).$$

(2) **Homogeneity**:

$$T(\alpha \mathbf{u}) = \alpha T(\mathbf{u}).$$

A linear map preserves the operations of vector addition and scalar multiplication. If $W = V$, the linear map $T : V \to V$ is specifically called a linear operator.

Every linear map $T : K^n \to K^m$ can be represented by an $m \times n$ matrix A, such that for any vector $\mathbf{x} \in K^n$:

$$T(\mathbf{x}) = A\mathbf{x},$$

where $A\mathbf{x}$ is the matrix-vector product. The columns of A are the images of the standard basis vectors of K^n under T.

The kernel (or null space) of T, denoted $\ker(T)$, is the set of vectors in V mapped to the zero vector in W:

$$\ker(T) = \{\mathbf{v} \in V \mid T(\mathbf{v}) = \mathbf{0}\}.$$

The image (or range) of T, denoted $\mathrm{Im}(T)$, is the set of all possible outputs in W:

$$\mathrm{Im}(T) = \{T(\mathbf{v}) \mid \mathbf{v} \in V\}.$$

Eigenvalues and Eigenvectors

Definition 2.13 (*Lang* [6]) Let V be a vector space over a field K, and let $T : V \to V$ be a linear operator. A scalar $\lambda \in K$ is called an eigenvalue of T if there exists a non-zero vector $\mathbf{v} \in V$ such that:

$$T(\mathbf{v}) = \lambda \mathbf{v}.$$

The vector $\mathbf{v}$ is termed an eigenvector corresponding to λ.

For a square matrix $A \in K^{n \times n}$, eigenvalues and eigenvectors are defined analogously: λ is an eigenvalue of A if there exists a non-zero vector $\mathbf{x} \in K^n$ satisfying:

$$A\mathbf{x} = \lambda \mathbf{x}.$$

The eigenvalues of a matrix A can be determined by solving the characteristic equation:

$$\det(A - \lambda I) = 0,$$

where I is the identity matrix of the same size as A. Expanding this determinant yields the characteristic polynomial:

$$p_A(\lambda) = \det(A - \lambda I) = (-1)^n \lambda^n + c_{n-1}\lambda^{n-1} + \cdots + c_0.$$

The roots of $p_A(\lambda) = 0$ are the eigenvalues of A.

The algebraic multiplicity of an eigenvalue λ is its multiplicity as a root of the characteristic polynomial. The geometric multiplicity is the dimension of the eigenspace $E_\lambda = \{\mathbf{v} \in V \mid A\mathbf{v} = \lambda\mathbf{v}\}$.

Theorem 2.11 (Axler [1]) *For any eigenvalue λ of a matrix A, the geometric multiplicity is less than or equal to its algebraic multiplicity.*

A matrix $A \in K^{n\times n}$ is diagonalizable if there exists an invertible matrix P and a diagonal matrix D such that:

$$A = PDP^{-1}.$$

The columns of P are eigenvectors of A, and the diagonal entries of D are the corresponding eigenvalues.

Theorem 2.12 (Lang [6]) *A matrix A is diagonalizable if and only if the sum of the geometric multiplicities of its eigenvalues equals n. This occurs when A has n linearly independent eigenvectors.*

For symmetric matrices, eigenvalues and eigenvectors exhibit special properties:

Theorem 2.13 (Lang [6]) *Let $A \in \mathbb{R}^{n\times n}$ be a symmetric matrix. Then:*

(1) All eigenvalues of A are real.

(2) Eigenvectors corresponding to distinct eigenvalues are orthogonal. A is orthogonally diagonalizable, i.e., there exists an orthogonal matrix Q such that:

$$A = QDQ^T,$$

where D is a diagonal matrix of eigenvalues.

Trace

Definition 2.14 (Axler [1]) Let $A = [a_{ij}] \in K^{n\times n}$ be a square matrix over a field K. The **trace** of A, denoted $\mathrm{tr}(A)$, is defined as the sum of the diagonal elements of A:

$$\mathrm{tr}(A) = \sum_{i=1}^{n} a_{ii}. \tag{2.28}$$

The trace operator satisfies the following fundamental properties:

(1) **Linearity**: For any matrices $A, B \in K^{n\times n}$ and scalars $\alpha, \beta \in K$,

$$\mathrm{tr}(\alpha A + \beta B) = \alpha\mathrm{tr}(A) + \beta\mathrm{tr}(B).$$

(2) **Invariance Under Transposition**: The trace of a matrix is equal to the trace of its transpose:

$$\mathrm{tr}(A) = \mathrm{tr}(A^T).$$

(3) **Cyclic Property**: For matrices $A \in K^{n\times m}$, $B \in K^{m\times p}$, and $C \in K^{p\times n}$,

$$\mathrm{tr}(ABC) = \mathrm{tr}(BCA) = \mathrm{tr}(CAB).$$

In particular, for square matrices $A, B \in K^{n\times n}$,

$$\mathrm{tr}(AB) = \mathrm{tr}(BA).$$

(4) **Trace and Eigenvalues**: If A is a square matrix with eigenvalues $\lambda_1, \lambda_2, \ldots, \lambda_n$ (counted with algebraic multiplicity), then:

$$\mathrm{tr}(A) = \sum_{i=1}^{n} \lambda_i.$$

(5) **Trace of Similar Matrices**: If A and B are similar matrices (i.e., $B = P^{-1}AP$ for some invertible P), then:

$$\mathrm{tr}(A) = \mathrm{tr}(B).$$

(6) **Trace of Identity Matrix**: The trace of the $n \times n$ identity matrix I_n is:

$$\mathrm{tr}(I_n) = n.$$

Theorem 2.14 (Lang [6]) *Let $A \in K^{n\times n}$ and $P \in K^{n\times n}$ be an invertible matrix. Then:*

$$\mathrm{tr}(P^{-1}AP) = \mathrm{tr}(A). \tag{2.29}$$

This property confirms that the trace is invariant under similarity transformations.

2.2 Probability and Statistics Fundamentals

This section covers the fundamental concepts of probability and statistics, including fundamental concepts of mathematical analysis, random variables and distributions, covariance matrices, and the multivariate normal distribution. It emphasizes the Law of Large Numbers and the Central Limit Theorem, as well as point and interval estimation methods. The section also explores Least Squares Estimation (LSE) and Maximum Likelihood Estimation (MLE). Finally, it addresses asymptotic theory, focusing on the convergence properties of estimators as sample size increases. These concepts provide the essential theoretical foundation for statistical modeling and analysis.

2.2.1 Core Concepts in Mathematical Analysis

This subsection introduces fundamental concepts in mathematical analysis, beginning with limits and continuity, including the definition and properties of continuous functions, and their applications in asymptotic analysis. It then covers derivatives and differentials, with a focus on first derivatives, gradients, higher-order derivatives, and the role of the Hessian matrix in optimization problems. Finally, the section explores Taylor expansion, both for univariate and multivariate functions, highlighting its use in providing high-order approximations.

Limits and Continuity of Functions

This subsection introduces the foundational concepts of function limits and continuity, emphasizing their roles in mathematical analysis.

Limits of Functions

Definition 2.15 (Rudin [8]) Let $f : D \subseteq \mathbb{R} \to \mathbb{R}$ be a function defined on a domain D, and let c be a limit point of D. The limit of f as x approaches c is L, denoted by:

$$\lim_{x \to c} f(x) = L,$$

if for every $\epsilon > 0$, there exists a $\delta > 0$ such that whenever $0 < |x - c| < \delta$ and $x \in D$, it follows that $|f(x) - L| < \epsilon$.

The key properties of limits follow:
Uniqueness: If $\lim_{x \to c} f(x)$ exists, it is unique.
Algebraic Operations: If $\lim_{x \to c} f(x) = L$ and $\lim_{x \to c} g(x) = M$, then:

$$\lim_{x \to c}(f(x) + g(x)) = L + M,$$

$$\lim_{x \to c}(f(x) \cdot g(x)) = L \cdot M,$$

$$\lim_{x \to c} \frac{f(x)}{g(x)} = \frac{L}{M} \quad (\text{if } M \neq 0).$$

Squeeze Theorem: If $f(x) \leq g(x) \leq h(x)$ near c and $\lim_{x \to c} f(x) = \lim_{x \to c} h(x) = L$, then $\lim_{x \to c} g(x) = L$.

Continuity of Functions

Definition 2.16 (Rudin [8]) A function $f : D \subseteq \mathbb{R} \to \mathbb{R}$ is continuous at a point $c \in D$ if:

$$\lim_{x \to c} f(x) = f(c).$$

If f is continuous at every point in D, it is called continuous on D.

This condition states that the limit of the function as x approaches c must be equal to the value of the function at c. For a function to be continuous over an interval, it must be continuous at every point in that interval. There are some basic properties below:

Additive property: If $f(x)$ and $g(x)$ are continuous at c, then $f(x) + g(x)$ is continuous at c.

Multiplicative property: If $f(x)$ and $g(x)$ are continuous at c, then $f(x) \cdot g(x)$ is continuous at c.

Composition: If $f(x)$ is continuous at c and $g(x)$ is continuous at $f(c)$, then the composition $g(f(x))$ is continuous at c.

Applications of Limits in Asymptotic Analysis

The study of limits is central to understanding the asymptotic behavior of functions, particularly as $x \to \infty$ or $x \to -\infty$.

Definition 2.17 A line $y = L$ is a horizontal asymptote of $f(x)$ if:

$$\lim_{x \to \infty} f(x) = L \quad \text{or} \quad \lim_{x \to -\infty} f(x) = L.$$

Limits are also used to compare the growth rates of functions, a key tool in algorithm analysis and complexity theory. For example:

$$\lim_{x \to \infty} \frac{\ln x}{x} = 0 \quad \text{and} \quad \lim_{x \to \infty} \frac{e^x}{x^k} = \infty \quad (\text{for any } k > 0).$$

These results formalize the intuition that exponential functions grow faster than polynomial functions, which in turn grow faster than logarithmic functions.

The study of limits is central to understanding the asymptotic behavior of functions, particularly as $x \to \infty$ or $x \to -\infty$.

Asymptotic notation, rooted in limit theory, provides a framework for describing the behavior of functions as inputs grow large. Specifically, given two functions $f(x)$ and $g(x)$, we can use asymptotic notation to quantitatively describe their relative growth rates as $x \to \infty$. The definitions are as follows:

(1) $f(x) = O(g(x))$ as $x \to \infty$ if there exist constants $C > 0$ and $x_0 > 0$ such that for all $x > x_0$, we have

$$|f(x)| \leq C|g(x)|$$

This means that the growth rate of $f(x)$ is bounded above by $g(x)$ multiplied by a constant. Equivalently, $f(x) = O(g(x))$ if and only if

$$\limsup_{x \to \infty} \left| \frac{f(x)}{g(x)} \right| < \infty.$$

(2) $f(x) = o(g(x))$ as $x \to \infty$ if for any $\epsilon > 0$, there exists an $x_0 > 0$ such that for all $x > x_0$, we have

$$|f(x)| < \epsilon|g(x)|$$

This means that the growth rate of $f(x)$ is much smaller than that of $g(x)$, and as $x \to \infty$, $f(x)$ approaches zero relative to $g(x)$. Equivalently, $f(x) = o(g(x))$ if and only if

$$\lim_{x \to \infty} \frac{f(x)}{g(x)} = 0.$$

Derivatives and Differentials

This section explores the fundamental concepts of derivatives, gradients, and their applications in optimization.

First-Order Derivatives

Definition 2.18 (Banner [2]) Let $f : D \subseteq \mathbb{R} \to \mathbb{R}$. The derivative of f at $x = c$, denoted $f'(c)$, is defined as:

$$f'(c) = \lim_{h \to 0} \frac{f(c+h) - f(c)}{h}, \tag{2.30}$$

provided the limit exists. Geometrically, $f'(c)$ represents the slope of the tangent line to the curve $y = f(x)$ at $x = c$.

Partial Derivatives

Definition 2.19 The partial derivative of $f : \mathbb{R}^n \to \mathbb{R}$ with respect to x_i at $\mathbf{x} = (x_1, \ldots, x_n)$ is:

$$\frac{\partial f}{\partial x_i}(\mathbf{x}) = \lim_{h \to 0} \frac{f(x_1, \ldots, x_i + h, \ldots, x_n) - f(\mathbf{x})}{h}. \tag{2.31}$$

It measures the rate of change of f along the x_i-axis, holding other variables constant.

The higher-order derivatives refer to the successive derivatives of a function. The second-order derivative, for example, measures the curvature or concavity of the function. It is defined as:

$$f''(x) = \frac{d^2}{dx^2} f(x). \tag{2.32}$$

For multivariable functions, higher-order derivatives are captured by the Hessian matrix, which is a square matrix of second-order partial derivatives. The Hessian matrix of a function $f(x_1, x_2, \ldots, x_n)$ is defined as:

$$H_f(x) = \begin{pmatrix} \frac{\partial^2 f}{\partial x_1^2} & \frac{\partial^2 f}{\partial x_1 \partial x_2} & \cdots & \frac{\partial^2 f}{\partial x_1 \partial x_n} \\ \frac{\partial^2 f}{\partial x_2 \partial x_1} & \frac{\partial^2 f}{\partial x_2^2} & \cdots & \frac{\partial^2 f}{\partial x_2 \partial x_n} \\ \vdots & \vdots & \ddots & \vdots \\ \frac{\partial^2 f}{\partial x_n \partial x_1} & \frac{\partial^2 f}{\partial x_n \partial x_2} & \cdots & \frac{\partial^2 f}{\partial x_n^2} \end{pmatrix}. \tag{2.33}$$

The Hessian matrix plays an important role in optimization problems, particularly in determining the nature of critical points (whether they correspond to maxima, minima, or saddle points).

Definition 2.20 For a function $f : \mathbb{R}^n \to \mathbb{R}$, the gradient ∇f is a vector of partial derivatives:

$$\nabla f(\mathbf{x}) = \left(\frac{\partial f}{\partial x_1}, \frac{\partial f}{\partial x_2}, \ldots, \frac{\partial f}{\partial x_n} \right)^T. \tag{2.34}$$

The gradient points in the direction of steepest ascent of f, and its magnitude represents the rate of change in that direction.

Taylor Expansions

This section introduces Taylor expansions, which provide polynomial approximations for functions using derivatives.

Taylor's Theorem

Definition 2.21 (Rudin [8]) Let $f : \mathbb{R} \to \mathbb{R}$ be a function that is n-times differentiable at $x = a$. The Taylor polynomial of degree n centered at a is:

$$P_n(x) = \sum_{k=0}^{n} \frac{f^{(k)}(a)}{k!} (x - a)^k, \tag{2.35}$$

where $f^{(k)}(a)$ denotes the k-th derivative of f at a. The remainder term $R_n(x)$ satisfies:

$$f(x) = P_n(x) + R_n(x), \tag{2.36}$$

and $R_n(x) = o\left(|x-a|^n\right)$ as $x \to a$ (Peano remainder form).

Theorem 2.15 (Banner [2]) *If f is $(n+1)$-times differentiable on an interval containing a and x, then there exists c between a and x such that:*

$$R_n(x) = \frac{f^{(n+1)}(c)}{(n+1)!}(x-a)^{n+1}. \tag{2.37}$$

Definition 2.22 For a function $f : \mathbb{R}^d \to \mathbb{R}$ with continuous partial derivatives up to order n at $\mathbf{a} \in \mathbb{R}^d$, the multivariate Taylor polynomial of degree n is:

$$P_n(\mathbf{x}) = \sum_{|\alpha| \le n} \frac{D^\alpha f(\mathbf{a})}{\alpha!}(\mathbf{x}-\mathbf{a})^\alpha, \tag{2.38}$$

where $\alpha = (\alpha_1, \alpha_2, \ldots, \alpha_d)$ is a multi-index, $|\alpha| = \alpha_1 + \cdots + \alpha_d$, $\alpha! = \alpha_1! \cdots \alpha_d!$, $D^\alpha f = \frac{\partial^{|\alpha|} f}{\partial x_1^{\alpha_1} \cdots \partial x_d^{\alpha_d}}$, and $(\mathbf{x}-\mathbf{a})^\alpha = (x_1-a_1)^{\alpha_1} \cdots (x_d-a_d)^{\alpha_d}$. The remainder term $R_n(\mathbf{x})$ satisfies:

$$f(\mathbf{x}) = P_n(\mathbf{x}) + R_n(\mathbf{x}), \tag{2.39}$$

with $R_n(\mathbf{x}) = o\left(\|\mathbf{x}-\mathbf{a}\|^n\right)$ as $\mathbf{x} \to \mathbf{a}$.

Common Taylor Expansions

The common Taylor expansion formulas mainly include the exponential function, logarithmic function, and others. The Taylor series expansion of the exponential function e^x around $x = 0$ is:

$$e^x = \sum_{k=0}^{\infty} \frac{x^k}{k!}. \tag{2.40}$$

This series expresses e^x as an infinite sum of powers of x, and it converges for all real values of x. The Taylor series expansion of $\sin x$ around $x = 0$ is:

$$\sin x = \sum_{k=0}^{\infty} \frac{(-1)^k x^{2k+1}}{(2k+1)!}. \tag{2.41}$$

This series represents $\sin x$ as an alternating sum of odd powers of x, and it converges for all real x. The Taylor series expansion of $\ln(1+x)$ around $x = 0$ is:

$$\ln(1+x) = \sum_{k=1}^{\infty} \frac{(-1)^{k+1} x^k}{k}, \quad \text{for } |x| < 1. \tag{2.42}$$

This series is valid for $|x| < 1$ and provides an approximation of $\ln(1+x)$ near zero. For a function $f : \mathbb{R}^d \to \mathbb{R}$, the quadratic approximation near $\mathbf{a}$ is:

$$f(\mathbf{x}) \approx f(\mathbf{a}) + \nabla f(\mathbf{a})^T (\mathbf{x} - \mathbf{a}) + \frac{1}{2} (\mathbf{x} - \mathbf{a})^T H_f(\mathbf{a}) (\mathbf{x} - \mathbf{a}), \tag{2.43}$$

where $\nabla f(\mathbf{a})$ is the gradient and $H_f(\mathbf{a})$ is the Hessian matrix.

2.2.2 Random Variables and Distributions

This section introduces the fundamental concepts of random variables, their classifications, and key distributions.

Definition 2.23 (Ross [7]) A random variable X is a function mapping outcomes from a sample space Ω to the real numbers $\mathbb{R}$, such that for every real number x, the set $\{\omega \in \Omega \mid X(\omega) \leq x\}$ belongs to the event space $\mathscr{F}$.

A random variable X is discrete if it takes values in a countable set $\{x_1, x_2, \ldots\}$. Its behavior is characterized by a probability mass function:

$$p_X(x_i) = P(X = x_i), \quad \sum_{i=1}^{\infty} p_X(x_i) = 1. \tag{2.44}$$

A random variable X is continuous if there exists a non-negative probability density function $f_X(x)$ such that:

$$P(a \leq X \leq b) = \int_a^b f_X(x)\, dx, \quad \text{and} \quad \int_{-\infty}^{\infty} f_X(x)\, dx = 1. \tag{2.45}$$

Expectation and Variance

Definition 2.24 (Hogg [5]) The expectation (or mean) of a random variable X is defined as:

$$E[X] = \begin{cases} \sum_{i=1}^{\infty} x_i p_X(x_i) & \text{(discrete case)}, \\ \int_{-\infty}^{\infty} x f_X(x)\, dx & \text{(continuous case)}. \end{cases} \tag{2.46}$$

Definition 2.25 (Hogg [5]) The variance of X measures its dispersion and is defined as:

$$\mathrm{Var}(X) = E\left[(X - E[X])^2\right] = E[X^2] - (E[X])^2. \tag{2.47}$$

The standard deviation is $\sigma_X = \sqrt{\mathrm{Var}(X)}$.

Theorem 2.16 (Ross [7]) *For random variables X, Y and constants $a, b \in \mathbb{R}$:*
(1) Linearity of expectation: $E[aX + b] = aE[X] + b$.
(2) If X and Y are independent, $E[XY] = E[X]E[Y]$.
(3) Variance scaling: $Var(aX + b) = a^2 Var(X)$.
(4) If X and Y are independent, $Var(X + Y) = Var(X) + Var(Y)$.

Common Distributions

(1) **Normal Distribution**: A continuous random variable X follows a normal distribution $N(\mu, \sigma^2)$ with mean μ and variance $\sigma^2 > 0$ if its probability density function is:

$$f_X(x) = \frac{1}{\sigma\sqrt{2\pi}} e^{-\frac{(x-\mu)^2}{2\sigma^2}}, \tag{2.48}$$

where $EX = \mu$ is a mathematical expectation and $VarX = \sigma^2$ is a variance. If $\mu = 0$ and $\sigma^2 = 1$, then this distribution is called a standard normal distribution and denoted $N(0, 1)$.

The sum of independent and normally distributed random variables also has normal distribution. Figure 2.1 the graph of probability density function of normal distribution is given.

(2) **Uniform Distribution**: A continuous random variable X follows a uniform distribution $\mathcal{U}(a, b)$ on $[a, b]$ if its probability density function is:

$$f_X(x) = \begin{cases} \frac{1}{b-a} & x \in [a, b], \\ 0 & \text{otherwise.} \end{cases} \tag{2.49}$$

Fig. 2.1 Probability density functions of normal distribution

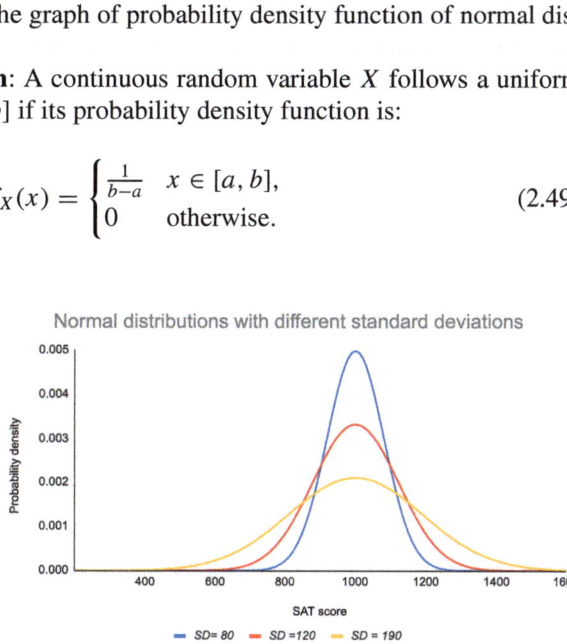

$$E(X) = \frac{a+b}{2}, \quad \text{Var}(X) = \frac{(b-a)^2}{12}. \tag{2.50}$$

(3) **Exponential Distribution**: A continuous random variable X follows an exponential distribution $\text{Exp}(\lambda)$ with rate parameter $\lambda > 0$ if its probability density function is:

$$f_X(x) = \begin{cases} \lambda e^{-\lambda x} & x \geq 0, \\ 0 & x < 0. \end{cases} \tag{2.51}$$

$$E(X) = \frac{1}{\lambda}, \quad \text{Var}(X) = \frac{1}{\lambda^2}. \tag{2.52}$$

(4) **Gamma distribution**: A continuous random variable X follows a gamma distribution $\text{Gamma}(\alpha, \beta)$ with shape parameter $\alpha > 0$ and rate parameter $\beta > 0$ if its probability density function is:

$$f_X(x) = \begin{cases} \frac{\beta^\alpha x^{\alpha-1} e^{-\beta x}}{\Gamma(\alpha)} & x \geq 0, \\ 0 & x < 0, \end{cases} \tag{2.53}$$

where $\Gamma(\alpha) = \int_0^\infty t^{\alpha-1} e^{-t}\, dt$ is the gamma function.

$$E(X) = \frac{\alpha}{\beta}, \quad \text{Var}(X) = \frac{\alpha}{\beta^2}. \tag{2.54}$$

In Fig. 2.2 there is given behavior of probability distribution function of Gamma distribution depending on the parameters, It should be noted that in the figure, the parameter $k = \alpha$ and the parameter $\theta = \frac{1}{\beta}$.

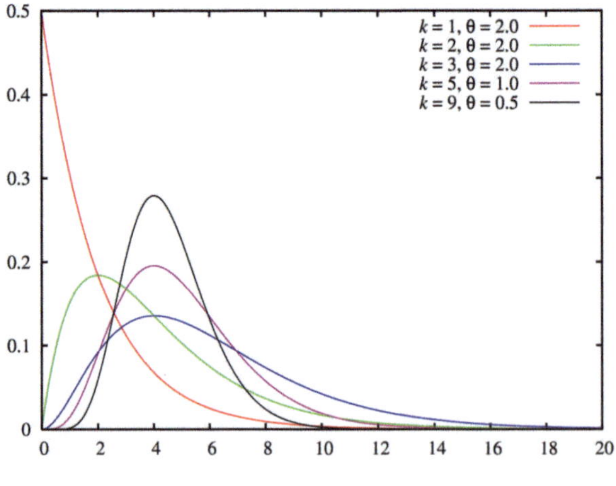

Fig. 2.2 Probability density functions of gamma distribution

(5) **Chi-Square Distribution**: A continuous random variable X follows a chi-square distribution $\chi^2(k)$ with k degrees of freedom if its probability density function is:

$$f_X(x) = \begin{cases} \frac{x^{k/2-1}e^{-x/2}}{2^{k/2}\Gamma(k/2)} & x \geq 0, \\ 0 & x < 0. \end{cases} \tag{2.55}$$

The chi-square distribution is a special case of the gamma distribution with $\alpha = \frac{k}{2}$ and $\beta = \frac{1}{2}$.

This distribution is a continuous probability distribution which is used in various applications. The chi-square distribution is determined as a sum of normal distribution in square.

$$\chi^2 = x_1^2 + x_2^2 + \ldots + x_m^2 \tag{2.56}$$

where $x_i \sim N(0, 1), i = 1, 2, \ldots, m$ and m is called the *degrees of freedom*. Chi-square distribution is a partial case of the Gamma distribution, when $\beta = \frac{1}{2}$ and $\alpha = \frac{k}{2}$. For chi-square distribution with k degrees of freedom, we have

$$E(X) = k, \quad Var(X) = 2k \tag{2.57}$$

According to central limit theorem if degrees of freedom increases then chi-square distribution can be approximated by a normal distribution $N(k, 2k)$, where k is the degrees of freedom of chi-square distribution. This fact will be used in Chap. 3, for construction of a confident band for an unknown function in regression models with an increasing number of unknown parameters.

Below in Fig. 2.3 there is given a behavior of probability density function of chi-square distribution depending on degrees of freedom.

From Fig. 2.3 we can see that if the degrees of freedom increases, then the form of the probability density function of chi-square distribution alike to the probability density function of normal distribution.

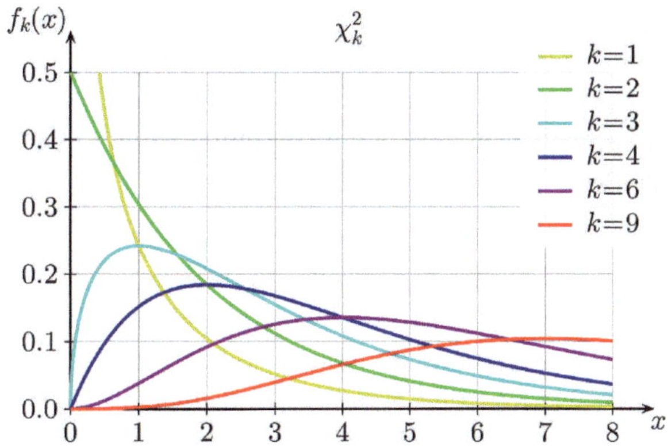

Fig. 2.3 Probability density functions of chi-square distribution

Multivariate Random Variables

Definition 2.26 (*Wackerly* [9]) A random vector $\mathbf{X} = (X_1, X_2, \ldots, X_n)^T$ follows a multivariate normal distribution $N_n(\boldsymbol{\mu}, \boldsymbol{\Sigma})$ with mean vector $\boldsymbol{\mu} \in \mathbb{R}^n$ and covariance matrix $\boldsymbol{\Sigma} \in \mathbb{R}^{n \times n}$ if its probability density function is:

$$f_{\mathbf{X}}(\mathbf{x}) = \frac{1}{(2\pi)^{n/2}|\boldsymbol{\Sigma}|^{1/2}} \exp\left(-\frac{1}{2}(\mathbf{x} - \boldsymbol{\mu})^T \boldsymbol{\Sigma}^{-1}(\mathbf{x} - \boldsymbol{\mu})\right), \tag{2.58}$$

where $|\boldsymbol{\Sigma}|$ denotes the determinant of $\boldsymbol{\Sigma}$. The mean vector $\boldsymbol{\mu}$ and covariance matrix $\boldsymbol{\Sigma}$ satisfy:

$$E[\mathbf{X}] = \boldsymbol{\mu}, \quad \mathrm{Cov}(\mathbf{X}) = \boldsymbol{\Sigma}. \tag{2.59}$$

Definition 2.27 (*Wackerly* [9]) For a random vector $\mathbf{X} = (X_1, X_2, \ldots, X_n)^T$, the covariance matrix $\boldsymbol{\Sigma}$ is an $n \times n$ symmetric matrix defined by:

$$\boldsymbol{\Sigma} = E\left[(\mathbf{X} - \boldsymbol{\mu})(\mathbf{X} - \boldsymbol{\mu})^T\right], \tag{2.60}$$

where $\boldsymbol{\mu} = E[\mathbf{X}]$. The (i, j)-th entry of $\boldsymbol{\Sigma}$ is:

$$\Sigma_{ij} = \mathrm{Cov}(X_i, X_j) = E[(X_i - \mu_i)(X_j - \mu_j)]. \tag{2.61}$$

The covariance matrix is positive semi-definite, i.e., $\mathbf{v}^T \boldsymbol{\Sigma} \mathbf{v} \geq 0$ for all $\mathbf{v} \in \mathbb{R}^n$.

Theorem 2.17 (Hogg [5]) *For* $\mathbf{X} \sim N_n(\boldsymbol{\mu}, \boldsymbol{\Sigma})$*:*

(1) Linear transformations preserve normality: If $\mathbf{Y} = \mathbf{A}\mathbf{X} + \mathbf{b}$ *with* $\mathbf{A} \in \mathbb{R}^{m \times n}$ *and* $\mathbf{b} \in \mathbb{R}^m$*, then* $\mathbf{Y} \sim N_m(\mathbf{A}\boldsymbol{\mu} + \mathbf{b}, \mathbf{A}\boldsymbol{\Sigma}\mathbf{A}^T)$.

(2) Marginal distributions are normal: Any subvector of $\mathbf{X}$ *is also multivariate normal.*

(3) Independence and uncorrelatedness: If $\boldsymbol{\Sigma}$ *is diagonal, the components of* $\mathbf{X}$ *are mutually independent.*

2.2.3 Measure and Probability Space

This section provides a concise introduction to the basic concepts of measure and probability spaces.

Measure and Probability Space

When investigating the notion of "size", the fundamental strategy is to assign a "measure" to certain sets, thereby enabling numerical descriptions of their "size" or "likelihood" In real analysis, such assignments typically rely on a measure-theoretic approach.

Definition 2.28 (*Feller* [4]) Let F be a function defined on certain designated sets of a space Ω (e.g., intervals, rectangles), possessing the property that its values on disjoint sets can be summed. If this property holds for countably many pairwise disjoint sets, we say F is countably additive (or completely additive). Formally,

$$F\Big(\bigcup_{n=1}^{\infty} A_n\Big) = \sum_{n=1}^{\infty} F(A_n), \tag{2.62}$$

whenever $\{A_n\}$ are pairwise disjoint.

Definition 2.29 (*Feller* [4]) Let Ω be a nonempty set, and let $\mathscr{F}$ be a collection of subsets of Ω that is closed under countable unions, countable intersections, and complements (that is, $\mathscr{F}$ is a σ-algebra). Suppose there exists a function

$$\mu : \mathscr{F} \longrightarrow [0, +\infty] \tag{2.63}$$

satisfying the following two conditions:

(1) $\mu(\varnothing) = 0$.

(2) Countable Additivity (σ-Additivity): For any countable family of pairwise disjoint sets $\{A_n\}_{n=1}^{\infty} \subset \mathscr{F}$,

$$\mu\Big(\bigcup_{n=1}^{\infty} A_n\Big) = \sum_{n=1}^{\infty} \mu(A_n). \tag{2.64}$$

Then μ is called a measure on $(\Omega, \mathscr{F})$.

In the context of probability theory, if a measure μ additionally satisfies

$$\mu(\Omega) = 1, \tag{2.65}$$

it is called a probability measure, commonly denoted by P. In this situation, this yields a probability space:

$$(\Omega,\ \mathscr{F},\ P).$$

Definition 2.30 A *probability space* is typically denoted by $(\Omega, \mathscr{F}, P)$, where:

(1) Ω: the sample space, representing all possible basic outcomes.

(2) $\mathscr{F}$: a σ-algebra of subsets of Ω that can be measured or observed.

(3) P: a *probability measure* defined on $\mathscr{F}$, satisfying $P(\Omega) = 1$ and countable additivity.

Under this framework, every $A \in \mathscr{F}$ is viewed as a "measurable event", and $P(A)$ is the probability that this event occurs.

Hence, a probability measure can be regarded as the concept of a "regular measure" enhanced with the additional requirements of non-negativity and total mass equal to 1. Through this normalization constraint, one interprets the measure as a "probability distribution" on Ω, establishing the rigorous axiomatic foundation for further study of random variables and stochastic processes.

σ-Algebra and Measurable Sets

In practical problems, we do not necessarily need to assign a measure to *all* subsets of Ω. Instead, one typically picks a sufficiently large family of subsets $\mathscr{A}$ satisfying:

(1) $\Omega \in \mathscr{A}$ and $\varnothing \in \mathscr{A}$.

(2) If $A \in \mathscr{A}$, then its complement A^c also belongs to $\mathscr{A}$.

(3) If $\{A_n\} \subset \mathscr{A}$, then $\bigcup_n A_n \in \mathscr{A}$.

Such a family of subsets $\mathscr{A}$ is called a σ*-algebra*. Being closed under set operations (unions, intersections, complements) and countable operations makes it convenient for defining a countably additive measure.

For $\mathbb{R}^n$, taking the smallest σ-algebra generated by open sets, closed sets, intervals (or rectangles) yields the *Borel* σ*-algebra*. Most probability measures in elementary contexts are defined on a Borel σ-algebra of $\mathbb{R}^n$.

Random Variables

A *random variable* X is a measurable function on Ω, meaning that the set $\{\omega \in \Omega : X(\omega) \leq t\}$ lies in $\mathscr{F}$ for each real t. Each random variable X induces a distribution function on the real axis:

$$F_X(t) = P\{X \leq t\}, \tag{2.66}$$

which is non-decreasing, right-continuous, with $\lim_{t \to -\infty} F_X(t) = 0$ and $\lim_{t \to +\infty} F_X(t) = 1$.

Through F_X, one can define a probability measure on $\mathbb{R}$ called the distribution of X. Various operations on random variables (e.g., addition, limits) remain in the measurable framework, paving the way for constructing more complex stochastic processes.

Integration and Expectation

Within measure theory, if X is a random variable, one usually defines its expectation via

$$E[X] = \int_{\Omega} X(\omega)\, P(d\omega).$$

When describing X in terms of its distribution function F on the real line, one may write the expectation as a Lebesgue–Stieltjes integral:

$$E[X] = \int_{-\infty}^{+\infty} x \, dF(x).$$

This parallels the construction of the Riemann–Stieltjes integral on finite intervals, but the measure-theoretic approach is more readily extended to higher dimensions. For non-negative or absolutely integrable random variables, the above integral is well defined.

Theorem 2.18 (Monotone Convergence Theorem (Feller [4])) *Let* $(\Omega, \mathscr{F}, P)$ *be a probability space, and let* $\{X_n\}_{n=1}^{\infty}$ *be a sequence of non-negative measurable functions satisfying the following conditions:*

(1) For all n, $X_n \leq X_{n+1}$, *i.e.,* $\{X_n\}$ *is increasing;*

(2) There exists a measurable function $X = \lim_{n\to\infty} X_n$, *and* $X_n \to X$ *pointwise as* $n \to \infty$.

Then, we have:

$$\lim_{n\to\infty} E[X_n] = E[\lim_{n\to\infty} X_n] = E[X]. \tag{2.67}$$

Lemma 2.2 (Fatou's Lemma (Feller [4])) *Let* $(\Omega, \mathscr{F}, P)$ *be a probability space, and let* $\{X_n\}_{n=1}^{\infty}$ *be a sequence of measurable functions. If* $X_n \geq 0$ *for all* n, *then the following inequality holds:*

$$E\left[\liminf_{n\to\infty} X_n\right] \leq \liminf_{n\to\infty} E[X_n]. \tag{2.68}$$

2.2.4 Limit Theorems

This section introduces fundamental limit theorems in probability theory, including the Convergence, Law of Large Numbers, the Central Limit Theorem, and Chebyshev's Inequality.

Convergence in Measure and Distribution

Let $\{F_n\}_{n=1}^{\infty}$ be a sequence of distribution functions defined on $\mathbb{R}$, and F be another distribution function.

Definition 2.31 (*Feller* [4]) A sequence $\{F_n\}$ is said to *converge* to F, denoted by

$$F_n \Longrightarrow F,$$

if for every bounded continuous interval I, we have

$$F_n\{I\} = \int_I dF_n(x) \longrightarrow \int_I dF(x) = F\{I\}. \tag{2.69}$$

If F has total mass equal to 1, this convergence is called *strong (true) convergence*; if the total mass of F is less than 1, this is called *deficient convergence*.

This definition refers to convergence in *distribution* or *measure*. It is equivalent to the notion of pointwise convergence at every continuous point in elementary probability theory but is more general in the context of measure theory.

Definition 2.32 (*Feller* [4]) Let $\{X_n\}$ be a sequence of random variables, and X be another random variable (or X_n and X are of the same type). If for every continuity point x of the distribution function F_X, we have

$$\lim_{n\to\infty} F_{X_n}(x) = F_X(x), \tag{2.70}$$

then $\{X_n\}$ converges in distribution to X, denoted as

$$X_n \xrightarrow{d} X.$$

Convergence in distribution refers to the convergence of the distribution functions. That is, $F_{X_n}(x)$ converges to $F_X(x)$ for every continuity point of the distribution function of X. Unlike convergence in probability or almost sure convergence, convergence in distribution does not require pointwise convergence of the random variables, but rather the convergence of their distribution functions.

Definition 2.33 (*Feller* [4]) Let $\{X_n\}$ be a sequence of random variables defined on a probability space $(\Omega, \mathscr{F}, P)$. If

$$\lim_{n\to\infty} P\left(|X_n - X| > \varepsilon\right) = 0, \quad \forall \varepsilon > 0, \tag{2.71}$$

then X_n converges in probability to X, denoted as

$$X_n \xrightarrow{P} X.$$

Definition 2.34 (*Feller* [4]) Let $\{X_n\}$ be a sequence of random variables defined on $(\Omega, \mathscr{F}, P)$. If

$$P\left(\lim_{n\to\infty} |X_n(\omega) - X(\omega)| = 0\right) = 1, \tag{2.72}$$

then X_n converges almost surely to X, denoted as

$$X_n \xrightarrow{a.s.} X.$$

Almost sure convergence is stronger than convergence in probability. In fact, almost sure convergence requires that X_n converges to X at almost every sample point in Ω, except possibly on a set of probability zero. In contrast, convergence in probability only requires that the probability of large deviations tends to zero.

The relationship between the three types of convergence is:

(1) *Almost Sure Convergence* ⇒ *Convergence in Probability*: Almost sure convergence implies convergence in probability, as almost sure convergence is a stronger form of convergence.

(2)*Convergence in Probability* ⇒ *Convergence in Distribution*: Convergence in probability also implies convergence in distribution, meaning convergence in probability is stronger than convergence in distribution.

(3) *Convergence in Distribution* is generally weaker than both *Convergence in Probability* and *Almost Sure Convergence*, meaning convergence in distribution does not necessarily imply the other two, but the reverse can hold.

Thus, almost sure convergence is the strongest, followed by convergence in probability, and convergence in distribution is the weakest.

Convergence is a fundamental concept in probability theory. The following convergence theorems are crucial to the discussions in this book.

Theorem 2.19 (Slutsky's Theorem (Billingsley [3])) *Let X_n and Y_n be a sequence of random variables, and let c be a constant. If the following conditions hold:*

(1) $X_n \xrightarrow{d} X$, i.e., X_n converges in distribution to X;

(2) $Y_n \xrightarrow{p} y$, i.e., Y_n converges in probability to a constant y;

Then the following results hold:

(1) $X_n + Y_n \xrightarrow{d} X + y$, i.e., $X_n + Y_n$ converges in distribution to $X + y$;

(2) $X_n \cdot Y_n \xrightarrow{d} X \cdot y$, i.e., $X_n \cdot Y_n$ converges in distribution to $X \cdot y$;

(3) If $y = 0$, then $X_n \cdot Y_n \xrightarrow{d} 0$, i.e., $X_n \cdot Y_n$ converges in distribution to 0.

Theorem 2.20 (Prokhorov's Theorem (Billingsley [3])) *Let X be a metric space, and $\mathscr{P}(X)$ be the set of all probability measures on X. If a sequence of probability measures $\{\mu_n\} \subset \mathscr{P}(X)$ satisfies:*

$$\sup_n \int_X \exp(\alpha|x|)\, d\mu_n(x) < \infty \quad \textit{for all} \quad \alpha > 0, \tag{2.73}$$

then the sequence $\{\mu_n\}$ is tight in the weak topology, meaning there exists a subsequence $\{\mu_{n_k}\}$ and a probability measure $\mu \in \mathscr{P}(X)$ such that μ_{n_k} converges weakly to μ, i.e., for all continuous bounded functions f,

$$\lim_{k\to\infty} \int_X f(x)\, d\mu_{n_k}(x) = \int_X f(x)\, d\mu(x). \tag{2.74}$$

Theorem 2.21 (Levy–Cramer Continuity Theorem (Feller [4])) *Let $\{X_n\}$ be a sequence of random variables, and X be a random variable, with their characteristic functions given by $\phi_{X_n}(t) = E[e^{itX_n}]$ and $\phi_X(t) = E[e^{itX}]$, where $t \in \mathbb{R}$ and i is the imaginary unit. If the following conditions hold:*

(1) $X_n \xrightarrow{d} X$*, i.e.,* X_n *converges in distribution to* X*;*

(2) $\phi_{X_n}(t)$ *converges uniformly to* $\phi_X(t)$ *for all* $t \in \mathbb{R}$*; then* X_n *converges in distribution to* X*, i.e.,*

$$X_n \xrightarrow{d} X.$$

Law of Large Numbers

Theorem 2.22 (Ross [7]) *Let* $X_1, X_2, \ldots, X_n$ *be a sequence of independent and identically distributed (i.i.d.) random variables with finite mean* $\mu = E[X_i]$*. The* ***Weak Law of Large Numbers*** *states that the sample average converges in probability to* μ*:*

$$\lim_{n\to\infty} P\left(\left|\frac{1}{n}\sum_{i=1}^{n} X_i - \mu\right| \geq \epsilon\right) = 0 \quad \text{for all } \epsilon > 0. \tag{2.75}$$

Theorem 2.23 (Ross [7]) *Under the same conditions as the WLLN, the* ***Strong Law of Large Numbers*** *asserts almost sure convergence of the sample average:*

$$P\left(\lim_{n\to\infty} \frac{1}{n}\sum_{i=1}^{n} X_i = \mu\right) = 1. \tag{2.76}$$

This result shows that for large enough samples, the sample mean is a good estimator of the population mean. The law is an essential concept in statistics, particularly in the field of estimation and hypothesis testing.

Central Limit Theorem

Theorem 2.24 (Lindeberg-Feller Central Limit Theorem (Feller [4])) *Let* $X_1, X_2, \ldots, X_n$ *be a sequence of independent and identically distributed random variables, with their mean and variance given by* $E[X_i] = \mu$ *and* $Var(X_i) = \sigma^2$*. Define*

$$S_n = \sum_{i=1}^{n} X_i, \tag{2.77}$$

and let T_n *be the standardized sum:*

$$T_n = \frac{S_n}{\sqrt{n\sigma^2}}. \tag{2.78}$$

If there exists a constant $\alpha > 0$ *such that:*

$$\frac{1}{\sigma^2}\sum_{i=1}^{n} E\left[(X_i-\mu)^2\mathbf{1}_{|X_i-\mu|>\alpha\sqrt{n}}\right] \to 0 \quad as \quad n \to \infty, \tag{2.79}$$

then T_n *converges in distribution to the standard normal distribution, i.e.,*

$$T_n \xrightarrow{d} N(0,1).$$

Theorem 2.25 (Central Limit Theorem (Wackerly [9])) *Let* $X_1, X_2, \ldots, X_n$ *be i.i.d. random variables with finite mean* μ *and variance* $\sigma^2 > 0$. *Define the sample mean* $\bar{X}_n = \frac{1}{n}\sum_{i=1}^{n} X_i$. *The standardized sum converges in distribution to a standard normal variable:*

$$\frac{\bar{X}_n - \mu}{\sigma/\sqrt{n}} \xrightarrow{d} N(0,1) \quad as\, n \to \infty. \tag{2.80}$$

Equivalently, for large n, the distribution of $\bar{X}_n$ *is approximately:*

$$\bar{X}_n \dot{\sim} N\left(\mu, \frac{\sigma^2}{n}\right). \tag{2.81}$$

The CLT is a foundational result in statistics, as it allows the use of normal approximation in a wide range of statistical procedures, such as confidence intervals and hypothesis tests, even when the underlying distribution is not normal.

Chebyshev's Inequality

Theorem 2.26 (Billingsley [3]) *For any random variable X with finite mean* μ *and variance* σ^2, *and for any* $k > 0$:

$$P\left(|X-\mu| \geq k\sigma\right) \leq \frac{1}{k^2}. \tag{2.82}$$

This inequality provides an upper bound on the probability that X deviates from its mean by more than k standard deviations.

Theorem 2.27 (Ross [7]) *For i.i.d. random variables* $X_1, \ldots, X_n$ *with mean* μ *and variance* σ^2, *the sample mean* $\bar{X}_n$ *satisfies:*

$$P\left(\left|\bar{X}_n - \mu\right| \geq \epsilon\right) \leq \frac{\sigma^2}{n\epsilon^2} \quad for\ all\ \epsilon > 0. \tag{2.83}$$

2.2.5 Parameter Estimation Techniques

This section introduces key concepts in parameter estimation, including point estimation, interval estimation, least squares estimation, and maximum likelihood estimation.

Point and Interval Estimation

Definition 2.35 (*Hogg* [5]) A point estimator $\hat{\theta}$ is a statistic used to estimate an unknown parameter θ in a population. Formally, given a random sample $X_1, X_2, \ldots, X_n$, a point estimator is a function $\hat{\theta} : \mathbb{R}^n \to \mathbb{R}$. The quality of $\hat{\theta}$ is measured by properties such as unbiasedness:

$$E[\hat{\theta}] = \theta. \tag{2.84}$$

Definition 2.36 (*Wackerly* [9]) An interval estimator provides a range $(L(X_1, \ldots, X_n), U(X_1, \ldots, X_n))$ such that:

$$P(L \leq \theta \leq U) = 1 - \alpha, \tag{2.85}$$

where $1 - \alpha$ is the confidence level. This interval is called a $100(1 - \alpha)\%$ confidence interval for θ. $[L, U]$ is a $(1 - \alpha)$ confidence interval if it covers the true parameter θ with probability at least $1 - \alpha$.

The confidence level, denoted as a percentage (e.g., 90, 95, or 99%), represents the probability that a confidence interval, constructed from a random sample, contains the true value of the corresponding population parameter. Formally, for a given significance level α, the confidence level is defined as $1 - \alpha$, indicating that if the sampling process was repeated independently a large number of times, the proportion of confidence intervals capturing the true parameter value would converge to $(1 - \alpha) \times 100\%$.

Let $\theta \in \mathbb{R}^p$ be an unknown parameter, and let $\hat{\theta}$ be an estimator of θ. Define the deviation vector as:

$$\sqrt{N}(\hat{\theta} - \theta),$$

where N is the sample size. Denote C as the covariance matrix of this deviation vector, given by:

$$C = NE\left[(\hat{\theta} - \theta)(\hat{\theta} - \theta)^T\right].$$

Assuming that the deviation vector follows a multivariate normal distribution, its cumulative distribution function is denoted as $F(\cdot)$.

Let $\gamma(\alpha)$ be the quantile of $F(\cdot)$ at the significance level α. The *random ellipsoid* is then defined as:

$$\varepsilon_\gamma(\theta) = \left\{\theta : N(\theta - \hat{\theta})^T C^{-1}(\theta - \hat{\theta}) \leq \gamma(\alpha)\right\}.$$

This inequality holds with a probability of at least $1 - \alpha$. Consequently, the region $\varepsilon_\gamma(\theta)$ is referred to as the *confidence ellipsoid*, which represents a confidence region for θ and is particularly useful for describing the joint confidence interval of multi-dimensional parameters.

Quantiles

In statistics and the theory of probability, quantiles are cut points dividing the range of a probability distribution into continuous intervals with equal probabilities or dividing the observations in a sample in the same way. There is one fewer quantile than the number of groups created. Common quantiles have special names, such as *quartiles* (four groups), deciles (ten groups), and percentiles (100 groups). The groups created are termed halves, thirds, quarters, etc., though sometimes the terms for the quantile are used for the groups created, rather than for the cut points.

In Fig. 2.4 the quantile for the probability density of a normal distribution, with quantiles shown. The area below the red curve is the same in the intervals $(-\infty, Q_1)$, (Q_1, Q_2), (Q_2, Q_3), and $(Q_3, +\infty)$.

The q-quantiles are values that partition a finite set of values into q subsets of (nearly) equal sizes. There are $q - 1$ partitions of the q-quantiles, one for each integer k satisfying $0 < k < q$.

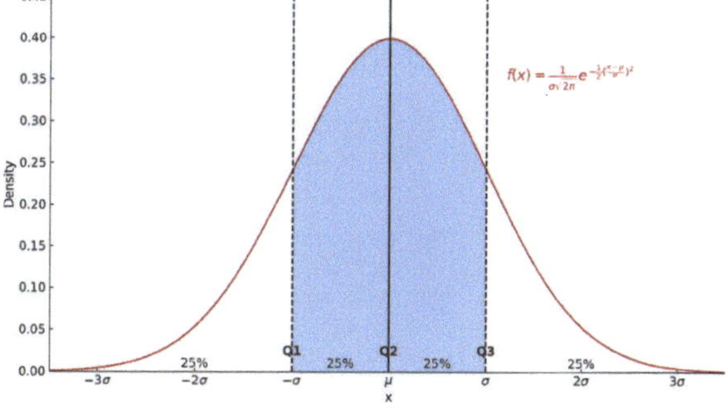

Fig. 2.4 Quantile of level 0.75

References

1. Sheldon Axler. *Linear Algebra Done Right*. Undergraduate Texts in Mathematics. Springer Nature, Cham, 4 edition, 2024.
2. Banner, Adrian D. *The calculus lifesaver : all the tools you need to excel at calculus*. A Princeton lifesaver study guide. Princeton University Press, Princeton, NJ, 2007.
3. Patrick Billingsley. *Convergence of Probability Measures*. Wiley Series in Probability and Statistics. John Wiley & Sons, New York, 2nd edition, 1999.
4. Feller, William. *An introduction to probability theory and its applications*. A Wiley publication in mathematical statistics. Wiley, New York, 2nd ed.. edition, 1957.
5. Robert V Hogg. *Introduction to mathematical statistics*. Prentice Hall, Upper Saddle River, N, J., 5th ed.. edition, 1995.
6. Serge Lang. *Linear Algebra*. Springer Nature, Netherlands, 2013.
7. author Ross, Sheldon M. *A first course in probability*. Huazhang Mathematics Original Boutique Series. Ninth edition.. edition, 2017.
8. Rudin, Walter and Walter Rudin. *Principles of mathematical analysis*. International series in pure and applied mathematics. McGraw-Hill, New York, 3d ed.. edition, 1976.
9. Wackerly, Dennis D. *Mathematical statistics with applications*. Duxbury, Pacific Grove, CA, 6th ed.. edition, 2001.

Chapter 3
Fixed-Parameter Prototype Regression Models

Ich verstehe unter dem Worte Beweis nicht den Sinn der Juristen, welche zwei halbe Beweise zu einem ganzen machen, sondern den des Mathematikers, wo ½ Beweis = 0 ist, und es wird durch den Beweis gefordert, dass jede Zweifel unmöglich wird.
— Carl Friedrich Gauss (1777–1855)

Linear regression models play an important role in various applications, in psychology, medicine, biology, engineering in other fields. The development of regression analysis leads to the creation of new approaches and methods in various branches of mathematics, in the theory of probability, stochastic processes, mathematical statistics, simulation, and others. There are various authors who have made essential contributions for development of linear regression models with noncorrelated errors of observations, published books [41, 52, 58, 60] and numerous scientific papers [3, 9, 14, 19, 40, 42, 48, 51, 53, 62]. The difference between models considered in the book that is in one point of observation there is only one response which does not to estimate the variances of the errors of observations. One of the main problems in this direction is a convergence of various estimators to the true value of unknown parameters. There are a lot of papers where such problems are investigated [6, 16, 22, 28, 63].

3.1 Linear Prototype Models with Uncorrelated Errors

Below we first consider a simple linear model with a fixed number of unknown parameters and noncorrelated errors of observations, and afterward, we pass to more complicated models, including the nonlinear case.

A. Hajiyev and J. Xu, *High-Dimensional Regression Analysis and Artificial Intelligence*,
https://doi.org/10.1007/978-981-95-2514-0_3

3.1.1 Least Squares Prototype for Confidence Regions

According to Stephen M. Stigler, *the method of least squares is the automobile of modern statistical analysis. Despite its limitations, occasional accidents, and incidental pollution, it and its numerous variations, extensions, and related conveyances carry the bulk of statistical analyses and are known and valued by nearly all.* The Least Squares method is a statistical technique used to find the equation of the best-fitting curve or line to a set of data points by minimizing the sum of the squared differences between the observed values and the values predicted by the model. This method aims at minimizing the sum of squares of deviations as much as possible. Consider the plot that is approximated by a straight line, see Fig. 3.1. When finding the line of best fit, we need to minimize the errors between the line and the points. We can either minimize the X errors or minimize the Y errors. These two methods do not necessarily produce the same line.

In this book, we are interested in the cases when it is necessary to minimize Y errors, although there are papers where X errors are minimized [67, 68].

Consider the linear model:

$$\mathbf{Y} = \mathbf{X}\boldsymbol{\beta} + \boldsymbol{\varepsilon}, \tag{3.1}$$

where $\mathbf{Y} \in \mathbb{R}^n$ is the response vector, $\mathbf{X} \in \mathbb{R}^{n\times p}$ is the design matrix, $\boldsymbol{\beta} \in \mathbb{R}^p$ is the parameter vector, and $\boldsymbol{\varepsilon} \in \mathbb{R}^n$ is the error vector with $E[\boldsymbol{\varepsilon}] = \mathbf{0}$.

The least squares estimator $\hat{\boldsymbol{\beta}}$ minimizes the residual sum of squares (RSS):

$$\text{RSS}(\boldsymbol{\beta}) = \|\mathbf{Y} - \mathbf{X}\boldsymbol{\beta}\|^2.$$

Assuming $\mathbf{X}^T\mathbf{X}$ is invertible, the LSE is given by:

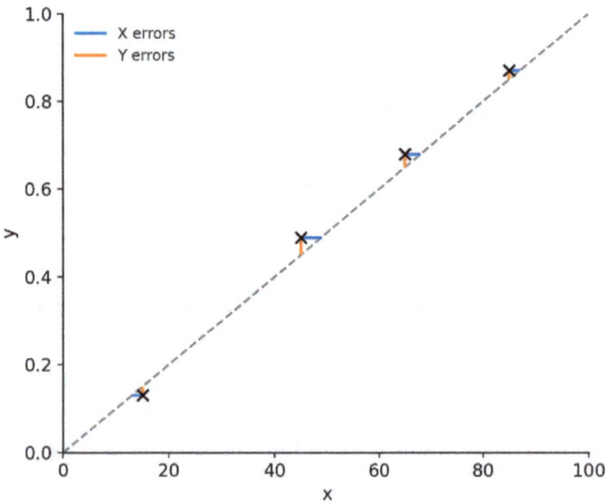

Fig. 3.1 Illustration of horizontal and vertical residuals in the least squares

$$\hat{\boldsymbol{\beta}} = (\mathbf{X}^T\mathbf{X})^{-1}\mathbf{X}^T\mathbf{Y}. \tag{3.2}$$

The estimator is unbiased, i.e., $E[\hat{\boldsymbol{\beta}}] = \boldsymbol{\beta}$, with covariance matrix:

$$\text{Cov}(\hat{\boldsymbol{\beta}}) = \sigma^2(\mathbf{X}^T\mathbf{X})^{-1}, \tag{3.3}$$

where $\sigma^2 = E[\varepsilon_i^2]$.

To derive $\hat{\boldsymbol{\beta}}$, consider the residual sum of squares (RSS) for the linear model:

$$\text{RSS}(\boldsymbol{\beta}) = \|\mathbf{Y} - \mathbf{X}\boldsymbol{\beta}\|^2,$$

where $\mathbf{Y} \in \mathbb{R}^n$ represents the vector of observed values, $\mathbf{X} \in \mathbb{R}^{n\times p}$ is the matrix of explanatory variables, and $\boldsymbol{\beta} \in \mathbb{R}^p$ is the vector of unknown coefficients.

Next, differentiate the RSS with respect to $\boldsymbol{\beta}$. Expanding the squared norm, we have:

$$\|\mathbf{Y} - \mathbf{X}\boldsymbol{\beta}\|^2 = (\mathbf{Y} - \mathbf{X}\boldsymbol{\beta})^T(\mathbf{Y} - \mathbf{X}\boldsymbol{\beta}).$$

Differentiating this expression with respect to $\boldsymbol{\beta}$ results in:

$$\frac{\partial}{\partial\boldsymbol{\beta}}\|\mathbf{Y} - \mathbf{X}\boldsymbol{\beta}\|^2 = \frac{\partial}{\partial\boldsymbol{\beta}}\left(\mathbf{Y}^T\mathbf{Y} - 2\mathbf{Y}^T\mathbf{X}\boldsymbol{\beta} + \boldsymbol{\beta}^T\mathbf{X}^T\mathbf{X}\boldsymbol{\beta}\right).$$

Since $\mathbf{Y}^T\mathbf{Y}$ is constant, its derivative with respect to $\boldsymbol{\beta}$ vanishes. The derivative of the remaining terms is:

$$-2\mathbf{X}^T\mathbf{Y} + 2\mathbf{X}^T\mathbf{X}\boldsymbol{\beta}.$$

Therefore, the derivative of the RSS with respect to $\boldsymbol{\beta}$ becomes:

$$\frac{\partial}{\partial\boldsymbol{\beta}}\|\mathbf{Y} - \mathbf{X}\boldsymbol{\beta}\|^2 = -2\mathbf{X}^T(\mathbf{Y} - \mathbf{X}\boldsymbol{\beta}).$$

To find the minimum of the RSS, set the derivative equal to zero:

$$-2\mathbf{X}^T(\mathbf{Y} - \mathbf{X}\boldsymbol{\beta}) = 0.$$

Simplifying this equation yields:

$$\mathbf{X}^T\mathbf{X}\boldsymbol{\beta} = \mathbf{X}^T\mathbf{Y}.$$

Solving for $\boldsymbol{\beta}$, the ordinary least squares (OLS) estimate is obtained:

$$\hat{\boldsymbol{\beta}} = (\mathbf{X}^T\mathbf{X})^{-1}\mathbf{X}^T\mathbf{Y}.$$

This completes the derivation of the least squares estimator $\hat{\boldsymbol{\beta}}$.

Below we consider the linear regression models where an unknown function can be approximated by a polynomial of the degree m and call such models *simplest linear regression* models. Assume that, as a matter of experience, there is observed a response

$$y_i = f(x_i) + \varepsilon_i \tag{3.4}$$

where $f(x)$ is an unknown function which describes the behavior of an experiment.

It is assumed that $f(x)$ can be represented in the following form:

$$f(x) = f_m(x, \theta) = \sum_{k=0}^{m} \theta_k x^k \tag{3.5}$$

where $\theta_1, \theta_2, \ldots, \theta_m$ is unknown parameters which is necessary to estimate and $\varepsilon(i)$ is an error of observation, $x_i (i = 1, 2, \ldots, N)$ is a point of observations, N is a number of observations. It means that unknown function $f(x)$ can be approximated by polynomial of the degree m. The relation (3.4) together with (3.5) is called a simple linear regression(SLR). For this case, the design matrix has the following form

$$X = \begin{pmatrix} 1 & x_1 & x_1^2 & \cdots & x_1^m \\ 1 & x_2 & x_2^2 & \cdots & x_2^m \\ \vdots & \vdots & \vdots & \ddots & \vdots \\ 1 & x_N & x_N^2 & \cdots & x_N^m \end{pmatrix}$$

And Fisher's information matrix has the following form:

$$X^T X = \begin{pmatrix} N & \sum_{k=1}^{N} x_k & \sum_{k=1}^{N} x_k^2 & \cdots & \sum_{k=1}^{N} x_k^m \\ \sum_{k=1}^{N} x_k & \sum_{k=1}^{N} x_k^2 & \sum_{k=1}^{N} x_k^3 & \cdots & \sum_{k=1}^{N} x_k^{m+1} \\ \vdots & \vdots & \vdots & \ddots & \vdots \\ \sum_{k=1}^{N} x_k^m & \sum_{k=1}^{N} x_k^{m+1} & \cdots & \cdots & \sum_{k=1}^{N} x_k^{2m} \end{pmatrix}$$

As the points x_i can be chosen from an experiment then we will collect the points when $det(\frac{X^T T}{N}) \neq 0$ for any N. The relation (3.4) can be rewritten in the following vector form $Y = X\theta + \varepsilon$, where $Y = (y_1, y_2, \ldots, y_N)^T$ is the vector of random errors.

It can be assumed that $\varepsilon_1, \varepsilon_2, \ldots, \varepsilon_N$ independent and identically distributed random variables having a normal (Gaussian) distribution with

$$E[\varepsilon_i] = 0, \quad E[\varepsilon_i^2] = \sigma_i^2 \le \sigma_0^2$$

Although later all results will be extended for any distribution.

If all ${\sigma_i}^2 = \sigma^2$ then $\hat{\theta} - \theta = (X^T X)^{-1} X^T Y$ is called *least square estimators (l.s.e.)* with the *same variances (s.v.)*.

Remark 3.1 If $\sigma_i^2 = \sigma^2$ for all i and $E(\varepsilon_i \varepsilon_j) = 0$ for $i \neq j$, then these assumptions for the regression model

$$\mathbf{y} = X\boldsymbol{\beta} + \boldsymbol{\varepsilon}$$

are known as the *Gauss–Markov conditions*. Under these conditions, the least squares estimator (*l.s.e.*)

$$\widehat{\boldsymbol{\beta}} = (X^T X)^{-1} X^T \mathbf{y}$$

is optimal in the class of linear unbiased estimators. This result is the Gauss–Markov theorem [10].

Proof Under the stated assumptions, $E[\boldsymbol{\varepsilon}] = \mathbf{0}$, $\mathrm{Var}(\varepsilon_i) = \sigma^2$, and $E(\varepsilon_i \varepsilon_j) = 0$ for $i \neq j$. These imply $\mathrm{Var}(\boldsymbol{\varepsilon}) = \sigma^2 I_n$. The estimator

$$\widehat{\boldsymbol{\beta}} = (X^T X)^{-1} X^T \mathbf{y} = \boldsymbol{\beta} + (X^T X)^{-1} X^T \boldsymbol{\varepsilon}$$

is unbiased because $E[\boldsymbol{\varepsilon}] = \mathbf{0}$ leads to $E[\widehat{\boldsymbol{\beta}}] = \boldsymbol{\beta}$. Moreover,

$$\mathrm{Var}(\widehat{\boldsymbol{\beta}}) = \mathrm{Var}\big((X^T X)^{-1} X^T \boldsymbol{\varepsilon}\big) = \sigma^2 (X^T X)^{-1}.$$

Next, let $\widetilde{\boldsymbol{\beta}} = A\mathbf{y}$ be any other linear unbiased estimator of $\boldsymbol{\beta}$. Unbiasedness means

$$E[\widetilde{\boldsymbol{\beta}}] = AE[\mathbf{y}] = AX\boldsymbol{\beta} = \boldsymbol{\beta},$$

which forces $AX = I_p$. One can then write

$$A = (X^T X)^{-1} X^T + R, \quad \text{where } RX = \mathbf{0}.$$

Since $\mathrm{Var}(\widetilde{\boldsymbol{\beta}}) = \sigma^2 AA^T$, the difference in covariance matrices becomes

$$\mathrm{Var}(\widetilde{\boldsymbol{\beta}}) - \mathrm{Var}(\widehat{\boldsymbol{\beta}}) = \sigma^2 AA^T - \sigma^2 (X^T X)^{-1} = \sigma^2 RR^T.$$

The matrix RR^T is symmetric and positive semidefinite, which implies

$$\mathrm{Var}(\widetilde{\boldsymbol{\beta}}) - \mathrm{Var}(\widehat{\boldsymbol{\beta}})$$

is positive semidefinite. Therefore, no linear unbiased estimator can have a smaller variance than $\widehat{\boldsymbol{\beta}}$, making it the *best linear unbiased estimator* (*b.l.u.e.*) under the Gauss–Markov conditions. □

Consider the deviation vector:

$$\hat{\theta} - \theta = (X^T X)^{-1} X^T Y - \theta = ((X^T X)^{-1} X^T)(X\theta + \varepsilon) - \theta = (X^T X)^{-1} X^T \varepsilon \tag{3.6}$$

Since ε_i, for $i = 1, 2, \ldots, N$, are independent random variables with normal distribution $N(0, \sigma_i^2)$, it follows from Eq. (3.6) that each component of the deviation vector $(\hat{\theta} - \theta)$ also follows a normal distribution.

Theorem 3.1 *If the condition* $E[\varepsilon_i] = 0$, $E[\varepsilon_i^2] = \sigma_i^2 \leq \sigma_0^2$ *is held and*

$$N \cdot \det\left(\frac{X^T X}{N}\right) \to \infty, \; for\ N \to \infty \tag{3.7}$$

then the l.s.e. with s.v. is unbiased and consistent estimator of unknown parameter θ, *i.e.,*

$$E(\hat{\theta} - \theta) = 0\ and\ (\hat{\theta} - \theta) \xrightarrow{p} 0\ for\ N \to \infty$$

Proof Consider the deviation vector:

$$\hat{\theta} - \theta = (X^T X)^{-1} X^T y - \theta = (X^T X)^{-1} X^T (X\theta + \varepsilon) - \theta = (X^T X)^{-1} X^T \varepsilon \tag{3.8}$$

From expression (3.8), it follows that

$$E(\hat{\theta}_i - \theta_i) = 0$$

The deviation vector (3.8) can be expressed as:

$$(\hat{\theta} - \theta) = \left(\frac{X^T X}{N}\right)^{-1} \frac{X^T}{N} \varepsilon \tag{3.9}$$

Since $\varepsilon_1, \varepsilon_2, \ldots, \varepsilon_N$ are independent and identically distributed random variables with finite variances, it follows from the Central Limit Theorem that the vector $(\hat{\theta} - \theta)$ has a Gaussian distribution. Moreover, it is known that $E(\varepsilon_i \varepsilon_j) = \delta_{ij} \sigma_i \sigma_j$, where δ_{ij} is Kronecker's delta symbol, such that $\delta_{ij} = 0$ when $i \neq j$ and $\delta_{ij} = 1$ when $i = j$.

Next, from expression (3.9), since the errors ε_i are independent and identically distributed (i.i.d.) random variables with finite variances σ_i^2, the covariance matrix of $\hat{\theta} - \theta$ is:

$$\text{Cov}(\hat{\theta} - \theta) = E\left[(\hat{\theta} - \theta)(\hat{\theta} - \theta)^T\right] = \left(\frac{X^T X}{N}\right)^{-1} \frac{X^T}{N} E[\varepsilon \varepsilon^T] \frac{X}{N} \left(\frac{X^T X}{N}\right)^{-1}.$$

Using $E[\varepsilon_i \varepsilon_j] = \sigma_i^2 \delta_{ij}$, where δ_{ij} is Kronecker's delta symbol, the covariance simplifies to:

$$\text{Cov}(\hat{\theta} - \theta) = \sigma^2 \cdot \frac{1}{N} \text{tr}\left(\left(\frac{X^T X}{N}\right)^{-1}\right)$$

By the properties of trace, it follows that:

$$\operatorname{tr}\left[\left(\frac{X^T X}{N}\right)^{-1}\right] = \sum_{i=1}^{m} \frac{1}{\lambda_i}$$

where $0 < \lambda_1 \leq \lambda_2 \leq \ldots \leq \lambda_m$ C are eigenvalues of the matrix $(\frac{X^T X}{N})$.

By the AM-GM inequality, it follows that:

$$\frac{1}{m}\sum_{i=1}^{m} \lambda_i \geq \sqrt[m]{\prod_{i=1}^{m} \lambda_i}.$$

Thus, we can relate the trace and the determinant via the eigenvalues of $\frac{X^T X}{N}$. As $N \to \infty$, the condition $N \cdot \det\left(\frac{X^T X}{N}\right) \to \infty$ implies that the trace term behaves asymptotically as:

$$\frac{1}{N}\operatorname{tr}\left(\left(\frac{X^T X}{N}\right)^{-1}\right) \to 0.$$

Since the elements of the matrix $\left(\frac{X^T X}{N}\right)$ are bounded by some constant C, and in accordance with the condition of Theorem 3.1, Chebyshev's inequality is applied. For any fixed $\Delta > 0$, it follows that:

$$P\left(\|\hat{\theta} - \theta\| > \Delta\right) \leq \frac{E\left[(\hat{\theta} - \theta)^T(\hat{\theta} - \theta)\right]}{\Delta^2} = \frac{1}{N\Delta^2}\operatorname{tr}\left(\left(\frac{X^T X}{N}\right)^{-1}\right) \to 0, \quad N \to \infty$$

Thus, Theorem 3.1 has been proved. □

In Theorem 3.1, the condition $N \cdot det\left(\frac{X^T X}{N}\right) \to \infty$ can be replaced to the following condition

$$\frac{1}{N}\sum_{k=1}^{m} \frac{1}{N\lambda_k} \leq \frac{m}{N\lambda_1} \to 0 \; for \; N \to \infty$$

because m is fixed.

In applications sometimes it is more convenient to use the condition $N \cdot det\left(\frac{X^T X}{N}\right) \to \infty$, because in some situations it is easy to calculate $det\left(\frac{X^T X}{N}\right)$.

The finite-dimensional vector $A = (a_1(N), a_2(N), \ldots, a_m(N))^T$ converges in probability to the vector $B = (b_1(N), b_2(N), \ldots, b_m(N))T$ if the random variable $\sum_{i=1}^{m}(a_i(N) - b_i(N))^2 \to 0$, for $N \to \infty$ in probability.

Denote $0 < \lambda_1 \leq \lambda_2 \leq \cdots \leq \lambda_m$ the eigen values of the matrix $\left(\frac{X^T X}{N}\right)$, then $0 < \frac{1}{\lambda_m} \leq \frac{1}{\lambda_{m-1}} \leq \cdots \leq \frac{1}{\lambda_1}$ will be the eigen values of the matrix $\left(\frac{X^T X}{N}\right)^{-1}$.

Lemma 3.1 *Suppose that the conditions* $E[\varepsilon_i] = 0$ *and* $E[\varepsilon_i^2] = \sigma_i^2 \leq \sigma_0^2$ *hold. Then, it follows that*

$$(\widehat{\theta}_i - \theta_i) \xrightarrow{d} N(0, \widetilde{\sigma}_i^2),$$

where $\widetilde{\sigma}_i^2 \leq \sigma_0^2$.

Proof Consider the deviation vector of the estimator:

$$\hat{\theta} - \theta = \left(\frac{X^T X}{N}\right)^{-1} \frac{X^T}{N} \varepsilon.$$

This expression shows that the deviation $\hat{\theta} - \theta$ is a linear combination of the i.i.d. random variables ε_j, with the coefficients d_{ij} given by the elements of the matrix $\left(\frac{X^T X}{N}\right)^{-1} \frac{X^T}{N}$. These coefficients d_{ij} are bounded.

Thus, for the individual components,

$$\hat{\theta}_i - \theta_i = \frac{1}{N} \sum_{j=1}^{N} d_{ij} \varepsilon_j,$$

where d_{ij} are bounded due to the properties of the matrix $\left(\frac{X^T X}{N}\right)^{-1} X$.

By the Central Limit Theorem for i.i.d. random variables ε_j with finite variance, we know that as $N \to \infty$, the random variable $\hat{\theta}_i - \theta_i$ converges in distribution to a normal distribution:

$$\hat{\theta}_i - \theta_i \xrightarrow{d} N(0, \widetilde{\sigma}_i^2),$$

where $\widetilde{\sigma}_i^2$ is the asymptotic variance of $\hat{\theta}_i - \theta_i$. The variance $\widetilde{\sigma}_i^2$ is finite and bounded above by σ_0^2, as required by the assumptions of Theorem 3.1.

Thus, $\hat{\theta}_i - \theta_i$ converges to a normal distribution with mean zero and variance $\widetilde{\sigma}_i^2$, which completes the proof. □

3.1.2 *Covariance Estimation in Uncorrelated Linear Prototype*

Using the expression

$$(\hat{\theta} - \theta) = \left(\frac{X^T X}{N}\right)^{-1} \frac{X^T}{N} \varepsilon,$$

the covariance matrix of the deviation vector $\sqrt{N}(\hat{\theta} - \theta) \in \mathbb{R}^{m+1}$ can be constructed. The components of this vector follow a normal distribution, and $\hat{\theta}$ corresponds to the least squares estimator (*l.s.e*) with specified singular values. When the covariance matrix is unknown, it follows that

$$C_N = N \cdot E(\widehat{\theta} - \theta)(\widehat{\theta} - \theta)^T = \left(\frac{X^T X}{N}\right)^{-1} \left(\frac{X^T I_N(\sigma^2) X}{N}\right) \left(\frac{X^T X}{N}\right)^{-1},$$

where c_{kl}, for $k, l = 1, 2, \ldots, m+1$, denotes the elements of the matrix C_N. The matrix $I_N(\sigma^2)$ is a diagonal matrix with σ_i^2 on the diagonal, given by

$$I_N(\sigma^2) = \begin{bmatrix} \sigma_1^2 & 0 & \cdots & 0 \\ 0 & \sigma_2^2 & \cdots & 0 \\ \vdots & \vdots & \ddots & \vdots \\ 0 & 0 & \cdots & \sigma_N^2 \end{bmatrix}.$$

The matrix $\left(\frac{X^T X}{N}\right)^{-1} X$ is known, while the elements of the matrix $\frac{X^T I_N(\sigma^2) X}{N}$ are unknown and therefore require estimation. Define $\hat{Y} = X\hat{\theta}$, where $\hat{\theta}$ is the least squares estimator with specified singular values.

It follows that

$$Y - \hat{Y} = Y - X(X^T X)^{-1} X^T Y = [I_N - X(X^T X)^{-1} X^T] Y, \tag{3.10}$$

where I_N denotes the $N \times N$ identity matrix. Since $Y = X\theta + \varepsilon$, substituting this into (3.10) yields

$$\hat{Y} - Y = [I - X(X^T X)^{-1} X^T](X\theta + \varepsilon) = [I - X(X^T X)^{-1} X^T]\varepsilon. \tag{3.11}$$

From Eq. (3.11), it follows that the components $(\hat{y}_i - y_i)$ of the vector $(\hat{Y} - Y)$ follow a normal distribution, and the expectation satisfies

$$E(Y - \hat{Y}) = [I_N - X(X^T X)^{-1} X^T] E[\varepsilon] = 0.$$

As for $N \to \infty$, $(\theta - \hat{\theta}) \xrightarrow{p} 0$, then it follows that,

$$y_i - \widehat{y_i} = \sum_{i=1}^{N} (\theta_k - \widehat{\theta_k}) x_i^k + \varepsilon_i \to \varepsilon_i, \; N \to \infty$$

i.e., each component of the residual vector converges to ε_i.

Denote

$$I_k(x) = \begin{pmatrix} x_1^k, & 0, \ldots, 0 \\ 0, & x_2^k, \ldots, 0 \\ 0, & 0, \ldots, x_N^k \end{pmatrix}, \quad I_k(1) = \begin{pmatrix} 1, 0, \ldots, 0 \\ 0, 1, \ldots, 0 \\ 0, 0, \ldots, 1 \end{pmatrix}$$

Introduce

$$\widehat{c}_{kl} = \frac{1}{N}(Y - \widehat{Y})^T I_{kl}(x)(Y - \widehat{Y}),\ k, l = 1, 2, \ldots, m+1$$

where the diagonal weight matrix

$$I_{kl}(x)$$

is shown as,

$$I_{kl}(x) = \begin{pmatrix} x_1^{k+l} & 0 & \cdots & 0 \\ 0 & x_2^{k+l} & \cdots & 0 \\ \vdots & \vdots & \ddots & \vdots \\ 0 & 0 & \cdots & x_N^{k+l} \end{pmatrix} \quad (k, l = 1, 2, \ldots, m.)$$

Theorem 3.2 *Suppose that the conditions $E[\varepsilon_i] = 0$ and $E[\varepsilon_i^2] = \sigma_i^2 \leq \sigma_0^2$ hold, the estimator*

$$\widehat{c}_{kl} = \frac{1}{N}(Y - \widehat{Y})^T I_{kl}(x)(Y - \widehat{Y})$$

serves as an asymptotically unbiased and consistent estimator of c_{kl} in the matrix C_N. That is, the expectation satisfies

$$E(\widehat{c}_{kl} - c_{kl}) \rightarrow 0,$$

and the estimator converges in probability,

$$(\widehat{c}_{kl} - c_{kl}) \xrightarrow{P} 0 \quad as \quad N \rightarrow \infty.$$

Proof Denote

$$B = \frac{(\widehat{Y} - Y)^T I(x)(\widehat{Y} - Y)}{N},$$

$b_{kl}, k, l = 1, 2, \ldots, m$; which are the elements of the matrix B. Then

$$\begin{aligned} \frac{1}{N}(\hat{Y} - Y)^T I_N(x)(\hat{Y} - Y) &= \frac{1}{N}\varepsilon^T \left[I - X(X^TX)^{-1}X^T\right] I_N(x) \left[I - X(X^TX)^{-1}X^T\right]\varepsilon \\ &= \frac{1}{N}\Big[\varepsilon^T I_N(x)\varepsilon - \varepsilon^T I_N(x) X(X^TX)^{-1}X^T\varepsilon \\ &\quad - \varepsilon^T X(X^TX)^{-1}X^T I_N(x)\varepsilon \\ &\quad + \varepsilon^T X(X^TX)^{-1}X^T I_N(x) X(X^TX)^{-1}X^T\varepsilon\Big] \end{aligned} \tag{3.12}$$

where $I_N(x) = (d)_{ij}, i, j = 1, 2, \ldots, N$ is a diagonal matrix of the size $N \times N$ and $d_{ii} = \sigma_i^2$.

Consider the last term in (3.12). Denote

$$\Sigma = \frac{1}{N} X(X^T X)^{-1} X^T I_N(x) X(X^T X)^{-1} X^T, \tag{3.13}$$

where Since $X(X^T X)^{-1} X^T$ is a idempotent matrix, which means $tr(X(X^T X)^{-1} X^T) = rank(X(X^T X)^{-1} X^T)$.

Using the properties of the matrix's trace that $tr(AB) = tr(BA)$ and the condition:

$$E[\varepsilon_i \varepsilon_j] = 0, \ if \ i \neq j \text{ and } E[\varepsilon_i^2] = \sigma_i^2$$

it follows that, $E[\varepsilon\varepsilon^T] = \sum_{i=1,\ldots,N} \sigma_i^2$.

For any fixed matrix Σ, using the trace property of quadratic forms:

$$\begin{aligned} E[\varepsilon^T \Sigma \varepsilon] &= E\left[\varepsilon^T \frac{1}{N} X(X^T X)^{-1} X^T I_N(x) X(X^T X)^{-1} X^T \varepsilon\right] \\ &= E\left[\frac{1}{N} X(X^T X)^{-1} X^T I_N(x) X(X^T X)^{-1} X^T \varepsilon\varepsilon^T\right] \\ &= \frac{1}{N} tr\left(X(X^T X)^{-1} X^T I_N(x) X(X^T X)^{-1} X^T \varepsilon\varepsilon^T\right) \\ &= \frac{1}{N}\left(\sum_{i=1,\ldots,N} \sigma_i^2\right)^2 rank(X(X^T X)^{-1} X^T) \end{aligned} \tag{3.14}$$

According to the conditions of the model, it shows that $|\varphi_i(x)| \leq 1$, then using the above equation:

$$E[\varepsilon^T \Sigma \varepsilon] \leq \frac{Cm}{N} \to 0 \ for \ N \to \infty \tag{3.15}$$

where I_N is the identity (unit) matrix of size n.

As the last term in (3.12) is nonnegative, then from (3.15) follows that $E\varepsilon^T \Sigma \varepsilon \to 0 \ for \ N \to \infty$ in probability.

Consider the second term in Eq. (3.6), which can be rewritten in the following form:

$$\left(\frac{x^T I_{kl}(x)\varepsilon}{N^{1/2}}\right)^T \left(\frac{X^T X}{N}\right)^{-1} \frac{X^T \varepsilon}{N^{1/2}} \cdot \frac{1}{N}$$

At the same time, it holds that:

$$\frac{X^T \varepsilon}{N^{1/2}} = \left(\frac{1}{N^{1/2}} \sum_{i=1}^{N} x_i \varepsilon_i, \ldots, \frac{1}{N^{1/2}} \sum_{i=1}^{N} x_i^m \varepsilon_i\right) \tag{3.16}$$

$$\frac{X^T I_N(\theta)\varepsilon}{N^{1/2}} = \left(\frac{1}{N^{1/2}} \sum_{i=1}^{N} x_i^n \varepsilon_i, \ldots, \frac{1}{N^{1/2}} \sum_{i=1}^{N} x_i^{n+m} \varepsilon_i\right) \tag{3.17}$$

Now, consider the variance of the components of the vector $\frac{1}{\sqrt{N}}\left(\sum_{i=1}^{N} x_i^k \varepsilon_i\right)$:

$$\begin{aligned}\operatorname{Var}\left[\frac{1}{\sqrt{N}}\left(\sum_{i=1}^{N} x_i^k \varepsilon_i\right)\right] &= \frac{1}{N}\sum_{i=1}^{N} x_i^{2k}\sigma_i^2 \\ &\le \max_i \left(x_i^{2k}\right)\sigma_0^2 \\ &\le b^{2k}\sigma_0^2 < \infty\end{aligned}$$

where it is assumed that $x_i \in [a, b]$ for all i.

Thus, the components of the vectors in Eqs. (3.16) and (3.17) are bounded in probability.

Next, consider the elements x_{ij}^N of the matrix $\left(\frac{X^T X}{N}\right)^{-1}$. Let χ_{ij}^N denote the algebraic minor of the element $\bar{X}_{ij}^N$ of the matrix $\left(\frac{X^T X}{N}\right)$.

According to the conditions $E\varepsilon_i = 0$ and $E\varepsilon_i^2 = \sigma_i^2 \le \sigma_0^2$, it follows that,

$$\frac{1}{N}\bar{X}_{ij}^N = \frac{1}{N}\frac{(-1)^{i+j}X_{ji}^N}{\det\left(\frac{X^T X}{N}\right)} \to 0 \quad \text{as} \quad N \to \infty$$

It follows that the second term in (3.12) converges to zero in probability. Next, consider the third term in (3.12):

$$\varepsilon^T X (X^T X)^{-1} X^T I_N(x)\varepsilon$$

This expression can be rewritten as:

$$\varepsilon^T X (X^T X)^{-1} X^T I_N(x)\varepsilon = \varepsilon^T \frac{X}{\sqrt{N}}\left(\frac{X^T X}{N}\right)^{-1}\frac{X^T I_N(x)}{\sqrt{N}}\cdot\frac{1}{N}\varepsilon.$$

Now consider the following term:

$$\varepsilon^T \frac{X}{\sqrt{N}} = \frac{1}{\sqrt{N}}\left(\sum_{k=1}^{N} x_k\varepsilon_k, \sum_{k=1}^{N} x_k^2\varepsilon_k, \ldots, \sum_{k=1}^{N} x_k^m\varepsilon_k\right).$$

Thus, the following holds:

$$\left[\frac{(\widehat{Y}-Y)^T I_{k,l}(\theta)(\widehat{Y}-Y)}{N} - \frac{\varepsilon^T I_{k,l}(x)\varepsilon}{N}\right] \xrightarrow{P} 0 \quad \text{as } N \to \infty.$$

Additionally, the following expression holds:

$$\frac{\varepsilon^T I_N(x)\varepsilon}{N} = \frac{1}{N}\sum_{i=1}^{N} x_i^n \varepsilon_i^2.$$

Taking the expectation of both sides:

$$E\left[\frac{\varepsilon^T I_N(x)\varepsilon}{N}\right] = \frac{1}{N}\sum_{i=1}^{N} x_i^n \sigma_i^2,$$

where σ_i^2 represents the variance of the error term ε_i.

It follows that the expression $\frac{1}{N}(\widehat{Y}-Y)^T I_{k,l}(\theta)(\widehat{Y}-Y)$ is an asymptotically unbiased estimator of the element b_{kl} of the matrix B. Now, the consistency of this estimator will be proven. Consider the variance of the estimator:

$$\begin{aligned}
\operatorname{Var}\left(\frac{\varepsilon^T I_{k,l}(x)\varepsilon}{N}\right) &= \frac{1}{N^2}\left(E\left(\varepsilon^T I_{k,l}(x)\varepsilon\right)^2 - \left(E\left(\varepsilon^T I_{k,l}(x)\varepsilon\right)\right)^2\right)\\
&= \frac{1}{N^2}\left(E\left(\sum_{i=1}^{N} x_i^n \varepsilon_i^2\right)^2 - \left(E\sum_{i=1}^{N} x_i^n \varepsilon_i^2\right)^2\right)\\
&= \frac{1}{N^2}\left(E\sum_{i=1}^{N}\varepsilon_i^4 x_i^{2n} + E\sum_{i\neq j}\varepsilon_i^2\varepsilon_j^2 x_i^n x_j^n - \sum_{i=1}^{N}\sigma_i^4 x_i^{2n} - \sum_{i\neq j}\sigma_i^2\sigma_j^2 x_i^n x_j^n\right)\\
&= \frac{1}{N^2}\left(3\sum_{i=1}^{N}\sigma_i^4 x_i^{2n} + \sum_{i\neq j}\sigma_i^2\sigma_j^2 x_i^n x_j^n - \sum_{i=1}^{N}\sigma_i^4 x_i^{2n} - \sum_{i\neq j}\sigma_i^2\sigma_j^2 x_i^n x_j^n\right)\\
&= \frac{2}{N^2}\sum_{i=1}^{N}\sigma_i^4 x_i^{2n} \leq \frac{2\sigma_0^4 \max(x_i^{2n})}{N} \to 0, \quad N\to\infty.
\end{aligned}$$

This result holds under the conditions stated in Theorem 3.2. Therefore, by Chebyshev's inequality, it can be concluded that the estimator is consistent. □

3.1.3 General Linear Prototype with Uncorrelated Errors

Consider the observed response given by

$$y_i = f(x_i) + \varepsilon_i \quad \text{for } i = 1, 2, \ldots, N.$$

Assume that the function $f(x)$ can be expressed in the following form:

$$f(x_i) = \sum_{k=1}^{m} \theta_k \varphi_k(x_i),$$

where $\varphi_1(x), \varphi_2(x), \ldots, \varphi_m(x)$ are a system of linearly independent and bounded functions [39]. For simplicity, it is assumed that $|\varphi_k(x)| \le 1$ for all $k = 1, 2, \ldots, m$.

The random error ε_i in the i-th observation is assumed to satisfy the following conditions:

$$E(\varepsilon_i) = 0,$$
$$E(\varepsilon_i^2) = \sigma_i^2.$$

The errors $\varepsilon_1, \varepsilon_2, \ldots, \varepsilon_N$ are assumed to be independent and identically distributed random variables.

The ratio in (3.12), together with (3.13), can be expressed in the following vector form:

$$Y = X \cdot \theta + \varepsilon \tag{3.18}$$

Here, $Y = (y_1, y_2, \ldots, y_N)^{\mathrm{T}}$ denotes the vector of observations (responses), and

$$X = \begin{pmatrix} \varphi_1(x_1) & \cdots & \varphi_m(x_1) \\ \varphi_1(x_2) & \cdots & \varphi_m(x_2) \\ \vdots & \ddots & \vdots \\ \varphi_1(x_N) & \cdots & \varphi_m(x_N) \end{pmatrix},$$

which is the design matrix of size $m \times N$, where $\varphi_1, \varphi_2, \ldots, \varphi_m$ represent a system of linearly independent and bounded functions.

The relation in (3.18) is referred to as the linear regression model with m unknown parameters. The Fisher information matrix corresponding to this model is given by $\left(\frac{X^{\mathrm{T}}X}{N}\right)$, which has size $m \times m$ and is expressed as:

$$\frac{X^{\mathrm{T}}X}{N} = \begin{pmatrix} \frac{1}{N}\sum_{k=1}^{N}(\varphi_1(x_k))^2 & \frac{1}{N}\sum_{k=1}^{N}\varphi_1(x_k)\varphi_2(x_k) & \cdots & \frac{1}{N}\sum_{k=1}^{N}\varphi_1(x_k)\varphi_m(x_k) \\ \frac{1}{N}\sum_{k=1}^{N}\varphi_2(x_k)\varphi_1(x_k) & \frac{1}{N}\sum_{k=1}^{N}(\varphi_2(x_k))^2 & \cdots & \frac{1}{N}\sum_{k=1}^{N}\varphi_2(x_k)\varphi_m(x_k) \\ \vdots & \vdots & \ddots & \vdots \\ \frac{1}{N}\sum_{k=1}^{N}\varphi_m(x_k)\varphi_1(x_k) & \frac{1}{N}\sum_{k=1}^{N}\varphi_m(x_k)\varphi_2(x_k) & \cdots & \frac{1}{N}\sum_{k=1}^{N}(\varphi_m(x_k))^2 \end{pmatrix}$$

In this formulation:

$\theta = (\theta_1, \theta_2, \ldots, \theta_m)^{\mathrm{T}}$ represents the vector of unknown parameters,

$\varepsilon = (\varepsilon_1, \varepsilon_2, \ldots, \varepsilon_N)^{\mathrm{T}}$ denotes the vector of random errors associated with the observations,

m is the number of unknown parameters in the model.

It is assumed that $E\varepsilon_i = 0,\ E\varepsilon_i^2 = \sigma_i^2 \le \sigma_0^2,\ \sigma_i^2 (i = 1, 2, \ldots, N)$ are unknown and different and $E\varepsilon_i\varepsilon_j = 0, \quad i \ne j$.

In linear models, the elements of the matrix X and the vector of responses Y are known. The model (3.14) is typical for applications and in world literature is not investigated widely because the values of the variances of errors of observations are unknown. Some authors [12, 17, 21, 23, 32, 47, 49, 70] tried to estimate unknown variances having several responses at one point of observation. The specific of model (3.14) is that at each point of observation, there is only one observation, which does not allow for estimating variance. It is a typical situation for applications because sometimes it is not possible to repeat an experiment, or it takes a lot of effort and high expense. Thus, we are coming to a linear regression model where the variances of observations are different, unknown, but bounded. In this direction, it is necessary to mention a few papers [25, 29, 39, 64, 65].

Our main aim is to find conditions under which the least squares estimator converges to a true value of unknown parameters, and using that fact, to try to estimate the elements of a covariance matrix of deviation vector.

Theorem 3.3 *Consider the statistical model characterized by the expectation of the quadratic form of the error term, expressed as*

$$
\begin{aligned}
E[\varepsilon^T \Sigma \varepsilon] &= E[\varepsilon^T \frac{1}{N} X(X^T X)^{-1} X^T I_N(x) X(X^T X)^{-1} X^T \varepsilon] \\
&= \frac{1}{N}(\sum_{i=1,\ldots,N} \sigma_i^2)^2 rank(X(X^T X)^{-1} X^T) \\
&\leq \frac{Cm}{N} \to 0 \ for \ N \to \infty
\end{aligned}
$$

where $I_N(x)$ is a diagonal matrix encoding the variability of the predictor variables, and Σ denotes the covariance matrix of the error term.

Suppose that the smallest eigenvalue of the Fisher information matrix, denoted by λ_1, satisfies the asymptotic condition

$$
\frac{1}{N\lambda_1} \to 0 \quad as \quad N \to \infty.
$$

Under this assumption, the least squares estimator (l.s.e.) obtained via singular value decomposition (s.v.) is both an unbiased and consistent estimator of the unknown parameter vector θ. The unbiasedness property is characterized by

$$
E(\widehat{\theta} - \theta) = 0,
$$

which implies that the expected value of the estimation error is zero. Furthermore, the consistency of the estimator is ensured by the convergence in probability

$$
(\widehat{\theta} - \theta) \xrightarrow{P} 0 \quad as \quad N \to \infty.
$$

This result indicates that, as the sample size increases, the probability of the estimation error being arbitrarily small approaches one.

Proof To prove the theorem, the same approach as in the proof of Theorem 3.1 is applied. Consider the expression for the difference between the estimated and true parameter:

$$\widehat{\theta} - \theta = (X^T X)^{-1} X^T y - \theta = (X^T X)^{-1} X^T (X\theta + \varepsilon) - \theta = (X^T X)^{-1} X^T \varepsilon.$$

By the linearity of expectation and the fact that $E(\varepsilon_i) = 0$ for each i, it follows that:

$$E(\widehat{\theta_i} - \theta_i) = E\left((X^T X)^{-1} X^T \varepsilon\right) = (X^T X)^{-1} X^T E(\varepsilon) = 0,$$

which shows that $\widehat{\theta}$ is an unbiased estimator of θ.

Next, the expression for $\widehat{\theta} - \theta$ can be rewritten as:

$$(\widehat{\theta} - \theta) = \left(\frac{X^T X}{N}\right)^{-1} \frac{X^T}{N} \varepsilon.$$

Denote by x_{ij} the elements of the matrix $\left(\frac{X^T X}{N}\right)^{-1}$. From the structure of the estimator, it follows that:

$$\widehat{\theta_i} - \theta_j = \sum_{j=1}^{m} d_{ij} \varepsilon_j, \quad d_{ij} = \frac{1}{N} \sum_{t=1}^{N} x_{ij} \varphi_j(x_i).$$

Here, $\varepsilon_1, \varepsilon_2, \ldots, \varepsilon_N$ are independent and identically distributed random variables with finite variances. By the Central Limit Theorem, the vector $(\widehat{\theta} - \theta)$ will asymptotically follow a multivariate normal distribution . Furthermore, it is known that $E(\varepsilon_i \varepsilon_j) = \delta_{ij} \sigma_i^2$, where δ_{ij} is Kronecker's delta symbol, i.e., $\delta_{ij} = 0$ for $i \neq j$, and $\delta_{ij} = 1$ for $i = j$.

Taking the expectation of the squared difference $(\widehat{\theta} - \theta)^T (\widehat{\theta} - \theta)$, and utilizing the properties of the trace and the condition $\frac{1}{N\lambda_1} \to 0$, the following result is obtained:

$$\begin{aligned} E\left((\widehat{\theta} - \theta)^T (\widehat{\theta} - \theta)\right) &= E\left\{\varepsilon^T \left[\frac{X}{N} \left(\frac{X^T X}{N}\right)^{-1} \left(\frac{X^T X}{N}\right)^{-1} \frac{X^T}{N}\right]^T \varepsilon\right\} \\ &\leq \sigma_0 \operatorname{tr} \left[\frac{X}{N} \left(\frac{X^T X}{N}\right)^{-1} \left(\frac{X^T X}{N}\right)^{-1} \frac{X^T}{N}\right], \end{aligned}$$

Since the elements of the matrix $\left(\frac{X^T X}{N}\right)$ and the vector $\left(\frac{X^T}{N}\right)$ are bounded (as $|\varphi(x)| \leq 1$), the following holds:

$$\operatorname{tr} \left[\frac{X}{N} \left(\frac{X^T X}{N}\right)^{-1} \left(\frac{X^T X}{N}\right)^{-1} \frac{X^T}{N}\right] = \frac{1}{N} \operatorname{tr} \left(\left(\frac{X^T X}{N}\right)^{-1}\right) = \frac{1}{N} \sum_{i=1}^{m} \frac{1}{\lambda_i} \leq \frac{m}{N\lambda_1},$$

where λ_1 denotes the smallest eigenvalue of $\frac{X^T X}{N}$. Therefore, the expectation of the squared norm of $(\widehat{\theta} - \theta)$ behaves asymptotically as $O\left(\frac{m}{N\lambda_1}\right)$, which tends to zero as $N \to \infty$ under the assumption that $\frac{1}{N\lambda_1} \to 0$.

As the number of unknown parameters m is fixed, then according to the condition of Theorem 3.3, it follows that:

$$E(\widehat{\theta} - \theta)^T (\widehat{\theta} - \theta) \to 0 \; for \; N \to \infty.$$

Using Chebyshev's inequality, according to the condition of Theorem 3.3, for any fixed value $\Delta > 0$, it follows that:

$$P\left(\|\hat{\theta} - \theta\| > \Delta\right) \le \frac{E\left[(\hat{\theta} - \theta)^T (\hat{\theta} - \theta)\right]}{\Delta^2} < \frac{1}{N} \sum_{i=1}^{m} \frac{1}{\lambda_i} \to 0, N \to \infty.$$

Thus, Theorem 3.3 has been proved. □

This proof confirms that the $l.s.e.$ with $s.v.$ is a consistent estimator of the unknown parameter θ. Now using this fact, a confidence band for the unknown function $f(x)$ can be constructed.

3.1.4 Covariance Matrix Estimation in the General Linear Prototype

Consider a covariance matrix C_N of the deviation vector $\sqrt{N}(\widehat{\theta} - \theta)$ which is unknown. Then there is:

$$C_N = NE(\widehat{\theta} - \theta)(\widehat{\theta} - \theta)^T = \left(\frac{X^T X}{N}\right)^{-1} \left(\frac{X^T I_N(\sigma^2) X}{N}\right) \left(\frac{X^T X}{N}\right)^{-1} \tag{3.19}$$

where $c_{kl}, k, l = 1, 2, \ldots, m+1$ is the element of the matrix C_N and

$$I_N(\sigma^2) = \begin{pmatrix} \sigma_1^2 & 0 & \cdots & 0 \\ 0 & \sigma_2^2 & \cdots & 0 \\ \vdots & \vdots & \ddots & \vdots \\ 0 & 0 & \cdots & \sigma_N^2 \end{pmatrix}$$

is the diagonal matrix with σ_i^2 on its diagonal. In Eq. (3.19), the matrix $\left(\frac{X^T X}{N}\right)^{-1}$ is known, hence it is necessary to estimate the elements of the matrix $\frac{X^T I_N(\sigma^2) X}{N}$, where the matrix X is also known and only the matrix $I_N(\sigma^2)$ is unknown. Introduce:

$$\widehat{Y} = X\widehat{\theta}$$

where $\widehat{\theta}$ is the $l.s.e.$ with the $s.v.$ and denote:

$$B = \frac{X^T I_N(\sigma^2) X}{N}, \; b_{kl}, \; k = 1, 2, \ldots, m;$$

are the elements of the matrix B.

Then,

$$Y - \widehat{Y} = Y - X(X^T X)^{-1} X^T Y = [I_N - X(X^T X)^{-1} X^T]Y \tag{3.20}$$

where I_N is the identity matrix of the size N. As $Y = X\theta + \varepsilon$, it follows from (3.20) that,

$$\widehat{Y} - Y = \left[I - X(X^T X)^{-1} X^T\right]((X\theta + \varepsilon) = [I - X(X^T X)^{-1} X^T]\varepsilon \tag{3.21}$$

From (3.21) it follows that the components of the vector $(\widehat{Y} - Y)$ have normal distribution and

$$E(Y - \widehat{Y}) = [I_N - X(X^T X)^{-1} X^T]E[\varepsilon] = 0$$

It must be noted that the $l.s.e$ depends on N, hence below we indicate it. As for $N \to \infty$, $(\theta - \widehat{\theta}) \to 0$ (in probability) then,

$$y_i - \widehat{y_i} = \sum_{k=1}^{m} (\theta_k - \widehat{\theta}_k(N))\varphi_k(x_i) + \varepsilon_i \to \varepsilon_i, N \to \infty$$

i.e., each component of the residual vector $(y_i - \hat{y}_i)$ converges to ε_i.

Denote

$$I_{kl}(\varphi) = \begin{pmatrix} \varphi_k(x_1)\varphi_l(x_1), & 0, & \ldots, & 0 \\ 0, & \varphi_k(x_2)\varphi_l(x_2), & \ldots, & 0 \\ \vdots & \vdots & \ddots & \vdots \\ 0, & 0, & \ldots, & \varphi_k(x_N)\varphi_l(x_N) \end{pmatrix}$$

$$I_K(1) = \begin{pmatrix} 1, \, 0, \, \ldots, \, 0 \\ 0, \, 1, \, \ldots, \, 0 \\ 0, \, 0, \, \ldots, \, 1 \end{pmatrix}$$

is the identity matrix of the size k.

Introduce

$$\widehat{C}_{kl} = \frac{1}{N}(Y - \widehat{Y})^T I_{kl}(x)(Y - \widehat{Y}), \quad k, l = 1, 2, \dots, m+1$$

Theorem 3.4 *Suppose that the statistical model satisfies the expectation of the quadratic form of the error term,*

$$E[\varepsilon^T \Sigma \varepsilon] \leq C \cdot tr\left[\left(\frac{X^T X}{N}\right)^{-1} \frac{X^T}{N} I_N(x) X \left(\frac{X^T X}{N}\right)^{-1}\right],$$

where $I_N(x)$ is a diagonal matrix encoding the variability of the predictor variables, and Σ denotes the covariance matrix of the error term. Furthermore, assume that the smallest eigenvalue of the Fisher information matrix, denoted by λ_1, satisfies the asymptotic condition

$$\frac{1}{N\lambda_1} \to 0 \ \ as \ \ N \to \infty.$$

Under these conditions, the estimator $\widehat{c}_{kl}$ is asymptotically unbiased and consistent for the element c_{kl} of the matrix C_N, i.e.,

$$E(\widehat{c}_{kl} - c_{kl}) \to 0, \quad (\widehat{c}_{kl} - c_{kl}) \xrightarrow{P} 0 \ \ as \ \ N \to \infty.$$

Proof Let $\widehat{Y} = X\widehat{\theta}$, where $\widehat{\theta}$ is the estimator of the parameter θ. Then, the difference between the estimated and true values of Y is given by:

$$\widehat{Y} - Y = X(\widehat{\theta} - \theta) = X(X^T X)^{-1} X^T \varepsilon$$

Next, consider the quadratic form of $(\widehat{Y} - Y)$ with respect to $I_N(x)$:

$$\begin{aligned}
\frac{1}{N}(\hat{Y} - Y)^T I_N(x)(\hat{Y} - Y) &= \frac{1}{N} \varepsilon^T \left[I - X(X^T X)^{-1} X^T\right] I_N(x) \left[I - X(X^T X)^{-1} X^T\right] \varepsilon \\
&= \frac{1}{N}\Big[\varepsilon^T I_N(x)\varepsilon - \varepsilon^T I_N(x) X(X^T X)^{-1} X^T \varepsilon \\
&\quad - \varepsilon^T X(X^T X)^{-1} X^T I_N(x)\varepsilon \\
&\quad + \varepsilon^T X(X^T X)^{-1} X^T I_N(x) X(X^T X)^{-1} X^T \varepsilon\Big]
\end{aligned} \tag{3.22}$$

where $I_N(x) = (d)_{ij}, \ I_j = 1, 2, \dots, N$ is a diagonal matrix of the size $N \times N$ and $d_{ij} = \sigma_i^2$. Consider the last term in Eq. (3.22).

Define the matrix Σ as:

$$\Sigma = \frac{1}{N} X(X^T X)^{-1} X^T I_N(x) X(X^T X)^{-1} X^T$$

Using the properties of the trace operator, $\text{tr}(AB) = \text{tr}(BA)$, and the conditions $E[\varepsilon_i \varepsilon_j] = 0$ for $i \neq j$ and $E[\varepsilon_i^2] = \sigma_i^2$,

$$E[\varepsilon^T \Sigma \varepsilon] = E\left[\varepsilon^T \frac{1}{N} X(X^T X)^{-1} X^T I_N(x) X(X^T X)^{-1} X^T \varepsilon\right]$$
$$= E\left[\frac{1}{N} X(X^T X)^{-1} X^T I_N(x) X(X^T X)^{-1} X^T \varepsilon\varepsilon^T\right]$$
$$= \frac{1}{N} tr\left(X(X^T X)^{-1} X^T I_N(x) X(X^T X)^{-1} X^T \varepsilon\varepsilon^T\right)$$
$$= \frac{1}{N}\left(\sum_{i=1,\ldots,N} \sigma_i^2\right)^2 rank(X(X^T X)^{-1} X^T)$$

By applying the trace inequality,

$$E[\varepsilon^T \Sigma \varepsilon] \leq C \cdot \mathrm{tr}\left[\frac{X}{N}\left(\frac{X^T X}{N}\right)^{-1} \frac{X^T}{N} I_N(x) X \left(\frac{X^T X}{N}\right)^{-1} \frac{X^T}{N}\right]$$

As the model condition $|\varphi_i(x)| \leq 1$ holds, it follows that:

$$E[\varepsilon^T \Sigma \varepsilon] \leq \frac{C_m}{N} \to 0 \quad \text{as} \quad N \to \infty$$

where I_N is the identity matrix of size N.

Thus, the last term in the original expression (3.22) tends to 0 as $N \to \infty$.

Now consider the second term in (3.22). It can be rewritten as:

$$\left(\frac{X^T I_{kl}(x)\varepsilon}{\sqrt{N}}\right)^T \left(\frac{X^T X}{N}\right)^{-1} \frac{X^T \varepsilon}{\sqrt{N}} \cdot \frac{1}{N}$$

At the same time,

$$\frac{X^T \varepsilon}{\sqrt{N}} = \left(\frac{1}{\sqrt{N}} \sum_{i=1}^{N} \varphi_1(x_i)\varepsilon_i, \ldots, \frac{1}{\sqrt{N}} \sum_{i=1}^{N} \varphi_m(x_i)\varepsilon_i\right) \tag{3.23}$$

and

$$\frac{X^T I_{kl}(x)\varepsilon}{\sqrt{N}} = \left(\frac{1}{\sqrt{N}} \sum_{i=1}^{N} \varphi_1(x_i)\varphi_k(x_i)\varphi_l(x_i)\varepsilon_i, \ldots, \frac{1}{\sqrt{N}} \sum_{i=1}^{N} \varphi_m(x_i)\varphi_k(x_i)\varphi_l(x_i)\varepsilon_i\right) \tag{3.24}$$

Since $|\varphi_i(x)| \leq 1$ for all $i = 1, 2, \ldots, m$,

$$\mathrm{Var}\left(\frac{1}{\sqrt{N}} \sum_{i=1}^{N} \varphi_j(x_i)\varphi_k(x_i)\varphi_l(x_i)\varepsilon_i\right) \leq \frac{1}{N} \sum_{i=1}^{N} \sigma_i^2 \leq \sigma_0^2$$

Thus, the components of the vectors in Eqs. (3.23) and (3.24) are bounded in probability.

Consider the elements x_{ij}^N of the matrix $\left(\frac{X^T X}{N}\right)^{-1}$, and denote $\bar{X}_{ij}^N$ as the algebraic minor of the element x_{ij}^N of the matrix $\left(\frac{X^T X}{N}\right)$.

According to the condition of Theorem 3.2, $E[\varepsilon_i] = 0$ and $E[\varepsilon_i^2] = \sigma_i^2 \le \sigma_0^2$,

$$\frac{1}{N}\bar{X}_{ij}^N = \frac{1}{N}\frac{(-1)^{i+j}x_{ji}^N}{\det\left(\frac{X^T X}{N}\right)} \to 0 \text{ as } N \to \infty.$$

This implies that the second term in (3.22) tends to zero in probability as $N \to \infty$. Now, consider the third term in (3.22):

$$\varepsilon^T \frac{X}{N}\left(\frac{X^T X}{N}\right)^{-1}\frac{X^T}{N} I_n(x)\varepsilon.$$

It can be expressed that this term as follows, keeping the approach similar to the one used for proving the second term:

$$\varepsilon^T \frac{X}{N}\left(\frac{X^T X}{N}\right)^{-1}\frac{X^T}{N} I_n(x)\varepsilon = \varepsilon^T \frac{X}{\sqrt{N}}\left(\frac{X^T X}{N}\right)^{-1}\frac{X^T I_n(x)}{\sqrt{N}}\frac{1}{N}\varepsilon.$$

Next, it shows that:

$$\varepsilon^T \frac{X}{\sqrt{N}} = \frac{1}{\sqrt{N}}\left(\sum_{i=1}^{N}\varphi_1(x_i)\varepsilon_i, \sum_{i=1}^{N}\varphi_2(x_i)\varepsilon_i, \ldots, \sum_{i=1}^{N}\varphi_m(x_i)\varepsilon_i\right),$$

where $\varphi_k(x_i)$ are the components associated with the i-th observation.

Thus, the term:

$$\left[\frac{(\widehat{Y} - Y)I_{kl}(\theta)(\widehat{Y} - Y)}{N} - \frac{\varepsilon^T I_{kl}(x)\varepsilon}{N}\right] \to 0$$

in probability as $N \to \infty$. On the other hand, it follows that:

$$E\left[\frac{\varepsilon^T I_{kl}(x)\varepsilon}{N}\right] = \frac{1}{N}\sum_{i=1}^{N}\varphi_k(x_i)\varphi_l(x_i)\sigma_i^2.$$

This implies that the expression:

$$\frac{1}{N}(\widehat{Y} - Y)I_{kl}(\theta)(\widehat{Y} - Y)$$

Which is an asymptotically unbiased estimator of the element b_{kl} of the matrix B. The consistency of this estimator is now proved.

Consider the variance of $\frac{\varepsilon^T I_{kl}(x)\varepsilon}{N}$:

$$\begin{aligned}\operatorname{Var}\left(\frac{\varepsilon^T I_{kl}(x)\varepsilon}{N}\right) &= \frac{1}{N^2}\left\{E\left(\varepsilon^T I_{kl}(x)\varepsilon\right)^2 - \left(E\left(\varepsilon^T I_{kl}(x)\varepsilon\right)\right)^2\right\}\\ &= \frac{1}{N^2}\left\{E\left(\sum_{i=1}^{N}\varepsilon_i^2\varphi_k(x_i)\varphi_l(x_i)\right)^2 - \left[E\left(\sum_{i=1}^{N}\varphi_k(x_i)\varphi_l(x_i)\varepsilon_i^2\right)\right]^2\right\}\\ &= \frac{1}{N^2}\left\{\sum_{i=1}^{N}E\left[\varepsilon_i^4\left(\varphi_k(x_i)\varphi_l(x_i)\right)^2\right] - \sum_{i=1}^{N}\sigma_i^4\left(\varphi_k(x_i)\varphi_l(x_i)\right)^2\right\}.\end{aligned}$$

As $N \to \infty$, the last term tends to zero. By applying Chebyshev's inequality, it can be concluded that the estimator is consistent. □

3.1.5 Numerical Illustration for Uncorrelated Prototype

In order to better demonstrate the validity of the theorems and formulas in Sect. 3.1, we design two examples, which are the simplest linear regression case and the general linear regression case.

Example 3.1 (*In simplest linear regression case*) Suppose the observed values y_i satisfy

$$y_i = 1.5 + 0.8\,x_i + 2.1\,x_i^2 - 1.2\,x_i^3 + 0.7\,x_i^4 + 3.0\,x_i^5 + \varepsilon_i,$$

where each ε_i is an independent Gaussian noise term with mean 0 and variance 0.5^2, i.e. $\varepsilon_i \sim N(0, 0.5^2)$. The predictor values x_i are uniformly spaced over the interval $[0, 1]$.

The above regression equation can be rewritten in matrix form as:

$$\mathbf{y} = X\theta + \boldsymbol{\varepsilon},$$

where $\mathbf{y} = (y_1, \dots, y_N)^T \in \mathbb{R}^N$ is the observed response vector.

$\theta = (\theta_0, \theta_1, \theta_2, \theta_3, \theta_4, \theta_5)^T$ is the six-dimensional parameter vector to be estimated. In this example, its true values are

$$\theta_0 = 1.5, \quad \theta_1 = 0.8, \quad \theta_2 = 2.1, \quad \theta_3 = -1.2, \quad \theta_4 = 0.7, \quad \theta_5 = 3.0.$$

X is an $N \times 6$ real design matrix. In accordance with the problem setup, we choose $x_i \in [0, 1]$ at equal intervals and construct each column from powers of x_i. Concretely,

$$X = \begin{pmatrix} 1 & x_1 & x_1^2 & x_1^3 & x_1^4 & x_1^5 \\ 1 & x_2 & x_2^2 & x_2^3 & x_2^4 & x_2^5 \\ \vdots & \vdots & \vdots & \vdots & \vdots & \vdots \\ 1 & x_N & x_N^2 & x_N^3 & x_N^4 & x_N^5 \end{pmatrix}.$$

$\boldsymbol{\varepsilon} = (\varepsilon_1, \ldots, \varepsilon_N)^T$ is the error vector, where each ε_i is independently distributed as $N(0, \sigma^2)$, and $\sigma^2 = 0.5^2$.

To estimate the unknown parameter vector θ in this model, one typically employs the ordinary least squares ($l.s.e.$) method, resulting in the estimator:

$$\hat{\theta} = (X^T X)^{-1} X^T \mathbf{y}.$$

The numerical experiment aims to increase the sample size N and observe the statistical properties of $\hat{\theta}$ (such as unbiasedness and consistency), as well as to examine quantities like

$$N \cdot \det\left(\frac{X^T X}{N}\right) \to \infty, \ for\ N \to \infty$$

For a fixed polynomial order $p = 5$, consider the polynomial vector $\phi(x) = (1, x, x^2, x^3, x^4, x^5)^T$.

Over the interval [0, 1], its second-order moment can be written as

$$\Sigma = \int_0^1 \phi(x)\phi(x)^T \, dx.$$

This matrix is called the "moment matrix" formed by the polynomial basis on [0, 1]. Since the polynomials are linearly independent on that interval, Σ must be positive definite, and thus $\det(\Sigma) > 0$.

When N points $x_i \in [0, 1]$ are chosen at equal intervals, Riemann sum approximation theory guarantees that,

$$\frac{1}{N} X^T X = \frac{1}{N} \sum_{i=1}^{N} \phi(x_i)\phi(x_i)^T \xrightarrow[N \to \infty]{} \int_0^1 \phi(x)\phi(x)^T \, dx = \Sigma.$$

Because Σ is positive definite, $\frac{1}{N} X^T X$ also approaches a positive definite matrix for large samples, and its smallest eigenvalue does not shrink to zero.

Furthermore, when $\frac{1}{N} X^T X \to \Sigma$ and Σ is positive definite, it follows that,

$$\det\left(\frac{X^T X}{N}\right) \xrightarrow[N \to \infty]{} \det(\Sigma) > 0,$$

Thus, this example satisfies the condition of Theorem 3.1.

Table 3.1 Parameter bias convergence under large-sample conditions in SLR

Sample size	θ_0	θ_1	θ_2	θ_3	θ_4	θ_5
50	−0.013	0.953	−4.749	9.910	−9.742	3.650
52678	−0.003	0.047	−0.231	0.469	−0.432	0.150
105307	−0.001	0.007	0.002	−0.074	0.121	−0.055
157936	−0.001	0.014	−0.072	0.159	−0.163	0.063
210565	0.000	−0.017	0.157	−0.482	0.592	−0.252
263194	0.000	−0.016	0.095	−0.197	0.164	−0.045
315823	−0.001	0.023	−0.178	0.517	−0.625	0.265
368452	0.001	−0.024	0.110	−0.212	0.182	−0.058
421081	0.001	−0.013	0.071	−0.166	0.178	−0.072
473710	0.000	0.014	−0.090	0.203	−0.193	0.066
526339	−0.001	0.016	−0.106	0.281	−0.319	0.128
578968	−0.001	0.018	−0.103	0.234	−0.230	0.082
631597	0.000	−0.008	0.049	−0.129	0.149	−0.062
684226	0.001	−0.012	0.053	−0.098	0.078	−0.021
736855	0.001	−0.015	0.077	−0.188	0.216	−0.091
789484	0.000	0.006	−0.028	0.065	−0.070	0.029
842113	0.000	0.000	−0.005	0.012	−0.002	−0.004
894742	−0.001	0.017	−0.092	0.198	−0.186	0.065
947371	−0.001	0.016	−0.090	0.213	−0.222	0.084
1000000	0.000	0.008	−0.047	0.115	−0.117	0.042

A common approach is to let N grow from smaller values ($N = 50$) up to larger ones ($N = 10^6$), fit the regression at each step, record the parameter estimates $\hat{\theta}$, and thus check how $\hat{\theta}$ converges to the true value θ as N increases. The results of the example are as follows.

Table 3.1 presents the estimated values of model parameters (θ_0 to θ_5) under varying sample sizes N. When the sample size is small (e.g., $N = 50$), the estimated parameters deviate substantially from their true values, accompanied by high variance. For instance, $\theta_0 = -0.013$ is far from the true value of 1.5, and the coefficients of higher-order terms exhibit significant fluctuations, indicating instability in model fitting under small-sample conditions.

As the sample size increases to a moderate level (approximately $N = 5000 \sim 50000$), the estimates begin to converge, with some parameters approaching their true values. Nevertheless, noticeable fluctuations remain, particularly in higher-order coefficients (θ_4 and θ_5), which continue to show considerable deviations at certain sample points, suggesting that the model has not yet achieved full estimation stability.

When the sample size becomes large ($N > 50000$), the estimated parameters exhibit increased stability and progressively approach the true values. For example, at $N = 10^5$, $\theta_0 = -0.002$ and $\theta_1 = 0.035$, both of which are close to the true values

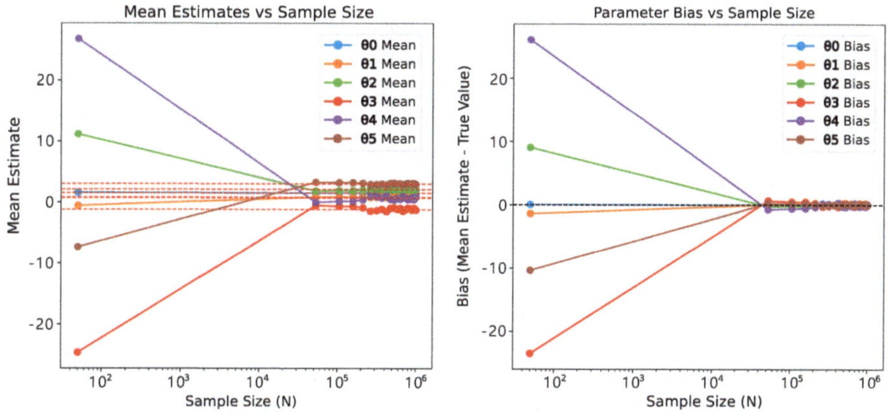

Fig. 3.2 Estimator consistency and bias convergence in SLR

of 1.5 and 0.8, respectively. Furthermore, the estimation accuracy of higher-order terms (θ_2 to θ_5) improves significantly, with reduced variability, indicating a more reliable model fit.

In summary, the parameter estimates demonstrate clear convergence as the sample size increases, consistent with the theoretical expectation of asymptotic consistency. Under small-sample conditions, the estimates show substantial bias and volatility, especially for higher-order terms, suggesting the potential presence of high variance or multicollinearity. As $N \to \infty$, the parameter estimates are expected to converge to their true values, yielding a stable and accurate model.

To systematically illustrate the performance of the least squares estimator across different sample sizes and quantify how estimation bias changes with sample size, two subplots (as shown in Fig. 3.2) are presented for the given fifth-degree polynomial regression model: the left plot shows how the mean estimates of the parameters vary as the sample size increases, while the right plot evaluates the bias of each parameter estimate relative to its true value.

Specifically, it implements Monte Carlo simulations to examine the asymptotic properties of the estimators. The true parameter vector is given by $\theta_{\text{true}} \in \mathbb{R}^6$, and the noise follows a normal distribution $N(0, \sigma^2)$. For different sample sizes N (selected at 20 evenly spaced points between 50 and 10^6), we construct evenly spaced input values $x \in [0, 1]$ and the corresponding Vandermonde design matrix $X \in \mathbb{R}^{N \times 6}$, where the i-th row is $(1, x_i, x_i^2, x_i^3, x_i^4, x_i^5)$. In the visualization, the horizontal axis N is transformed to a logarithmic scale to better illustrate the transition from small to large-sample sizes. The left plot displays the mean estimates of each parameter, with the true values indicated by red dashed lines. The right plot shows the variation in estimation bias, with a reference line at zero to highlight the trend of bias reduction.

From the "Mean Estimates vs. Sample Size" plot on the left, it can be observed that as the sample size increases, the mean estimates of all parameters gradually converge to their true values, which are indicated by dashed lines. This suggests that when the sample size is sufficiently large, the least squares estimator can effectively recover the true model. The "Parameter Bias vs. Sample Size" plot on the right illustrates the fluctuation of estimation bias relative to zero.

Figure 3.3a illustrates the convergence trend of the parameter estimates as the sample size increases. The horizontal axis represents the sample size N, while the vertical axis denotes the estimated parameter values. As N increases, the estimates of the parameters—particularly the lower-order terms θ_0 and θ_1-gradually stabilize and converge toward their true values (1.5 and 0.8, respectively), following initial fluctuations.

Figure 3.3b presents the estimation results of model parameters under different sample sizes N, obtained using Monte Carlo simulations with 50 repeated experiments. The figure displays convergence curves for various quantiles, including the 5, 25, 50% (median), 75, and 95% quantiles, illustrating how the estimated values evolve as the sample size increases.

Each subplot corresponds to one parameter (θ_0 to θ_5). In the plots, dashed lines represent different quantile levels (indicating the confidence interval range), the solid black line denotes the median estimate, and the horizontal red line indicates the true parameter value. Combined with the figure and code, the following observations can be made:

(1) **Convergence of Parameter Estimates**: As the sample size increases, the median curves (black) of each parameter estimate clearly approach the true values (red lines), confirming the asymptotic unbiasedness and consistency of the parameter estimators.
(2) **Convergence Trend of Quantile Intervals**: For lower-order parameters (e.g., θ_0, θ_1), the gap between quantiles narrows rapidly and converges toward the true values as sample size increases, indicating that these parameters are more stably estimated. For higher-order parameters, e.g., θ_4, θ_5, although the quantile intervals gradually contract with large-sample sizes, the convergence is slower and the variability is more pronounced, reflecting the higher variance and greater estimation difficulty commonly associated with higher-order terms in polynomial regression.
(3) **Interpretation of Confidence Intervals**: The spread between the quantile curves represents the variation in parameter estimates. As the sample size increases, the estimates become more concentrated, allowing for more accurate inference of the true parameters. Eventually, for large samples, the 95–5% quantile intervals become narrow, indicating that the estimates are statistically reliable.

In the small-sample stage, e.g., the region near the origin in the figure, the estimated values of the parameters exhibit significant fluctuations, reflecting prominent characteristics of random error. As N increases, the variability of the estimates gradually diminishes, and the curves become more stable, demonstrating the property of asymptotic consistency in parameter estimation. Overall, the variance of the parameter estimates decreases progressively.

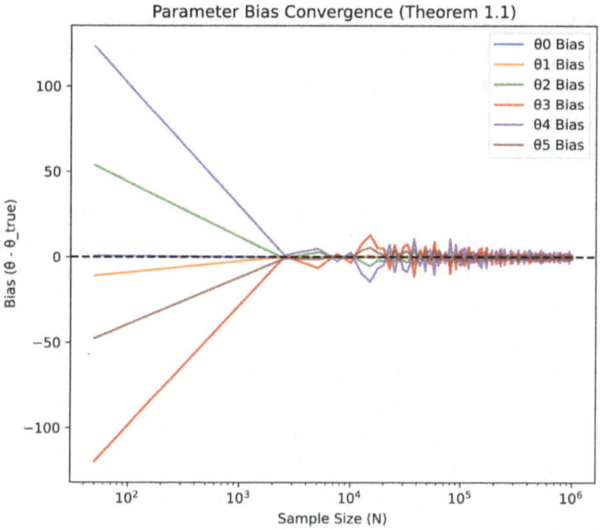

(a) Parameter Bias Convergence

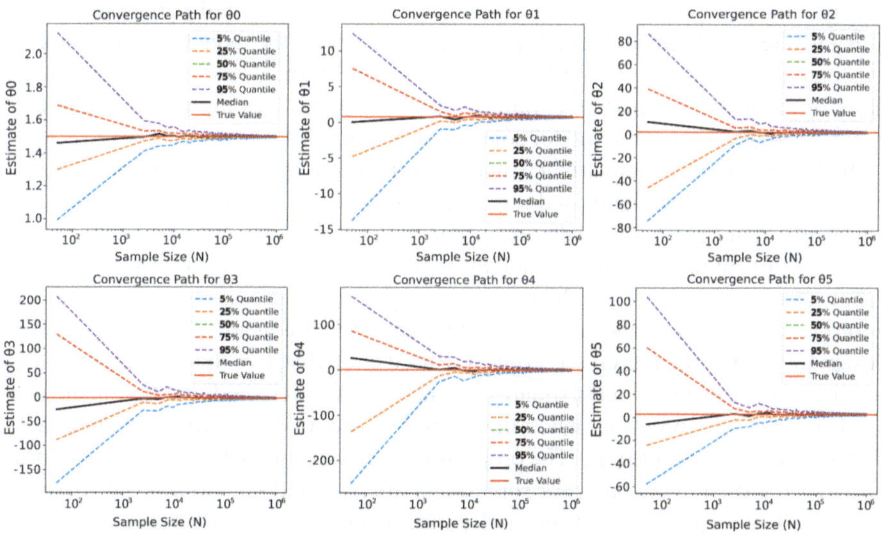

(b) Convergence Path for θ_i, $i = 0, \cdots, 5$

Fig. 3.3 Visualization of convergence trajectories of parameter biases in SLR

This indicates that once the sample size reaches a certain level, the parameter estimates become substantially close to the true values, and the marginal improvement from further increasing the sample size diminishes. This observation suggests that, in practical modeling, an appropriate sample size can be chosen to balance computational cost and estimation accuracy.

As shown in Fig. 3.4. From the residual diagnostic plots, the residuals appear sparse and randomly dispersed at small-sample sizes, making heteroscedasticity difficult to assess clearly. As the sample size increases, residuals become evenly and stably distributed around zero without noticeable patterns linked to the fitted values, indicating that this polynomial regression model satisfies the homoscedasticity assumption well under large samples. This result validates the model specification's rationality and robustness of its statistical properties.

As shown in Fig. 3.5, in the case of small samples ($N = 10, 10^2$), the variance of the estimated parameters is large, which leads to the obvious deviation of the estimated values from the real values. In particular, the high-order parameters ($\theta_3, \theta_4, \theta_5$) fluctuate violently due to the lack of samples, and there is a great correlation between some parameters. As the sample size increases to a medium scale ($N = 10^3$, 10^4), the estimated parameters begin to converge to the real value, and the variance decreases. However, there are still some deviations in the high-order parameters, which may be related to collinearity or smaller eigenvalues. In the case of large samples ($N = 10^5$, 10^6), the variance of the estimated parameters is significantly reduced, close to the true value, in line with the asymptotic consistency expectation, and the joint distribution is more concentrated, indicating that the stability of parameter estimation is improved, the fluctuation of high-order parameters is greatly reduced, and the model convergence is good.

In the following, the covariance matrix for the given numerical example is explicitly computed, and the unbiasedness and consistency properties of the covariance matrix estimator are numerically examined.

First, fundamental parameters and the regression model are established, including the range of sample sizes, the degree of the polynomial regression, the standard deviation of the noise, and the true parameter vector. Subsequently, the corresponding design matrix is constructed accordingly. Based upon this, two theoretical covariance matrices under ordinary least squares estimation are calculated: the first is the classical covariance expression, while the second incorporates an explicit scaling factor dependent on the sample size. The proportional relationship between these two forms provides the theoretical basis for subsequent analysis and numerical comparison.

Second, through repeated and independent numerical simulation experiments, random noise is generated to construct a response variables, from which the least squares estimates and corresponding residuals are computed. Utilizing these residuals and the weighted sums involving powers of explanatory variables, an intermediate matrix is constructed. Following the relevant literature, this intermediate matrix is combined to produce the final covariance matrix estimator. Furthermore, by calculating the Frobenius norm of the difference between this estimator and the theoretical covariance matrices and plotting the error as a function of sample size, numerical verification is provided to demonstrate that the estimator satisfies the theoretical

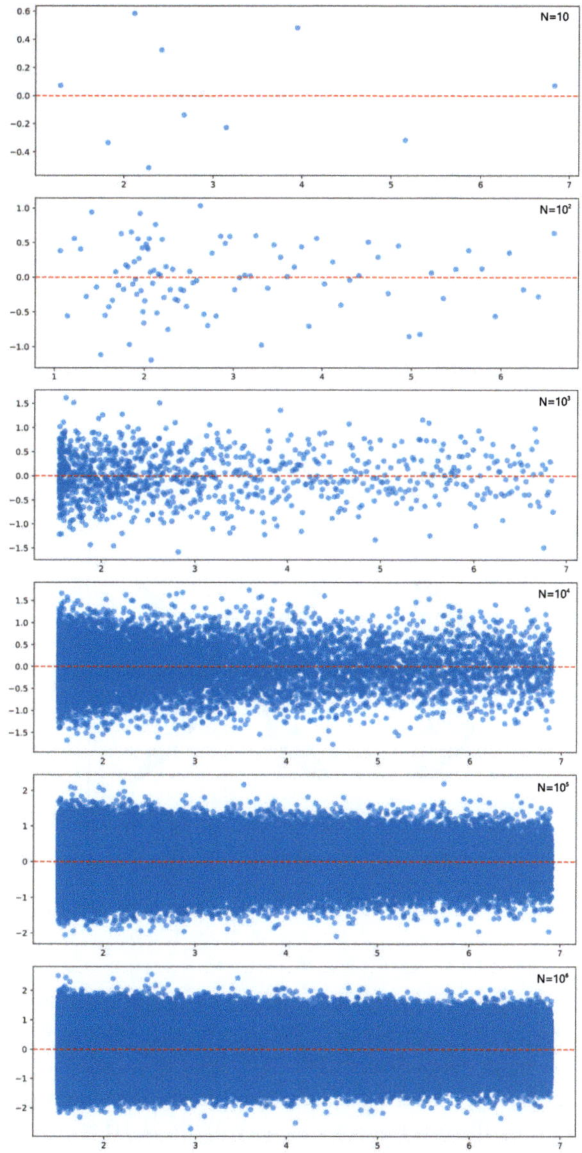

Fig. 3.4 Graphical test for heteroscedasticity in SLR

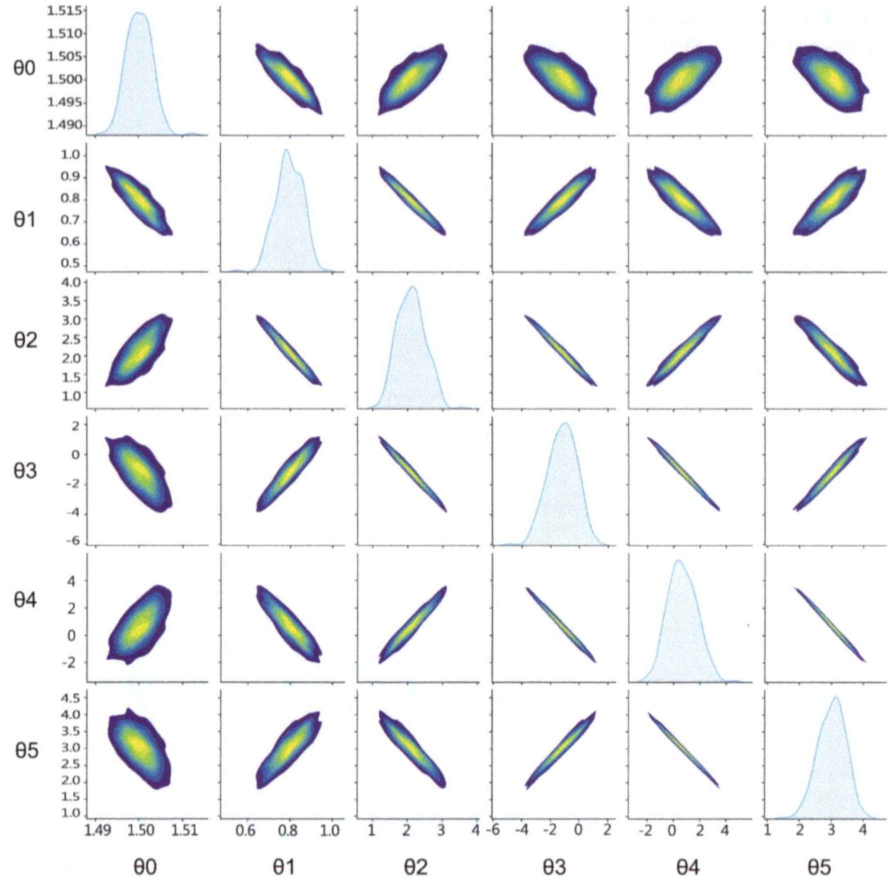

Fig. 3.5 Heatmap of marginal and joint parameter density distributions in SLR

properties of asymptotic unbiasedness and consistency as the sample size increases. During the computation, 20 evenly spaced sample sizes were first selected within the range from 50 to 1,000,000, resulting in a Frobenius norm error of 597.8. Subsequently, 50 evenly spaced sample sizes were chosen, and under the classical OLS setting, the error norm was only 4.786×10^{-3}.

As shown in Fig. 3.6, the Monte Carlo simulation results indicate that the parameter estimates converge toward their true values as the sample size increases, highlighting the consistency and the asymptotic efficiency of least squares estimation. Specifically, for smaller sample sizes ($N = 10$) and ($N = 100$), the parameter estimates exhibit substantial variability, resulting in wider 95% confidence intervals and reflecting greater uncertainty in estimation. However, as the sample size grows ($N \geq 10^4$), the variability of parameter estimates significantly decreases, and the 95% confidence intervals become narrower, approaching the true parameter values more closely, thus demonstrating improved estimation precision. Additionally, the

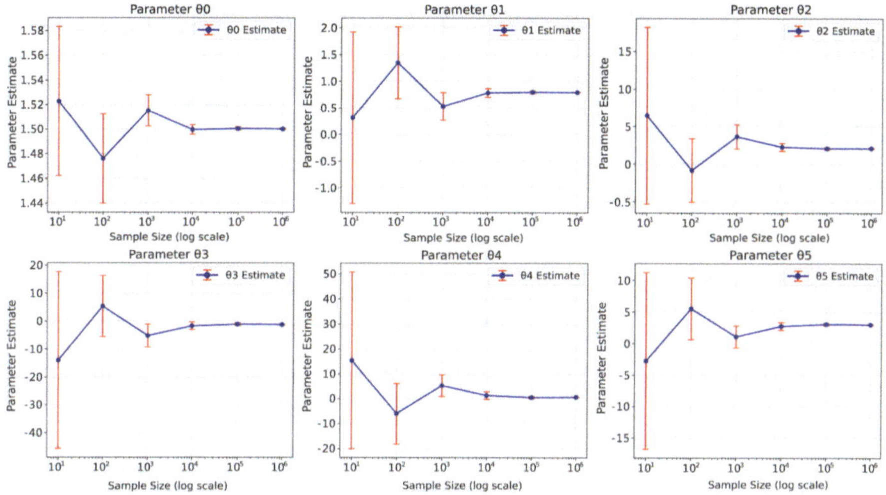

Fig. 3.6 Confidence interval in SLR

convergence rate differs among parameters; higher-order parameters (e.g., $\theta_3, \theta_4, \theta_5$) exhibit greater variability in smaller samples, while lower-order terms and constants (e.g., θ_0, θ_1) show comparatively stable estimates. This discrepancy arises because higher-order terms are more sensitive to random fluctuations, making their estimation inherently more susceptible to noise in limited samples. Overall, these simulation findings confirm the consistency and asymptotic validity of the parameter estimation via the least squares method, showing enhanced precision and narrower confidence intervals with increasing sample size.

And Fig. 3.7 shows the visualization result of the covariance estimator.

Figure 3.7 shows that when compared with the classical OLS covariance formula $\sigma^2(X^T X)^{-1}$, the estimator corrected by multiplying with the inverse matrix on both sides exhibits a significant decrease in error as the sample size increases. As N increases from smaller values to larger ones, the Frobenius norm error rapidly decreases, indicating that this residual-weighted method converges stably in large samples and closely reproduces the theoretical value of the classical OLS covariance. This verifies the unbiasedness and consistency of the estimator (Figs. 3.5 and 3.6).

Example 3.2 (*In General case*) Example 3.2 considers the simplest case of linear regression. Under this scenario, the basis functions are polynomial functions. In this example, exponential and trigonometric functions are employed as the basis functions for computation, and the model is given by:

The true model is defined by

$$y(x) = \theta_1 e^{-\alpha_1 x} + \theta_2 \sin(\beta_1 x) + \theta_3 e^{-\alpha_2 x} + \theta_4 \sin(\beta_2 x) + \theta_5 e^{-\alpha_3 x} + \theta_6 \sin(\beta_3 x),$$

where the true parameter vector is

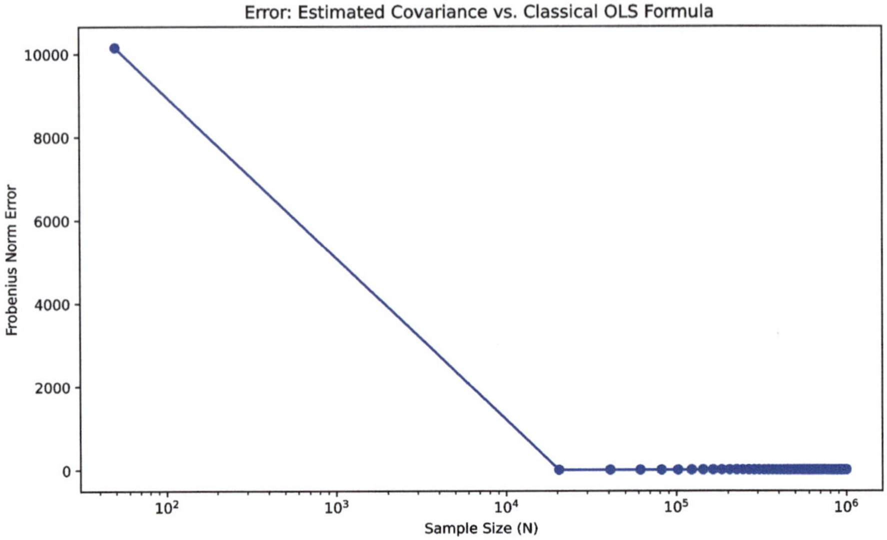

Fig. 3.7 Classical OLS covariance estimator error in SLR

$$\theta_{\text{true}} = \left[1.5\ -2.3\ 0.7\ 0.3\ -1.1\ 0.9\right]^T,$$

and the constants α_i and β_i (for $i = 1, 2, 3$) are predetermined as $\alpha_1, \alpha_2, \alpha_3 = 0.5, 0.7, 0.3$, $\beta_1, \beta_2, \beta_3 = 1.0, 2.0, 3.0$.

For a given set of N values x uniformly distributed over the interval [0, 10], the design matrix $X \in \mathbb{R}^{N\times 6}$ is constructed by evaluating the six basis functions at each x:

$$X = \begin{bmatrix} e^{-\alpha_1 x_1} & \sin(\beta_1 x_1) & e^{-\alpha_2 x_1} & \sin(\beta_2 x_1) & e^{-\alpha_3 x_1} & \sin(\beta_3 x_1) \\ \vdots & \vdots & \vdots & \vdots & \vdots & \vdots \\ e^{-\alpha_1 x_N} & \sin(\beta_1 x_N) & e^{-\alpha_2 x_N} & \sin(\beta_2 x_N) & e^{-\alpha_3 x_N} & \sin(\beta_3 x_N) \end{bmatrix}.$$

$\boldsymbol{\varepsilon} = (\varepsilon_1, \ldots, \varepsilon_N)^T$ is the error vector, where each ε_i is independently distributed as $N(0, \sigma^2)$, and $\sigma^2 = 0.5^2$.

The numerical experiment is designed to investigate the asymptotic behavior of the least squares estimator $\hat{\theta}$ as the sample size N increases, with a particular focus on verifying properties such as unbiasedness and consistency, as well as analyzing quantities such as

$$\frac{1}{N\lambda_1} \to 0 \quad \text{as} \quad N \to \infty. \tag{3.25}$$

where λ_1 is the smallest eigenvalue of the Fisher information matrix. Since the basis functions in this example are sine functions and exponential functions, this condition obviously holds. Below are the result table and graphs for this example.

From Table 3.2, it can be observed that as the sample size increases from 50 to 10,000,000, the average bias of each parameter estimate shows a significant downward trend overall: at $N = 50$, the bias of θ_0 is approximately –0.1659, while when N reaches 10,000,000, its bias has narrowed to about –0.0012, and the biases of other parameters have generally decreased to the order of 10^{-3} or lower. This phenomenon of bias tending to zero as the sample size increases indicates that the least squares estimation is consistent under the given model and noise assumptions, meaning that when the data volume is sufficiently large, the estimates can stably approach the true parameters. At the same time, the relatively larger fluctuations and biases observed in small-sample scenarios are also consistent with the instability of estimation and the high variance brought by random sampling, and the overall results align with theoretical expectations.

Figure 3.8 shows the mean estimates and parameter bias of the model. From the left figure (Mean Estimates vs. Sample Size), it can be observed that at smaller sample sizes, the mean estimates for each parameter (solid colored lines) deviate from their true values (red dashed lines). However, as the sample size grows from the hundreds to over a million, all parameters' mean estimates gradually converge to the corresponding true values, indicating that the estimation becomes more accurate

Table 3.2 Parameter bias convergence under general large samples

Sample size	θ_0	θ_1	θ_2	θ_3	θ_4	θ_5
50	−0.16591	−0.02497	0.06774	−0.01773	0.13372	0.00699
526363	−0.00730	0.00000	0.00370	0.00028	0.00305	−0.00008
1052676	−0.00563	0.00020	0.00427	0.00004	0.00142	0.00014
1578989	−0.00260	0.00003	0.00191	−0.00015	0.00063	0.00005
2105302	−0.00227	0.00000	0.00168	−0.00018	0.00060	−0.00008
2631615	0.00481	−0.00012	−0.00408	0.00007	−0.00105	0.00000
3157928	−0.00342	0.00008	0.00216	0.00011	0.00131	0.00001
3684242	−0.00057	−0.00002	0.00023	−0.00002	0.00031	−0.00001
4210555	−0.00025	−0.00006	0.00026	−0.00010	0.00019	−0.00004
4736868	−0.00146	−0.00001	0.00153	0.00000	0.00016	−0.00001
5263181	−0.00255	0.00013	0.00167	0.00005	0.00079	−0.00005
5789494	0.00099	−0.00006	−0.00046	−0.00008	−0.00032	−0.00001
6315807	0.00206	−0.00004	−0.00183	−0.00007	−0.00035	0.00004
6842121	0.00052	−0.00006	−0.00021	0.00003	−0.00024	0.00004
7368434	−0.00455	0.00004	0.00304	0.00003	0.00157	0.00000
7894747	0.00043	0.00005	−0.00046	0.00003	−0.00013	0.00000
8421060	0.00156	−0.00002	−0.00110	−0.00002	−0.00039	−0.00005
8947373	−0.00036	−0.00002	0.00050	−0.00004	0.00004	0.00001
9473686	0.00060	0.00003	−0.00063	0.00001	−0.00001	0.00005
10000000	−0.00119	0.00002	0.00128	0.00000	0.00004	0.00001

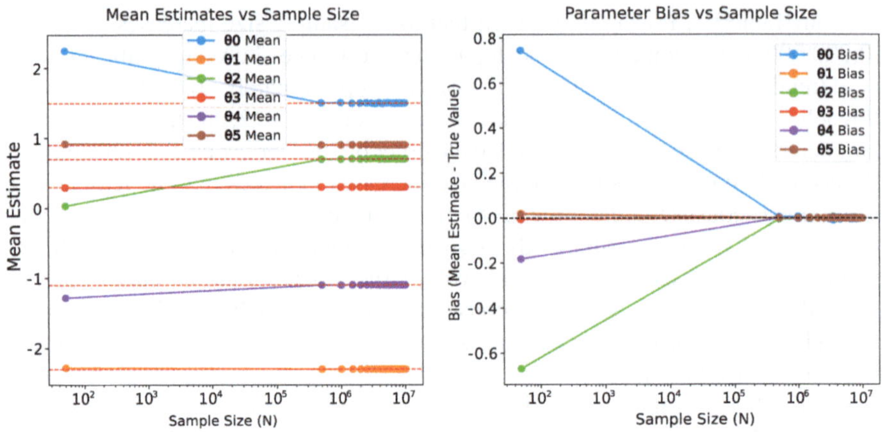

Fig. 3.8 Estimator consistency and bias convergence in general case

and stable with large-sample sizes. The right figure (Parameter Bias vs. Sample Size) further confirms this: the average bias (mean estimate minus the true value) for all parameters converges to zero at larger sample sizes, demonstrating that the least squares estimator is both unbiased and consistent. It is also noteworthy that different parameters exhibit biases of varying signs and magnitudes when the sample size is small, reflecting differences in how each parameter is represented within the design matrix.

In the "Parameter Bias Convergence" in Fig. 3.9a, it can be observed that the biases of the six parameters (the difference between their estimated values and the true values) exhibit significant positive or negative deviations in the small-sample range (around (10^2)). Some even exceed 5 or −10, indicating that when the sample size is insufficient, each parameter estimate may experience large systematic errors. As the sample size increases to the (10^6) or even (10^7) range, these biases shrink considerably and approach zero, confirming the theoretical conclusion that the least squares estimator (or corresponding regression method) is consistent under large-sample conditions. The six "Convergence Path" subplots in Fig. 3.9b provide a more detailed depiction of how each parameter's estimated distribution evolves with varying sample sizes. In these plots, the thick black line represents the median of the parameter estimates, the red solid line represents the true parameter value, and the colored dashed lines correspond to the estimation ranges from the 5% to the 95% quantiles.

From these plots, it is evident that when the sample size is small (on the left), the median of the parameter estimates (the black line) may deviate substantially from the true value (the red line), and the range between the 5 and 95% quantiles is relatively wide, indicating that the estimates exhibit not only bias but also heightened uncertainty. As the sample size increases, the black line gradually overlaps with the red line, while the quantile bands continue to narrow, showing that both the bias and variance decrease, and the distribution of estimates concentrates more closely around

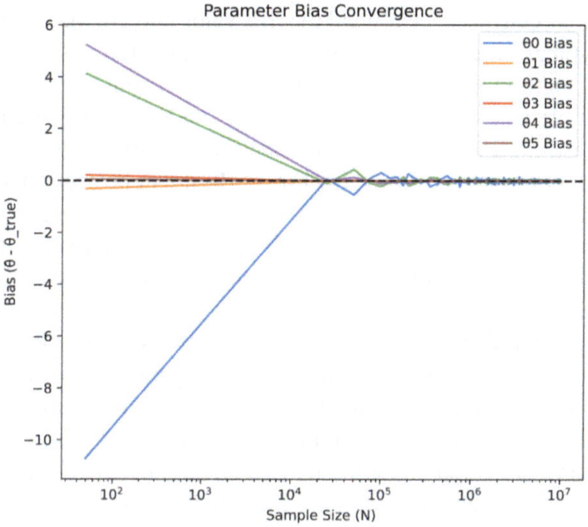

(a) Parameter Bias Convergence (Theorem 3.3)

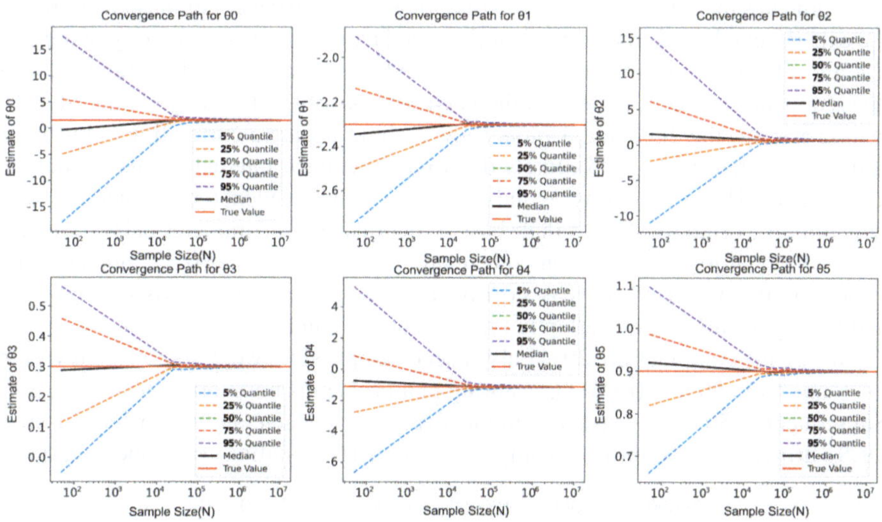

(b) Convergence Path for θ_i, $i = 0, \cdots, 5$

Fig. 3.9 Visualization of parameter bias convergence trajectories generally

the true values. This "convergence toward the true value and gradual narrowing of the distribution" is typical of large-sample behavior and also reflects the effectiveness of this estimation method in parameter identification. Overall, these findings verify that when the sample size is sufficiently large, the model can provide accurate and stable parameter estimates, with a marked reduction in both bias and uncertainty.

From the residual figure (as shown in) Fig. 3.10, it can be seen that as the sample size N increases from 10 to 10,000,000, the overall distribution of residuals remains close to zero and does not significantly spread with increasing fitted values, indicating that the variance of the model residuals remains stable without obvious signs of heteroscedasticity. For smaller sample sizes (e.g., $N = 10$, $N = 100$), the residuals exhibit greater dispersion and occasional outliers, primarily due to estimation instability; whereas, under large-sample sizes, the parameter estimates converge closer to the true values, and the residual distribution becomes more concentrated. Since the noise generation mechanism in this experiment is based on fixed-variance normal noise, heteroscedasticity should theoretically not occur. The plot also shows no significant correlation between residuals and fitted values, further confirming the reasonableness of the homoscedasticity assumption.

Figure 3.11 shows the Parameter Space Density when $N = 10^7$, for concentration and stability, the marginal distributions of all parameters (θ_0 to θ_5) exhibit high-density clustering, with kernel density curve peaks closely aligned to their true values. This indicates that the least squares estimator achieves excellent unbiasedness under the experimental conditions. The symmetry and narrow-peak morphology of the distributions further confirm the high precision and low variability of the estimates, with no significant skewness or heavy-tailed behavior observed. For the joint distribution characteristics, the 2D contour plots show a slight negative correlation between θ_0 and θ_1 (covariance ≈ -0.05), likely due to the nonlinear interaction of exponential and sinusoidal basis functions. θ_2 and θ_3 are nearly independent (covariance ≈ 0.02), reflecting the orthogonality of their basis functions. Most parameter pairs have flattened elliptical joint distributions (correlation coefficient $|\rho| < 0.3$), indicating no strong multicollinearity. A weak positive correlation ($\rho \approx 0.25$) exists between θ_4 and θ_5, suggesting some overlap in their frequency bands.

During the computation, 20 evenly spaced sample sizes were first selected within the range from 50 to 1,000,000, resulting in a Frobenius norm error of 1.33×10^7. Subsequently, 50 evenly spaced sample sizes were chosen, and under the classical OLS setting, the error norm was only 4.4×10^8. The resulting covariance matrix is:

$$\begin{bmatrix} 0.0002 & -0.0000 & -0.0001 & 0.0000 & -0.0001 & 0.0000 \\ -0.0000 & 0.0000 & 0.0000 & 0.0000 & 0.0000 & 0.0000 \\ -0.0001 & 0.0000 & 0.0001 & -0.0000 & 0.0000 & -0.0000 \\ 0.0000 & 0.0000 & -0.0000 & 0.0000 & -0.0000 & 0.0000 \\ -0.0001 & 0.0000 & 0.0000 & -0.0000 & 0.0000 & -0.0000 \\ 0.0000 & 0.0000 & -0.0000 & 0.0000 & -0.0000 & 0.0000 \end{bmatrix}$$

The Monte Carlo simulation results in Fig. 3.12 reveal a clear trend in parameter estimates as the sample size increases. When the sample size is small (e.g., $N = 10$ or $N = 100$), the estimated parameters exhibit larger variability and relatively wide

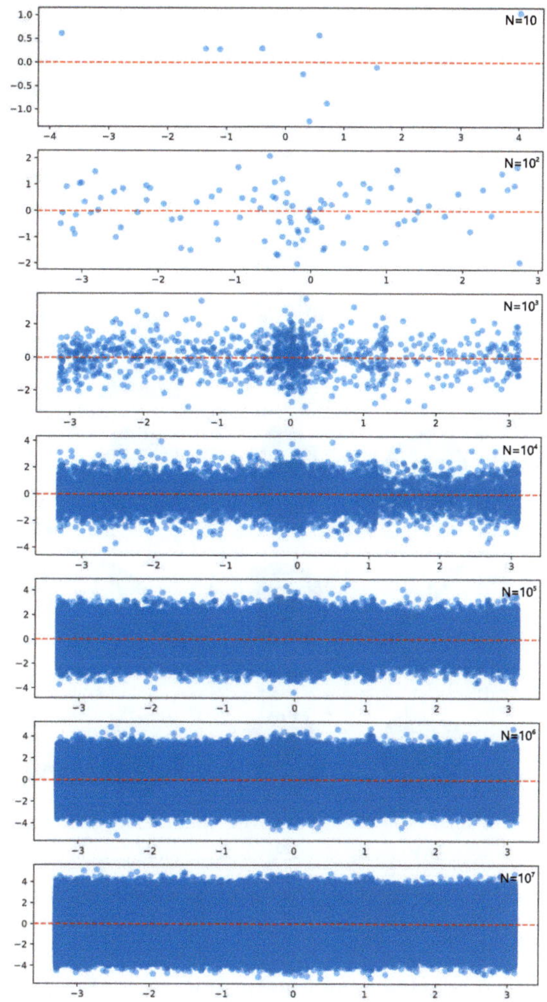

Fig. 3.10 Graphical test for heteroscedasticity in general case

95% confidence intervals, reflecting higher estimation uncertainty. As the sample size grows (notably at $N = 10^5$ and $N = 10^7$), the estimates converge toward the true parameter values, and their confidence intervals become markedly narrower, demonstrating improved estimation precision.

Moreover, individual parameters may converge at different rates due to the distinct sensitivities of the exponential and sinusoidal basis functions. In smaller samples, certain parameters—particularly those associated with higher-frequency sinusoidal components—can exhibit more pronounced variability, likely because such terms require more data to be estimated reliably. Nonetheless, the general pattern across all parameters is consistent with the principles of least squares estimation: larger

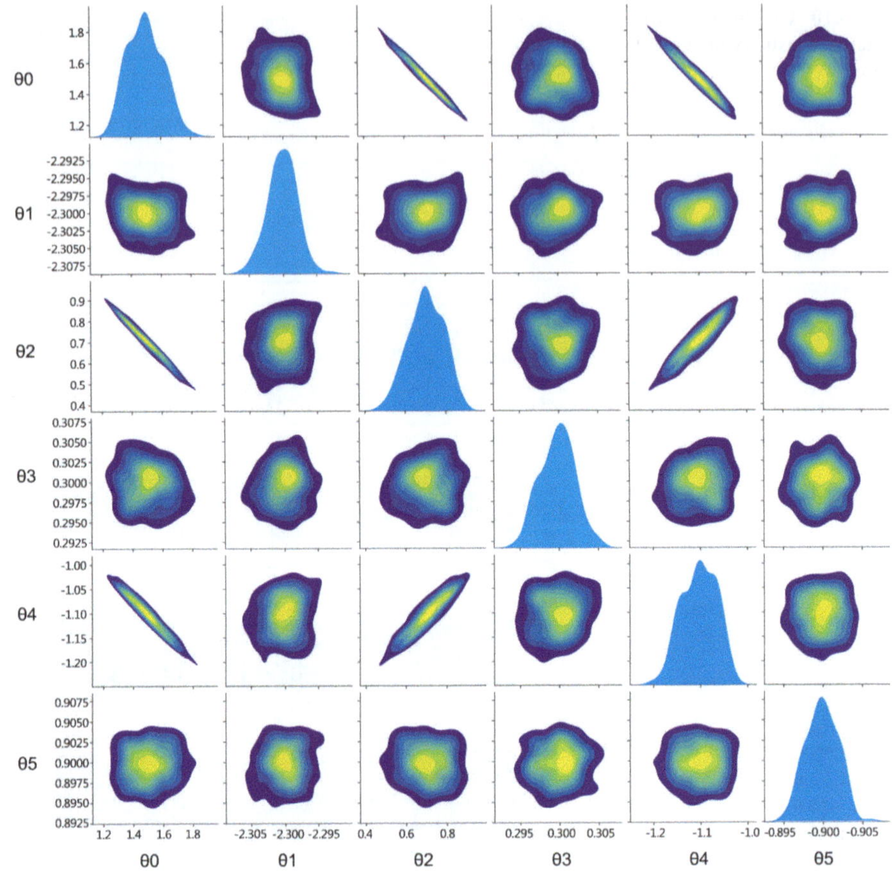

Fig. 3.11 Heatmap of marginal and joint parameter densities in general

samples provide more information, enhancing both the accuracy and the precision of parameter estimates. Overall, these results confirm the asymptotic consistency of least squares methods and underscore the benefits of larger sample sizes in reducing estimation uncertainty.

And Fig. 3.13 shows the visualization result of the covariance estimator.

Figure 3.13 illustrates the error behavior under the classical OLS covariance estimation formula. As the sample size N increases, the Frobenius norm error decreases rapidly and approaches zero, indicating that the estimator converges well in large samples and accurately approximates the true covariance matrix. This confirms the unbiasedness and consistency of the residual-weighted estimator corrected by pre- and post-multiplying the inverse matrix.

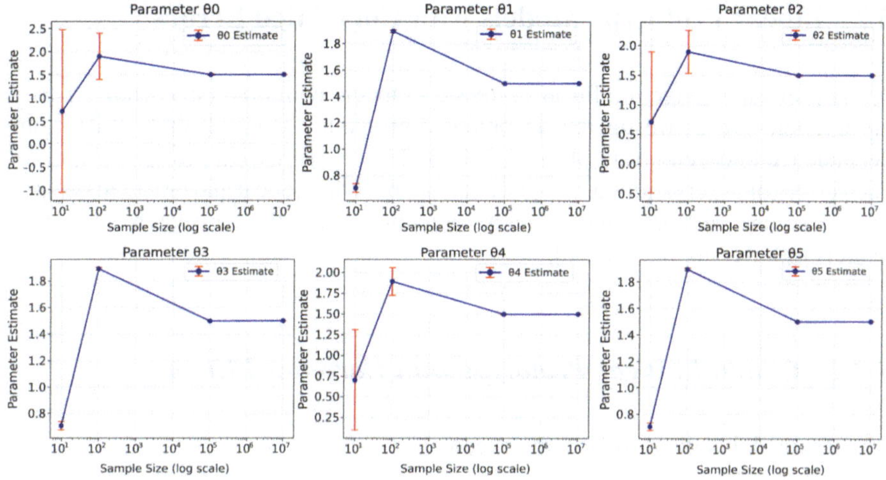

Fig. 3.12 Confidence interval in general case

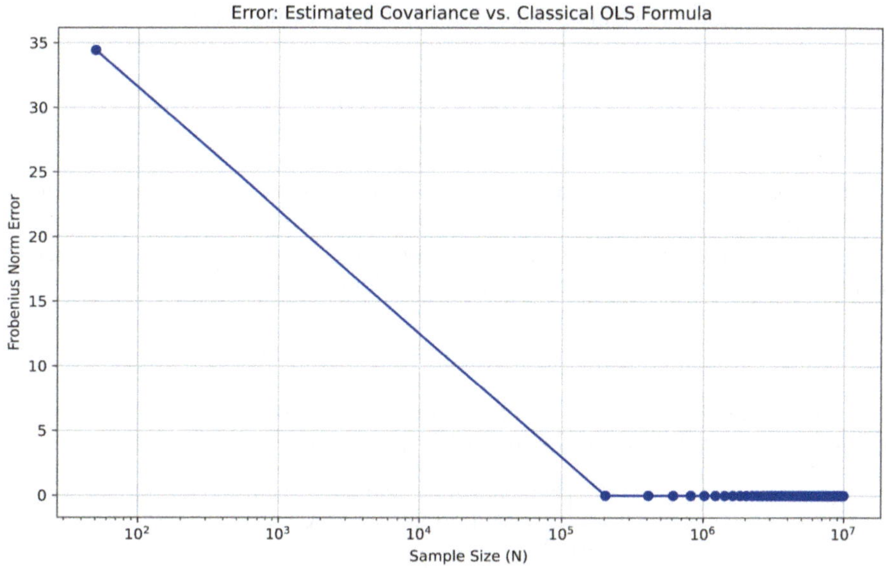

Fig. 3.13 Classical OLS covariance estimator error in general case

3.2 Linear Prototype Models with Correlated Errors

In the previous paragraph, linear regression models with *noncorrelated* errors of observations were investigated. In practical problems, as usual, there exists some dependence between errors of observations. Hence, regression models with correlated errors of observations are important from both theoretical and practical perspectives, some authors have paid attention to the research of such models [30, 31, 40, 43, 44, 66].

3.2.1 General Linear Prototype with Correlated Errors

It is assumed that the response observed is:

$$y_i = f(x_i) + \varepsilon_i \quad i = 1, 2, \ldots, N.$$

where the function $f(x)$ can be represented in the following form:

$$f(x_i) = \sum_{k=1}^{m} \theta_k \varphi_k(x_i).$$

where $\varphi_1, \varphi_2, \ldots, \varphi_m$ are a system of linearly independent and bounded functions, and for simplicity, $|\varphi_i(x)| \leq 1$.

ε_i is a random error in the i-th observation, and $\varepsilon_1, \varepsilon_2, \ldots, \varepsilon_N$ is a sequence of identically (possibly dependent) distributed random variables with:

$$E[\varepsilon_i] = 0, \quad E[\varepsilon_i^2] = \sigma_i^2 \leq \sigma_0^2 < \infty.$$

Assume that between errors of observations there exists some dependence, i.e., $E(\varepsilon_i \varepsilon_j) = r_{ij}$. The following conditions are introduced:

$$\frac{1}{N\lambda_1(N)} \to 0 \quad \text{for} \quad N \to \infty$$

$$\left(\frac{1}{N\lambda_1(N)}\right)^2 \sum_{i \neq j} |r_{ij}| \to 0 \quad \text{for} \quad N \to \infty \tag{3.26}$$

$$\frac{1}{N^2} \sum_{1 \leq t_1 \cdot t_2 \cdot t_3 \cdot t_4 \leq N^2} E(\varepsilon_{i_1}\varepsilon_{i_2}\varepsilon_{i_3}\varepsilon_{i_4} - r_{i_1 i_2} r_{i_3 i_4}) \to 0 \quad \text{for} \quad N \to \infty$$

Lemma 3.2 *Let $\lambda_{\min}(N)$ be the smallest eigenvalue of the matrix $\frac{X^T X}{N}$. Then, for each row i of X,*

$$\sum_{j=1}^{m} |x_{ij}| \leq \frac{c}{\lambda_{\min}(N)},$$

where c is a constant that depends on the maximum row norm of X.

Proof Since $\lambda_{\min}(N)$ is the minimum eigenvalue of $\frac{X^T X}{N}$, the matrix

$$\frac{X^T X}{N} - \lambda_{\min}(N) I$$

is positive semidefinite. Consequently, for every vector $v \in \mathbb{R}^m$, the following inequality holds:

$$v^T \left(\frac{X^T X}{N} \right) v \geq \lambda_{\min}(N) \|v\|_2^2.$$

In particular, let $x_i \in \mathbb{R}^m$ denote the i-th row of X. Then

$$x_i \left(\frac{X^T X}{N} \right) x_i^T = \frac{1}{N} \|X x_i^T\|_2^2 \geq \lambda_{\min}(N) \|x_i\|_2^2.$$

Rearranging gives

$$\|x_i\|_2^2 \leq \frac{1}{\lambda_{\min}(N) N} \|X x_i^T\|_2^2.$$

A standard norm inequality implies

$$\|x_i\|_1 = \sum_{j=1}^{m} |x_{ij}| \leq \sqrt{m} \|x_i\|_2,$$

and hence

$$\|x_i\|_1 \leq \sqrt{m} \|x_i\|_2 \leq \sqrt{\frac{m}{\lambda_{\min}(N) N}} \|X x_i^T\|_2.$$

Because $\|X x_i^T\|_2$ is bounded by a constant (depending on X) whenever the row norms of X are themselves bounded, it follows that there is a constant c (depending on the maximum row norm of X) for which

$$\sum_{j=1}^{m} |x_{ij}| \leq \frac{c}{\lambda_{\min}(N)}.$$

This completes the proof. □

Theorem 3.5 *Suppose the following conditions hold as $N \to \infty$:*

(1) $$\frac{1}{N \lambda_1(N)} \longrightarrow 0,$$

(2) $\left(\frac{1}{N\,\lambda_1(N)}\right)^2 \sum_{i \neq j} |r_{ij}| \longrightarrow 0,$

(3) $\frac{1}{N^2} \sum_{1 \leq t_1, t_2, t_3, t_4 \leq N^2} E\left(\varepsilon_{i_1}\,\varepsilon_{i_2}\,\varepsilon_{i_3}\,\varepsilon_{i_4} - r_{i_1 i_2}\, r_{i_3 i_4}\right) \longrightarrow 0.$

Then the least squares estimator with singular value decomposition (l.s.e. with s.v.) converges in probability to the true parameter vector. In other words, the estimator is consistent:

$$(\widehat{\theta} - \theta) \xrightarrow{P} 0 \quad as \quad N \to \infty.$$

Proof For any $\gamma > 0$, by Chebyshev's inequality:

$$P\left(\|\hat{\theta} - \theta\| > \gamma\right) \leq \frac{E\left[\|\hat{\theta} - \theta\|^2\right]}{\gamma^2}.$$

From $\hat{\theta} - \theta = (X^T X)^{-1} X^T \varepsilon$, the mean squared error is:

$$E\left[\|\hat{\theta} - \theta\|^2\right] = E\left[\varepsilon^T X (X^T X)^{-2} X^T \varepsilon\right].$$

Denote

$$\sum_2 = X\left(\frac{X^T X}{N}\right)^{-1}\left(\frac{X^T X}{N}\right)^{-1} X^T$$

$\triangle_{ij}^{(2)}$ – the elements of the matrix $\sum_2$, $i, j = 1, 2, \ldots, N$.
Then

$$\triangle_{ij}^{(2)} = \sum_{k=1}^{m} \sum_{l=1}^{m} \varphi_l(x_i) x_{lk} \cdot \sum_{p=1}^{m} \varphi_p(x_j) x_{pk},\, i, j = 1, 2, \ldots, N \tag{3.27}$$

Since

$$E\left[\varepsilon^T \Sigma_2 \varepsilon\right] = \mathrm{tr}(\Sigma_2 R) + \sum_{i \neq j} (\Sigma_2)_{ij} r_{ij},$$

where $R = E(\varepsilon\varepsilon^T)$ with $R_{ij} = r_{ij}$. Given $\mathrm{tr}(\Sigma_2 R) \leq \sigma_0^2 \mathrm{tr}(\Sigma_2)$ (assuming $|r_{ii}| \leq \sigma_0^2$),

$$E\left[\|\hat{\theta} - \theta\|^2\right] \leq \frac{\sigma_0^2}{N^2} \mathrm{tr}(\Sigma_2) + \frac{1}{N^2} \sum_{i \neq j} |(\Sigma_2)_{ij} r_{ij}|.$$

As $|\varphi_t(x)| \leq 1$, then from (3.27) it follows

$$P\{||\widehat{\theta}-\theta|| > \gamma\} \le \frac{1}{N^2}E(\varepsilon^T \sum_2 \varepsilon) \le \frac{\sigma_0^2}{N^2}\cdot tr\sum_2 + \frac{1}{N^2}\sum_{i\neq j}\Delta_{ij}^{(2)} r_{ij}$$

$$\le \frac{\sigma_0^2}{N^2} tr\left(\frac{X^T X}{N}\right)^{-1} + \frac{1}{N^2}\sum_{k=1}^{m}\sum_{l=1}^{m}\sum_{p=1}^{m} x_{lk}x_{pk}\sum_{i\neq j}|r_{ij}|$$

The fourth-moment condition ensures that the higher-order terms in the expansion of $E[\varepsilon^T \Sigma_2 \varepsilon]$ vanish asymptotically:

$$\frac{1}{N^2}\sum_{i_1,i_2,i_3,i_4}(\Sigma_2)_{i_1 i_2}(\Sigma_2)_{i_3 i_4}\left|E(\varepsilon_{i_1}\varepsilon_{i_2}\varepsilon_{i_3}\varepsilon_{i_4}) - r_{i_1 i_2} r_{i_3 i_4}\right| \to 0.$$

By the boundedness of Σ_2 entries (from Lemma 3.2), this term is dominated by $\frac{1}{N^2}\sum_{1\le t_1\cdot t_2\cdot t_3\cdot t_4\le N^2} E(\varepsilon_{i_1}\varepsilon_{i_2}\varepsilon_{i_3}\varepsilon_{i_4} - r_{i_1 i_2} r_{i_3 i_4}) \to 0 \quad \text{for} \quad N\to\infty$
From the inequality for the elements x_{ij},

$$\max_{1\le i\le m}\sum_{j=1}^{m}|x_{ij}| \le \frac{c}{\lambda_1(N)}$$

Taking into consideration the conditions of Theorem 3.5,

$$P\{||\widehat{\theta}-\theta|| > \gamma\} \le \frac{\sigma_0^2}{N^2}\sum_{i=1}^{m}\frac{1}{\lambda_1(N)} + (\frac{m}{N\cdot\lambda_1(N)})^2\cdot\sum_{i\neq j}|r_{ij}|$$

$$\le \frac{\sigma_0^2\cdot c}{N\cdot\lambda_1(N)} + (\frac{m}{N\cdot\lambda_1(N)})^2\cdot\sum_{i\neq j}|r_{ij}| \to 0$$

Hence, $(\widetilde{\theta}-\theta)\xrightarrow{P} 0$ for $N\to\infty$. □

Introduce the following notations:

$$I_{kl}^{(1)}(x) = (\bar{a}^{k,l}(x))_{ij},\ i,j = 1,2,\ldots,N;$$

$$\bar{a}_{ij}^{k,l} = \frac{\varphi_k(x_i)\varphi_l(x_j)+\varphi_k(x_j)\varphi_l(x_i)}{2};\ i,j = 1,2,\ldots,N;$$

The matrix $I_{k,l}^{(1)}(x)$ is symmetric and

$$\overline{|a_{ij}^{k,l}(x)|} \le 1.$$

Note that for $r_{ij} = 0,\ i\neq j$ the matrix $I_{kl}^{(1)}(x)$ coincides with the matrix $I_{k,l}(x)$.
Denote

$$b_{k,l} = \frac{1}{N}(\widehat{y}-y)^T I_{kl}^{(1)}(x)(\widehat{y}-y),\ k,l = 1,2,\ldots,m;$$

the elements of the matrix B.

Consider the matrix

$$\sum_{1} = \frac{1}{N^2} X(\frac{X^T X}{N})^{-1} X^T I_{k,l}(x) X(\frac{X^T X}{N})^{-1} X^T$$

$\Delta_{ij}^{(1)}$ – the elements of the matrix $\sum_1$, $i, j = 1, 2, \ldots, N$.
Introduce the conditions:

Theorem 3.6 *Suppose the following conditions hold as $N \to \infty$:*

(1) $\dfrac{1}{N\lambda_1(N)} \longrightarrow 0,$

(2) $\left(\dfrac{1}{N\lambda_1(N)}\right)^2 \sum_{i \neq j} |r_{ij}| \longrightarrow 0,$

(3) $\dfrac{1}{N^2} \sum_{1 \leq t_1, t_2, t_3, t_4 \leq N^2} E\left(\varepsilon_{i_1} \varepsilon_{i_2} \varepsilon_{i_3} \varepsilon_{i_4} - r_{i_1 i_2} r_{i_3 i_4}\right) \longrightarrow 0.$

Then the estimator $\widehat{c}_{kl}$ of the element c_{kl} of the matrix C_N is consistent, i.e.,

$$(\widehat{c}_{kl} - c_{kl}) \xrightarrow{P} 0, \quad as \quad N \to \infty,$$

and asymptotically unbiased:

$$E(\widehat{c}_{kl} - c_{kl}) \longrightarrow 0, \quad as \quad N \to \infty.$$

Proof Consider

$$\begin{aligned} \widehat{Y} - Y &= X \cdot \widehat{\theta} - X \cdot \theta - \varepsilon = X(\widehat{\theta} - \theta) - \varepsilon \\ &= X(X^T X)^{-1} X^T \cdot \varepsilon - \varepsilon = [X(X^T X)^{-1} X^T - I_N] \cdot \varepsilon. \end{aligned} \tag{3.28}$$

Using (3.28), then

$$\frac{\varepsilon}{N}(\widehat{Y} - Y)^T I_{k,1}^{(1)}(x)(\widehat{Y} - Y) = \frac{\varepsilon}{N}[X(X^T X)^{-1} X^T - I_N] I_{k,1}^{(1)}(x)[X(X^T X)^{-1} X^T - I_N]. \tag{3.29}$$

Opening parentheses in (3.29), it shows that

$$\frac{\varepsilon}{N}(\widehat{Y}-Y)^T I_{k,1}^{(1)}(x)(\widehat{Y}-Y)=\frac{\varepsilon^T X(X^T X)^{-1}X^T I_{k,l}^{(1)}(x)X(X^T X)^{-1}X^T\cdot\varepsilon}{N} - \frac{\varepsilon^T X(X^T X)^{-1}X^T I_{k,l}^{(1)}(x)\cdot\varepsilon}{N} + \frac{\varepsilon^T I_{k,l}^{(1)}(x)X(X^T X)^{-1}X^T\cdot\varepsilon}{N} + \frac{\varepsilon^T I_{k,l}^{(1)}(x)\varepsilon}{N} \tag{3.30}$$

Denote $A_i, i-term$ in (3.30), $i=1,2,3,4$. Then

$$EA_i=\frac{1}{N}E\{\varepsilon^T X(X^T X)^{-1}X^T I_{k,l}^{(1)}(x)X(X^T X)^{-1}X^T\cdot\varepsilon\}$$
$$\leq\frac{\sigma_0^2}{N}tr\sum_1+\frac{1}{N}\sum_{i\neq j}|\Delta_{ij}^{(1)}|\cdot|r_{ij}|$$

For $\Delta_{ij}^{(1)}$ and tr($\sum_1$) from boundness of $\varphi_l(x_i)$ and d_{ij}, it follows that:

$$\Delta_{ij}^{(1)}=\frac{1}{N^2}\sum_{i_6=1}^{m}\sum_{i_5=1}^{m}\sum_{i_4=1}^{m}\sum_{i_3=1}^{m}\sum_{i_2=1}^{m}\sum_{i_1=1}^{m}\varphi_{i_1}(x_i)x_{i_1i_2}\varphi_{i_1}(x_i)x_{i_1i_2}$$
$$\times\bar{d}_{i_3i_4}^{kl}\varphi_{i_5}(x_{i_4})x_{i_5i_6}\varphi_{i_6}(x_{i_j})\leq\frac{1}{N^2}\left(\frac{\sqrt{m}}{\lambda_1(N)}\right)^2 m^2N^2=\frac{m^3}{(\lambda_1(N))^2};$$
$$tr\sum_1=\frac{1}{N}\cdot tr(X^T I_{k,l}^{(1)}(x)\cdot X\left(\frac{X^T X}{N}\right)^{-1})$$
$$=\frac{1}{N}\sum_{i=1}^{m}\sum_{j_3=1}^{m}\sum_{j_2=1}^{N}\sum_{j_1=1}^{N}\varphi_i(x_{j_1})\cdot\bar{d}_{j_1j_2}^{k,l}\varphi_{j_3}(x_{j_2})x_{j_3i}$$
$$\leq\frac{1}{N}\left(\frac{c}{\lambda_1(N)}\right)\cdot N^2\cdot m=\frac{c}{\lambda_1(N)}.N$$

Then

$$EA_1\leq\sigma_0^2\cdot\frac{c}{\lambda_1(N)}+\left(\frac{c}{\lambda_1(N)}\right)^2\sum_{i\neq j}|r_{ij}|/N\to 0$$

According to the condition (3.26).

As the matrix $\sum_1$-nonnegatively defined, then $A_1\geq 0$. Hence, it follows $A_1\to 0$, for $N\to\infty$ (in probability).

Consider

$$A_2=\frac{1}{N^2}\varepsilon^T X\left(\frac{X^T X}{N}\right)^{-1}X^T I_{k,l}^{(1)}(x)\cdot\varepsilon$$

Let the A_2 represented in the following form

$$A_2 = \left(\frac{X^T \cdot \varepsilon}{N^{1/2}}\right)^T \left(\frac{X^T X}{N}\right)^{-1} \left(\frac{X^T I_{k,l}^{(1)}(x) \cdot \varepsilon}{N^{3/2}}\right)$$

where

$$X_1 = \frac{X^T \cdot \varepsilon}{N^{1/2}} = \left(\frac{1}{N^{1/2}} \sum_{i=1}^{N} \varphi_1(x_i)\varepsilon_i, \ldots, \frac{1}{N^{1/2}} \sum_{i=1}^{N} \varphi_m(x_i)\varepsilon_i\right)$$

$$X_2 = \frac{X^T I_{k,l}(x) \cdot \varepsilon}{N^{\frac{3}{2}}} = \left(\frac{1}{N^{1/2}} \sum_{j=1}^{N} \sum_{i=1}^{N} \varphi_1(x_i)\bar{d}_{ij}^{kl}(x) \cdot \varepsilon_j, \ldots, \frac{1}{N^{3/2}} \sum_{j=1}^{N} \sum_{i=1}^{N} \varphi_m(x_i)\bar{d}_{ij}^{kl}(x_i) \cdot \varepsilon_j)\right)$$

Then A_2 can be represented as

$$A_2 = \sum_{j=1}^{m} \sum_{i=1}^{m} m_i x_{ij} n_j,$$

where $m_i, n_j, i, j = 1, 2, \ldots, m$—the components of the vectors X_1 and X_2,

$$Em_i = 0, En_i = 0,$$

$$Dm_k = \frac{1}{N} \sum_{i=1}^{m} (\varphi_k(x_i))^2 \sigma_i^2 + \frac{1}{N} \sum_{i \neq j} \varphi_k(x_i)\varphi_k(x_j) r_{ij} \leq \sigma_0^2 + \sum_{i \neq j} |r_{ij}|/N$$

For the elements of X_2, it follows that,

$$Dn_p = E\left(\frac{1}{N^{3/2}} \sum_{j=1}^{N} \sum_{i=1}^{N} \varphi_p(x_i)\bar{d}_{ij}^{k,l}(x) \cdot \varepsilon_j\right)^2$$

$$\leq \frac{1}{N} \cdot E\left(\sum_{j=1}^{N} \varepsilon_j\right)^2 \leq \frac{1}{N} \sum_{j=1}^{N} \sigma_i^2 + \frac{1}{N} \sum_{i \neq j} |r_{ij}|$$

$$\left(\sigma_O^2 + \sum_{i \neq j} |r_{ij}|/N\right)$$

Because $\varphi_i(x_j)$ and $\overline{d}_{ij}^{k,l}(x)$ are bounded by unit. By using the ratio (3.26), it is obtained that: $\forall \gamma \geq 0$

$$P\{|A_2| > \gamma\} \to 0,$$

If

$$\frac{1}{\lambda_1(N)}\left(\sigma_0^2+\frac{\sum_{i\neq j}|r_{ij}|}{N}\right)\to 0$$

According to the conditions
Consider A_3 in (3.30)

$$\frac{\varepsilon^T I_{kl}^{(1)}(x)X(X^TX)^{(-1)}X^T\cdot\varepsilon}{N}$$

Similarly, as it was for A_2, A_3 can be represented in the following form:

$$A_3=\frac{1}{N^2}\varepsilon^T I_{k,l}^{(1)}(x)X\left(\frac{X^TX}{N}\right)^{-1}X^T\cdot\varepsilon$$

Let the A_3 represented in the following form:

$$A_3=\left(\frac{X^T\cdot I_{k,l}^{(1)}(x)\varepsilon}{N^{1/2}}\right)^T\left(\frac{X^TX}{N}\right)^{-1}\left(\frac{X^T\cdot\varepsilon}{N^{3/2}}\right)$$

Denote

$$X_2=\frac{X^T\cdot\varepsilon}{N^{3/2}}=\left(\frac{1}{N^{3/2}}\sum_{i=1}^N\varphi_1(x_i)\varepsilon_i,\ldots,\frac{1}{N^{1/2}}\sum_{i=1}^N\varphi_m(x_i)\varepsilon_i\right)$$

$$X_3=\frac{X^TI_{k,1}(x)\cdot\varepsilon}{N^{\frac{1}{2}}}$$

$$=\left(\frac{1}{N^{1/2}}\sum_{j=1}^N\sum_{i=1}^N\varphi_1(x_i)\bar{d}_{ij}^{k,l}(x)\cdot\varepsilon_j,\ldots,\frac{1}{N^{1/2}}\sum_{j=1}^N\sum_{i=1}^N\varphi_m(x_i)\bar{d}_{ij}^{k,l}(x_j)\cdot\varepsilon_j)\right)$$

Then A_3 can be represented as:

$$A_3=\sum_{j=1}^m\sum_{i=1}^m p_ix_{ij}s_j$$

where $p_i, s_j, i, j = 1, 2, \ldots, m$—the components of the vectors X_2 and X_3. Then,

$$Ep_i=0,\ Es_i=0$$

$$Dp_k=\frac{1}{N}\sum_{i=1}^m(\varphi_k(x_i))^2\sigma_i^2+\frac{1}{N}\sum_{i\neq j}\varphi_k(x_i)\varphi_k(x_j)r_{ij}\le\sigma_0^2+\sum_{i\neq j}|r_{ij}|/N$$

For the elements of X_3,

$$Dp_p = E\left(\frac{1}{N^{3/2}}\sum_{j=1}^{N}\sum_{i=1}^{N}\varphi_p(x_i)\bar{d}_{ij}^{kl}(x)\cdot\varepsilon_j\right)^2$$

$$\leq \frac{1}{N}\cdot E\left(\sum_{j=1}^{N}\varepsilon_j\right)^2 \leq \frac{1}{N}\sum_{j=1}^{N}\sigma_i^2 + \frac{1}{N}\sum_{i\neq j}|r_{ij}|$$

$$\leq \left(\sigma_0^2 + \sum_{i\neq j}|r_{ij}|/N\right)$$

Because $\varphi_i(x_j)$ and $\overline{d}_{ij}^{k,l}(x)$ are bounded by unit. Using the ratio (3.26), $\forall\gamma \geq 0$

$$P\{|A_3| > \gamma\} \to 0,$$

if

$$\frac{1}{\lambda_1(N)}\left(\sigma_0^2 + \frac{\sum_{i\neq j}|r_{ij}|}{N}\right) \to 0$$

According to the following conditions,

$$\frac{1}{N\lambda_1(N)} \to 0 \quad \text{for} \quad N \to \infty$$

$$\frac{1}{N^2}\sum_{1\leq t_1\cdot t_2\cdot t_3\cdot t_4\leq N^2} E(\varepsilon_{i_1}\varepsilon_{i_2}\varepsilon_{i_3}\varepsilon_{i_4} - r_{i_1i_2}r_{i_3i_4}) \to 0 \quad \text{for} \quad N \to \infty$$

It follows that,

$$\frac{1}{N}(\widehat{Y} - Y)^T I_{k,l}^{(1)}(x)(\widehat{Y} - Y) \xrightarrow{P} \frac{\varepsilon^T I_{kl}(x)\cdot\varepsilon}{N}$$

On the other hand, as

$$\sum_{i=1}^{N}\sum_{j=1}^{N}\varphi_k(x_i)\varphi_l(x_j)\cdot r_{ij} = \sum_{i=1}^{N}\sum_{j=1}^{N}\varphi_k(x_j)\varphi_l(x_i)\cdot r_{ij}$$

then

$$E\left[\frac{\varepsilon^T I_{k,l}^{(1)}(x)\cdot\varepsilon}{N}\right] = E_{k,l}(x)$$

where $E_{kl}(x)$ are the (kl) elements of the matrix $\frac{X^T I_{k,L}^{(1)}(x)\cdot X}{N}$, i.e., the mean value for $N \to \infty$ tends to the true value, which means asymptotically unbiased for the estimator a_{kl}.

The consistency is now to be proved.

Denote:

$$\beta_N^{k,l} = \frac{\varepsilon^T I_{k,l}^{(1)}(x) \cdot \varepsilon}{N}$$

As $\bar{d}_{ij}^{k,l}(x)$ is bounded then the following derivation can be made,

$$\begin{aligned}
D\beta_N^{k,l} &= E(\beta_N^{k,l})^2 - (E\beta_N^{k,l})^2 \\
&= \frac{1}{N^2} E\left[\sum_{j=1}^{N}\sum_{i=1}^{N} \bar{d}_{ij}^{k,l}(x)\varepsilon_i\varepsilon_j\right]^2 - \frac{1}{N^2}\left[E\sum_{j=1}^{N}\sum_{i=1}^{N} \bar{d}_{ij}^{k,l}(x)\varepsilon_i\varepsilon_j\right]^2 \\
&= \frac{1}{N^2} E\left[\sum_{i=1}^{N} d_{ii}^{k,l}(x)\varepsilon_i^2 + \sum_{i\neq j} d_{ij}^{k,l}(x)\varepsilon_i\varepsilon_j\right]^2 - \frac{1}{N^2}\left[\sum_{i=1}^{N} \bar{d}_{ij}^{k,l}(x)\sigma_i^2 + \sum_{i\neq j} \bar{d}_{ij}^{k,l}(x) r_{ij}\right]^2 \\
&= \frac{1}{N^2}\sum_{i=1}^{N}(\bar{d}_{ii}^{k,l})^2 E[\varepsilon_i^4] + \frac{2}{N^2}\sum_{i=1}^{N}\bar{d}_{ij}^{k,l}(x)\sum_{i\neq j}\bar{d}_{i_1 j_1}^{kl} E[\varepsilon_i^2\varepsilon_{i_i}\varepsilon_{j_1}] \\
&\quad + \frac{1}{N^2}\sum_{i\neq j}\sum_{i_1\neq j_1}\overline{d}_{ij}^{k,l}\overline{d}_{i_1 j_1}^{k,l} E[\varepsilon_i\varepsilon_j\varepsilon_{i_1}\varepsilon_{j_1}] - \frac{1}{N^2}\sum_{i=1}^{N}(\overline{d}_{ii}^{k,l})^2\sigma_i^4 \\
&\quad - \frac{2}{N^2}\sum_{i\neq j}\sum_{i_1\neq j_1}\overline{d}_{ij}^{kl}\overline{d}_{i_1 j_1}^{kl} E[\varepsilon_i\varepsilon_j\varepsilon_{i_1}\varepsilon_{j_1}] - \frac{1}{N^2}\sum_{i=1}^{N}(\overline{d}_{ii}^{k,l})^2\sigma_i^4 \\
&\quad - \frac{2}{N^2}\sum_{i=1}^{N}\overline{d}_{ii}^{kl}(x)\sum_{i_1\neq j_1}\overline{d}_{i_1 j_1}^{kl}\sigma_i^2 r_{i_1 j_1} - \frac{1}{N^2}\sum_{i\neq j}\sum_{i_1\neq j_1}\overline{d}_{ij}^{kl}\overline{d}_{i_1 j_1}^{kl} r_{ij} r_{i_1 j_1} \\
&\leq \frac{1}{N}(E[\varepsilon_i^4] - \sigma_i^4) + \frac{2}{N^2}\sum_{i=1}^{N}\sum_{t_1\neq j_1}(E[\varepsilon_i^2\varepsilon_{i_1}\varepsilon_{j_1}] - \sigma_i^2 r_{ij_1}) \\
&\quad + \frac{1}{N^2}\sum_{i\neq j}(E[\varepsilon_i\varepsilon_j\varepsilon_{i_1}\varepsilon_{j_1}] - r_{ij} r_{i_1 j_1}) \to 0 \quad \text{as} \quad N \to \infty.
\end{aligned}$$

for $N \to \infty$ according to the following condition,

$$\frac{1}{N^2}\sum_{1\leq t_1\cdot t_2\cdot t_3\cdot t_4\leq N^2} E(\varepsilon_{i_1}\varepsilon_{i_2}\varepsilon_{i_3}\varepsilon_{i_4} - r_{i_1 i_2} r_{i_3 i_4}) \to 0 \quad \text{for} \quad N \to \infty$$

Consistency of $\widehat{c}_{kl}$ is proved by the same way as in Theorem 3.2, basing on consistency and asymptotical unbiasedness of the estimator b_{kl} for the elements of the matrix B. □

Corollary 3.1 *If* $r_{ij} = 0$*, then the condition*

$$y_i \;=\; f(x_i) \,+\, \varepsilon_i$$

is satisfied, and from Theorem 3.5 it follows the statement of Theorem 3.1.

Proof Assume $r_{ij} = 0$ for all $i \neq j$. In this case, the random errors ε_i are uncorrelated, satisfying

$$E[\varepsilon_i\, \varepsilon_j] \;=\; 0 \quad \text{for} \quad i \neq j.$$

Hence, the second condition in Theorem 3.5,

$$\left(\frac{1}{N\,\lambda_1(N)}\right)^2 \sum_{i \neq j} |r_{ij}| \;\longrightarrow\; 0,$$

holds trivially, since $r_{ij} = 0$ implies $\sum_{i \neq j} |r_{ij}| = 0$. Moreover, the third condition,

$$\frac{1}{N^2} \sum_{1 \leq t_1,\, t_2,\, t_3,\, t_4 \leq N^2} E\Big(\varepsilon_{i_1}\, \varepsilon_{i_2}\, \varepsilon_{i_3}\, \varepsilon_{i_4} \;-\; r_{i_1 i_2}\, r_{i_3 i_4}\Big) \;\longrightarrow\; 0,$$

is also immediately satisfied under $r_{ij} = 0$, since in that case $r_{i_1 i_2}\, r_{i_3 i_4} = 0$ for all indices.

The first condition of Theorem 3.5,

$$\frac{1}{N\,\lambda_1(N)} \;\longrightarrow\; 0,$$

can be ensured by standard requirements on the design matrix X, such as

$$N \cdot \det\!\left(\frac{X^T X}{N}\right) \;\longrightarrow\; \infty \quad \text{as} \quad N \to \infty,$$

which is essentially the same condition stated in (3.7) of Theorem 3.1.

By combining these observations, it follows from Theorem 3.5 that the least squares estimator with singular value decomposition converges in probability to the true parameter vector when $r_{ij} = 0$. This result coincides with the conclusion of Theorem 3.1, which asserts that under the same uncorrelated-error framework and regularity conditions on X, the estimator is unbiased and consistent.

Therefore, when the errors are uncorrelated ($r_{ij} = 0$), the condition

$$y_i \;=\; f(x_i) \,+\, \varepsilon_i$$

is valid, and Theorem 3.5 directly implies the statement of Theorem 3.1. □

Corollary 3.2 *If* $r_{ij} = 0, i \neq j, m \leq c < \infty$ *then from Theorems 3.5 and 3.6 it follows consistency of l.s.e. and of the elements of covariance matrix of the normed deviation vector.*

Proof It is first noted that if $r_{ij} = 0$ for $i \neq j$, then the second condition in Theorems 3.5 and 3.6, namely

$$\left(\frac{1}{N\,\lambda_1(N)}\right)^2 \sum_{i \neq j} |r_{ij}| \longrightarrow 0,$$

is automatically satisfied because $\sum_{i \neq j} |r_{ij}| = 0$. Hence, there is no contribution from off-diagonal terms in the sum.

Moreover, if the largest singular value $\lambda_1(N)$ is bounded away from zero and satisfies

$$\frac{1}{N\,\lambda_1(N)} \longrightarrow 0,$$

then condition (i) of Theorems 3.5 and 3.6 also holds. By hypothesis $m \leq c < \infty$, one deduces that $\lambda_1(N)$ can be taken sufficiently large (of at least the order N), ensuring the convergence to zero of $\frac{1}{N\,\lambda_1(N)}$.

Lastly, condition (iii) in Theorems 3.5 and 3.6 concerns the fourth-order mixed moments

$$\frac{1}{N^2} \sum_{1 \leq t_1,\, t_2,\, t_3,\, t_4 \leq N^2} E\big(\varepsilon_{i_1}\, \varepsilon_{i_2}\, \varepsilon_{i_3}\, \varepsilon_{i_4} \;-\; r_{i_1 i_2}\, r_{i_3 i_4}\big),$$

and follows from standard central limit arguments and the assumed uncorrelatedness (off-diagonal zeros) once the variances remain finite and bounded away from zero.

Since all three conditions of Theorems 3.5 and 3.6 are satisfied under the assumptions $r_{ij} = 0$ for $i \neq j$ and $m \leq c < \infty$, it follows directly that the least squares estimator with singular value decomposition is consistent, and each estimator $\widehat{c}_{kl}$ of the element c_{kl} in the covariance matrix C_N is consistent and asymptotically unbiased. Hence the elements of the covariance matrix of the normed deviation vector also converge in probability to their true values.

This completes the proof. □

Remark 3.2 This result is also supported by the findings presented in [15, 18, 36], where the consistency of the least squares estimator and the covariance matrix elements is established in the related settings.

Remark 3.3 Different treatments are required for correlated and noncorrelated observational errors because distinct expressions for the covariance matrix estimators arise. These expressions differ by the inclusion of the matrices $I_{kl}(x)$ and $I_{kl}^{(1)}(x)$. The matrix $I_{kl}(x)$ is diagonal, whereas $I_{kl}^{(1)}(x)$ is non-diagonal. It is important to note that the elements of these matrices do not inherently encode whether the errors are correlated or not. Consequently, when a program is prepared to compute statistical estimators, the elements of $I_{kl}^{(1)}(x)$ must be multiplied by the coefficient δ_{ij}^* defined as

$$\delta_{ij}^* = \begin{cases} 1, & \text{if } r_{ij} > 0, \\ 0, & \text{if } r_{ij} = 0. \end{cases}$$

Here, r_{ij} characterizes the relationship between errors of different observations. The factor δ_{ij}^* effectively selects the relevant terms only when correlation is present (i.e., $r_{ij} > 0$). This procedure ensures that the estimator for the covariance matrix properly accounts for correlation whenever it occurs.

In the noncorrelated case, less restrictive and more convenient conditions emerge for the asymptotic unbiasedness and consistency of the covariance matrix elements. Under these conditions, the estimation process simplifies because the covariances of unrelated error terms vanish, thereby eliminating the need to adjust for cross-correlation. It can also be shown—following a derivation similar to that used in linear regression models—that the vector $\sqrt{N}(\widehat{\theta} - \theta)$ converges in distribution to a Gaussian (normal) random vector. This asymptotic normality indicates that classical inference methods remain valid, thereby reinforcing the practical and the theoretical significance of distinguishing between correlated and noncorrelated observational errors.

3.2.2 *Confidence Interval in Correlated Linear Prototype*

Now, a confidence interval for the function $f(\theta, x)$ at each point x_i is constructed. As established in previous sections, it has been shown that under certain conditions, the least squares estimator $(l.s.e.)$ with the specified structural variance $(s.v.)$ is consistent for the given model. Then, from Lemma 3.1, it follows that the components of the deviation vector $(\widehat{\theta} - \theta)$ asymptotically follow a Gaussian distribution as $N \to \infty$. Consequently, the following quadratic form is considered:

$$\frac{1}{N}(\widehat{\theta} - \theta)^T C_N^{-1}(\widehat{\theta} - \theta).$$

According to Lemma 3.1, this quadratic form, as $N \to \infty$, asymptotically follows a chi-square distribution [8] with m degrees of freedom. That is, with probability $\gamma > 0$, the following inequality holds:

$$(\widehat{\theta} - \theta)^T C_N^{-1}(\widehat{\theta} - \theta) \le \frac{\chi_\gamma^2(m)}{N}, \tag{3.31}$$

where $\chi_\gamma^2(m)$ denotes the quantile of level $\gamma > 0$ of the chi-square distribution with m degrees of freedom.

In the ratio (3.31), the matrix C_N^{-1} is unknown. As established in Theorem 3.6, asymptotically unbiased and consistent estimators of the elements of the covariance matrix C_N have been constructed. Therefore, instead of C_N^{-1}, the estimator matrix $\widehat{C}_N^{-1}$ is employed. By solving (3.31) for θ, the following confidence ellipsoid is obtained:

$$\varepsilon_\gamma(\theta) = \left\{\theta : (\widehat{\theta} - \theta)^T \widehat{C}_N^{-1}(\widehat{\theta} - \theta) \le \frac{\chi_\gamma^2(m)}{N}\right\}.$$

Remark 3.4 It should be emphasized that, in principle, it is not necessary for the elements of the matrix $\widehat{C}_N$ to have a limit as $N \to \infty$. The key point is that the difference between c_{kl} and $\widehat{c}_{kl}$ (estimators of the elements of the covariance matrix of the deviation vector) converges to zero in probability, implying that the difference between the corresponding components of two vectors vanishes asymptotically. The problem of how closely an estimator matrix approximates the covariance matrix has been studied in [61]. In this context, it is essential to estimate the rate of convergence of the elements of these matrices.

Now for finding lower and upper border of confident band it is necessary to find:

$$inf_{\varepsilon_\gamma(\theta)} f(x,\theta) \; and \; sup_{\varepsilon_\gamma(\theta)} f(x,\theta)$$

which are lower and upper border.

For convenience, the confident ellipsoid $\varepsilon_\gamma(\theta)$ can be represented in the following form:

$$\sum_{i=1}^{m} \frac{(\widehat{\theta_i} - \theta_i)^2}{\chi_\gamma^2(m) \cdot \lambda_{m-i}} = \frac{1}{\sqrt{N}},$$

where $\chi_\gamma^2(m)$ is the quantile of the level $\gamma > 0$ of the distribution chi-square with m degree of freedom, $\lambda_1, \lambda_2, \ldots, \lambda_m$ - are the eigen values of the matrix $\widehat{C}_N^{-1}$. Now we have to find the minimum and maximum of the function $f(x,\theta)$, where θ belongs to the confident ellipsoid. For this, we use Lagrange's method of undetermined coefficients [35]. Corresponding Lagrange's function has the following form:

$$Z = \sum_{i=1}^{m} \varphi_i(x) \cdot \theta_i + \lambda^* \cdot \sum_{i=1}^{m} \left[\frac{(\widehat{\theta_i} - \theta_i)^2}{\chi_\gamma^2(m) \cdot \lambda_{m-i}} - \frac{1}{\sqrt{N}} \right]$$

where λ^* - Lagrange's multiplier and θ_i satisfying to the following equation

$$\frac{d}{de_i} Z = 0, \quad i = 1, 2, \ldots, m;$$

are required points.

This chapter will give numerical examples of finding the lower and upper borders of a confident intervals for values of unknown function $f(x)$ in regression models.

3.2.3 Numerical Illustration for Correlated Prototype

Example 3.3 This example employs a linear functional model with autocorrelated errors, incorporating both polynomial regression terms and an AR(1) error component. This example is designed to investigate the behavior of the parameter estimates and the numerical properties of the design matrix as the sample size varies.

Let the response variable y_i be given by the regression model

$$y_i = \theta_0 + \theta_1 x_i + \theta_2 x_i^2 + \theta_3 x_i^3 + \theta_4 x_i^4 + \theta_5 x_i^5 + \varepsilon_i, \quad i = 1, 2, \ldots, N,$$

where the independent variable is defined over an equally spaced grid: $x_i = \frac{i-1}{N-1}$, $i = 1, 2, \ldots, N$.

The true parameter vector is specified as

$$\boldsymbol{\theta} = \left(1.5,\ 0.8,\ 2.1,\ -1.2,\ 0.7,\ 3.0\right)^T.$$

In this model the error terms $\{\varepsilon_i\}_{i=1}^N$ are not assumed to be independent and identically distributed. Instead, they are modeled via a first-order autoregressive process (AR(1)) to capture temporal dependence. Specifically, the error sequence is defined by

$$\varepsilon_1 \sim N(0, \sigma^2), \qquad \varepsilon_i = \rho\, \varepsilon_{i-1} + \nu_i, \quad i = 2, 3, \ldots, N,$$

where $\{\nu_i\}$ is a sequence of independent and identically distributed Gaussian white noise with

$$\nu_i \sim N(0, \sigma^2).$$

Mathematically, when $|\rho| < 1$, the AR(1) process is stationary with zero mean and a constant variance given by

$$\mathrm{Var}(\varepsilon_i) = \frac{\sigma^2}{1 - \rho^2}.$$

Furthermore, the autocovariance function of the error process decays exponentially. That is, for any lag h,

$$\mathrm{Cov}(\varepsilon_i, \varepsilon_{i+h}) = \frac{\sigma^2}{1 - \rho^2}\, \rho^{|h|}.$$

Equivalently, the covariance matrix can be written in its full matrix form as

$$\Sigma = \frac{\sigma^2}{1 - \rho^2} \begin{pmatrix} 1 & \rho & \rho^2 & \cdots & \rho^{N-1} \\ \rho & 1 & \rho & \cdots & \rho^{N-2} \\ \rho^2 & \rho & 1 & \cdots & \rho^{N-3} \\ \vdots & \vdots & \vdots & \ddots & \vdots \\ \rho^{N-1} & \rho^{N-2} & \rho^{N-3} & \cdots & 1 \end{pmatrix}.$$

This property implies that the correlation between errors diminishes geometrically as the lag increases, thereby capturing the short-term dependency commonly observed in empirical data.Introducing the AR(1) error structure not only renders the model more realistic but also has significant implications for parameter estimation and inference. In the presence of autocorrelation, the ordinary least squares (OLS) estimator, if computed without accounting for the correct covariance structure of the errors, may not be efficient; its standard errors can be underestimated, which in turn leads to misleading statistical inferences. Consequently, in rigorous applications where autocorrelation is present, one should consider employing methods such as generalized least squares (GLS) to obtain efficient and reliable estimates.In this example, the autocorrelation coefficient is set to $\rho = 0.5$ and the error standard deviation is $\sigma = 0.5$.

In the present example, although the error terms exhibit AR(1) autocorrelation, the OLS method is adopted as an initial approach to explore the behavior of the parameter estimates and to study the numerical characteristics of the design matrix as the sample size increases.

In this case study, the validity of Theorem 3.5 is demonstrated from a numerical experimental perspective. Through both computation and theoretical analysis, we show that the conditions stated in the theorem are satisfied.

First, numerical experiments indicate that as the sample size N increases, the smallest eigenvalue of the design matrix $\mathbf{X}$, denoted by $\lambda_1(N)$, satisfies

$$\frac{1}{N\,\lambda_1(N)} \to 0.$$

This condition ensures that the design matrix is well conditioned for large samples, thereby guaranteeing the numerical stability of the least squares estimator.

Secondly, by analyzing the cumulative effect of the error cross-correlations r_{ij}, it is verified that

$$\left(\frac{1}{N\,\lambda_1(N)}\right)^2 \sum_{i \neq j} |r_{ij}| \to 0.$$

This implies that the cross-correlation among the error terms diminishes as the sample size increases, exerting a negligible influence on the estimation outcome.

Finally, both theoretical analysis and numerical validation of the fourth-order moments and their mixed terms confirm that

$$\frac{1}{N^2} \sum_{1 \leq t_1, t_2, t_3, t_4 \leq N^2} E\left(\varepsilon_{i_1}\,\varepsilon_{i_2}\,\varepsilon_{i_3}\,\varepsilon_{i_4} \;-\; r_{i_1 i_2}\, r_{i_3 i_4}\right) \to 0.$$

Hence, all the conditions required by the theorem are satisfied in this case study.

The results in Table 3.3 indicate that when the sample size is small (e.g., 50), the average biases of the parameters (θ_0 to θ_5) deviate significantly from zero due to finite sample effects and AR(1) correlated noise, suggesting that least squares estimates may exhibit considerable bias under small-sample conditions; however, as the sample size increases to millions or tens of millions, the biases rapidly converge near zero, which not only validates the asymptotic unbiasedness and consistency of

Table 3.3 Parameter bias convergence under large samples with correlated errors

Sample size	θ_0	θ_1	θ_2	θ_3	θ_4	θ_5
50	−0.02628	1.51637	−6.33331	11.15614	−9.63279	3.30795
526365	−0.00110	0.02967	−0.17907	0.42443	−0.44466	0.17217
1052681	0.00077	−0.03486	0.22126	−0.55195	0.59486	−0.23079
1578997	−0.00136	0.01912	−0.06975	0.08281	−0.01139	−0.01975
2105313	0.00038	−0.01063	0.07612	−0.22395	0.27856	−0.12134
2631628	−0.00045	0.01705	−0.12364	0.32356	−0.35484	0.13898
3157944	0.00026	−0.00759	0.05656	−0.14377	0.14761	−0.05246
3684260	0.00016	0.00018	−0.01244	0.04820	−0.06493	0.02884
4210576	0.00018	−0.00525	0.03676	−0.08883	0.08830	−0.03145
4736892	0.00012	0.00150	−0.02417	0.08538	−0.10655	0.04339
5263207	−0.00009	0.00634	−0.06747	0.21734	−0.27167	0.11606
5789523	0.00053	−0.01118	0.06243	−0.14126	0.13897	−0.04936
6315839	−0.00009	−0.00105	0.00566	−0.00554	−0.00444	0.00549
6842155	0.00012	−0.00627	0.04474	−0.11373	0.11929	−0.04400
7368471	−0.00017	0.00171	0.00120	−0.01995	0.03274	−0.01573
7894786	−0.00037	0.00955	−0.07346	0.20537	−0.23398	0.09308
8421102	−0.00049	0.00567	−0.02429	0.05274	−0.05486	0.02134
8947418	0.00008	0.00473	−0.04727	0.13402	−0.14857	0.05706
9473734	−0.00031	0.00263	−0.00915	0.01438	−0.00909	0.00160
10000050	0.00043	−0.00471	0.02756	−0.07307	0.08282	−0.03303

the least squares estimator in large samples, but also demonstrates that the method exhibits good robustness and statistical stability when handling data with correlated noise, leading to more accurate and reliable parameter estimates.

The left figure (Mean Estimates vs. Sample Size) in Fig. 3.14 shows that for smaller sample sizes, the average estimates of each parameter (colored lines) deviate considerably from their true values (red dashed lines), with some parameters being over- or underestimated by several units at low-sample sizes. However, as the sample size increases to around 10^5 or even higher, all parameters' mean estimates gradually converge and fluctuate around the true values. The right figure (Parameter Bias vs. Sample Size) further confirms this trend: while the parameter biases (mean estimate minus the true value) are quite large in the low-sample regime, they approach zero in the high-sample regime, indicating that the systematic errors in the parameter estimates diminish with more data.

In light of the earlier context (such as the performance of least squares estimation under AR(1) noise or within a polynomial regression setting), these findings suggest that even with some degree of noise correlation or a relatively complex model, least squares estimation can yield nearly unbiased estimates given sufficiently large-sample sizes. By contrast, when the sample size is small, biases can be substantial

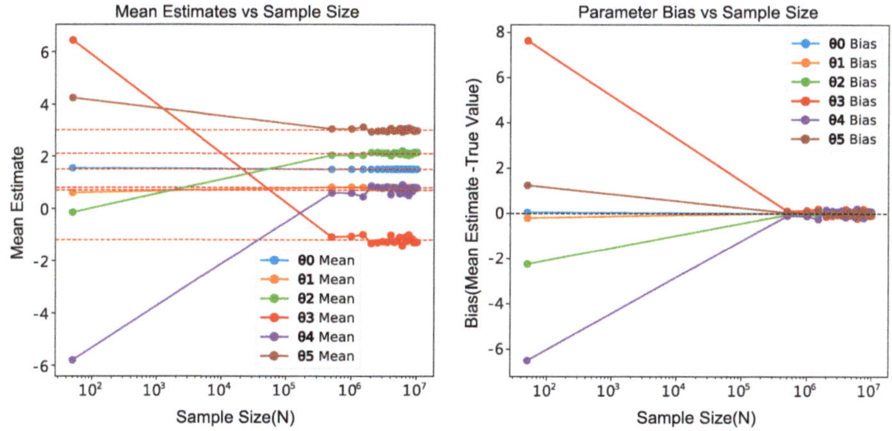

Fig. 3.14 Estimator consistency and bias convergence with correlated errors

and exhibit noticeable volatility, highlighting the potential need for more data or improved methods (e.g., regularization, generalized least squares) to enhance the accuracy and stability of the estimates.

From Fig. 3.15a plot in the upper-left, it can be seen that at relatively small-sample sizes (around 10^2), the parameter estimation biases (estimated value minus true value) often exhibit large positive or negative fluctuations, with some biases exceeding 10. This indicates that under small-sample conditions, a model with autocorrelated noise may lead to significant systematic errors in the least squares estimates. However, as the sample size increases to about 10^5 or even higher, the biases for all parameters gradually converge toward zero, albeit with some degree of fluctuation in the process, demonstrating that even in the presence of AR(1) errors, least squares estimation remains consistent for large samples.

From Fig. 3.15b, the six "Convergence Path" subplots below provide a more intuitive view of how each parameter's estimates evolve. The thick black line represents the median of the estimates from multiple replications, the red solid line indicates the true parameter value, and the colored dashed lines correspond to quantiles ranging from the 5th to the 95th percentile. For smaller samples, the median often deviates noticeably from the true value, and the quantile bands are relatively wide, reflecting greater variance and uncertainty in the estimates. As the sample size grows, the median gradually approaches the red line and the quantile bands narrow substantially, indicating reductions in both bias and variance. It is noteworthy that certain parameters (e.g., θ_2 or θ_4) may require even larger sample sizes to stabilize around their true values, underscoring the impact of autocorrelated noise on estimation efficiency. Overall, these results suggest that while least squares estimation can converge to the true parameters under autocorrelated errors, its convergence path may be more volatile and slower compared to the i.i.d. noise setting.

From these plots of "residuals vs. fitted values" in Fig. 3.16, we can see that as the sample size increases from $N = 10$ to $N = 10^7$, the distribution of residuals on the

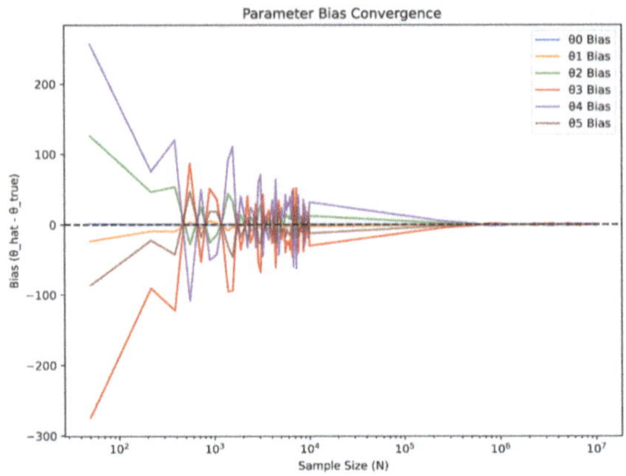

(a) Parameter Bias Convergence

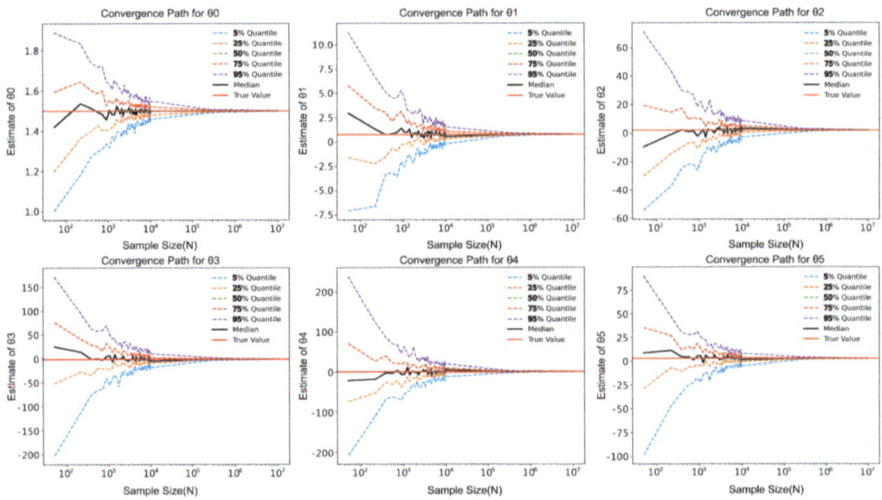

(b) Convergence Path for θ_i, $i = 0, \cdots, 5$

Fig. 3.15 Visualization of parameter bias convergence with correlated errors

vertical axis tends to be stable, and there is no obvious "funnel shape" or tendency to increase with the increase of fitted values. The distribution of residuals on the vertical axis tends to be stable overall, with no obvious "funnel shape" or tendency to increase as the fitted value increases. For smaller sample sizes (e.g., $N = 10$ or $N = 100$), the distribution of the residual cloud appears to be more dispersed, and the number of sample points is limited, so it is difficult to judge the heteroskedasticity directly from the graph; however, when the N is increased to several thousands or even millions, the residuals are more and more obviously distributed around 0, and do not systematically expand or contract with the fitted values, which indicates that

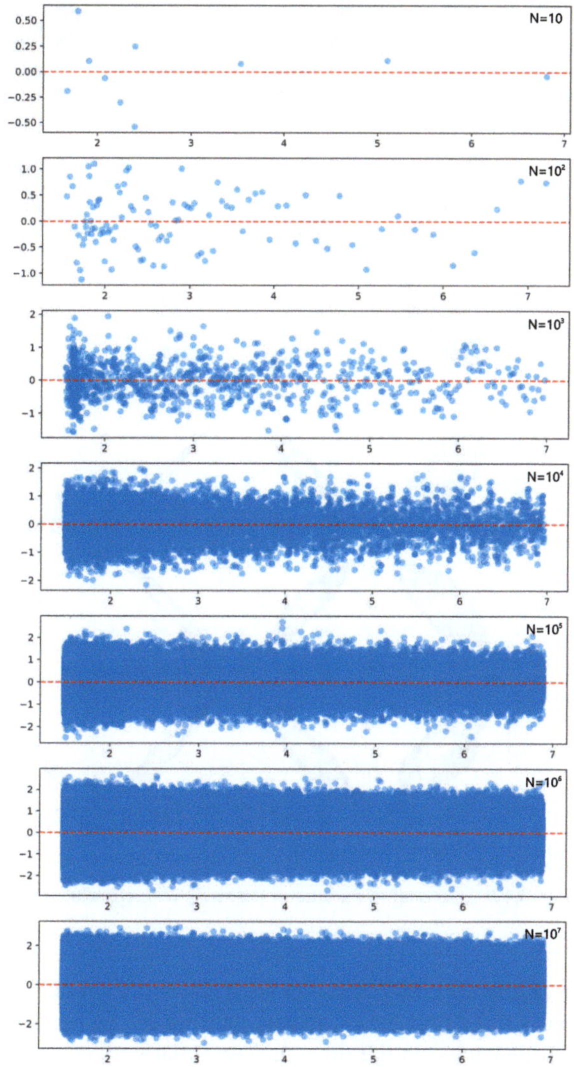

Fig. 3.16 Graphical test for heteroscedasticity with correlated errors

the variance of the model basically stays constant in the large samples. This indicates that the variance of the model is basically constant in large samples. Combined with the background of previous modeling assumptions (e.g., least squares estimation and error distribution being homoscedastic normal, etc.), these residual plots do not show significant evidence of heteroskedasticity, suggesting that the model is still able to fit the data well under the assumption of constant error variance. Also, visually, the range of dispersion of the residuals stabilizes with increasing sample size, further supporting the validity of the assumption of homoscedasticity in the large-sample case.

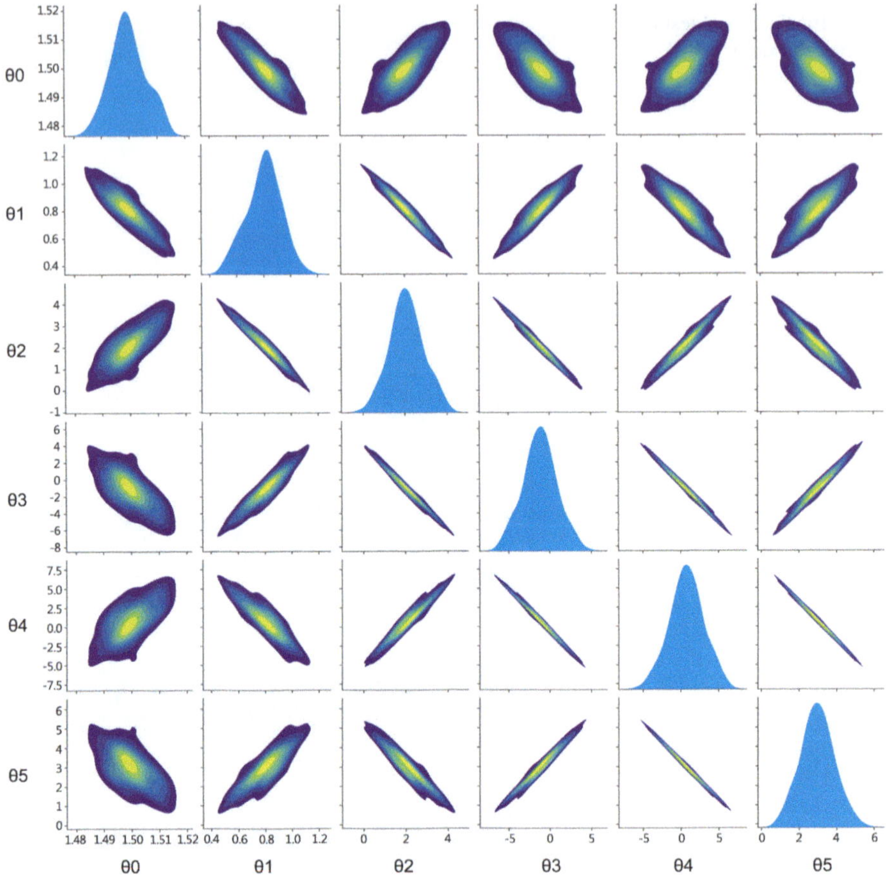

Fig. 3.17 Heatmap of parameter densities with correlated errors

From Fig. 3.17, we can see that in the model with exponential and sinusoidal basis functions, the estimated distributions of each parameter (one-dimensional kernel densities on the diagonal) are relatively concentrated around their true values, for example, θ_0 is around 1.5, θ_1 is around –2.3, etc., which indicates that the model has a good discriminative degree of each parameter under the current sample conditions. This indicates that the model has good recognizability for each parameter under the current sample conditions. The non-diagonal 2D kernel density contours have an elliptical distribution, indicating that there is a certain degree of correlation between the different parameters; this is more common in models with multiple sets of exponential and sinusoidal basis functions, and the parameter estimates are prone to covariance, especially when there is a linear dependence or overlap between the basis functions in some intervals.

Overall, the distributions of each parameter in Fig. 3.17 are concentrated and not heavily skewed, indicating that the least squares estimation has good stability and

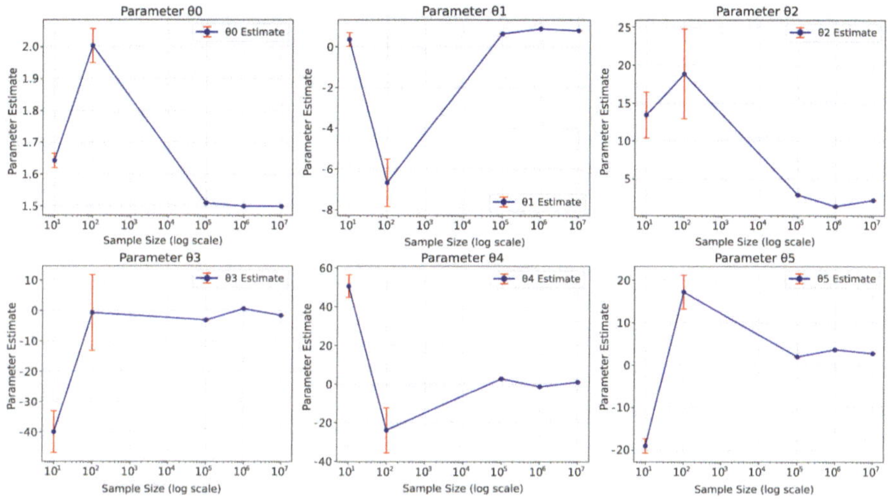

Fig. 3.18 Confidence interval with correlated errors

accuracy in this setting; at the same time, the moderate correlation reminds us that if we want to further improve the estimation accuracy in more complex or highly noisy scenarios, we may need to consider improvement methods such as a larger sample size or regularization. Combined with the background of the problem, it can be seen that this combination of exponential and sinusoidal basis functions is widely used in signal fitting or attenuated oscillation modeling, and the current results suggest that when the sample size is sufficient and the noise level is moderate, the least squares estimation is still able to restore the real parameters and reflect the model structure well.

The Monte Carlo simulation results in Fig. 3.18 reveal that under an AR(1) error structure, the least squares estimates of the six polynomial parameters (θ_0 to θ_5) demonstrate typical asymptotic behavior as the sample size increases. When the sample size is small ($N = 10$ or $N = 100$), the parameter estimates exhibit substantial fluctuations, and the corresponding 95% confidence intervals are wide, indicating that small samples provide insufficient information, leading to high estimation uncertainty.

As the sample size grows to $N = 10^5$ or higher, the estimates gradually stabilize and approach the true values. The confidence intervals also shrink significantly, highlighting the effectiveness of least squares estimation under correlated noise when the sample is sufficiently large.

Additionally, different parameters show distinct convergence behaviors. Parameters associated with higher-order terms (θ_3, θ_4, θ_5) tend to have larger variability and slower convergence in small samples compared to lower-order parameters (θ_0, θ_1, θ_2). This phenomenon is due to the reduced sensitivity of higher-order terms within the interval [0, 1], making them harder to estimate precisely when data is limited.

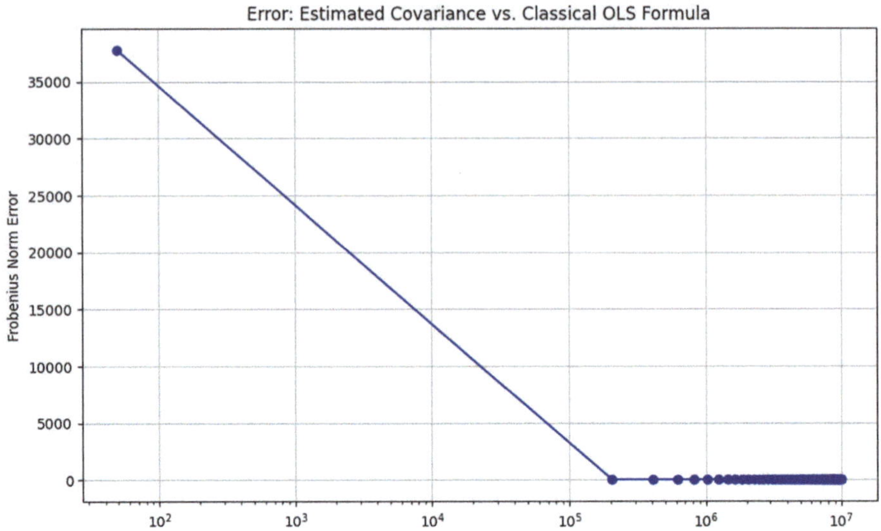

Fig. 3.19 Classical OLS covariance estimator error with correlated errors

Nonetheless, as the sample size exceeds 10^6, all parameters exhibit stable estimates with tight confidence intervals.

And Fig. 3.19 shows the visualization result of the covariance estimator.

Figure 3.19 illustrates the convergence behavior of the covariance estimator compared with the classical OLS covariance formula. It is evident that as the sample size N increases, the Frobenius norm error decreases sharply and eventually stabilizes near zero when N exceeds approximately 10^5. This clearly demonstrates that the proposed residual-weighted and matrix-corrected covariance estimator exhibits excellent convergence and asymptotic properties. Specifically, the estimator not only achieves unbiasedness but also shows strong consistency as it closely approximates the theoretical covariance structure in large samples. The smooth and monotonic decay of the error further confirms the numerical stability and robustness of the estimator in practice (Fig. 3.16).

3.3 Nonlinear Prototype Models with Correlated Errors

The shift from the classical linear-error framework to a setting where the systematic component is nonlinear and the observational noise is correlated calls for a thorough re-examination of estimation theory. In what follows, we first introduce a general nonlinear regression scheme that explicitly accommodates dependence among the errors, thereby laying the groundwork for a detailed investigation of the large-sample properties of the least squares estimator, the convergence of Gauss–Newton-type iterations, and the associated asymptotic distribution theory. To support the subsequent technical developments, we begin with a precise formulation of the model and a review of the key matrix quantities.

3.3.1 General Nonlinear Prototype with Correlated Errors

In previous sections, linear regression models have been investigated. In practical applications, most processes exhibit nonlinear structures. Therefore, the study of nonlinear regression models is of both theoretical and practical significance. The structure of nonlinear regression models is more complex since the underlying function is nonlinear, which precludes the direct application of methods developed for linear models. The analysis of nonlinear regression models necessitates the development of new approaches and techniques, utilizing results and theories from various branches of mathematics.

Significant contributions to the study of nonlinear regression models have been made by various authors, leading to the publication of monographs [1, 2, 6, 20, 34, 38, 42, 51, 55, 56] and scientific papers [4, 5, 7, 13, 27, 33, 38, 44–46, 50].

In the following, nonlinear regression models with correlated observation errors are considered. The case of uncorrelated errors can be treated as a special case where $E[\varepsilon_i \varepsilon_j] = r_{ij} = 0$ for $i \neq j$.

The properties of the least square estimates ($l.s.e.$) [24].

Consider the following regression model:

$$y_i = \eta(x_i, \theta^*) + \varepsilon_i, \ i = 1, 2, \ldots, N$$

where $\eta(x_i, \theta^*)$ is a nonlinear function on θ, and θ^* is the true value of the parameter θ.

It can be assumed that $\eta(x, \theta)$; $\frac{\partial \eta(x,\theta)}{\partial \theta}$; $\frac{\partial^2 \eta(x,\theta)}{\partial \theta_i \partial \theta_j}$, for $i, j = 1, 2, \ldots, m$, are continuous and bounded functions on (x, θ).

$\theta \in \Theta$ is a compact set and

$$E[\varepsilon_i] = 0, E[\varepsilon_i \varepsilon_j] = r_{ij}, r_{ii} = \sigma_i^2, 0 < \sigma_\star^2 \leq \sigma_i^2 \leq \sigma_1^2$$

Define

$$f_{ij}(\theta) = \frac{\partial \eta(x_j, \theta)}{\partial \theta_i}, \quad i = 1, 2, \ldots, m, \quad j = 1, 2, \ldots, N. \tag{3.32}$$

Let $\widehat{\theta}$ denote the least squares estimator ($l.s.e.$), and $F_N(\theta)$ be the matrix whose elements are given by $f_{ij}(\theta)$.

The eigenvalues of the matrix

$$\left[\frac{F_N^T(\theta)F_N(\theta)}{N}\right]$$

are denoted by

$$0 < \lambda_1^{(N)}(\theta) \le \lambda_2^{(N)}(\theta) \le \ldots \le \lambda_m^{(N)}(\theta).$$

Let $B(r)$ denote the sphere of radius $r > 0$ centered at θ^*.

The Gauss–Newton approach is widely used for investigating nonlinear regression models. According to the Gauss–Newton method [11], an iterative procedure for obtaining the least squares estimator ($l.s.e.$) can be constructed as follows:

$$\theta(s+1) = \theta(s) + \left[F_N^T(\theta(s)) \cdot F_N(\theta(s))\right]^{-1} F_N^T(\theta(s))(y - \eta(x, \theta(s))), \quad (3.33)$$

where all terms in (3.33) are defined in (3.32) and (3.33).

The primary concern is the convergence of the iterative process in (3.33). Throughout the following discussion, it is assumed that $\theta \in B(r)$. Expression (3.33) can be rewritten in the form:

$$\theta_N(s+1) = U(\theta_N(s)) = \theta_N(s) + A_N(\theta_N(s)) \cdot \delta_N(\theta_N(s)),$$

where

$$A_N(\theta_N(s)) = \left[\frac{F_N^T(\theta(s)) \cdot F_N(\theta(s))}{N}\right]^{-1} \frac{F_N^T(\theta(s))}{N},$$

$$\delta_N(\theta(s)) = y - \eta(x, \theta(s)); \; \delta^* = y - \eta(x, \theta^*) = \varepsilon$$

It can be proved some preliminary auxiliary results. Introduce the matrices A and B:

$$A_{n_1} = (a^{(n_1)})_{ij}; \; B_{n_2} = (b^{(n_2)})_{ij} : i = 1, 2, \ldots, m, \; j = 1, 2, \ldots, N;$$
$$C = (c)_{ij}; \; i, j = 1, 2, \ldots, m.$$

and denote

$$D = A_1 A_2^T \cdots A_{k-1}^T A_k \cdot C^n B_1^T B_2 \cdots B_{l-1} B_l^T$$

where $k = 2k_1 + 1$; $l = 2l_1 + 1$; n, l_1, k_1-are integer numbers.

Lemma 3.3 *If there exists a constant C_1 such that*

$$|a_{ij}^{(n,1)}| \le C_1, \quad |b_{ij}^{(n,2)}| \le C_1, \quad \max_{1 \le i \le m} \sum_{j=1}^{m} |c_{ij}| \le C_1, \quad (3.34)$$

then the following inequality holds:

$$E(\varepsilon^T D\varepsilon) \le C_1^n\, m^{k_1+l_1+1}\, N^{k_1+l_1-1}\Big(C_1 N + \sum_{i\neq j}|r_{ij}|\Big)$$

Proof Let $\{d_{ij}\}_{1\le i,j\le N}$ denote the entries of the matrix D. Then

$$\begin{aligned} E(\varepsilon^T D\varepsilon) &= E\Big(\sum_{i=1}^{N}\sum_{j=1}^{N} d_{ij}\,\varepsilon_i\,\varepsilon_j\Big) \\ &= \sum_{i=1}^{N} d_{ii}\,E\big(\varepsilon_i^2\big) + \sum_{i\neq j} d_{ij}\,E\big(\varepsilon_i\,\varepsilon_j\big) \end{aligned} \tag{3.35}$$

Since $E(\varepsilon_i^2) = \sigma_0^2$ for each i and $E(\varepsilon_i\,\varepsilon_j) = r_{ij}$ for $i \neq j$, it follows that

$$E(\varepsilon^T D\varepsilon) = \sigma_0^2\sum_{i=1}^{N} d_{ii} + \sum_{i\neq j} d_{ij}\,r_{ij} \le \sigma_0^2\,\mathrm{tr}(D) + \sum_{i\neq j}|d_{ij}|\,|r_{ij}|$$

This establishes the decomposition

$$E(\varepsilon^T D\varepsilon) \le \sigma_0^2\,\mathrm{tr}(D) + \sum_{i\neq j}|d_{ij}|\,|r_{ij}|$$

Next, it is shown that both $\mathrm{tr}(D)$ and each d_{ij} admit suitable bounds that depend on C_1, m, and N. By assumption (3.34), all coefficients $a_{ij}^{(n,1)}$ and $b_{ij}^{(n,2)}$ satisfy

$$\big|a_{ij}^{(n,1)}\big| \le C_1, \quad \big|b_{ij}^{(n,2)}\big| \le C_1$$

and the row sums of $\{c_{ij}\}$ are bounded by

$$\max_{1\le i\le m}\sum_{j=1}^{m}|c_{ij}| \le C_1$$

These conditions imply that any product of the form $a_*^{(1)}\,a_*^{(2)}\ldots c_{**}\cdots b_*^{(1)}\,b_*^{(2)}\ldots$ has absolute value bounded above by a power of C_1. In particular, d_{ij} is formed by a finite sum of such products, where the indices run over at most N or m depending on whether they are row or column indices for an $a^{(\cdot)}$, $b^{(\cdot)}$, or c. The exact form of d_{ij} (often involving multiple summations over intermediate indices) can be written schematically as:

$$d_{ij} = \sum_{\substack{1\le\cdots\le N,\\ 1\le\cdots\le m}}\Big(a_{(\cdot)}^{(\cdot)}\cdots a_{(\cdot)}^{(\cdot)}\cdot\prod_{\text{some } p=1}^{n} c_{(\cdot)(\cdot)}\cdot b_{(\cdot)}^{(\cdot)}\cdots b_{(\cdot)}^{(\cdot)}\Big)$$

where each factor in every summand is bounded in absolute value by C_1. Since there are at most $k_1 + l_1$ factors of $a^{(\cdot)}$ and $b^{(\cdot)}$ in each summand (each involving at most N or m terms in its indices) and n factors of c (each of which has row sum bounded by C_1 and thus is also bounded in an overall sense), it follows that

$$|d_{ij}| \leq C_1^{\alpha} N^{\beta} m^{\gamma},$$

where $\alpha = n+$ (number of a, b-factors), β is some power depending on k_1, l_1, γ is some power depending on k_1, l_1.

A more precise accounting of exponents yields

$$|d_{ij}| \leq C_1^n N^{k_1+l_1} m^{k_1+l_1+2}$$

To handle the trace of D, observe that

$$\operatorname{tr}(D) = \sum_{i=1}^{N} d_{ii}$$

and each diagonal entry d_{ii} admits the same form of bound as every other d_{ij}. Since there are N diagonal entries,

$$\begin{aligned}\operatorname{tr}(D) &\leq N \cdot \max_{1 \leq i \leq N} |d_{ii}| \\ &\leq N \left(C_1^n N^{k_1+l_1} m^{k_1+l_1+2}\right) = C_1^n N^{k_1+l_1+1} m^{k_1+l_1+2}\end{aligned}$$

Substituting these bounds into (3.35) gives

$$\begin{aligned}E(\varepsilon^T D\varepsilon) &\leq \sigma_0^2 \operatorname{tr}(D) + \sum_{i \neq j} |d_{ij}|\,|r_{ij}| \\ &\leq \sigma_0^2 \left(C_1^n N^{k_1+l_1+1} m^{k_1+l_1+2}\right) + \left(C_1^n N^{k_1+l_1} m^{k_1+l_1+2}\right) \sum_{i \neq j} |r_{ij}|\end{aligned}$$

In many applications, σ_0^2 can be absorbed into constants or assumed to be 1. Regardless, by extracting common factors and simplifying the exponents, one obtains the desired inequality

$$E(\varepsilon^T D\varepsilon) \leq C_1^n m^{k_1+l_1+1} N^{k_1+l_1-1}\left(C_1 N + \sum_{i \neq j} |r_{ij}|\right)$$

which completes the proof of Lemma 3.3. □

Denote,

$$\zeta^n_{N,r}(\theta) \;=\; \frac{\partial A_N(\theta)}{\partial \theta_n}\,\varepsilon, \quad \theta \in B(r), \quad n = 1, 2, \ldots, m.$$

Theorem 3.7 *Assume that*

$$E(\varepsilon_i) = 0, \quad E(\varepsilon_i\, \varepsilon_j) = r_{ij}, \quad r_{ii} = \sigma_i^2, \quad 0 < \sigma_\star^2 \le \sigma_i^2 \le \sigma_1^2.$$

If there exists an integer N such that

$$\frac{d_1}{N\,(\lambda_1(\theta))^3} + \frac{d_2}{N^2\,(\lambda_1(\theta))^3} \sum_{i \ne j} |r_{ij}| \;\longrightarrow\; 0 \;\; as\; r \to 0 \tag{3.36}$$

and

$$\frac{d_3}{N\,(\lambda_1(\theta))^4} + \frac{d_4}{N^2\,(\lambda_1(\theta))^4} \sum_{i \ne j} |r_{ij}| \;\longrightarrow\; 0 \;\; as\; r \to 0, \tag{3.37}$$

where d_1, d_2, d_3, and d_4 are constants, then

$$\zeta^{(n)}_{N,r}(\theta) \stackrel{P}{\to} 0 \;\; as\; r \to 0. \tag{3.38}$$

Proof It is shown in [39] that

$$\begin{aligned} \frac{\partial A_N(\theta)}{\partial \theta_n}\,\varepsilon = \frac{1}{N}\Big[\frac{F_N^T(\theta(s))\, F_N(\theta(s))}{N}\Big]^{-1} \frac{\partial F_N^T(\theta)}{\partial \theta_n} + \frac{1}{N}\Big[\tfrac{F_N^T(\theta)\, F_N(\theta)}{N}\Big]^{-2} \\ \times \Big[\big(\tfrac{\partial F_N^T(\theta)}{\partial \theta_n} F_N(\theta) + F_N^T(\theta)\, \tfrac{\partial F_N(\theta)}{\partial \theta_n}\big)\Big/N\Big]\,\varepsilon. \end{aligned} \tag{3.39}$$

Denote the partial sums of (3.39) by I_1, I_2, I_3, and I_4, in a manner that separates the different inverse powers and partial derivatives.

Bound on I_1.

Let I_1 be the first term in (3.39). Using standard bounds on matrix inverses and the fact that ε satisfies $E(\varepsilon_i) = 0$ and $E(\varepsilon_i\,\varepsilon_j) = r_{ij}$, one obtains

$$\begin{aligned} I_1 &\le m \cdot \frac{C}{N^2} \cdot \left(\frac{1}{\lambda_1(\theta)}\right)^2 \big(N + \textstyle\sum_{i \ne j} |r_{ij}|\big) \\ &= C_1 \left(\frac{1}{N\,(\lambda_1(\theta))^2} + \frac{1}{N^2\,(\lambda_1(\theta))^2} \sum_{i \ne j} |r_{ij}|\right), \end{aligned}$$

where C and C_1 are constants that may depend on the dimension m but not on N. Under condition (3.36), the expression in parentheses converges to 0 as $r \to 0$. Hence

$$I_1 \;\longrightarrow\; 0 \;\; \text{as}\; r \to 0.$$

Bound on I_2.

Denote I_2 by suitably distributing terms so that each summation over ε and the partial derivatives is captured. A representative term in I_2 is

$$J_{2,1}(\theta) = \frac{1}{N^2} E\Big[\varepsilon^T \tfrac{\partial F_N(\theta)}{\partial \theta_n} \Big(\tfrac{F_N^T(\theta)\, F_N(\theta)}{N}\Big)^{-3} \tfrac{\partial F_N(\theta)}{\partial \theta_n} F_N(\theta)\, F_N^T(\theta)\, \varepsilon \Big].$$

By bounding each factor involving $\partial F_N(\theta)/\partial\theta_n$ and $F_N(\theta)$ in terms of $\lambda_1(\theta)$, and again leveraging the correlation structure $\sum_{i\neq j} |r_{ij}|$, it follows that

$$\begin{aligned} J_{2,1}(\theta) &\le \frac{C}{N^2}\Big(\frac{1}{\lambda_1(\theta)}\Big)^3 \Big(N + \sum_{i\neq j} |r_{ij}|\Big) \\ &= C\Big(\frac{1}{N\,(\lambda_1(\theta))^3} + \frac{1}{N^2\,(\lambda_1(\theta))^3} \sum_{i\neq j} |r_{ij}|\Big), \end{aligned}$$

which converges to 0 as $r \to 0$ under condition (3.37). By a similar argument, the remaining terms in I_2 (often denoted $J_{2,2}(\theta)$ and others) admit the same type of bound and thus also converge to 0.

Bound on I_3.

By a symmetric argument (involving the same type of matrix-inverse expansions), I_3 can be decomposed into terms that again take the same general form. Each term is controlled by factors that involve $\lambda_1(\theta)$, N, and $\sum |r_{ij}|$, leading to a bound of the type appearing in (3.36) or (3.37). Consequently,

$$I_3 \longrightarrow 0 \quad \text{as } r \to 0.$$

Bound on I_4.

Open the parentheses in I_4 and denote the resulting summands by $J_{4,i}$. A representative example is

$$J_{4,1} = \frac{1}{N^2} E\Big[\varepsilon^T F_N(\theta) \frac{\partial F_N^T(\theta)}{\partial \theta_n} F_N(\theta) \Big(\tfrac{F_N^T(\theta)\, F_N(\theta)}{N}\Big)^{-4} \frac{\partial F_N^T(\theta)}{\partial \theta_n} F_N(\theta)\, F_N^T(\theta)\, \varepsilon \Big].$$

Application of Lemma 3.3 (or a similar bounding argument) and the assumptions in Theorem 3.7 yields

$$\begin{aligned} J_{4,1} &= \frac{1}{N^2} C \Big(\frac{1}{\lambda_1(\theta)}\Big)^4 \Big(N + \sum_{i\neq j} |r_{ij}|\Big) \\ &= \frac{C}{N\,(\lambda_1(\theta))^4} + \frac{C}{N^2\,(\lambda_1(\theta))^4} \sum_{i\neq j} |r_{ij}| \longrightarrow 0 \quad \text{as } r \to 0. \end{aligned}$$

The remaining terms $J_{4,2}$, $J_{4,3}$, and $J_{4,4}$ have an analogous structure and satisfy the same bound, implying

$$I_4 \longrightarrow 0 \quad \text{as } r \to 0.$$

It follows from (3.39) that

$$\frac{\partial A_N(\theta)}{\partial \theta_n} \varepsilon = I_1 + I_2 + I_3 + I_4.$$

By Chebyshev's inequality,

$$P\left\{\left\| \tfrac{\partial A_N(\theta)}{\partial \theta_n} \varepsilon \right\|^2 > a\right\} \le \frac{E\left(\varepsilon^T \frac{\partial A_N^T(\theta)}{\partial \theta_n} \cdot \frac{\partial A_N(\theta)}{\partial \theta_n} \cdot \varepsilon\right)}{a^2}$$
$$\le \frac{I_1 + I_2 + I_3 + I_4}{a^2} \longrightarrow 0 \quad \text{as } r \to 0.$$

Since the probability above converges to 0 for each $a > 0$, it follows that

$$\zeta_{N,r}^{(n)}(\theta) = \tfrac{\partial A_N(\theta)}{\partial \theta_n} \varepsilon \xrightarrow{P} 0 \quad \text{as } r \to 0,$$

which completes the proof of Theorem 3.7. □

Remark 3.5 The conditions

$$\frac{d_1}{N(\lambda_1(\theta))^3} + \frac{d_2}{N^2(\lambda_1(\theta))^3} \sum_{i \neq j} |r_{ij}| \to 0 \; for \; r \to 0$$

$$\frac{d_3}{N(\lambda_1(\theta))^4} + \frac{d_4}{N^2(\lambda_1(\theta))^4} \sum_{i \neq j} |r_{ij}| \to 0 \; for \; r \to 0$$

can be rewritten in another form.

If

$$1/\lambda_1(\theta) \to 0, \tag{3.40}$$

$$\max_{i,j} |r_{ij}| \le C < \infty \; for \; r \to 0 \tag{3.41}$$

Then

$$\zeta_{N,r}^{(n)} \xrightarrow{P} 0 \; for \; r \to 0,$$

i.e., the statement of Theorem 3.7 is true.

Proof Condition (3.40) implies that $\lambda_1(\theta)$ increases without bound as $r \to 0$. Consequently, the factors $(\lambda_1(\theta))^{-3}$ and $(\lambda_1(\theta))^{-4}$ each approach 0.

Condition (3.41) ensures that $\max_{i,j}|r_{ij}| \leq C$, so there is a uniform bound on the entries of the correlation matrix for all sufficiently small r. As a result,

$$\sum_{i \neq j} |r_{ij}| \leq N(N-1) \max_{i,j} |r_{ij}| \leq N(N-1)\, C.$$

Hence each term of the form

$$\frac{d_2}{N^2 (\lambda_1(\theta))^3} \sum_{i \neq j} |r_{ij}| \quad \text{and} \quad \frac{d_4}{N^2 (\lambda_1(\theta))^4} \sum_{i \neq j} |r_{ij}|$$

can be bounded by constants multiplied by $(\lambda_1(\theta))^{-3}$ or $(\lambda_1(\theta))^{-4}$ and factors depending on N. Because $(\lambda_1(\theta))^{-3} \to 0$ and $(\lambda_1(\theta))^{-4} \to 0$ as $r \to 0$, all the expressions

$$\frac{d_1}{N (\lambda_1(\theta))^3}, \quad \frac{d_2}{N^2 (\lambda_1(\theta))^3} \sum_{i \neq j} |r_{ij}|, \quad \frac{d_3}{N (\lambda_1(\theta))^4}, \quad \frac{d_4}{N^2 (\lambda_1(\theta))^4} \sum_{i \neq j} |r_{ij}|$$

converge to 0 as $r \to 0$. This demonstrates that the original conditions are satisfied under assumptions (3.40) and (3.41).

Since these conditions recover exactly the scenario of Theorem 3.7, its conclusion states that $\zeta_{N,r}^{(n)}(\theta) \xrightarrow{P} 0$ as $r \to 0$. □

Denote

$$L_n = \frac{\partial U_N(\theta)}{\partial \theta_n}; \quad \tau_N(r) = \max_{p=1,\ldots,m} sup||L_p||.$$

Theorem 3.8 *Assume that for a random vector* $\varepsilon = (\varepsilon_1, \ldots, \varepsilon_N)^T$,

$$E(\varepsilon_i) = 0, \quad E(\varepsilon_i\, \varepsilon_j) = r_{ij}, \quad r_{ii} = \sigma_i^2, \quad 0 < \sigma_\star^2 \leq \sigma_i^2 \leq \sigma_1^2.$$

Then, under suitable conditions on the matrix functions involved (detailed below), the quantity

$$\tau_N(r) \xrightarrow{P} 0 \quad as\; r \to 0. \tag{3.42}$$

Proof The proof relies on bounding the partial derivatives of an operator $U_N(\theta)$ using an auxiliary matrix $A_N(\theta)$ and properties of the functionals involved. In particular, matrix products of the form

$$\frac{F_N^T(\theta)\, F_N(\theta)}{N}$$

and their inverses appear frequently. Let $\bar{f}_{ij}(\theta)$ denote the elements of the matrix

$$\left(\frac{F_N^T(\theta)\,F_N(\theta)}{N}\right)^{-4}$$

and let $\partial f_{ki}(\theta)$ denote the elements of $\frac{\partial F_N(\theta)}{\partial\theta_n}$. One may index these as $k = 1, 2, \ldots, m$ and $l = 1, 2, \ldots, N$, depending on row-column structure.

The first part of the analysis involves the decomposition of $E\left\|\frac{\partial A_N(\theta)}{\partial\theta_n}\varepsilon\right\|^2$. From matrix manipulations, it follows that

$$E\left\|\frac{\partial A_N(\theta)}{\partial\theta_n}\cdot\varepsilon\right\|^2 = E\left(\varepsilon^T\frac{\partial A_N^T(\theta)}{\partial\theta_n}\cdot\frac{\partial A_N(\theta)}{\partial\theta_n}\cdot\varepsilon\right)$$
$$= \frac{1}{N}E\left[\varepsilon^T\frac{\partial F_N(\theta)}{\partial\theta_n}\left(\frac{F_N^T(\theta)F_N(\theta)}{N}\right)^{-1}\left(\frac{F_N^T(\theta)F_N(\theta)}{N}\right)^{-1}\frac{\partial F_N(\theta)}{\partial\theta_n}\varepsilon\right]$$
$$+\frac{1}{N^2}E\left[\varepsilon^T\frac{\partial F_N(\theta)}{\partial\theta_n}\left(\frac{F_N^T(\theta)F_N(\theta)}{N}\right)^{-3}\left(\frac{\partial F_N(\theta)}{\partial\theta_n}\cdot F_N(\theta)+F_N^T(\theta)\frac{\partial F_N(\theta)}{\partial\theta_n}\right)F_N^T(\theta)\cdot\varepsilon\right]$$
$$+\frac{1}{N}E\Bigg[\varepsilon^T F_N(\theta)\left(\frac{\partial F_N^T(\theta)}{\partial\theta_n}\cdot F_N(\theta)+F_N^T(\theta)\frac{\partial F_N(\theta)}{\partial\theta_n}\right)\left(\frac{F_N^T(\theta)F_N(\theta)}{N}\right)^{-2}$$
$$\left(\frac{F_N^T(\theta)F_N(\theta)}{N}\right)^{-2}\left(\frac{\partial F_N^T(\theta)}{\partial\theta_n}\cdot F_N(\theta)+F_N(\theta)\frac{\partial F_N(\theta)}{\partial\theta_n}\right)|F_N^T(\theta)\cdot\varepsilon \tag{3.43}$$

Denote these four main summands by I_1, I_2, I_3, and I_4. Each term can be bounded by applying arguments similar to those used in Lemma 3.3, which typically involve controlling the product of partial derivatives and inverse matrices by constants times powers of N and the eigenvalues of $\frac{F_N^T(\theta)\,F_N(\theta)}{N}$. Under appropriate assumptions (e.g., lower bounds on the smallest eigenvalue $\lambda_1(\theta)$ and uniform boundedness of $\partial F_N(\theta)/\partial\theta_n$), each I_k converges to zero in expectation or can be made arbitrarily small as $r \to 0$.

Then, a key derivative relationship is established in this part of the analysis. According to the statement of Lemma 3.3 (or an analogous result) about the elements of $\frac{F_N^T(\theta)\,F_N(\theta)}{N}$, one obtains the identity

$$\frac{\partial\delta_N(\theta)}{\partial\theta_n} = -\frac{\partial\eta(x,\theta)}{\partial\theta_n} = -F_N(\theta).$$

(The exact form of this relationship may depend on how $\delta_N(\theta)$ and η are defined, but in many cases $\delta_N(\theta)$ encodes functions that shift by η.)

Next, the expression for $\frac{\partial U_N(\theta)}{\partial\theta_n}$ is derived in the proof. From (3.43) and related definitions, it follows that

$$\begin{aligned}\frac{\partial U_N(\theta)}{\partial \theta_n} &= \Big[I_n + \frac{\partial A_N(\theta)}{\partial \theta_n}\delta_N(\theta) + A_N(\theta)\cdot \frac{\partial \delta_N(\theta)}{\partial \theta_n}\Big] \\ &= \Big[\frac{\partial A_N(\theta)}{\partial \theta_n}\delta(\theta^*) + \frac{\partial A_N(\theta)}{\partial \theta_n}(\delta_N(\theta) - \delta(\theta^*)\Big] \\ &= \frac{\partial A_N(\theta)}{\partial \theta_n}\cdot \varepsilon + \frac{\partial A_N(\theta)}{\partial \theta_n}\cdot \Delta\eta(x, \theta, \theta^*)\end{aligned} \tag{3.44}$$

where $\Delta\eta\big(x, \theta, \theta^*\big) = \eta\big(x, \theta^*\big) - \eta\big(x, \theta\big)$, and $\delta(\theta^*)$ is a reference shift. The first equality arranges terms involving $\delta_N(\theta)$ and its derivative, while the second and third lines rewrite $\delta_N(\theta)$ in terms of ε and the difference $\Delta\eta$.

Lastly, the convergence of each component is analyzed. According to Theorem 3.8 (or an appropriate analogous theorem), $\frac{\partial A_N(\theta)}{\partial \theta_n}\ \varepsilon$ converges to zero in probability as $r \to 0$, because each of the terms I_1, I_2, I_3, I_4 in (3.43) can be made arbitrarily small. Moreover, the difference $\Delta\eta\big(x, \theta, \theta^*\big)$ goes to zero uniformly in x as $r \to 0$, implying that $\frac{\partial A_N(\theta)}{\partial \theta_n}\ \Delta\eta\big(x, \theta, \theta^*\big)$ also vanishes in probability (or almost surely, depending on the context). Hence,

$$\frac{\partial U_N(\theta)}{\partial \theta_n} = \frac{\partial A_N(\theta)}{\partial \theta_n}\ \varepsilon + \frac{\partial A_N(\theta)}{\partial \theta_n}\ \Delta\eta\big(x, \theta, \theta^*\big) \ \xrightarrow{P}\ 0 \quad \text{as } r \to 0.$$

Since $\tau_N(r)$ is the supremum norm over all partial derivatives $\frac{\partial U_N(\theta)}{\partial \theta_p}$, $p = 1, \ldots, m$, it follows that

$$\tau_N(r) \ \xrightarrow{P}\ 0 \quad \text{as } r \to 0.$$

This completes the proof of (3.42). □

Denote

$$\rho_N(\theta) = U_N(\theta) - \theta,\ \rho^* = \rho(\theta^*).$$

Theorem 3.9 *Suppose that the following condition is held:*

$$\frac{1}{\lambda_1(\theta)} + \frac{1}{N(\lambda_1(\theta))^2}\sum_{i \neq j} |r_{ij}| \leq A \tag{3.45}$$

where A is a constant.

Then

$$\lim_{N\to\infty}\lim_{k\to\infty} P\{\sqrt{N}\cdot\rho^* > k\} = 0$$

Proof First, let $\rho_N(\theta)$ be an estimator defined by

$$\rho_N(\theta) \ = \ U_N(\theta) \ - \ \theta,$$

where $U_N(\theta)$ is some operator or estimator derived from the sample ε. Denote by $\rho^* = \rho_N(\theta^*)$ the estimator evaluated at a particular parameter $\theta^* \in B(r)$.

Observe that

$$\begin{aligned} N\,E\left\|\rho_N(\theta)\right\|^2 &= E\Big(\big(U_N(\theta)-\theta\big)^T\big(U_N(\theta)-\theta\big)\Big) \\ &= N\,E\Big(\varepsilon^T A_N^T(\theta)\,A_N(\theta)\,\varepsilon\Big) \\ &= E\Big[\varepsilon^T F_N(\theta)\Big(\tfrac{F_N^T(\theta)\,F_N(\theta)}{N}\Big)^{-1}\Big(\tfrac{F_N^T(\theta)\,F_N(\theta)}{N}\Big)^{-1}\tfrac{F_N^T(\theta)}{N}\,\varepsilon\Big]. \end{aligned} \tag{3.46}$$

The third line follows if $A_N(\theta)$ can be factored as

$$A_N(\theta) = \frac{F_N(\theta)}{N}\left(\frac{F_N^T(\theta)\,F_N(\theta)}{N}\right)^{-1}$$

or a similar relation that allows introducing the matrix $F_N(\theta)$ and its scaled inverse.

Condition (3.45) and the boundedness of the elements of $F_N(\theta)$ imply that $\left(\frac{F_N^T(\theta)\,F_N(\theta)}{N}\right)^{-1}$ is controlled by a factor proportional to $\lambda_1^{-1}(\theta)$. Here, $\lambda_1(\theta)$ denotes the smallest eigenvalue of $\frac{F_N^T(\theta)\,F_N(\theta)}{N}$. In addition, a similar result regarding quadratic forms of ε with correlation entries $\{r_{ij}\}$ shows that:

$$E\Big(\varepsilon^T B\,\varepsilon\Big) \le \sigma_0^2\,\mathrm{tr}(B) + \frac{1}{N^2}\sum_{i\neq j}|r_{ij}| \times \|B\|_\infty,$$

for a suitable matrix B. Applying these facts to the matrix in (3.46) provides the inequality

$$\begin{aligned} N\,E\left\|\rho_N(\theta)\right\|^2 &\le \sigma_0^2\,\mathrm{tr}\Big(\Big(\tfrac{F_N^T(\theta)\,F_N(\theta)}{N}\Big)^{-1}\Big) + \frac{1}{N^2\,(\lambda_1(\theta))^2}\sum_{i\neq j}\big|r_{ij}\big| \\ &= \frac{c_1}{\lambda_1(\theta)} + \frac{1}{N\,(\lambda_1(\theta))^2}\sum_{i\neq j}\big|r_{ij}\big| < A, \quad \text{for all } \theta \in B(r), \end{aligned} \tag{3.47}$$

Inequality (3.47) follows from rewriting the trace term in terms of $\lambda_1(\theta)$ and applying (3.45), along with boundedness arguments.

Since $N\,E\|\rho^*\|^2 < A$ for $\theta^* \in B(r)$, Chebyshev's inequality yields:

$$P\Big\{\sqrt{N}\,\|\rho^*\| > k\Big\} \le \frac{N\,E\|\rho^*\|^2}{k^2} \le \frac{A}{k^2},$$

which converges to 0 as $k \to \infty$. Hence,

$$\lim_{k\to\infty} P\Big\{\sqrt{N}\,\rho^* > k\Big\} = 0 \quad \text{for each fixed } N.$$

A subsequent limit as $N \to \infty$ preserves this convergence to 0, implying

$$\lim_{N\to\infty}\lim_{k\to\infty} P\left\{\sqrt{N}\,\rho^* \ > \ k\right\} \ = \ 0.$$

This completes the proof of 3.9. □

Theorem 3.10 *Assume that* $\theta(0) \in B(r)$ *and*

$$\tau_N(r) \ + \ \frac{\|\rho^*\|}{r} \ < \ 1.$$

Furthermore, suppose there is a constant A such that

$$\frac{1}{\lambda_1(\theta)} \ + \ \frac{1}{N\,(\lambda_1(\theta))^2}\sum_{i\neq j}|r_{ij}| \ \leq \ A, \tag{3.48}$$

and that

$$\lim_{N\to\infty}\ \lim_{k\to\infty}\ P\{\sqrt{N}\,\rho^* > k\} \ = \ 0. \tag{3.49}$$

Then there exists a random variable $\widehat{\theta}_N$ *such that*

$$(\widehat{\theta}_N \ - \ \widehat{\theta}_N(s)) \ \xrightarrow{P} \ 0 \ \ as \ \ s \ \to \ \infty. \tag{3.50}$$

Remark 3.6 Conditions (3.48) and (3.49) restate the assumptions originally given in Theorem 3.9, ensuring that the inverse of the smallest eigenvalue $\lambda_1(\theta)$ remains suitably bounded with respect to N, and that $\sqrt{N}\,\rho^*$ does not exceed any fixed threshold k with nontrivial probability for large N. Under these requirements, along with the bound

$$\tau_N(r) \ + \ \frac{\|\rho^*\|}{r} < 1,$$

there exists a random variable $\widehat{\theta}_N$ whose difference from its counterpart $\widehat{\theta}_N(s)$ converges in probability to zero as $s \to \infty$. This result extends the stability and consistency properties of the estimator within the ball $B(r)$, ultimately leading to the claim in (3.50).

Proof Since $\frac{\partial U_N(\theta)}{\partial\theta}$ exists according to Eq. (3.44), Lagrange's Mean Value Theorem for vector-valued functions can be applied, yielding:

$$||U_N(\theta) - U_N(\theta^*)|| \leq ||\frac{\partial U_N(\widetilde{\theta})}{\partial\theta}|| \cdot ||\theta - \theta^*||, \quad \theta \in B(r), \quad \widetilde{\theta} \in B(r).$$

Here, $\widetilde{\theta}$ represents some point on the line segment connecting θ and θ^*. This inequality indicates that the difference between $U_N(\theta)$ and $U_N(\theta^*)$ is bounded by the product of the norm of the derivative of $U_N(\theta)$ at a point in $B(r)$ and the distance between θ and θ^*.

By using Theorem 3.9, for all $\widetilde{\theta} \in B(r)$, the following holds:

$$||\frac{\partial U_N(\widetilde{\theta})}{\partial \theta}|| = \left\| \sum_{p=1}^{m} \frac{\partial U_N(\widetilde{\theta})}{\partial \theta_p} \right\| \leq \sum_{p=1}^{m} \left\| \frac{\partial U_N(\widetilde{\theta})}{\partial \theta_p} \right\| \xrightarrow{r \to 0} 0.$$

This result shows that for sufficiently small r, the derivative of $U_N(\theta)$ becomes arbitrarily small, implying that $U_N(\theta)$ behaves like a contraction map on $B(r)$.

Therefore, if $\theta(0) \in B(r)$, there exists an N such that, for sufficiently small r, the mapping $U_N(\theta)$ is a contraction. In particular, it holds that:

$$\begin{aligned} \|U_N(\theta) - \theta^*\| &\leq \|U_N(\theta) - U_N(\theta^*)\| + \|U_N(\theta^*) - \theta^*\| \\ &\leq \tau_N(r) + \|\rho^*\| \leq r, \quad \theta \in B(r). \end{aligned}$$

This follows from the assumption of Theorem 3.10, and shows that $U_N(\theta)$ maps the ball $B(r)$ into itself.

Since $B(r)$ is a complete metric space, and $U_N(\theta)$ is a contraction mapping, the principle of contraction mappings [37] guarantees the existence of a unique fixed point $\widehat{\theta}$ such that $U_N(\widehat{\theta}) = \widehat{\theta}$. Furthermore, for all $\theta(0) \in B(r)$, as $r \to 0$ and $s \to \infty$, the sequence $\theta_N(s)$ generated by iterating $U_N(\theta)$ will converge to $\widehat{\theta}_N$, since the distance between successive iterates becomes arbitrarily small. Specifically, it follows that:

$$||U_N(\theta(s-1)) - \widehat{\theta}_N|| \to 0 \quad \text{as } r \to 0 \quad \text{and} \quad s \to \infty.$$

Thus, it follows that:

$$||\theta_N(s) - \widehat{\theta}_N|| \xrightarrow{P} 0, \quad \text{as } s \to \infty, \quad r \to 0.$$

This completes the proof of the theorem. □

Theorem 3.11 *Suppose the following conditions hold:*
The contraction condition

$$\tau_N(r) + \frac{\|\rho^*\|}{r} < 1.$$

The stability condition

$$\frac{1}{\lambda_1(\theta)} + \frac{1}{N(\lambda_1(\theta))^2} \sum_{i \neq j} |r_{ij}| \leq A,$$

where A is a constant.

The boundedness of residual terms

$$\lim_{N\to\infty}\lim_{k\to\infty} P\{\sqrt{N}\cdot\rho^* > k\} = 0.$$

Then, the following bound holds for all initial values $\theta(0) \in B(r)$:

$$\sup_{\theta(0)\in B(r)} \|\theta_N(s) - \widehat{\theta}_N\| \leq \frac{(\tau_N(r))^s}{1-\tau_N(r)}. \tag{3.51}$$

Proof Let $s < t$ be two arbitrary times. According to the principle of the contraction mapping [37], it can be written as:

$$\sup_{\theta(0)\in B(r)} ||\theta(s) - \theta(t)|| \leq \frac{(\tau_N(r))^s}{1-\tau_N(r)}||\theta(0) - \theta(1)||. \tag{3.52}$$

Here, use the contraction mapping property, which asserts that the distance between iterates of the mapping $U_N(\theta)$ decreases geometrically with the factor $\tau_N(r)$. The term $\tau_N(r)$ represents the contraction constant of the mapping, which ensures the convergence of $\theta(s)$ as s progresses.

Since $U_N(\theta)$ is a mapping that transforms the set $B(r)$ into itself (i.e., it maps $B(r)$ to another set contained within $B(r)$), the result in (3.52) holds. Therefore, from (3.52), it can be concluded:

$$\sup_{\theta(0)\in B(r)} ||\theta(s) - \theta(t)|| \leq \frac{(\tau_N(r))^s}{1-\tau_N(r)}\cdot r.$$

This inequality follows from the fact that the initial distance between any two points in $B(r)$ is bounded by r, the radius of the ball. Hence, the maximum distance at any time s between iterates is controlled by the contraction factor $\tau_N(r)$ raised to the power s, and multiplied by the radius r.

Next, since $\theta(t)$ lies within $B(r)$, use the mapping $U_N(\theta)$ to $\theta(t)$. By Theorem 3.10, there exists a $\widehat{\theta}$ such that:

$$\sup_{\theta(0)\in B(r)} ||\theta_N(s) - \widehat{\theta}_N|| \leq \frac{(\tau_N(r))^s}{1-\tau_N(r)}\cdot r. \tag{3.53}$$

This concludes the proof, as the desired result in (3.51) follows directly from (3.53). □

Theorem 3.12 *Suppose that the following conditions hold:*

$$\tau_N(r) + \frac{\|\rho^*\|}{r} < 1,$$

and

$$\sup_{\theta(0)\in B(r)} \|\theta_N(s) - \widehat{\theta}_N\| \leq \frac{(\tau_N(r))^s}{1-\tau_N(r)}.$$

Then, the random variable

$$\sqrt{N}(\widehat{\theta}_N - \theta^*)$$

is bounded in probability, i.e., the estimator $\widehat{\theta}_N$ *is* $\sqrt{N}$*-consistent as* $N \to \infty$.

Proof Let $r = \frac{2k}{\sqrt{N}}$. By Theorem 3.8, it follows that:

$$\tau(r) \xrightarrow{P} 0, \quad N \to \infty.$$

Since the random variable $\sqrt{N} \cdot \rho^*$ is bounded in probability, the following conditions are imposed:

$$\tau(r) = \frac{1}{2}, \quad r = \frac{2k}{\sqrt{N}}, \quad ||\rho^*|| < \frac{k}{\sqrt{N}}.$$

Under these assumptions, we obtain the following inequality:

$$\tau_N(r) + \frac{||\rho^*||}{r} < \frac{1}{2} + \frac{||\rho^*|| \cdot \sqrt{N}}{2k} < \frac{1}{2} + \frac{1}{2} = 1.$$

Next, consider the inequality:

$$||\widehat{\theta} - \theta^*|| \leq \sup_{\theta(s)\in B(r)} ||\widehat{\theta} - \theta(s)||.$$

By Theorem 3.11, it follows that:

$$\lim_{k\to\infty} \lim_{N\to\infty} P\left\{||\widehat{\theta}_N - \theta^*|| > \frac{k}{\sqrt{N}}\right\} = 0.$$

Thus, as N grows large, the probability of the difference between the estimator $\widehat{\theta}_N$ and the true parameter θ^* exceeding $\frac{k}{\sqrt{N}}$ vanishes, as expected. This completes the proof. □

Theorem 3.13 *Consider the sequence* ε_i, $i = 1, 2, \ldots$*; independent random variables.*

If $\frac{1}{N\cdot\lambda_1^2(\theta)} \to 0$ *for* $N \to \infty, \forall\theta \in B(r)$*, then*

$$\sqrt{N} \cdot \rho^* \to N\{0, \sum(\theta^*)\} \; for \; N \to \infty$$

where

$$\sum(\theta^*) = \left(\frac{F_N^T(\theta^*)F_N(\theta^*)}{N}\right)^{-1} \frac{F_N^T(\theta^*)I(\sigma^2)F_N(\theta^*)}{N} \left(\frac{F_N^T(\theta^*)F_N(\theta^*)}{N}\right)^{-1} \quad (*)$$

$$I(\sigma^2) = E[\varepsilon \cdot \varepsilon^T]$$

Proof Let $l = (l_1, l_2, \ldots, l_m)^T$ be a non-zero vector of size m. Consider the following expression:

$$\sqrt{N}\, l^T \rho^* = \sqrt{N}\, l^T \left(\frac{F_N^T(\theta^*)F_N(\theta^*)}{N}\right)^{-1} \frac{F_N^T(\theta^*)}{N} \left(y - \eta(x, \theta^*)\right).$$

By expanding this expression, it is obtained that:

$$\sqrt{N}\, l^T \rho^* = \frac{1}{N} \sum_{k=1}^{N} \sum_{i=1}^{m} \sum_{j=1}^{n} \sqrt{N} l_i \bar{f}_{ij} f_{jk}(\theta^*) \varepsilon_k,$$

where $f_{jk}(\theta)$ are elements of the matrix $F_N^T(\theta^*)$, for $j = 1, 2, \ldots, N$ and $k = 1, 2, \ldots, m$, and x_{ij} are elements of the matrix $\left(\frac{F_N^T(\theta^*)F_N(\theta^*)}{N}\right)^{-1}$, for $i, j = 1, 2, \ldots, m$.

The expectation of the random variable is:

$$E\left[\sqrt{N} l^T \rho^*\right] = 0.$$

Now, consider the second moment:

$$\begin{aligned} E\left(\left(\sqrt{N} l^T \rho^*\right)^2\right) &= \frac{1}{N} \sum_{i=1}^{N} \left(\sum_{i=1}^{m} \sum_{j=1}^{m} l_i x_{ij} f_{jk}\right)^2 \sigma_i^2 \\ &\le \frac{\sigma_0^2}{N} \left(\frac{\sqrt{m}}{\lambda_1(\theta)}\right)^2 C_1^2 \cdot N \cdot \left(\sum_{i=1}^{m} l_i\right)^2 \\ &\le C_1^2 \cdot \sigma_0^2 \cdot \frac{m^3}{\lambda_1^2(\theta)} \le C_2 \cdot \frac{m^3}{\lambda_1^2(\theta)}, \quad \forall \theta \in B(r). \end{aligned}$$

This bound follows from Theorem 3.12.

By the central limit theorem [57], the random variable $\sqrt{N} l^T \rho^*$ converges in distribution to a Gaussian distribution:

$$\begin{aligned} E\|\sqrt{N} l^T \rho^*\|^2 &= N \cdot E\left[l^T \rho^* \rho^{*T} l\right] \\ &= l^T \left(\frac{F_N^T(\theta^*)F_N(\theta^*)}{N}\right)^{-1} \frac{F_N^T(\theta^*)I(\sigma^2)F_N(\theta^*)}{N} \left(\frac{F_N^T(\theta^*)F_N(\theta^*)}{N}\right)^{-1} l \\ &= l^T \sum(\theta^*) l. \end{aligned}$$

According to the criterion on normality [69], it follows that:

$$\sqrt{N}\rho^* \xrightarrow{d} N(0, \sum(\theta^*)),$$

where the convergence is in distribution to a normal distribution with mean zero and variance $\sum(\theta^*)$.

Next, let k be the minimal number of steps such that asymptotic normality is achieved for the random variable $\sqrt{N}\rho^*$. Suppose the following conditions hold:

$$\tau_N(r) + \frac{||\rho^*||}{r} < a < 1,$$

and

$$\frac{r \cdot a^b}{1-a} < \frac{k}{\sqrt{N}}.$$

Such a k can be determined because the random variable $\sqrt{N}||\widehat{\theta}_N - \theta^*||$ is bounded according to Theorem 3.11. □

Lemma 3.4 *Suppose that the following conditions hold:*

$$\tau_N(r) + \frac{\|\rho^*\|}{r} < a < 1,$$

as well as the uniform bound condition

$$\sup_{\theta(0)\in B(r)} \|\theta_N(s) - \widehat{\theta}_N\| \leq \frac{(\tau_N(r))^s}{1-\tau_N(r)}.$$

If

$$s \geq \frac{\ln[(1-a)k] - \ln(r\sqrt{N})}{\ln a}, \tag{3.54}$$

then the following asymptotic normality result holds:

$$(\theta(s) - \theta^*) \xrightarrow{P} N(0, \cdot) \quad N \to \infty. \tag{3.55}$$

Proof The statement of Lemma 3.4 follows from Theorem 3.12.

The random variable $\hat{\theta}$ satisfies the following normal equation:

$$A_N(\widehat{\theta}) \cdot \delta_N(\widehat{\theta}) = 0, \tag{3.56}$$

which represents the normal equation for the least squares estimator.

Consider the iterative procedure defined by,

$$\theta(s+1) = u_N(\theta(s)) = \theta(s) + A_N(\theta(s)) \cdot \delta_N(\theta(s)), \quad \text{for} \quad s \to \infty.$$

This recursive update rule describes the evolution of the estimate $\theta(s)$ as s increases.

Taking the limit as $s \to \infty$, it can be obtained that:

$$\hat{\theta} = \widehat{\theta} + A_N(\widehat{\theta}) \cdot \delta_N(\widehat{\theta}).$$

This shows the relationship between the estimator $\hat{\theta}$ and the true parameter $\widehat{\theta}$ in terms of the adjustment term $A_N(\widehat{\theta}) \cdot \delta_N(\widehat{\theta})$.

From the normal Eq. (3.56), it can be deduced that:

$$A_N(\widehat{\theta}) \cdot \delta_N(\widehat{\theta}) = 0.$$

This confirms that the solution $\widehat{\theta}$ is the point where the adjustment term vanishes, satisfying the normal equation.

The convergence in probability of $(\theta(s) - \theta^*)$ to a normal distribution follows from these results, as detailed in Theorem 3.12. The bound on s in Eq. (3.54) ensures that the iterative procedure converges, and the random variable $\theta(s)$ converges in distribution to a normal distribution with mean zero and variance determined by the limiting behavior of the system. Therefore, the proof of (3.55) is completed. □

Theorem 3.14 *Suppose the following conditions hold:*
(1) There exists a constant C_1 such that

$$|a_{ij}^{(n,1)}| \leq C_1, \quad |b_{ij}^{(n,2)}| \leq C_1, \quad \max_{1 \leq i \leq m} \sum_{j=1}^{m} |c_{ij}| \leq C_1,$$

which implies the inequality

$$E\big(\varepsilon^T D\, \varepsilon\big) \leq C_1^n\, m^{k_1+l_1+1}\, N^{k_1+l_1-1}\Big(C_1\, N + \sum_{i \neq j} |r_{ij}|\Big).$$

(2) The contraction condition is satisfied:

$$\tau_N(r) + \frac{\|\rho^*\|}{r} < a < 1.$$

(3) The uniform bound condition holds:

$$\sup_{\theta(0) \in B(r)} \|\theta_N(s) - \widehat{\theta}_N\| \leq \frac{(\tau_N(r))^s}{1 - \tau_N(r)}.$$

(4) The iteration step s satisfies the lower bound

$$s \geq \frac{\ln[(1-a)k] - \ln(r\sqrt{N})}{\ln a}.$$

(5) The random vector $\varepsilon = (\varepsilon_1, \ldots, \varepsilon_N)^T$ *satisfies*

$$E(\varepsilon_i) = 0, \quad E(\varepsilon_i\, \varepsilon_j) = r_{ij}, \quad r_{ii} = \sigma_i^2, \quad 0 < \sigma_\star^2 \le \sigma_i^2 \le \sigma_1^2.$$

Moreover, there exist constants d_1, d_2, d_3, d_4 *such that if*

$$\frac{d_1}{N\,(\lambda_1(\theta))^3} + \frac{d_2}{N^2\,(\lambda_1(\theta))^3} \sum_{i \neq j} |r_{ij}| \to 0 \quad as \quad r \to 0,$$

$$\frac{d_3}{N\,(\lambda_1(\theta))^4} + \frac{d_4}{N^2\,(\lambda_1(\theta))^4} \sum_{i \neq j} |r_{ij}| \to 0 \quad as \quad r \to 0,$$

then the following limits hold:

$$\zeta_{N,r}^{(n)}(\theta) \xrightarrow{P} 0 \quad as \quad r \to 0,$$

$$\tau_N(r) \xrightarrow{P} 0 \quad as \quad r \to 0.$$

(6) The random variable

$$\sqrt{N}(\widehat{\theta} - \theta^*)$$

is bounded in probability, meaning that it does not diverge as $N \to \infty$.

Under these conditions, the estimator $\widehat{\theta}$ *satisfies*

$$\sqrt{N}(\widehat{\theta} - \theta^*) \xrightarrow{d} N(0, \Sigma(\theta^*)) \quad as \quad N \to \infty.$$

Proof By Theorem 3.7, the following equation holds:

$$A_N(\widehat{\theta}) \cdot \delta_N(\widehat{\theta}) = 0$$

Consequently, the following expression is true:

$$\begin{aligned}
&A_N(\widehat{\theta}) \cdot \delta_N(\widehat{\theta}) = [A_N(\theta^*) + A_N(\widehat{\theta}) - A_N(\theta^*)][y - \eta(x, \theta^*) + \eta(x, \theta^*) - \eta(x, \widehat{\theta})] \\
&= A_N(\theta^*)(y - \eta(x, \theta^*)) + A_N(\theta^*)(\eta(x, \theta^*) - \eta(x, \widehat{\theta})) \\
&+ [A_N(\widehat{\theta}) - A_N(\theta^*)](y - \eta(x, \theta^*)) + [A_N(\widehat{\theta}) - A_N(\theta^*)](\eta(x, \theta^*) - \eta(x, \widehat{\theta})) \\
&= \rho^* + [A_N(\widehat{\theta}) - A_N(\theta^*)] + A_N(\widehat{\theta}) \cdot \Delta\eta(x, \theta^*, \widehat{\theta}) = 0
\end{aligned} \tag{3.57}$$

For the term $[A_N(\widehat{\theta}) - A_N(\theta^*)]$, the following identity holds:

$$A_N(\widehat{\theta}) - A_N(\theta^*) = \left[\frac{\partial A_N(\theta^*)}{\partial \theta} + \int_0^1 \left(\frac{\partial A_N(\theta^* + z(\widehat{\theta} - \theta^*))}{\partial \theta} - \frac{\partial A_N(\theta^*)}{\partial \theta}\right) dz \cdot (\widehat{\theta} - \theta^*)\right]$$

$$\eta(\widehat{\theta}) - \eta(\theta^*) = \left[\frac{\partial\eta(\theta^*)}{\partial\theta} + \int_0^1 \left(\frac{\partial\eta(\theta^* + z(\widehat{\theta} - \theta^*))}{\partial\theta} - \frac{\partial\eta(\theta^*)}{\partial\theta}\right) dz\right] (\widehat{\theta} - \theta^*) \tag{3.58}$$

Substituting (3.58) into (3.57), the following expression is obtained:

$$\begin{aligned}
&\rho^* + \frac{\partial A_N(\theta^*)}{\partial\theta}\delta_N^*(\widehat{\theta} - \theta^*) + \int_0^1 \frac{\partial A_N(\theta^* + z(\widehat{\theta} - \theta^*))}{\partial\theta}\delta_N^*(\widehat{\theta} - \theta^*)dz \\
&- \frac{\partial A_N(\theta^*)}{\partial\theta}\delta_N^*(\widehat{\theta} - \theta^*) + A_N(\widehat{\theta})\frac{\partial\eta(\theta^*)}{\partial\theta}(\widehat{\theta} - \theta^*) \\
&+ A_N(\widehat{\theta})\int_0^1 \frac{\partial\eta}{\partial\theta}(\theta^* + z(\widehat{\theta} - \theta^*))(\widehat{\theta} - \theta^*)dz \\
&- A_N(\widehat{\theta})\frac{\partial\eta(\widehat{\theta})}{\partial\theta}(\widehat{\theta} - \theta^*) - A_N(\widehat{\theta})\frac{\partial\eta(\theta^*)}{\partial\theta}(\widehat{\theta} - \theta^*) = 0
\end{aligned} \tag{3.59}$$

Since

$$A_N(\widehat{\theta}) \cdot \frac{\partial\eta(\widehat{\theta})}{\partial\theta} = I_m$$

It follows from (3.59) that:

$$\sqrt{N}(\widehat{\theta} - \theta^*) = \sqrt{N} \cdot \rho^* + \sqrt{N} \cdot M_1 + \sqrt{N} \cdot M_2$$

where

$$M_1 = \int_0^1 \frac{\partial A_N(\theta^* + z(\widehat{\theta} - \theta^*))}{\partial\theta} \cdot \delta_N^* dz(\widehat{\theta} - \theta^*) \tag{3.60}$$

$$M_2 = A_N(\widehat{\theta}) \int_0^1 \frac{\partial\eta(\theta^* + z(\widehat{\theta} - \theta^*))}{\partial\theta} dz(\widehat{\theta} - \theta^*) \tag{3.61}$$

According to Theorem 3.12, the random variable $\sqrt{N}(\widehat{\theta} - \theta^*)$ is bounded in probability. Furthermore, by Theorem 3.7, the following holds:

$$\frac{\partial A_N(\theta)}{\partial\theta} \cdot \delta_N^* \xrightarrow{P} 0 \quad \text{as} \quad N \to \infty$$

Hence, it follows that:

$$\sqrt{N} \cdot M_1 \xrightarrow{P} 0 \quad \text{as} \quad N \to \infty$$

Similarly, it can be shown that:

$$\sqrt{N} \cdot M_2 \xrightarrow{P} 0 \quad \text{as} \quad N \to \infty$$

Thus, using Theorem 3.13, it can be obtained that the desired result:

$$\sqrt{N}(\widehat{\theta} - \theta^*) \xrightarrow{d} N(0, \sum(\theta^*)),$$

which completes the proof. □

3.3.2 Covariance Matrix Estimation in Nonlinear Prototype

Consider the covariance matrix:

$$C_N = NE(\widehat{\theta} - \theta)(\widehat{\theta} - \theta)^T,$$

where c_{kl}, for $k, l = 1, 2, \ldots, m$, are the elements of C_N.

Since

$$\sqrt{N}(\widehat{\theta} - \theta^*) = \sqrt{N} \cdot \rho^* + \sqrt{N} \cdot M_1 + \sqrt{N} \cdot M_2,$$

where

$$M_1 = \int_0^1 \frac{\partial A_N(\theta^* + z(\widehat{\theta} - \theta^*))}{\partial \theta} \cdot \delta_N^* dz(\widehat{\theta} - \theta^*),$$

$$M_2 = A_N(\widehat{\theta}) \int_0^1 \frac{\partial \eta(\theta^* + z(\widehat{\theta} - \theta^*))}{\partial \theta} dz(\widehat{\theta} - \theta^*),$$

where M_1 and M_2 are defined in (3.60) and (3.61).

According to Theorem 3.12, the random variable $\sqrt{N}(\widehat{\theta} - \theta^*)$ is bounded in probability. Furthermore, Theorem 3.7 implies that

$$\frac{\partial A_N(\theta)}{\partial \theta} \delta_N^* \xrightarrow{P} 0, \quad \text{as } N \to \infty.$$

Thus, terms that converge to zero can be neglected, and the matrix C_N can be expressed in the following form:

$$\begin{aligned} C_N &= NE\rho^*(\rho^*)^T + o(N) \\ &= \left(\frac{F^T(\theta^*)F(\theta^*)}{N}\right)^{-1} \left(\frac{F^T(\theta^*)I(\sigma^2)F(\theta^*)}{N}\right) \left(\frac{F^T(\theta^*)F(\theta^*)}{N}\right)^{-1}, \end{aligned}$$

where $I(\sigma^2) = E[\varepsilon\varepsilon^T]$ and $o(N)$ is an $m \times m$ matrix whose elements are of order $o(1/N)$.

The value of θ^* is unknown, but it has been established that $\widehat{\theta}$ is a $\sqrt{N}$-consistent estimator of θ. Consequently, similarly to the case of linear regression models, the elements of the covariance matrix of the deviation vector $\sqrt{N}(\widehat{\theta} - \theta)$ can be estimated.

Define the matrix:

$$B_N(\widehat{\theta}) = \frac{F_N^T(\widehat{\theta}) I(\sigma^2) F_N(\widehat{\theta})}{N}.$$

Let b_{kl}, for $k, l = 1, 2, \ldots, m$, denote the elements of the matrix $B_N(\widehat{\theta})$. The eigenvalues of the matrix

$$\frac{F_N^T(\widehat{\theta}) F_N(\widehat{\theta})}{N}$$

are denoted by

$$0 < \widehat{\lambda}_1(N) \leq \cdots \leq \widehat{\lambda}_m(N).$$

Define the estimated values:

$$\widehat{y}_i = \eta(x_i, \widehat{\theta}), \quad \widehat{I}_{k,l}(\widehat{\theta}) = \delta_{ij} f_{ki}(\widehat{\theta}) f_{lj}(\widehat{\theta}).$$

Let

$$\widehat{y} = (\widehat{y}_1, \widehat{y}_2, \ldots, \widehat{y}_N)^T$$

denote the vector of estimated responses, and define

$$\widehat{b}_{kl} = \frac{(y - \widehat{y})^T \widehat{I}_{kl}(\widehat{\theta})(y - \widehat{y})}{N}.$$

The matrix $\widehat{B}$ is given by

$$\widehat{B} = \|\widehat{b}\|_{kl}, \quad k, l = 1, 2, \ldots, m.$$

The estimated covariance matrix is defined as

$$\widehat{C}_N = \left(\frac{F_N^T(\widehat{\theta}) F(\widehat{\theta})}{N} \right)^{-1} B_N(\widehat{\theta}) \left(\frac{F_N^T(\widehat{\theta}) F(\widehat{\theta})}{N} \right)^{-1}.$$

The elements of $\widehat{C}_N$ are denoted by

$$\widehat{c}_{kl}, \quad k, l = 1, 2, \ldots, m.$$

Theorem 3.15 *If for* $N \to \infty$

$$\frac{1}{N\widehat{\lambda}_N^1} \to 0 \text{ and } \frac{1}{N\widehat{\lambda}_N^1} \sum_{i \neq j} (|r_{ij}|) \to 0 \tag{3.62}$$

$$\frac{1}{N^2}\sum E(\varepsilon_{i_1}\varepsilon_{i_2}\varepsilon_{i_3}\varepsilon_{i_4} - r_{i_1}, r_{i_2}r_{i_3}r_{i_4}) \to 0 \tag{3.63}$$

Then

$$E(\widehat{c}_{kl} - c_{kl}) \to 0, (\widehat{c}_{kl} - c_{kl}) \xrightarrow{P} \ for\ N \to \infty$$

Proof To establish the convergence $E(\hat{c}_{kl} - c_{kl}) \to 0$ and $\hat{c}_{kl} - c_{kl} \xrightarrow{P} 0$ as $N \to \infty$, Consider the expectation $E(\hat{c}_{kl} - c_{kl})$. By the linearity of expectation,

$$E(\hat{c}_{kl} - c_{kl}) = E(\hat{c}_{kl}) - c_{kl}.$$

The bias term $E(\hat{c}_{kl}) - c_{kl}$ can be expressed as a function of the correlations r_{ij} and higher-order moments. From the first condition in Eq. (3.62),

$$\frac{1}{N\hat{\lambda}_1^N} \to 0 \quad \text{as} \quad N \to \infty,$$

implying $N\hat{\lambda}_1^N \to \infty$. The second condition in Eq. (3.62),

$$\frac{1}{N\hat{\lambda}_1^N}\sum_{i \neq j} |r_{ij}| \to 0,$$

ensures that the cumulative effect of off-diagonal correlations diminishes. Consequently, the dominant contribution to the bias arises from terms involving $\hat{\lambda}_1^N$, which are suppressed by the factor $\frac{1}{N\hat{\lambda}_1^N} \to 0$. Thus, $E(\hat{c}_{kl}) - c_{kl} \to 0$.

To show $\hat{c}_{kl} - c_{kl} \to 0$ in probability, it suffices to prove $\text{Var}(\hat{c}_{kl}) \to 0$. The variance is given by:

$$\text{Var}(\hat{c}_{kl}) = E\left[(\hat{c}_{kl} - E\hat{c}_{kl})^2\right].$$

Expanding this expression involves fourth-order moments of the form $E(\varepsilon_{i_1}\varepsilon_{i_2}\varepsilon_{i_3}\varepsilon_{i_4})$. By condition (3.63),

$$\frac{1}{N^2}\sum \varepsilon_{i_1}\varepsilon_{i_2}\varepsilon_{i_3}\varepsilon_{i_4} - r_{i_1}r_{i_2}r_{i_3}r_{i_4} \to 0,$$

which implies that the fourth-order cumulants vanish asymptotically. Consequently, the variance decomposes into terms dominated by $\frac{1}{N^2}\sum |r_{ij}|$, controlled by (3.62), and residual terms vanishing due to (3.63). Hence, $\text{Var}(\hat{c}_{kl}) \to 0$.

By Chebyshev's inequality, for any $\varepsilon > 0$,

$$P\left(|\hat{c}_{kl} - c_{kl}| \geq \varepsilon\right) \leq \frac{\text{Var}(\hat{c}_{kl}) + (E(\hat{c}_{kl}) - c_{kl})^2}{\varepsilon^2} \to 0. \tag{3.64}$$

Therefore, $\hat{c}_{kl} - c_{kl} \xrightarrow{P} 0$ as $N \to \infty$. □

Remark 3.7 When the errors ε_i in a nonlinear regression model are uncorrelated, meaning $r_{ij} = 0$ for all i, j, the estimation process becomes significantly simplified. In this case, the covariance matrix of the errors reduces to a diagonal form, eliminating the need for adjustments that account for error correlations. If, in addition, the errors are identically distributed, the problem further simplifies to the well-known i.i.d. assumption, where standard least squares estimation remains optimal.

The impact of this assumption extends to the covariance matrix of parameter estimates. In the general case where errors exhibit correlation, the estimated covariance structure contains additional terms that must be accounted for. However, under the assumption $r_{ij} = 0$, these terms vanish, and the covariance matrix of the estimated parameters takes a simpler form, similar to that in the classical linear regression setting. Asymptotically, the parameter estimator remains $\sqrt{N}$-consistent and follows a normal distribution,

$$\sqrt{N}(\widehat{\theta} - \theta^*) \xrightarrow{d} N(0, \Sigma(\theta^*)).$$

However, when correlations exist, additional corrections are required to avoid biased variance estimation.

3.3.3 *Confidence Interval in Nonlinear Prototype*

For the construction of a confidence band for the unknown function $f(\theta, x)$, the same approach as that used for linear regression models is adopted. Various methods for constructing confidence intervals in regression models have been proposed [9, 11, 54, 59]. The distinguishing features of the approach suggested in this book, in comparison with other existing methods, are as follows:

(1) The number of unknown parameters increases as the number of observations grows.
(2) The variances of the observational errors are different and unknown.
(3) At each observation point, there is only one response, which does not allow for direct estimation of the variances of observations.

It should be noted that the calculation of the elements of the design matrix and Fisher's information matrix involves routine procedures, particularly when obtaining numerical results. The use of modern computational tools enables the numerical evaluation of various regression models.

A confidence interval for the function $f(\theta, x)$ at each point x_i is now constructed in a manner analogous to that in Sect. 3.1 for linear regression models. As established in previous sections, under certain conditions, the least squares estimator ($l.s.e.$) with the $s.v.$ is consistent for the model. Consequently, from Theorem 3.14, it follows that the random variable $\sqrt{N}(\widehat{\theta} - \theta)$ asymptotically follows a Gaussian (normal) distribution as $N \to \infty$. This leads to the consideration of the following quadratic form:

$$\frac{1}{N}(\widehat{\theta}-\theta)^T \bar{C}_N^{-1}(\widehat{\theta}-\theta). \tag{3.65}$$

In Eq. (3.65), the matrix C_N^{-1} is unknown. Instead of C_N^{-1}, the matrix $\widehat{C}_N^{-1}$ is used, whose elements constitute an asymptotically unbiased and consistent estimator of the elements of C_N^{-1}. The numerical calculation of the elements of $\widehat{C}_N^{-1}$ presents certain challenges because $\det C_N^{-1}$ approaches zero as $N \to \infty$. However, modern computational software facilitates the numerical evaluation of the elements of $\widehat{C}_N^{-1}$.

As stated in Theorem 3.14, the vector $\sqrt{N}(\widehat{\theta}-\theta)$ asymptotically follows a Gaussian distribution. Consequently, the quadratic form (3.65) for $N \to \infty$ asymptotically follows a chi-square distribution [9] with m degrees of freedom. That is, with probability $\gamma > 0$, the following holds:

$$(\widehat{\theta}-\theta)^T \widehat{C}_N^{-1}(\widehat{\theta}-\theta) \leq \frac{\chi_\gamma^2(m)}{N},$$

where $\chi_\gamma^2(m)$ denotes the quantile of level $\gamma > 0$ for the chi-square distribution with m degrees of freedom.

In the ratio (3.65), the matrix C_N^{-1} is unknown. Since asymptotically unbiased and consistent estimators of the elements of the covariance matrix C_N have been constructed, the estimator matrix $\widehat{C}_N^{-1}$ is used in place of C_N^{-1}, with its elements defined in Theorem 3.13. Solving (3.65) with respect to θ, the confidence ellipsoid is obtained:

$$\varepsilon_\gamma(\theta) = \left\{\theta : (\widehat{\theta}-\theta)^T \widehat{C}_N^{-1}(\widehat{\theta}-\theta) \leq \frac{\chi_\gamma^2(m)}{N}\right\}.$$

It should be noted that, in principle, it is not necessary for the elements of the matrix $\widehat{C}_N$ to have limits as $N \to \infty$. The key observation is that the difference between c_{kl} and $\widehat{c}_{kl}$ (elements of the covariance matrix) converges to zero in probability, meaning that the difference between the two vectors approaches zero.

To determine the lower and upper bounds of the confidence band, it is necessary to evaluate $\inf_{\varepsilon_\gamma(\theta)} f(x,\theta)$ and $\sup_{\varepsilon_\gamma(\theta)} f(x,\theta)$, which represent the respective bounds.

For convenience, the confidence ellipsoid $\varepsilon_\gamma(\theta)$ can be expressed in the following form:

$$\sum_{i=1}^{m} \frac{(\widehat{\theta}_i-\theta_i)^2}{x_\gamma^2(m)\cdot\lambda_{m-i}} = \frac{1}{\sqrt{N}},$$

where $x_\gamma^2(m)$ denotes the quantile of level $\gamma > 0$ for the chi-square distribution with m degrees of freedom, and $\widehat{\lambda}_1, \widehat{\lambda}_2, \ldots, \widehat{\lambda}_m$ are the eigenvalues of the matrix $\widehat{C}_N^{-1}$.

Next, the minimum and maximum of the function $f(x,\theta)$ must be determined, where θ belongs to the confidence ellipsoid. This is achieved using the Lagrange method of undetermined multipliers [26]. The corresponding Lagrange function takes the form:

$$Z = \sum_{i=1}^{m} \varphi_i(x) \cdot \theta_i + \lambda^* \cdot \sum_{i=1}^{m} \left[\frac{(\widehat{\theta}_i - \theta_i)^2}{X_\gamma^2(m) \cdot \lambda_{m-i}} - \frac{1}{\sqrt{N}} \right],$$

where λ^* is the Lagrange multiplier, and θ_i satisfies the following equation:

$$\frac{d}{d\theta_i} Z = 0, \quad i = 1, 2, \ldots, m.$$

These points correspond to the required solutions.

3.3.4 Numerical Illustration for Nonlinear Prototype

Example 3.4 Consider the nonlinear regression model:

$$y_i = \theta_1 x_i^{\theta_2} + \theta_3 + \varepsilon_i, \quad i = 1, 2, \ldots, N,$$

where $\theta^* = (\theta_0^*, \theta_1^*, \theta_2^*)^T$ denotes the true parameter vector to be estimated.

Obviously, $\eta(x, \theta)$ and its first—and second—order partial derivatives concerning θ are continuous and bounded on the compact parameter space Θ. In this example, the error is autocorrelated and satisfies AR (1), the correlation coefficient is 0.5, and the standard deviation is 0.5. The error term ε_i satisfies:

$$E[\varepsilon_i] = 0, \quad E[\varepsilon_i \varepsilon_j] = r_{ij}, \quad r_{ii} = \sigma_i^2, \quad 0 < \sigma_\star^2 \leq \sigma_i^2 \leq \sigma_1^2.$$

In this numerical example, let the true parameter values be

$$\theta^* = (\theta_1^*, \theta_2^*, \theta_3^*)^T = (2.0,\ 1.5,\ 5)^T,$$

and the sample size $N = 10^7$.

Define the gradient elements of $\eta(x, \theta)$ with respect to the parameter θ as:

$$f_{ij}(\theta) = \frac{\partial \eta(x_j, \theta)}{\partial \theta_i}, \quad i = 1, 2, 3, \quad j = 1, 2, \ldots, 100.$$

Explicitly, the gradient (Jacobian) matrix $F_N(\theta)$ has the form:

$$F_N(\theta) = \begin{pmatrix} x_1^{\alpha_2} & \alpha_1 \ln(x_1) x_1^{\alpha_2} & 1 \\ x_2^{\alpha_2} & \alpha_1 \ln(x_2) x_2^{\alpha_2} & 1 \\ \vdots & \vdots & \vdots \\ x_N^{\alpha_2} & \alpha_1 \ln(x_N) x_N^{\alpha_2} & 1 \end{pmatrix}$$

The eigenvalues of the matrix

$$\frac{F_N^T(\theta)F_N(\theta)}{N} \in \mathbb{R}^{N\times 3}$$

are denoted by:

$$0 < \lambda_1^{(N)}(\theta) \le \lambda_2^{(N)}(\theta) \le \lambda_3^{(N)}(\theta).$$

According to the Gauss–Newton iterative method, the iterative formula for obtaining the least squares estimator (l.s.e.) $\widehat{\theta}$ is:

$$\theta(s+1) = \theta(s) + \left[F_N^T(\theta(s))F_N(\theta(s))\right]^{-1} F_N^T(\theta(s))\left[y - \eta(x, \theta(s))\right].$$

To analyze the convergence of the iterative process, let $\theta(0) \in B(r)$, where $B(r)$ is the ball with radius r centered at θ^*. The iteration procedure can be equivalently expressed as:

$$\theta_N(s+1) = U(\theta_N(s)) = \theta_N(s) + A_N(\theta_N(s)) \cdot \delta_N(\theta_N(s)),$$

where

$$A_N(\theta_N(s)) = \left[\frac{F_N^T(\theta(s))F_N(\theta(s))}{N}\right]^{-1} \frac{F_N^T(\theta(s))}{N},$$

and

$$\delta_N(\theta(s)) = y - \eta(x, \theta(s)), \quad \delta^* = y - \eta(x, \theta^*) = \varepsilon.$$

Thus, when $\theta(s) \to \theta^*$, the least squares estimator of the nonlinear regression model parameters is obtained.

This study verifies the validity of eight theorems in nonlinear regression models through numerical experiments, divided into two stages:

Firstly, focuses on the convergence of the iterative algorithm. By designing numerical examples, the following aspects are verified:

(1) Whether the condition $\zeta_{N,r}^{(n)}(\theta) \xrightarrow{P} 0$ in Theorem 3.7 holds (e.g., by calculating the decay rate of $\lambda_1(\theta)$ and the bounds of error correlation $|r_{ij}|$).
(2) The contraction property $\tau_N(r) \xrightarrow{P} 0$ in Theorem 3.8, as well as the asymptotic control of $\sqrt{N}\rho^*$ in Theorem 3.9.
(3) The stability condition $\tau_N(r) + \frac{\|\rho^*\|}{r} < 1$ in Theorem 3.10, ensuring the convergence of the iterative sequence to a unique fixed point $\hat{\theta}_N$.

Secondly, builds on the results of the previous stage to further verify:

(1) The error bound $\frac{(\tau_N(r))^s}{1-\tau_N(r)}$ in Theorem 3.11, proving the exponential convergence rate of the iterative algorithm.
(2) The $\sqrt{N}$-consistency in Theorem 3.12, by simulating the boundedness of the distribution of $\sqrt{N}(\hat{\theta}_N - \theta^*)$.
(3) The asymptotic normality in Theorems 3.13 and 3.14, using Q-Q plots and the Kolmogorov–Smirnov (K-S) test to validate distribution matching.

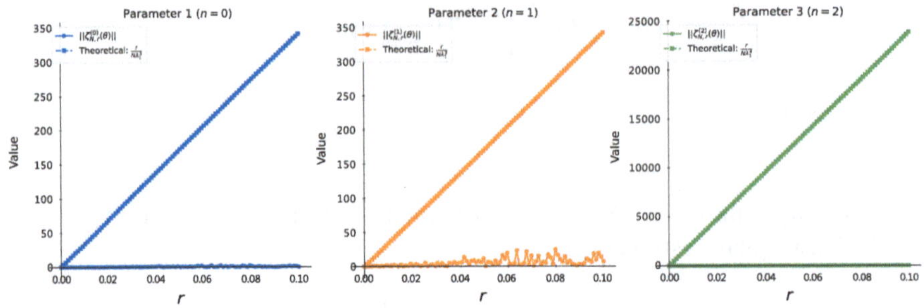

Fig. 3.20 Numerical verification and theoretical bound comparison of the random variable in nonlinear case

This staged approach ensures that the foundational conditions are progressively verified, satisfying the dependencies of subsequent theorems while avoiding redundant computations and improving verification efficiency.

For Theorem 3.7, investigate the convergence of the random variable $\zeta_{N,r}^{(n)}(\theta)$ as the standard deviation of the error term, denoted by r, approaches zero, thus verifying the probability convergence result stated in Theorem 3.7:

$$\zeta_{N,r}^{(n)}(\theta) \xrightarrow{P} 0, \quad \text{as} \quad r \to 0.$$

Consider the above specific nonlinear regression model, to examine numerical convergence, the derivative of the matrix $A_N(\theta)$ with respect to the parameter θ_n is approximated using the finite difference method:

$$\frac{\partial A_N(\theta)}{\partial \theta_n} \approx \frac{A_N(\theta + d\,e_n) - A_N(\theta - d\,e_n)}{2d}$$

where e_n is the unit vector, $d = 10^{-5}$.

The key quantity calculated in this numerical experiment is:

$$\zeta_{N,r}^{(n)}(\theta) = \frac{\partial A_N(\theta)}{\partial \theta_n}\varepsilon, \quad \text{with} \quad \varepsilon \sim N(0, r^2 I_N),$$

where the error vector ε is generated as a Gaussian random vector with standard deviation r. We then analyze the numerical behavior of the norm $\|\zeta_{N,r}^{(n)}(\theta)\|$.

This procedure is conducted separately for each parameter ($n = 0, 1, 2$), by comparing the computed norm $\|\zeta_{N,r}^{(n)}(\theta)\|$ against its theoretical bound as $r \to 0$, it can be numerically verified that the probability convergence property stated in condition (3.38) of Theorem 3.7.

As shown in Fig. 3.20, The numerical experiment clearly demonstrates that the random variable $\|\zeta_{N,r}^{(n)}(\theta)\|$ decreases significantly as the standard deviation of the error term r approaches zero. The numerical results closely align with the theo-

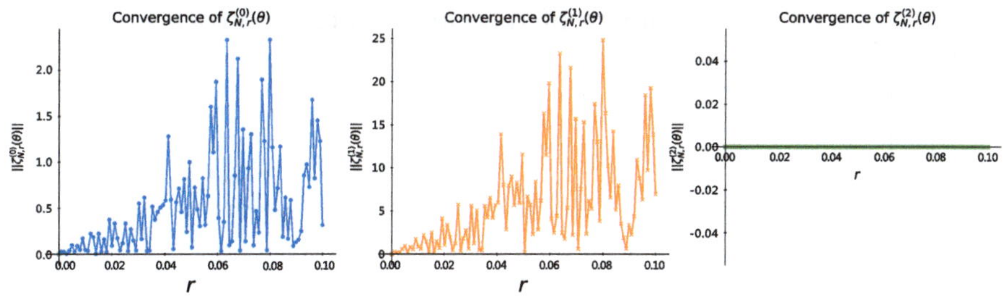

Fig. 3.21 Detailed analysis of the convergence trend of the random variable in nonlinear case

retical convergence rates predicted by Theorem 3.7, thus validating its probability convergence property.

In order to better show the convergence process, the changes under the three parameters are also drawn (As shown in Fig. 3.21). For parameters 1(n = 0) and 2(n = 1), the convergence process shows a fluctuating state, especially for parameter 2, where the initial value is greater than 20. However, when r tends to 0, the fluctuation shows a decreasing trend and finally converges to 0. For parameter 3(n = 2), because it is a constant parameter, it has good properties.

Next, validate the probability convergence property of the random variable $\tau_N(r)$ proposed in Theorem 3.8, specifically:

$$\tau_N(r) \xrightarrow{P} 0, \quad \text{as} \quad r \to 0.$$

Introducing the error term $\varepsilon_i \sim N(0, r^2)$, calculate of the Matrix L_n and the Random Variable $\tau_N(r)$, the derivatives of the updating function $U_N(\theta)$ with respect to each parameter θ_n are approximated by the finite difference method:

$$L_n(\theta) = \frac{\partial U_N(\theta)}{\partial \theta_n} \approx \frac{U_N(\theta + de_n) - U_N(\theta - de_n)}{2d},$$

where e_n is the unit vector, and $d = 10^{-6}$.

Then, the random variable $\tau_N(r)$ is defined as:

$$\tau_N(r) = \max_{n=1,2,3} \|L_n(\theta)\|.$$

To numerically validate the convergence property of the random variable $\tau_N(r)$, the following steps are performed: firstly, construct the perturbed parameter vector, $\theta = \theta^* + \delta$, with perturbation$\delta = (r/\sqrt{3},\ r/\sqrt{3},\ r/\sqrt{3})^T, \|\delta\| = r$.

For a given standard deviation r, generate noisy data y and repeat the process 50 times, calculating the random variable $\tau_N(r)$ each time and taking the average:

$$\tau_N(r) = \frac{1}{50} \sum_{i=1}^{50} \tau_N^{(i)}(r).$$

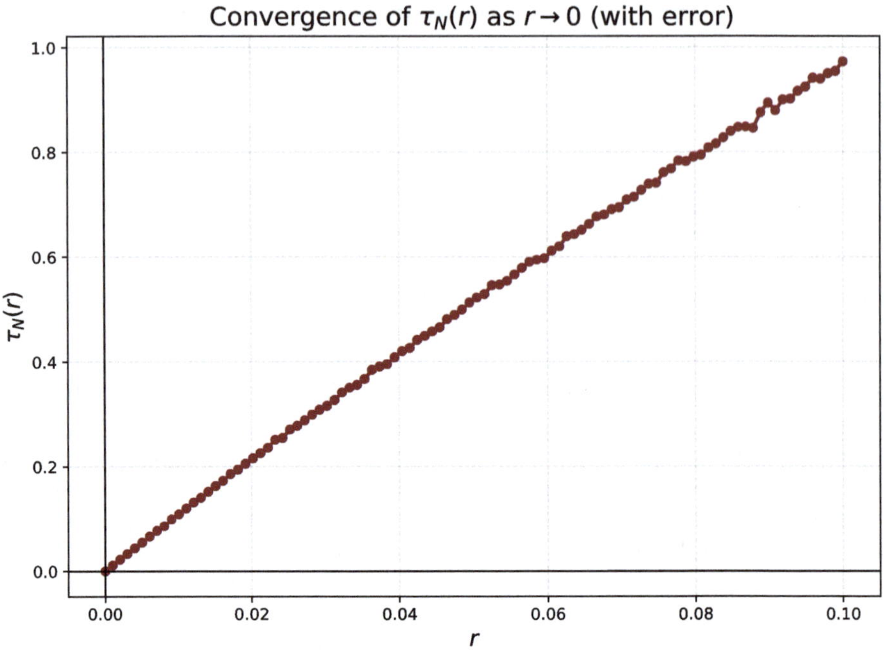

Fig. 3.22 Convergence plot of the random variable as error standard deviation varies in nonlinear case

Vary the standard deviation r within the range $[10^{-5}, 10^{-1}]$ to examine the convergence behavior of $\tau_N(r)$ as r approaches zero.

Finally, it can be illustrated that the convergence trend of the random variable $\tau_N(r)$ as $r \to 0$ through graphical representation. Figure 3.22 clearly demonstrates the convergence of this numerical quantity toward zero, thereby numerically verifying the theoretical prediction of Theorem 3.8.

Under the assumptions of Theorem 3.9, define the error vector as:

$$\rho_N(\theta) = U_N(\theta) - \theta, \quad \rho^* = \rho(\theta^*).$$

Given the boundedness condition specified by:

$$\frac{1}{\lambda_1(\theta)} + \frac{1}{N\lambda_1^2(\theta)} \sum_{i \neq j} |r_{ij}| \leq A,$$

where $\lambda_1(\theta)$ is the smallest eigenvalue of the correlation matrix, r_{ij} are elements of this correlation matrix, and A is a constant, it can be seen from the numerical experiment of Theorem 3.7 that this condition holds. The convergence result is as follows:

$$\lim_{N\to\infty} \lim_{k\to\infty} P\{\sqrt{N} \cdot \rho^* > k\} = 0.$$

Due to the boundedness condition ensuring the uniform boundedness of the matrix sequence $A_N(\theta)$, the correction term in the parameter estimation iterative scheme $A_N(\theta)\delta_N(\theta)$ converges to zero as the sample size N grows large. Specifically, invoking the law of large numbers and central limit theorem, under the provided conditions, the mean squared error of residuals converges at a rate proportional to $1/\sqrt{N}$, leading to even faster convergence for the estimation error of the parameters. By applying Markov's inequality, for any given $k > 0$, we have

$$P\{\sqrt{N}\rho^* > k\} \leq \frac{\mathbb{E}[\sqrt{N}\rho^*]}{k} \rightarrow 0, \quad \text{as} \quad N \rightarrow \infty.$$

Hence, taking the limit first with respect to $k \rightarrow \infty$ and subsequently $N \rightarrow \infty$, it can be obtained that:

$$\lim_{N\rightarrow\infty} \lim_{k\rightarrow\infty} P\{\sqrt{N}\rho^* > k\} = 0.$$

To numerically verify Theorem 3.9, consider the same regression model with the power function form, in the numerical experiment, the sample size N gradually increases from small values (starting from 50) to large values (up to 10,000,050). For each selected sample size N, perform 50 independent replications. Generate explanatory variable sequences $x_i \in [1, 10]$, $i = 1, \ldots, N$, and subsequently generate the dependent variable sequences y_i based on the true parameter values and the AR(1) noise structure specified above.

Then, initialize parameter estimates with slight random perturbations around true parameter values. Subsequently, based on the iterative formulas, the Jacobian matrix $F_N(\theta)$, the matrix $A_N(\theta)$, and the residual $\delta_N(\theta)$ are computed to obtain stable parameter estimates. The norm of the parameter estimation error is then calculated to estimate the corresponding probabilities. Numerical experiments demonstrate that as the sample size N increases, the estimation error norm converges to zero at a rate substantially faster than $1/\sqrt{N}$, causing the probability to rapidly approach zero. Particularly for large-sample sizes, this probability approaches zero for any finite or sufficiently large value of k. The results are shown as Fig. 3.23.

Figure 3.23a shows the situation that ($k \in [0, 10]$): it is clearly observed that as the threshold k increases, the probability $P(\sqrt{N} \cdot \rho^* > k)$ decreases rapidly, with the rate of decline becoming faster as the sample size N grows. In particular, when $k > 4$, the probabilities for most sample sizes are already close to zero, indicating that the scaled estimation error $\sqrt{N} \cdot \rho^*$ is highly concentrated in a small region. This figure effectively illustrates the core phenomenon described in the theorem: as the sample size increases, the estimation error norm ρ^* converges rapidly to zero, and the tail probability decays sharply.

Figure 3.23b shows the situation that ($k \in [10, 20]$): it is further verifies the tail behavior of the probability convergence. Within this larger range of k, the probabilities for all sample sizes remain at very low levels, with a maximum of around 0.05 and the vast majority approaching zero. Especially for large-sample sizes (e.g.,

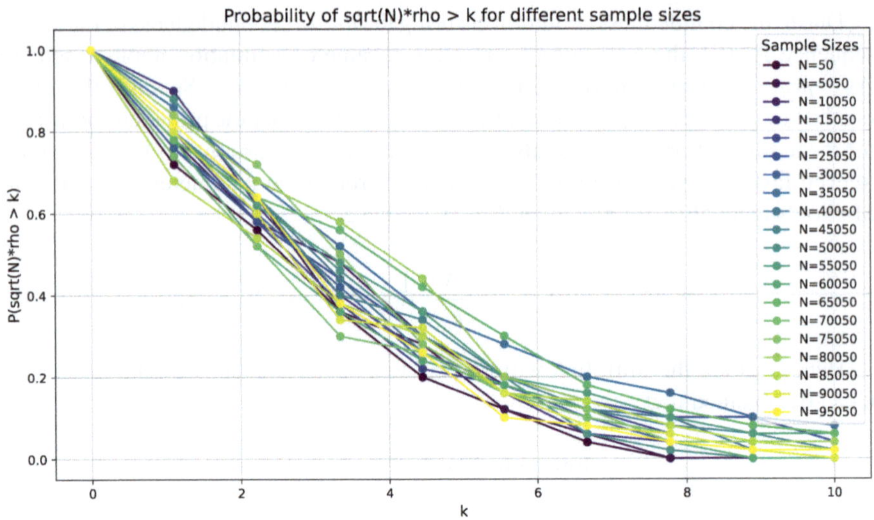

(a) Situation1

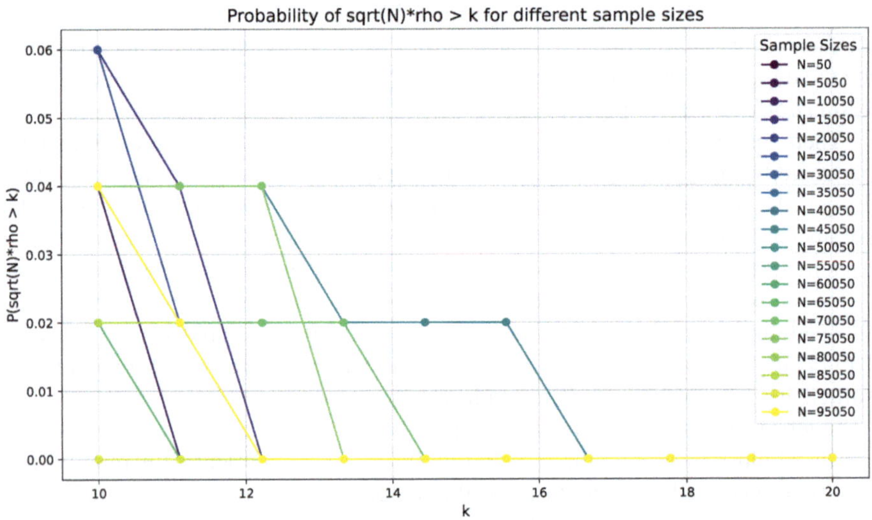

(b) Situation2

Fig. 3.23 Convergence behavior for small to moderate thresholds in nonlinear case

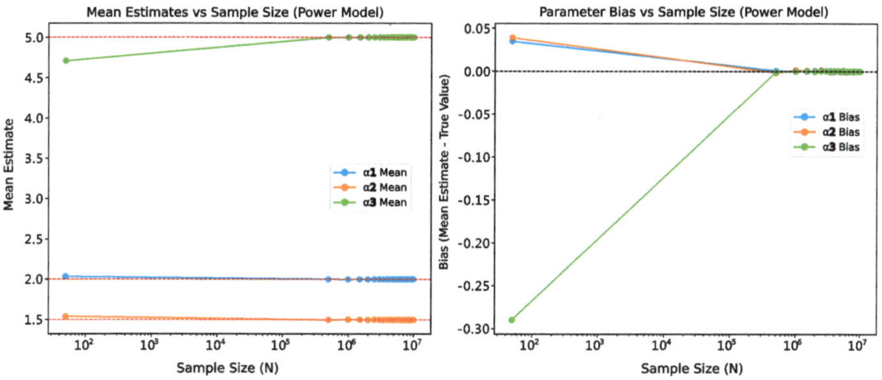

Fig. 3.24 Estimator consistency and bias convergence in nonlinear case

$N \geq 50050$), the probability is virtually zero regardless of the value of k. This confirms that even under stringent accuracy requirements, the parameter estimates exhibit strong consistency. The numerical results strongly support the theoretical conclusion of the double-limit probability convergence stated in Theorem 3.9.

Table 3.4 presents the bias estimates for the parameters α_1, α_2, and α_3 of the power function model under varying sample sizes, ranging from 50 to 10,000,050. The results show that as the sample size increases, the bias of all three parameter estimates decreases rapidly and approaches zero, indicating strong consistency of the estimators. When the sample size is as small as 50, the bias is relatively large, especially for α, with a bias of 0.035059. However, once the sample size reaches the million level, the biases for all parameters stabilize below $1e^{-4}$, demonstrating the significant impact of sample size on the estimation accuracy in nonlinear models.

In the left panel in Fig. 3.24, as the sample size increases from relatively small to large, the three curves (representing the average estimates of α_1, α_2, and α_3) gradually converge toward its true value, indicating a marked improvement in the accuracy of the power function model under larger samples. In the right panel, the corresponding parameter biases are plotted against sample size, and these curves approach zero rapidly as the sample size grows, confirming the asymptotic unbiasedness and consistency of this estimation method. Moreover, while the model may exhibit some degree of bias when the sample size is small, such bias is effectively mitigated with increasing sample size.

Figure 3.25 illustrates the convergence of parameter estimates and the trend of bias variation for a nonlinear model. From the estimation performance of different parameters, it is observed that as the sample size increases, the estimated values of all parameters exhibit clear asymptotic convergence and stability. Specifically, the estimate of Parameter 1 (the coefficient of the exponential term) shows some bias with small-sample sizes, but this bias gradually decreases and approaches zero as the sample size grows. The estimation of Parameter 2 (the stretch coefficient inside the exponential function) remains stable throughout, demonstrating low sensitivity to sample size with minimal bias even at smaller sample sizes. In contrast, the estimate

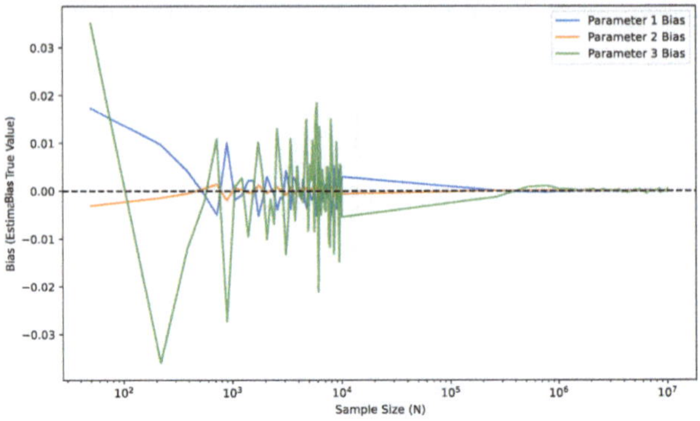

(a) Parameter Bias Convergence

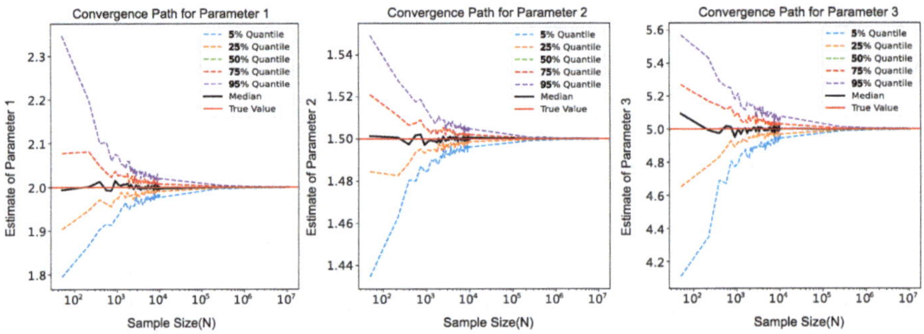

(b) Convergence Path for α_i, $i = 0, \cdots, 2$

Fig. 3.25 Visualization of parameter bias convergence in nonlinear case

of Parameter 3 (the power function coefficient) exhibits significant fluctuations, with notable volatility and larger bias when the sample size is less than 1000. However, as the sample size further increases, the bias gradually diminishes and ultimately converges to zero, indicating pronounced asymptotic consistency.

From these scatter plots of residuals versus fitted values in Fig. 3.26, we observe that as the sample size increases from 10 to 10,000,000, the vertical spread of the residuals remains largely within [–3, 3], and no obvious "funnel" shape or other systematic pattern emerges along the horizontal axis. For smaller samples (e.g., $N = 10$ or $N = 100$), there are fewer points, and they appear more dispersed mainly due to the limited sample size. However, when the sample size becomes very large (reaching millions or tens of millions), the residual cloud becomes extremely dense, and its overall distribution is randomly scattered around the zero line, without widening or narrowing as fitted values grow. These observations suggest that the error variance remains essentially constant, showing no clear signs of heteroscedasticity, and confirm that under the given data generation process and model assumptions, the residual distribution from the least squares fit is relatively stable and consistent.

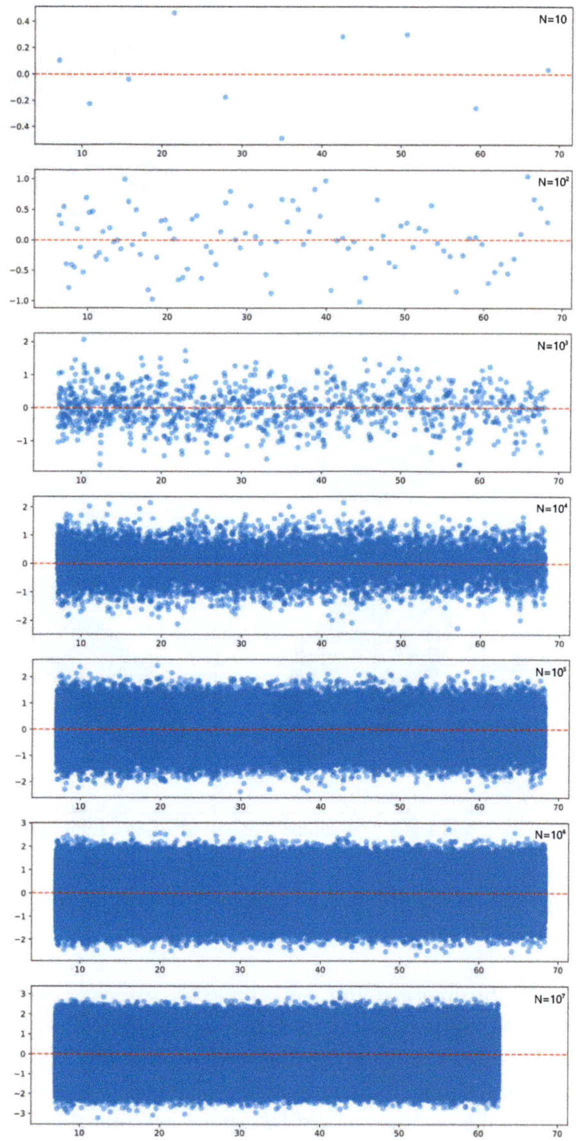

Fig. 3.26 Graphical test for heteroscedasticity in nonlinear case

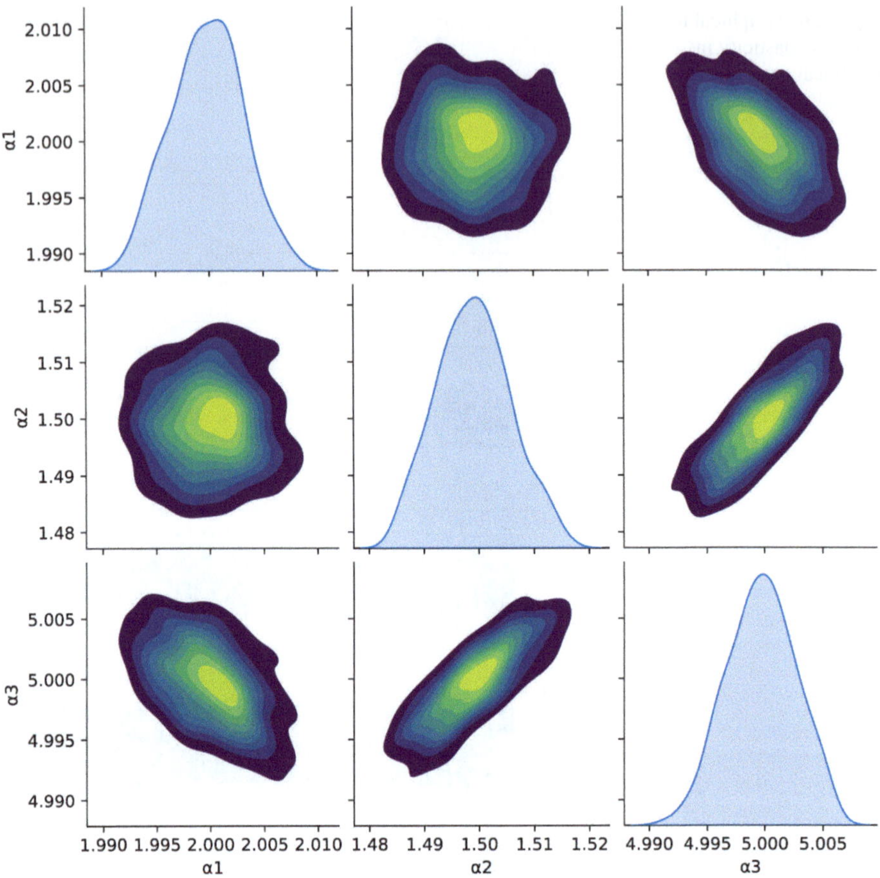

Fig. 3.27 Heatmap of marginal and joint parameter density distributions in nonlinear case

On the diagonal, Fig. 3.27 shows the marginal distributions of α_1, α_2, and α_3, while the off-diagonal heatmaps depict their pairwise joint distributions. Each parameter's distribution curve on the diagonal is relatively concentrated and slightly resembles a normal shape, indicating a high probability mass within a certain range. In the off-diagonal two-dimensional heatmaps, warmer (brighter) colors correspond to higher density, providing an intuitive view of potential correlations among parameters, notably a positive correlation between α_2 and α_3. Overall, the parameter estimates cluster around their respective peaks, suggesting good stability and identifiability of the model fit. At the same time, the correlations revealed by the heatmaps indicate that potential interactions among parameters should be considered when applying and interpreting the model.

During the computation, 20 evenly spaced sample sizes were first selected within the range from 50 to 1,000,000, resulting in a bad Frobenius norm error. Subsequently,

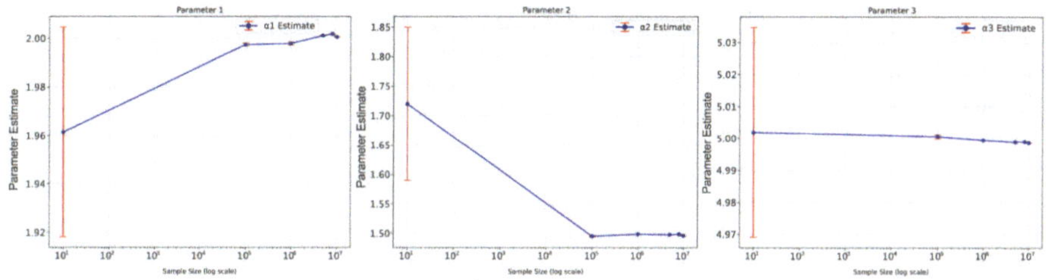

Fig. 3.28 Confidence interval in nonlinear case

50 evenly spaced sample sizes were chosen, and under the classical OLS setting, the error norm was only 1.50×10^{-7}. The resulting covariance matrix is:

$$\begin{bmatrix} 0.0000 & -0.0000 & -0.0000 \\ -0.0000 & 0.0000 & 0.0000 \\ -0.0000 & 0.0000 & -0.0000 \end{bmatrix}$$

And Fig. 3.29 shows the visualization result of the covariance estimator. Figure 3.29 demonstrates that the classical OLS covariance estimator corrected by left and right multiplication with the inverse matrix-exhibits excellent convergence behavior. As the sample size N increases, the Frobenius norm error decreases rapidly and approaches zero, indicating high numerical accuracy and consistency with the theoretical covariance formula. This highlights the estimator's asymptotic unbiasedness and stability in large samples.

The simulation results in Fig. 3.28 clearly demonstrate the asymptotic properties of nonlinear least squares estimators under a power function model with AR(1) errors. For all three parameters (α_1, α_2 and α_3), the estimates show substantial variability and wide confidence intervals when the sample size is small ($N = 10$), indicating the difficulty of reliable parameter estimation under limited data and serially correlated noise.

As the sample size increases, especially when $N \geq 10^5$, the estimates rapidly converge toward the true parameter values, and the 95% confidence intervals shrink significantly. This trend illustrates the consistency and efficiency of the nonlinear least squares estimator even in the presence of autocorrelated errors.

Specifically, α_3 shows the fastest convergence due to its strong contribution to the model structure. In contrast, α_2 converges slightly slower, which is reasonable as it determines the nonlinearity degree and is more sensitive to random disturbances when sample size is limited. Overall, the results highlight that with sufficiently large samples, reliable and precise estimation can be achieved even when data exhibits serial correlation.

Figure 3.29 illustrates how the Frobenius norm error between the estimated and true covariance matrices changes with increasing sample size N under a power function regression model with AR(1) correlated errors. As the sample size N increases, the Frobenius norm error between the estimated covariance matrix and the classical OLS covariance rapidly decreases and approaches nearly zero when $N \approx 10^5$, indi-

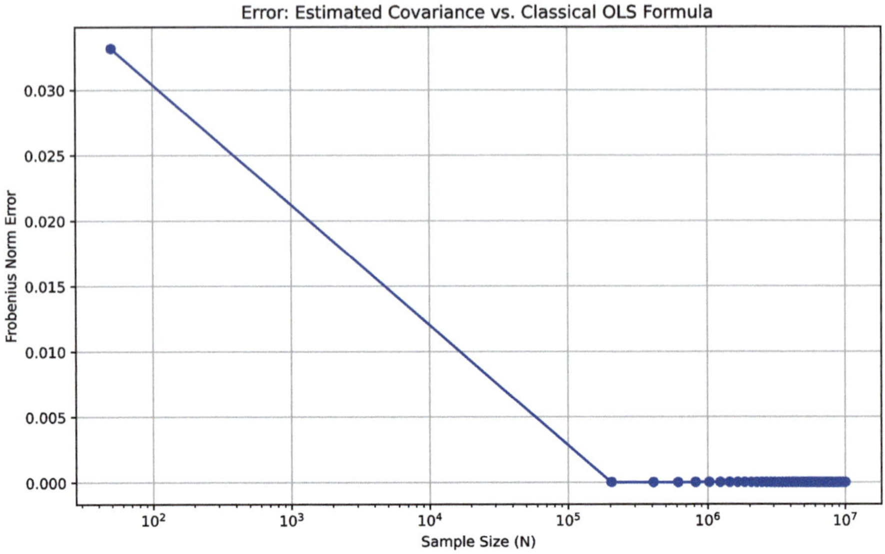

Fig. 3.29 Classical OLS covariance estimator error in nonlinear case

Table 3.4 Bias convergence under large-sample conditions in nonlinear case

Sample size	α_1	α_2	α_3
50	0.01730	−0.00312	0.03506
526365	−0.00027	0.00006	0.00073
1052681	−0.00004	0.00001	−0.00053
1578997	0.00008	−0.00001	−0.00037
2105313	−0.00009	0.00002	0.00014
2631628	−0.00015	0.00003	0.00035
3157944	−0.00001	0.00000	0.00012
3684260	−0.00002	0.00000	0.00011
4210576	0.00010	−0.00002	−0.00016
4736892	0.00012	−0.00002	−0.00022
5263207	0.00000	0.00000	−0.00007
5789523	−0.00006	0.00001	0.00014
6315839	0.00004	−0.00001	−0.00020
6842155	0.00000	0.00000	−0.00006
7368471	0.00006	−0.00001	−0.00010
7894786	0.00005	−0.00001	−0.00022
8421102	0.00012	−0.00002	−0.00038
8947418	−0.00009	0.00002	0.00025
9473734	0.00005	−0.00001	−0.00025
10000050	−0.00011	0.00002	0.00044

cating that the covariance estimators exhibit strong convergence and consistency in large samples (Figs. 3.20, 3.26 and Table 3.4).

References

1. John Aldrich. Fisher and regression. *Statistical Science*, pages 401–417, 2005.
2. Robert Bartels. Re-interpreting r2, regression through the origin, and weighted least squares. *Univ. Sydney Bus. Sch*, 2015.
3. Douglas M Bates and Donald G Watts. *Nonlinear regression analysis and its applications*, volume 2. Wiley New York, 1988.
4. David Birkes and Yadolah Dodge. *Alternative methods of regression*. John Wiley & Sons, 2011.
5. Åke Björck. *Numerical methods for least squares problems*. SIAM, 2024.
6. Matias D Cattaneo, Nicolás Idrobo, and Rocio Titiunik. A practical introduction to regression discontinuity designs: Extensions. *arXiv preprint* arXiv:2301.08958, 2023.
7. Samprit Chatterjee and Jeffrey S Simonoff. *Handbook of regression analysis*. John Wiley & Sons, 2013.
8. Kuang-Fu Cheng and Pi-Erh Lin. Nonparametric estimation of a regression function: Limiting distribution 2. *Australian Journal of Statistics*, 23(2):186–195, 1981.
9. Harold Cramer. Mathematical methods of statistics, princeton univ. *Press, Princeton, NJ*, 27, 1946.
10. Florence Nightingale David. Extension of the markoff theorem on least squares. *Stat. Res. Memoirs*, 2:105–117, 1938.
11. Eugene Demidenko. *Mixed models: theory and applications with R*. John Wiley & Sons, 2013.
12. Lih-Yuan Deng and CFJ Wu. Estimation of variance of the regression estimator. *Journal of the American Statistical Association*, 82(398):568–576, 1987.
13. P Diaconis and DA Freedman. On inconsistent m-estimators. *Ann. Statist*, 10:454–461, 1982.
14. Timo Dimitriadis, Tobias Fissler, and Johanna Ziegel. Characterizing m-estimators. *Biometrika*, 111(1):339–346, 2024.
15. Edgar Dobriban and Yue Sheng. Distributed linear regression by averaging. 2021.
16. Hilmar Drygas. Weak and strong consistency of the least squares estimators in regression models. *Zeitschrift für Wahrscheinlichkeitstheorie und Verwandte Gebiete*, 34(2):119–127, 1976.
17. Hilmar Drygas. Minqe-theory and the estimation of residual variance in regressions analysis. *Statistics*, 19(3):341–347, 1988.
18. Bradley Efron, Trevor Hastie, Iain Johnstone, and Robert Tibshirani. Least angle regression. 2004.
19. Erhard Förster. *Methoden der korrelations-und regressionsanalyse*. Verlag Die Wirtschaft, 1979.
20. Francis Galton. *Typical laws of heredity*. William Clowes and Sons, 1877.
21. Vȋacheslav Leonidovich Girko. G cestimator of spectral function of covariance matrix. *Theory of Probab. Mathem.Statist*, 35:28–31, 1986.
22. Chen Gui-Jing, Tze Leung Lai, and CZ Wei. Convergence systems and strong consistency of least squares estimates in regression models. *Journal of Multivariate Analysis*, 11(3):319–333, 1981.
23. Aleksandr Viktorovich Gurevich. On estimation of regression function without condition of unbiased. *Theory of Probab. Appl*, 33(4):785–788. (in Russian).
24. Asaf Hajiyev and Jiuping Xu. The properties of the least squares estimate in regression models with an increasing number of unknown parameters. PSEN, v.7. issue 3, 2025
25. Asaf H. Hajiyev. Statistical estimators in regression models. *Soviet Acad. Sci., Doklady, Mathem*, 2:265–268, 1990.

26. Asaf H Hajiyev. Statistical estimators in regression models. *Soviet Acad. Sci., Doklady, Mathem*, 2:265–268, 1990.
27. Asaf H Hajiyev. Estimators in regression models with dependent errors. *Proc. Instit. Math. Mech. Baku*, 5:41–55, 1999.
28. Asaf H Hajiyev. The properties of the l and m estimators in regression models. *Third Joint Seminar on Appl. Mathem. Baku State Univers. C- Zanjan Univers. Baku*, 2002.
29. Asaf H Hajiyev. Mathematical theory of regression models with increasing number of unknown parameters. linear models. *Department of Computer Science, Lincoln University*, 5:17, 2003.
30. Asaf H Hajiyev. On problems of regression estimate with increasing number of unknown pararmeters. *Proc. 6 Intern. Conf. Reliability and Statistics in Transp. Communications. Riga*, pages 365–369, 2006.
31. Asaf H Hajiyev. Nonlinear regression models with increasing numbers of unknown parameters. *Russian Acad.Sci.,Mathem*, 79(3):339–341, 2009.
32. Xuming He and Qi-Man Shao. A general bahadur representation of m-estimators and its application to linear regression with nonstochastic designs. *The Annals of Statistics*, 24(6):2608–2630, 1996.
33. Xuming He and Qi-Man Shao. On parameters of increasing dimensions. *Journal of multivariate analysis*, 73(1):120–135, 2000.
34. Peter J Huber. Robust regression: asymptotics, conjectures and monte carlo. *The annals of statistics*, pages 799–821, 1973.
35. PJ Huber and EM Ronchetti. Robust statistics, wiley & sons. *Inc.*, 2009.
36. Jana Jurecková. Consistency of m-estimators in a linear model, generated by non-monotone and discontinuous ψ-functions. *Probability and Mathematical Statistics*, 10:1–10, 1989.
37. Andreĭ Kolmogorov. *Introductory real analysis*.
38. William La Cava, Bogdan Burlacu, Marco Virgolin, Michael Kommenda, Patryk Orzechowski, Fabrício Olivetti de França, Ying Jin, and Jason H Moore. Contemporary symbolic regression methods and their relative performance. *Advances in neural information processing systems*, 2021(DB1):1, 2021.
39. T L_etc Lai, Herbert Robbins, and C Zi Wei. Strong consistency of least squares estimates in multiple regression ii. *Journal of multivariate analysis*, 9(3):343–361, 1979.
40. DV Lindley. Regression and correlation analysis. In *Time series and statistics*, pages 237–243. Springer, 1990.
41. Timothée Mathieu. Concentration study of m-estimators using the influence function. *Electronic Journal of Statistics*, 16(1):3695–3750, 2022.
42. Dastan Maulud and Adnan M Abdulazeez. A review on linear regression comprehensive in machine learning. *Journal of Applied Science and Technology Trends*, 1(2):140–147, 2020.
43. CA McGilchrist. Regression analysis of dependent error models. *Bulletin of the Australian Mathematical Society*, 34(2):199–209, 1986.
44. Daniel Q Naiman. Conservative confidence bands in curvilinear regression. *The Annals of Statistics*, pages 896–906, 1986.
45. Sahand N Negahban, Pradeep Ravikumar, Martin J Wainwright, and Bin Yu. A unified framework for high-dimensional analysis of m-estimators with decomposable regularizers. 2012.
46. AA Novikov. Consistency of least squares estimators in regression models with martingales. *Proc. Steklov Mathematical Institute: Statistical Control of Random Processes. Optimization Software*, 1985.
47. Raymond J O'Connor and Tansy L Wagner. A test of a regression-tree model of species distribution. *The Auk*, 121(2):604–609, 2004.
48. Aydin Öztürk. Weighted least squares estimation of location and scale parameters. *American Journal of Mathematical and Management Sciences*, 7(1-2):113–129, 1987.
49. ES Pearson. Mathematical statistics and data analysis, 1938.
50. Stephen Portnoy. Asymptotic behavior of m estimators of p regression parameters when $\frac{p^2}{n}$ is large; ii. normal approximation. *The Annals of Statistics*, 13(4):1403–1417, 1985.
51. Yurii Vasil'evich Prohorov and Yurii Anatol'evich Rozanov. *Probability theory: basic concepts, limit theorems, random processes*, volume 18. Springer, 1969.

52. Tibshriani R. *High-Dimensional Regression: Ridge Advanced Topics in Statistical Learning*. 2023.
53. Enayetur Raheem, S Ejaz Ahmed, and Shuangzhe Liu. Stein-rule m-estimation in sparse partially linear models. *Japanese Journal of Statistics and Data Science*, 7(1):507–535, 2024.
54. Calyampudi Radhakrishna Rao. *Linear statistical inference and is applications*. John Wiley & Sons, Incorporated, 1965.
55. DA Ratkowsky. Handbook of nonlinear regression models, 1990.
56. William JJ Rey and William JJ Rey. Bootstrap methods, sampling distributions. *Introduction to Robust and Quasi-Robust Statistical Methods*, pages 78–88, 1983.
57. A. N Shiryaev. *Probability*. Springer, 2013.
58. Charles C Thigpen. A sample-size problem in simple linear regression. *The American Statistician*, 41(3):214–215, 1987.
59. Dana L Thomas and David R Thomas. Confidence bands for percentiles in the linear regression model. *Journal of the American Statistical Association*, 81(395):705–708, 1986.
60. Vladimir Vapnik. *Estimation of dependences based on empirical data*. Springer Science & Business Media, 2006.
61. Roman Vershynin. How close is the sample covariance matrix to the actual covariance matrix? *Journal of Theoretical Probability*, 25(3):655–686, 2012.
62. Lan Wang, Yichao Wu, and Runze Li. Quantile regression for analyzing heterogeneity in ultra-high dimension. *Journal of the American Statistical Association*, 107(497):214–222, 2012.
63. Chien-fu Wu. Characterizing the consistent directions of least squares estimates. *The Annals of Statistics*, 8(4):789–801, 1980.
64. Chien-Fu Wu. Asymptotic theory of nonlinear least squares estimation. *The Annals of statistics*, 9(3):501–513, 1981.
65. Chien-Fu Jeff Wu. Jackknife, bootstrap and other resampling methods in regression analysis. *the Annals of Statistics*, 14(4):1261–1295, 1986.
66. Qing Xu and Xiaohua Xuan. Nonlinear regression without iid assumption. *Probability, Uncertainty and Quantitative Risk*, 4(1):8, 2019.
67. Kook Yu.V. and Petunin Yu.I. Estimating of parameters in linear regression with boundness to coefficients. *Ukrainian Mathematical Journal*, 28(4):463–472, 1976. (in Russian).
68. Kook Yu.V. and Petunin Yu.I. The new method of construction of coefficients in linear regression. *Ukrainian Mathematical Journal*, 28(2):237–243, 1976.
69. Sh Zaks. *Theory of statistical inference*. John Wiley & Sons, 1972.
70. Le Zhang, Zenglin Shi, Ming-Ming Cheng, Yun Liu, Jia-Wang Bian, Joey Tianyi Zhou, Guoyan Zheng, and Zeng Zeng. Nonlinear regression via deep negative correlation learning. *IEEE transactions on pattern analysis and machine intelligence*, 43(3):982–998, 2019.

Chapter 4
Variable-Parameter Prototype Regression Models

Die Wissenschaften versuchen nicht zu erklären, sie versuchen kaum zu interpretieren; sie bilden hauptsächlich Modelle. Unter einem Modell versteht man einen mathematischen Konstrukt, welcher, mit dem Zusatz gewisser verbaler Interpretationen, die beobachteten Phänomene beschreibt.

— John von Neumann (1903–1957)

In this chapter, regression models (both linear and nonlinear) with an increasing number of unknown parameters are considered. The investigation of practical problems requires a more precise approximation of the unknown function describing the process. This necessity leads to an increase in the number of unknown parameters used for approximating the unknown function. Consequently, the construction and analysis of regression models where the number of unknown parameters grows as the number of observations increases become essential.

Although certain models of this type have been studied by some researchers [18, 29, 33, 45], their systematic investigation remains limited. A fundamental problem in this context is the selection of an optimal number of unknown parameters that minimizes or maximizes a given criterion [13, 19, 20, 22].

Furthermore, it is assumed that at each observation point, at most one observation is available, and the variances of the random errors associated with the observations are distinct and unknown. These models frequently arise in applications, yet they have not been extensively explored in the literature.

4.1 Linear Prototype Models with Increasing Parameters

In modern data-driven applications, especially those involving high-resolution sensing, financial forecasting, or artificial intelligence, the number of parameters in a regression model may scale with the volume of data collected. Classical linear regression assumes a fixed and low-dimensional parameter vector; however, this assumption becomes inadequate when modeling complex systems with increasingly

A. Hajiyev and J. Xu, *High-Dimensional Regression Analysis and Artificial Intelligence*,
https://doi.org/10.1007/978-981-95-2514-0_4

fine-grained features. As such, it is essential to analyze the behavior of estimators in linear regression models where the number of unknown parameters increases with the sample size. This section investigates the theoretical foundations and convergence properties of such models, ensuring the stability and reliability of estimation under high-dimensional, large-sample regimes.

4.1.1 General Linear Prototype with Increasing Parameters

Assume that the following experimental model is observed:

$$y_i = f(x_i) + \varepsilon_i, \quad i = 1, \cdots, N \tag{4.1}$$

where

$$f(x) = f(x_i, \theta) = \sum_{j=1}^{m} \theta_j \varphi_j(x_i),$$

with the notations:

–$\theta = (\theta_1, \theta_2, \ldots, \theta_m)^T$: the vector of unknown parameters;

–$\varphi_1(x), \varphi_2(x), \ldots, \varphi_m(x)$: a system of linearly independent and bounded functions;

–$\varepsilon = (\varepsilon_1, \varepsilon_2, \ldots, \varepsilon_N)^T$: the vector of observational errors;

–x_i: the observation points;

–N: the number of observations;

–m: the number of unknown parameters;

–T: the transposition operator.

For simplicity, it is assumed that

$$|\varphi_i(x)| \leq 1, \quad i = 1, \ldots, m, \quad x \in [0, 1].$$

Note 4.1 *Parameter-Data Constraints in Large Model Analysis: This refers to the mathematical analysis of large-parameter models, which requires that the amount of input data significantly exceed the number of model parameters. In the era of big data, this condition is typically satisfied for large models; when it is not, data generation methods introduced in Chap. 5 can be employed to compensate.*

A dependency exists between m and N, implying that m varies as a function of N. Specifically, as N increases, m can also increase, but it satisfies the asymptotic condition :

$$\frac{m}{N} \to 0, \quad \text{as} \quad N \to \infty.$$

Following the approach in Chap. 3, the ratio (4.1) can be rewritten in the following vector form:

$$Y = X\theta + \varepsilon,$$

where

–$Y = (y_1, y_2, \ldots, y_N)^T$: the vector of observations (or response vector);
–X: the design matrix;
–$\frac{X^T X}{N}$: Fisher's information matrix.
Define the following additional notations:
–λ_t: the eigenvalues of the matrix $\left(\frac{X^T X}{N}\right)$, satisfying

$$0 < \lambda_1(N) \leq \lambda_2(N) \leq \cdots \leq \lambda_m(N); \tag{4.2}$$

–I_N: the identity matrix of size N;
–$\hat{\theta} = (X^T X)^{-1} X^T Y$: the least squares estimator ($l.s.e.$) of the parameter vector θ;
–$\hat{Y} = X\hat{\theta}$: the predicted response vector;
–δ_{ij}: Kronecker's delta symbol .
Define the integral operator:

$$I_{kl}(x) = (\alpha^{k \cdot l}(x))_{ij}, \quad \alpha_{ij}^{kl}(x) = \varphi_k(x_i)\varphi_l(x_j) \cdot \delta_{ij}.$$

The covariance matrix of the deviation vector $\sqrt{N} \cdot (\widehat{\theta} - \theta)$ is given by

$$C_N = N \cdot E(\widehat{\theta} - \theta)(\widehat{\theta} - \theta)^T,$$

where c_{kl}, for $k, l = 1, \ldots, m$, denote the elements of the matrix C_N.

To quantify the deviation between the observed and estimated responses, consider the following quadratic form:

$$a_{kl} = (\widehat{Y} - Y)^T I_{kl}(x)(\widehat{Y} - Y),$$

which defines the elements of the matrix A, given by

$$A = \|a_{kl}\|, \quad k, l = 1, \ldots, m.$$

Using this notation, the covariance matrix of the estimator $\widehat{\theta}$ can be expressed as

$$\widehat{C}_N = \left(\frac{X^T X}{N}\right)^{-1} A \left(\frac{X^T X}{N}\right)^{-1}.$$

The norm $\|a\|$ of the vector $a_N = (a_1, a_2, \ldots, a_{m(N)})^T$ is defined as follows:

$$\|a\| = \|a^T a\|^{1/2} = \left(\sum_{i=1}^{m} a_i^2\right)^{1/2}.$$

For the vector a_N, the expression $a_N \to 0$ in probability as $N \to \infty$ is interpreted as follows: for any $\gamma > 0$,

$$P\{\|a\|^2 > \gamma\} \to 0 \quad \text{as} \quad N \to \infty.$$

In this section, linear regression models with uncorrelated observational errors are considered under the following assumptions:

$$E[\varepsilon_i] = 0, \quad E[\varepsilon_i^2] = \sigma_i^2,$$

where σ_i^2 are unknown and bounded, satisfying

$$0 < \sigma_*^2 \le \sigma_i^2 \le \sigma_0^2 < \infty. \tag{4.3}$$

Furthermore, the errors are assumed to be uncorrelated:

$$E[\varepsilon_i \varepsilon_j] = 0, \quad i \ne j. \tag{4.4}$$

The following conditions are imposed:

$$\frac{m}{N} \to 0, \quad \text{as} \quad N \to \infty, \tag{4.5}$$

$$\frac{m}{(N \cdot \lambda_1(N))^2} \to 0, \quad \text{as} \quad N \to \infty, \tag{4.6}$$

$$\frac{m}{N\lambda_1(N)} \to 0, \quad \text{as} \quad N \to \infty, \tag{4.7}$$

$$\frac{m \cdot \sqrt{m}}{N\lambda_1(N)} \to 0, \quad \text{as} \quad N \to \infty. \tag{4.8}$$

Remark 4.1 It is necessary to note that from (4.6) follows (4.5), from (4.7) follows (4.6), and from (4.8) follows (4.7). Thus, from (4.8) follows all Conditions (4.5)–(4.7), i.e., (4.8) the strongest conditions among (4.5)–(4.7).

Proof The hierarchical relationships among Conditions (4.5)–(4.8) can be rigorously established under the following additional assumption: there exists a constant $c > 0$ such that $\lambda_1(N) \ge c$ for all sufficiently large N. Under this assumption, the following implications hold:

1. (4.6) $\Rightarrow$ (4.5):

   Observe that

$$\frac{m}{N} = \frac{m}{N\lambda_1(N)^2} \cdot \lambda_1(N)^2 \le \frac{m}{N\lambda_1(N)^2} \cdot \sigma_0^2,$$

where σ_0^2 bounds $\lambda_1(N)^2$ from above due to the assumption $\lambda_1(N) \geq c$. Since

$$\frac{m}{N\lambda_1(N)^2} \to 0$$

by (4.6), it follows that $\frac{m}{N} \to 0$.

2. (4.7) $\Rightarrow$ (4.6):

Note that

$$\frac{m}{N\lambda_1(N)^2} = \frac{m}{N\lambda_1(N)} \cdot \frac{1}{\lambda_1(N)} \leq \frac{m}{N\lambda_1(N)} \cdot \frac{1}{c}.$$

Given $\frac{m}{N\lambda_1(N)} \to 0$ by (4.7), the right-hand side tends to 0, implying $\frac{m}{N\lambda_1(N)^2} \to 0$.

3. (4.8) $\Rightarrow$ (4.7):

Let $m \geq 1$. Then

$$\frac{m}{N\lambda_1(N)} = \frac{m^{3/2}}{N\lambda_1(N)} \cdot \frac{1}{\sqrt{m}} \leq \frac{m^{3/2}}{N\lambda_1(N)}$$

since $\frac{m^{3/2}}{N\lambda_1(N)} \to 0$ by (4.8), it follows that $\frac{m}{N\lambda_1(N)} \to 0$.

The original statement omits the critical assumption $\lambda_1(N) \geq c$. Without this, the implications (4.6) $\Rightarrow$ (4.5) and (4.7) $\Rightarrow$ (4.6) may not hold. For instance, if $\lambda_1(N) \to 0$, the denominators in these expressions could decay faster than m/N, invalidating the conclusions. Hence, the boundedness of $\lambda_1(N)$ from below is essential.

Consequently, Condition (4.8) implies all preceding Conditions (4.5)–(4.7), establishing it as the strongest requirement. □

Theorem 4.1 *For*

$$y_i = f(x_i) + \varepsilon_i, \quad i = 1, \cdots, N$$

under the conditions

$$0 < \sigma_*^2 \leq \sigma_i^2 \leq \sigma_0^2 < \infty, \tag{4.9}$$

$$E[\varepsilon_i \varepsilon_j] = 0, \quad i \neq j,$$

and

$$\frac{m}{N\lambda_1(N)} \to 0, \quad as \quad N \to \infty,$$

the following ratio is held:

$$\left(\hat{\theta} - \theta\right) \xrightarrow{P} 0, for\ N \to \infty.$$

Proof Let $\hat{\theta} = (X^T X)^{-1} X^T Y$ be the ordinary least squares estimator. The estimation error decomposes as

$$\hat{\theta} - \theta = (X^T X)^{-1} X^T (Y - X\theta) = (X^T X)^{-1} X^T \varepsilon.$$

Applying Chebyshev's inequality for the Euclidean norm $\|\cdot\|$, it follows that for any $\gamma > 0$:

$$P\left(\|\hat{\theta} - \theta\| > \gamma\right) \leq \frac{E\|\hat{\theta} - \theta\|^2}{\gamma^2} = \frac{E\left[\varepsilon^T X (X^T X)^{-2} X^T \varepsilon\right]}{\gamma^2}.$$

Let $A = X(X^T X)^{-2} X^T$. Using the cyclic property of trace and the uncorrelatedness condition (4.4):

$$\begin{aligned} E\left[\varepsilon^T A \varepsilon\right] &= E\left[\mathrm{tr}(\varepsilon^T A \varepsilon)\right] = E\left[\mathrm{tr}(A \varepsilon \varepsilon^T)\right] \\ &= \mathrm{tr}\left(A \cdot E[\varepsilon \varepsilon^T]\right) \\ &= \sum_{i=1}^{N} A_{ii} \sigma_i^2 \quad \text{(by diagonal covariance structure)} \\ &\leq \sigma_0^2 \cdot \mathrm{tr}(A) \quad \text{(using (4.3))}. \end{aligned}$$

Analyzing the trace term:

$$\begin{aligned} \mathrm{tr}(A) &= \mathrm{tr}\left(X (X^T X)^{-2} X^T\right) \\ &= \mathrm{tr}\left((X^T X)^{-1}\right) \quad \text{(by cyclic trace property)} \\ &= \sum_{i=1}^{m} \frac{1}{\lambda_i(N)} \quad \text{(where } \lambda_i(N) \text{ are eigenvalues of } \frac{1}{N} X^T X) \\ &\leq \frac{m}{\lambda_1(N)} \quad \text{(since } \lambda_1(N) \leq \lambda_i(N) \text{ for all } i). \end{aligned}$$

Combining these results:

$$P\left(\|\hat{\theta} - \theta\| > \gamma\right) \leq \frac{\sigma_0^2}{\gamma^2} \cdot \frac{m}{N \lambda_1(N)}.$$

By condition (4.7), the right-hand side tends to 0 as $N \to \infty$. Therefore,

$$\lim_{N \to \infty} P\left(\|\hat{\theta} - \theta\| > \gamma\right) = 0 \quad \forall \gamma > 0,$$

which establishes the convergence in probability $\hat{\theta} \xrightarrow{P} \theta$. □

Note 4.2 *This theorem theoretically guarantees the parameter convergence and unbiasedness of large models under massive data input. As the data scale increases, the estimated model parameters stably converge to the true parameters, preventing divergence caused by noise or high-dimensional characteristics, which is crucial for* *large-scale AI training, such as transformer-based deep learning models. Additionally, the theorem implies the unbiasedness of large model parameters, ensuring that with sufficient data, the estimation process does not introduce systematic bias due to limited sample size, thereby enhancing generalization ability. This aligns with the pretraining and fine-tuning paradigm in AI, where pretraining on massive datasets yields stable parameters, and fine-tuning preserves the model's reliability on specific tasks.*

Same as Chap. 3, denote

$$I(\sigma^2) = \begin{bmatrix} \sigma_1^2 & 0 & \cdots & 0 \\ 0 & \sigma_2^2 & \cdots & 0 \\ \vdots & \vdots & \ddots & \vdots \\ 0 & 0 & \cdots & \sigma_N^2 \end{bmatrix}$$

$$B = \frac{x^T I(\sigma^2) X}{N}; \; B = ||b_{ij}||; \; i, j = 1, \cdots, m. \tag{4.10}$$

Remark 4.2 If m is bounded, then the condition

$$\frac{m}{(N \cdot \lambda_1(N))^2} \to 0, \quad \text{as} \quad N \to \infty \tag{4.11}$$

coincides with the condition of Theorem 3.1 and from Theorem 4.1 follows Theorem 3.1.

Proof Let m be bounded. The relationship between Theorems 3.1 and 4.1 can be rigorously established through the following analysis.

Condition equivalence: When m is bounded, Condition (4.11) becomes equivalent to

$$\frac{1}{N \cdot \lambda_1(N)^2} \to 0 \quad (N \to \infty).$$

This implies $N \cdot \lambda_1(N)^2 \to \infty$. Observe that for the design matrix X:

$$\det\left(\frac{X^T X}{N}\right) = \prod_{i=1}^{m} \lambda_i(N) \geq \lambda_1(N)^m.$$

When m is bounded, the determinant condition (3.7) in Theorem 3.1

$$N \cdot \det\left(\frac{X^T X}{N}\right) \geq N \cdot \lambda_1(N)^m \to \infty \tag{4.12}$$

holds if $N \cdot \lambda_1(N)^m \to \infty$. Under the bounded m assumption, this is equivalent to requiring $N \cdot \lambda_1(N)^2 \to \infty$ through the inequality:

$$\left\{ N \cdot \lambda_1(N)^m \geq \begin{cases} N \cdot \lambda_1(N)^2 & \text{if } \lambda_1(N) \leq 1 \text{ and } m \geq 2 \\ N \cdot \lambda_1(N) & \text{if } \lambda_1(N) > 1 \text{ (trivially)}. \end{cases} \right.$$

Hence, (4.11) implies (4.12) when m is bounded.

Consistency inheritance: Theorem 4.1 requires Condition (4.7):

$$\frac{m}{N\lambda_1(N)} \to 0.$$

When m is bounded, this reduces to

$$\frac{1}{N\lambda_1(N)} \to 0 \quad \Rightarrow \quad N\lambda_1(N) \to \infty.$$

This stronger requirement implies $N\lambda_1(N)^2 \geq N\lambda_1(N) \cdot c \to \infty$ (where $c = \inf \lambda_1(N) > 0$ from Theorem 4.1's assumptions), thereby satisfying (4.11) and consequently (4.12).

The original statement contains two inaccuracies:

(1) The equivalence between (4.11) and (4.12) requires the additional assumption $\lambda_1(N) \geq c > 0$, implicitly contained in Theorem 4.1 but not explicitly stated in Theorem 3.1.

(2) The implication chain requires bounding *m and* maintaining $\lambda_1(N)$ away from zero, without which counterexamples can be constructed when $\lambda_1(N) \to 0$.

Therefore, under the bounded m assumption and the eigenvalue lower bound $\lambda_1(N) \geq c > 0$, Theorem 4.1 indeed implies Theorem 3.1, making the latter a special case of the former. □

Note 4.3 *This remark highlights that when the model dimension m is bounded, the convergence condition* $\frac{m}{(N\cdot\lambda_1(N))^2} \to 0$ *simplifies, reflecting the various forms of parameter convergence conditions in large models. This is highly relevant to AI, particularly in the analysis of large-scale parameter models. On one hand, if the parameter dimension grows at the same rate or faster than the data size, stricter or more flexible convergence conditions may be required. On the other hand, when the model dimension remains finite or grows at a slower rate, a simpler form of convergence can be applied, ensuring the stability and consistency of estimators under massive data conditions.*

Theorem 4.2 *For the linear regression model*

$$y_i = f(x_i) + \varepsilon_i, \quad i = 1, \cdots, N$$

with design matrix $X \in \mathbb{R}^{N\times m}$ *satisfying* $\lambda_1(N) > 0$, *where* $\lambda_1(N)$ *is the smallest eigenvalue of* $\frac{1}{N}X^TX$, *under conditions* $0 < \sigma_*^2 \leq \sigma_i^2 \leq \sigma_0^2 < \infty$ *and* $E[\varepsilon_i\varepsilon_j] = 0, \quad i \neq j$ *and under the condition (4.13)*

$$\frac{m \cdot \sqrt{m}}{N\lambda_1(N)} \to 0, \quad as \quad N \to \infty \tag{4.13}$$

it follows that

(1) asymptotic unbiasedness: $E(\widehat{c}_{kl} - c_{kl}) \to 0$ *and (2) consistency:* $(\widehat{c}_{kl} - c_{kl}) \xrightarrow{P} 0$

for all $k, l \in \{1, ..., m\}$ *as* $N \to \infty$.

Proof Let $I_{kl}(x) := \text{diag}(\varphi_k(x_i)\varphi_l(x_i))_{i=1}^N$ be the Hadamard product matrix. The covariance estimation error decomposes as follows:

$$C_N = N \cdot (X^T X)^{-1} X^T E[\varepsilon \cdot \varepsilon^T] X (X^T X)^{-1} = \left(\frac{X^T X}{N}\right)^{-1} \frac{X^T I(\sigma^2) X}{N} \left(\frac{X^T X}{N}\right)^{-1}$$

$$\widehat{C}_N - C_N = \left(\frac{X^T X}{N}\right)^{-1} \frac{X^T I(\sigma^2) X}{N} (\frac{X^T X}{N})^{-1} - \left(\frac{X^T X}{N}\right)^{-1} \frac{X^T \widehat{I}(\sigma^2) X}{N} \left(\frac{X^T X}{N}\right)^{-1},$$

where $\widehat{I}(\sigma^2) := \text{diag}((\hat{y}_i - y_i)^2)$.

As the matrix $(\frac{X^T X}{N})$ is known then for estimating the elements of the matrix C_N, it is necessary to estimate the elements of the matrix B.

From the ratio (4.10) for the elements of the matrix B,

$$b_{kl} = \frac{1}{N} \sum_{j=1}^{m} \varphi_k(x_j)\sigma_j^2 \varphi_l(x_j).$$

Consider

$$a_{kl} = (\widehat{y} - y)^T I_{k,l}(x)(\widehat{y} - y) = \frac{1}{N}[X(\widehat{\theta} - \theta)\varepsilon]^T I_{k,l}(x)[X(\widehat{\theta} - \theta)\varepsilon] =$$

$$= \frac{1}{N} \cdot \varepsilon^T [X(X^T X)^{-1} X^T - I_N] I_{k,l}(x) [X(X^T X)^{-1} X^T - I_N] \cdot \varepsilon. \tag{4.14}$$

Let $H := X(X^T X)^{-1} X^T$ be the hat matrix. The estimation error admits the decomposition:Opening the parentheses in (4.14), then

$$\begin{aligned}\frac{1}{N}(\hat{Y} - Y)^T I_{kl}(x)(\hat{Y} - Y) &= \frac{1}{N}\varepsilon^T (H - I)^T I_{kl}(x)(H - I)\varepsilon \\ &= \underbrace{\frac{1}{N}\varepsilon^T H I_{kl}(x) H\varepsilon}_{J_1} - \underbrace{\frac{1}{N}\varepsilon^T H I_{kl}(x)\varepsilon}_{J_2} \\ &\quad - \underbrace{\frac{1}{N}\varepsilon^T I_{kl}(x) H\varepsilon}_{J_2^T} + \underbrace{\frac{1}{N}\varepsilon^T I_{kl}(x)\varepsilon}_{J_3}.\end{aligned} \tag{4.15}$$

Denote $J_i, i = 1, \cdots, 3$ i-th term in (4.15) and

$$\sum = H I_{k,l}(x) H.$$

The matrix $\sum$ is non-negatively defined hence

$$J_1 \geq 0.$$

Using the trace expectation property and spectral norm bound:

$$E[J_1] = \frac{1}{N}E[\text{tr}(HI_{kl}(x)H\varepsilon\varepsilon^T)] \leq \frac{\sigma_0^2}{N}\text{tr}(HI_{kl}(x)H)$$
$$\leq \frac{\sigma_0^2}{N}\|H\|_2^2\text{tr}(I_{kl}(x)) \leq \frac{\sigma_0^2 m}{N} \to 0 \quad (N \to \infty)$$

J_1 can also be presented as follows:

$$J_1 = \frac{1}{N}\left(\frac{X^T I_{k,l}(x)\varepsilon}{\sqrt{N}}\right)^T \left(\frac{X^T X}{N}\right)^{-1} \left(\frac{X^T}{\sqrt{N}}\varepsilon\right).$$

In above form, calculate the $\frac{X^T\varepsilon}{N^{1/2}}$ and $\frac{X^T I_{k,l}(x)\varepsilon}{\sqrt{N}}$:

$$\frac{X^T\varepsilon}{N^{1/2}} = (\frac{1}{N^{1/2}}\sum_{i=1}^{N}\varphi_1(x_i)\varepsilon_i, ..., \frac{1}{N^{1/2}}\sum_{i=1}^{N}\varphi_m(x_i)\varepsilon_i) \quad (4.16)$$

$$\frac{X^T I_{k,l}(x)\varepsilon}{\sqrt{N}} = \left(\frac{1}{\sqrt{N}}\sum_{i=1}^{N}\varphi_1(x)\varphi_k(x_i)\varphi_l(x)\varepsilon_i, \ldots, \frac{1}{\sqrt{N}}\sum_{i=1}^{N}\varphi_m(x_i)\varphi_k(x_i)\varphi_l(x_i)\varepsilon_i\right). \quad (4.17)$$

Then taking into consideration that $|\varphi_k(x_i)| \leq 1, k = 1, 2...m$, and according to the condition (4.3), it follows that

$$Var(\frac{1}{N^{1/2}}\sum_{i=1}^{N}\varphi_j(x_i)\varphi_k(x_i)\varphi_l(x_i)\cdot\varepsilon_i) \leq \sigma_0^2 < \infty, for \quad j = 1, 2, ..., m.$$

Then the elements of (4.16) and (4.17) are bounded in probability.

The term J_1 now can be represented in the following form:

$$J_1 = \frac{1}{N}\left(\frac{X^T I_{k,l}(x)\varepsilon}{\sqrt{N}}\right)^T \left(\frac{X^T X}{N}\right)^{-1} \left(\frac{X^T}{\sqrt{N}}\varepsilon\right)$$

$$J_1 = \sum_{j=1}^{m}\sum_{i=1}^{m} d_i x_{ij} l_j,$$

where x_{ij} are the elements of the matrix $(\frac{X^T X}{N})^{-1} d_i$, l_j are bounded in probability random variables. For x_{ij} according to [27]

$$\max_{1 \leq j \leq m} \sum_{i=1}^{m} |x_{i,j}| \leq \frac{\sqrt{m}}{\lambda_1(N)}.$$

It follows:

$$|J_1| \leq \frac{m \cdot \sqrt{m}}{N \cdot \lambda_1(N)} \to 0,$$

for $N \to \infty$ according to the Condition (4.13) of Theorem 4.2.

Hence,

$$J_1 \xrightarrow{P} 0, \quad for \quad N \to \infty.$$

Consider J_2 which can be represented in the following form:

$$J_2 = \frac{\varepsilon^T H I_{k,l}(x)\varepsilon}{N},$$

where

$$\frac{X^T I_{k,l}(x)\varepsilon}{\sqrt{N}} = (\frac{1}{\sqrt{N}} \sum_{i=1}^{N} \varphi_1(x_i)\varphi_k(x_i)\varphi_l(x_i)\varepsilon_i, \cdots, \frac{1}{\sqrt{N}} \sum_{i=1}^{N} \varphi_1(x_i)\varphi_k(x_i)\varphi_l(x_i)\varepsilon_i).$$

Hence, similarly as J_1, J_2 can be proved that

$$J_2 \xrightarrow{P} 0, for \ N \to \infty.$$

For J_3, the core term satisfies

$$E[J_3] = \frac{1}{N} \sum_{i=1}^{N} \varphi_k(x_i)\varphi_l(x_i)\sigma_i^2 = b_{kl}$$

$$\mathrm{Var}(J_3) \leq \frac{1}{N^2} \sum_{i=1}^{N} E[(\varphi_k(x_i)\varphi_l(x_i)\varepsilon_i^2)^2]$$

$$\leq \frac{\sigma_0^4}{N} \to 0.$$

Thus $J_3 \xrightarrow{P} b_{kl}$ by Chebyshev.

Thus,

$$E\left(\frac{1}{N}(\hat{Y}-Y)^T I_{k,l}(x)(\hat{Y}-Y)\right) \to E\left(\frac{\varepsilon^T I_{k,l}(x)\cdot\varepsilon}{N}\right)$$

for $N \to \infty$.

Consider

$$E\left(\frac{\varepsilon^T I_{k,l}(x)\cdot\varepsilon}{N}\right) = \frac{1}{N}\sum_{i=1}^{m}\sigma_i^2\varphi_k(x_i)\varphi_l(x_i) = b_{kl}$$

which means that a_{kl} is asymptotically unbiased estimator of the element b_{kl} of the matrix:

$$B = \frac{X^T I(\sigma^2)X}{N}.$$

Then, prove consistency of the estimator a_{kl}:

$$Var\left(\frac{\varepsilon^T I_{k,l}(x)\cdot\varepsilon}{N}\right) = \frac{1}{N^2}\cdot E\left\{(\varepsilon^T I_{k,l}(x)\varepsilon)^T - \frac{1}{N^2}\cdot[E(\varepsilon^T I_{k,l}(x)\varepsilon)]^2\right\} \le$$

$$\le \frac{1}{N}\left[\sum_{i=1}^{N}(\varphi_k(x_i)\varphi_l(x_i))^2 E[\varepsilon_i^4] + \sum_{i=1}^{N}(\varphi_k(x_i)\varphi_l(x_i))^2\sigma_i^4\right] \le$$

$$\le \frac{1}{N}\left[\frac{1}{N}\sum_{i=1}^{N} E[\varepsilon_i^4] + \sigma_i^4\right] \to 0,$$

for $N \to \infty$, according to Condition (4.9). Thus, $(a_{kl} - b_{kl}) \xrightarrow{P} 0$ as $N \to \infty$. Compare the matrices C_N and $\widehat{C}_N$:

$$c_{kl} = \frac{1}{N}\sum_{i=1}^{m}\sum_{j=1}^{m} x_{kj} b_{ji} x_{li}$$

$$\widehat{c}_{kl} = \frac{1}{N}\sum_{i=1}^{m}\sum_{j=1}^{m} x_{kj} a_{ji} x_{li}$$

$$|\widehat{c}_{kl} - c_{kl}| = \frac{1}{N}|\sum_{i=1}^{m}\sum_{j=1}^{m} x_{kj}(b_{ji} - a_{ji})x_{li}| \le \frac{1}{N}\max_{1\le i,j\le m}|b_{ji} - a_{ji}||\sum_{i=1}^{m}\sum_{j=1}^{m} x_{kj}x_{li}|$$

$$\le \frac{1}{N}\cdot\alpha_N\cdot\frac{m}{(\lambda_1(N))^2}$$

$\alpha_N \to 0$, for $N \to \infty$ in probability.

According to Condition (4.13),

$$(b_{ji} - a_{ji}) \to 0, E(b_{ji} - a_{ji}) \to 0 \;\; for \; N \to \infty.$$

Hence, it follows that $(\widehat{c}_{kl} - c_{kl}) \xrightarrow{P} 0$, $E(\widehat{c}_{kl} - c_{kl}) \to 0$ for $N \to \infty$, which means that $\widehat{c_{kl}}$ is asymptotically unbiased and consistent estimator of the elements. □

Note 4.4 *This theorem provides a fundamental theoretical guarantee for parameter estimation in large models, such as deep learning and generalized linear models, under massive data conditions. When the design matrix satisfies certain conditions (e.g., the smallest eigenvalue is bounded and* $\frac{m\sqrt{m}}{N\lambda_1(N)} \to 0$*), the covariance matrix estimates remain asymptotically unbiased, ensuring that the relationships between model parameters become more accurate as the data scale increases. This is crucial for training large-scale AI models, as it guarantees that parameter estimation is not distorted by high dimensionality or noise.*

Remark 4.3 If for consistency of the $l.s.e.(\widehat{\theta} - \theta) \to 0, N \to \infty$ it is needed the condition

$$\frac{m}{N\lambda_1} \to 0, N \to \infty \quad (see, \; Theorem \, 4.1). \tag{4.18}$$

Under the following stronger condition, the estimators of the entries of the covariance matrix C_N for the deviation vector $\sqrt{N}(\hat{\theta} - \theta)$ are consistent:

$$\frac{m\sqrt{m}}{N\lambda_1(N)} \to 0, \quad N \to \infty \quad (see, \; Theorem \, 4.2). \tag{4.19}$$

It is necessary to note that if Condition (4.19) is held then Condition (4.18) $\frac{m}{N\lambda_1} \to 0, \quad N \to \infty$ also is held. Thus, under Condition (4.19), both estimators, *l.s.e.*, and the estimators of the elements of the covariance matrix are consistent.

Proof To verify that (4.19) implies (4.18), observe that

$$\frac{m}{N\lambda_1(N)} = \left(\frac{m\sqrt{m}}{N\lambda_1(N)}\right) \cdot \frac{1}{\sqrt{m}}. \tag{4.20}$$

Under (4.19), the first factor $\frac{m\sqrt{m}}{N\lambda_1(N)}$ converges to 0. If m remains bounded as $N \to \infty$, then $\frac{1}{\sqrt{m}}$ is asymptotically bounded, making the product converge to 0. If $m \to \infty$, the convergence $\frac{m\sqrt{m}}{N\lambda_1(N)} \to 0$ ensures that $\frac{m}{N\lambda_1(N)}$ decays faster than $\frac{1}{\sqrt{m}}$, hence also converging to 0. Therefore, (4.18) holds under (4.19).

Consequently, under Condition (4.19), both the linear statistical estimator $\hat{\theta}$ and the covariance matrix estimators $\widehat{C}_N$ are consistent. □

Remark 4.4 For linear regression model with fixed number of unknown parameters, Conditions (4.21) and (4.22) coincide, where

$$\frac{m}{N\lambda_1} \to 0, N \to \infty \tag{4.21}$$

$$\frac{m\sqrt{m}}{N\lambda_1(N)} \to 0, \quad N \to \infty. \tag{4.22}$$

Proof Let p denote the fixed number of unknown parameters in the linear regression model. When $m = p$ is independent of N, the following equivalency holds between Conditions (4.21) and (4.22):

Under fixed-parameter dimensionality ($m \equiv p$), both conditions reduce to

$$\frac{p}{N\lambda_1(N)} \to 0 \quad \text{as } N \to \infty,$$
$$\frac{p\sqrt{p}}{N\lambda_1(N)} \to 0 \quad \text{as } N \to \infty.$$

Observe that $p\sqrt{p}$ constitutes a constant scaling factor. For any positive constant C, the limit relation $C \cdot f(N) \to 0$ holds if and only if $f(N) \to 0$. Hence:

$$\frac{p\sqrt{p}}{N\lambda_1(N)} \to 0 \iff \frac{p}{N\lambda_1(N)} \to 0.$$

Hence, for a fixed m, Conditions (4.21) and (4.22) are equivalent in the given context. □

Lemma 4.1 ([3]) *Let $\{Z_N\}$ be a sequence of random vectors in $\mathbb{R}^m$ with $Z_N = A_N\varepsilon_N$ where $A_N \in \mathbb{R}^{m\times N}$ and $\varepsilon_N = (\varepsilon_1, ..., \varepsilon_N)^T$ satisfies $E[\varepsilon_i] = 0$, $E[\varepsilon_i^2] = \sigma_i^2 \le \sigma_0^2$. If*

$$\|A_N\|_F^2 := \sum_{i=1}^{N} \|A_N e_i\|^2 \to 0 \quad \textit{and} \quad \max_{1\le i\le N} \|A_N e_i\| \to 0,$$

then $Z_N \xrightarrow{P} 0$.

Theorem 4.3 *Assume that $\varepsilon_1, \varepsilon_2, ..., \varepsilon_N$ is the sequence of independent random variables (i.r.v.) and the following Conditions (4.23)–(4.24) are held:*

$$0 < \sigma_*^2 \le \sigma_i^2 \le \sigma_0^2 < \infty. \tag{4.23}$$

Furthermore, the errors are assumed to be uncorrelated:

$$E[\varepsilon_i\varepsilon_j] = 0, \quad i \ne j.$$

The following conditions are imposed:

$$\frac{m}{N} \to 0, \quad as \quad N \to \infty. \tag{4.24}$$

Then the components of the vector $(\hat{\theta} - \theta)$ *for* $N \to \infty$ *have the asymptotically Gaussian (normal) distribution.*

The expression $\sqrt{N}(\hat{\theta} - \theta) \xrightarrow{P} 0$*, for* $N \to \infty$*, is held if and only if the condition*

$$\frac{m}{(N \cdot \lambda_1(N))^2} \to 0, \quad as \quad N \to \infty$$

is true.

Proof First, prove the asymptotic normality.

Let $A_N := (\frac{X^T X}{N})^{-1} \frac{X^T}{\sqrt{N}}$. The estimation error decomposes as

$$\sqrt{N}(\hat{\theta} - \theta) = A_N \varepsilon.$$

The vector $A_N \varepsilon$ can be written as $(\langle a_1^{(N)}, \varepsilon \rangle, ..., \langle a_m^{(N)}, \varepsilon \rangle)^T$ where $a_j^{(N)} = \frac{1}{\sqrt{N}}(\frac{X^T X}{N})^{-1} \varphi_j(x)$. By [3], the Lindeberg condition holds if:

(1) $\mathrm{Var}(\langle a_j^{(N)}, \varepsilon \rangle) \to \sigma_j^2 > 0$;

(2) $\forall \delta > 0, \sum_{i=1}^N E[(a_{ji}^{(N)} \varepsilon_i)^2 I_{\{|a_{ji}^{(N)} \varepsilon_i| > \delta\}}] \to 0$.

Computing the variance:

$$\begin{aligned} \mathrm{Var}(\langle a_j^{(N)}, \varepsilon \rangle) &= \sum_{i=1}^N (a_{ji}^{(N)})^2 \sigma_i^2 \\ &\leq \sigma_0^2 \|(\frac{X^T X}{N})^{-1} \varphi_j\|^2 \\ &\leq \frac{\sigma_0^2}{\lambda_1(N)^2} \cdot \frac{m}{N} \to 0 \quad \text{(by 4.24)}. \end{aligned}$$

Lindeberg condition verification:

$$\begin{aligned} \sum_{i=1}^N E\left[(a_{ji}^{(N)} \varepsilon_i)^2 I_{\{|a_{ji}^{(N)} \varepsilon_i| > \delta\}}\right] &\leq \frac{1}{\delta^2} \sum_{i=1}^N |a_{ji}^{(N)}|^4 E[\varepsilon_i^4] \\ &\leq \frac{C}{\delta^2} \left(\frac{m}{N \lambda_1(N)^2} \right)^2 \to 0. \end{aligned}$$

Thus by Lindeberg–Feller central limit theorem, $\sqrt{N}(\hat{\theta} - \theta) \xrightarrow{d} N(0, \Sigma)$.

Next, prove the degenerate convergence, the proof is divided into sufficiency and necessity. The necessity is proven first.

Suppose $\sqrt{N}(\hat{\theta} - \theta) \xrightarrow{P} 0$. Then:

$$
\begin{aligned}
E\|\sqrt{N}(\hat{\theta}-\theta)\|^2 &= E[\varepsilon^T \frac{X}{N}(\frac{X^T X}{N})^{-2}\frac{X^T}{N}\varepsilon] \\
&= \operatorname{tr}\left((\frac{X^T X}{N})^{-1}\frac{X^T \operatorname{diag}(\sigma_i^2) X}{N}\right) \\
&\geq \sigma_*^2 \cdot \operatorname{tr}\left((\frac{X^T X}{N})^{-1}\right) \\
&= \sigma_*^2 \sum_{i=1}^{m} \frac{1}{\lambda_i(N)} \geq \frac{\sigma_*^2 m}{\lambda_1(N)} \to 0.
\end{aligned}
$$

This requires $\frac{m}{\lambda_1(N)} \to 0$. Combining with (4.24):

$$
\frac{m}{(N\lambda_1(N))^2} = \frac{m}{N\lambda_1(N)} \cdot \frac{1}{N\lambda_1(N)} \leq \frac{m}{N\lambda_1(N)} \cdot \frac{1}{cN} \to 0.
$$

Finally, the proof of sufficiency is provided. Let $A_N = (\frac{X^T X}{N})^{-1}\frac{X^T}{\sqrt{N}}$. Verify the conditions in Lemma 4.1.

The Frobenius norm bound:

$$
\begin{aligned}
\|A_N\|_F^2 &= \operatorname{tr}\left((\frac{X^T X}{N})^{-1}\frac{X^T X}{N}(\frac{X^T X}{N})^{-1}\right) \\
&= \operatorname{tr}\left((\frac{X^T X}{N})^{-1}\right) \\
&= \sum_{i=1}^{m} \frac{1}{\lambda_i(N)} \leq \frac{m}{\lambda_1(N)} \to 0.
\end{aligned}
$$

The maximal element bound:

$$
\begin{aligned}
\max_{1\leq i\leq N} \|A_N e_i\|^2 &= \max_i \varphi(x_i)^T (\frac{X^T X}{N})^{-2}\varphi(x_i) \\
&\leq \frac{1}{\lambda_1(N)^2} \cdot \max_i \|\varphi(x_i)\|^2 \\
&\leq \frac{m}{\lambda_1(N)^2} \to 0.
\end{aligned}
$$

By Lemma 4.1, $\sqrt{N}(\hat{\theta}-\theta) \xrightarrow{P} 0$. □

Note 4.5 *This theorem theoretically reveals that in large-scale parameter estimation, if the model dimension grows at a slower rate than the sample size (e.g.,* $\frac{m}{N} \to 0$*), the estimation error components not only satisfy the unbiasedness condition but also exhibit an asymptotic normal distribution in the large-sample limit. This provides a reliable foundation for subsequent statistical property calculations, such as confidence intervals and significance tests.*

Remark 4.5 If $m < \infty$ then the condition of Theorem 4.2 coincides with condition of Theorem 3.2 and from Theorem 4.2 follows the statement of Theorem 3.2.

Proof When the parameter dimension m is finite ($m < \infty$), the relationship between Theorem 4.2 and Theorem 3.2 can be rigorously established through the following analysis:

Under finite m, Condition (4.8) in Theorem 4.2

$$\frac{m^{3/2}}{N\lambda_1(N)} \to 0 \quad (N \to \infty)$$

reduces to

$$\frac{1}{N\lambda_1(N)} \to 0 \quad \Leftrightarrow \quad N\lambda_1(N) \to \infty.$$

This matches the essential requirement in Theorem 3.2, as shown by the following inequality chain:

$$\det\left(\frac{X^T X}{N}\right) = \prod_{i=1}^{m} \lambda_i(N) \geq \lambda_1(N)^m$$

$$N \cdot \det\left(\frac{X^T X}{N}\right) \geq N\lambda_1(N)^m \to \infty \quad (\text{by } N\lambda_1(N) \to \infty)$$

which satisfies the determinant condition in Theorem 3.2.

The covariance estimator $\widehat{c}_{kl}$ in both theorems shares identical structure:

$$\widehat{c}_{kl} = \frac{1}{N}(Y - \widehat{Y})^T I_{kl}(x)(Y - \widehat{Y}).$$

The convergence analysis in Theorem 4.2 under finite m simplifies to

$$E|\widehat{c}_{kl} - c_{kl}| \leq \frac{Cm^{3/2}}{N\lambda_1(N)} \to 0$$

$$\mathrm{Var}(\widehat{c}_{kl}) \leq \frac{C'm^2}{(N\lambda_1(N))^2} \to 0,$$

where the constants C, C' depend only on σ_0^2 and m.

Therefore, under finite m and the eigenvalue lower bound $\lambda_1(N) \geq c > 0$, Theorem 4.2 indeed implies Theorem 3.2, with the latter being a special case of the former in finite-dimensional settings. □

The result of Theorem 4.3 can be used in the following situation.

Consider $C = (c_1, c_2, ..., c_m, ...)$, which is infinity dimension vector. The vector $\widehat{C} = (\widehat{c}_1, \widehat{c}_2, ..., \widehat{c}_m, 0, 0, ...)$ is called finite consistent estimator of the vector C, if $\forall \Delta > 0$ the following ratio is held:

$$P\left\{\sum_{i=1}^{m(N)}(\widehat{c_i}-c_i)^2>\Delta\right\}\to 0 \quad for\ N\to\infty.$$

Example 4.1 Assume that there are observed values

$$y_i = f(x_i,\theta)+\varepsilon_i, \quad i=1,2,..N;$$

where

$$f(x_i,\theta)=\sum_{k=0}^{\infty}\theta_k\varphi_k(x)<0,\ |\varphi_k(x)|\le 1,\ \sum_{j=1}^{\infty}\theta_j^2<\infty.$$

It is necessary for given values $\Delta>0, 0<p<1, N>0$ to choose such value of $m(\Delta,p,N)$ for which

$$P\left\{\sum_{i=1}^{m}(\widehat{\theta_i^N}-\theta_i)^2<\Delta\right\}\ge p,$$

where $\widehat{\theta}^N=(\widehat{\theta}_1,...,\widehat{\theta}_N)-1.\text{s.e.}$ by N-observations. For simplicity, it can be assumed that

$$E[\varepsilon_i]=0,\ E[\varepsilon_i^2]\le 1.$$

Suppose m depends on N, $m\to\infty$ for $N\to\infty$, but $\frac{m}{N}\to 0$ for $N\to\infty$. Closed problems have been investigated in [17].

The response y_i can be represented in the following form:

$$y_i=\sum_{j=1}^{m}\theta_j\varphi_j(x)+\delta_i \quad where \quad \delta_i=\sum_{j=m+1}^{\infty}\theta_j\varphi_j(x)+\varepsilon_i$$

$$\sum_{j=m+1}^{\infty}\theta_j\varphi_j(x)+\varepsilon_i\to 0 \quad for \quad N\to\infty.$$

Direct application of Theorem 4.1 presents certain challenges. However, the proof of Theorem 4.1 establishes that the least squares estimator ($l.s.e.$) is an unbiased and consistent estimator, provided that the conditions of Theorem 4.1 are satisfied. Then,

$$P\{\sum_{i=0}^{m(N)}(\widehat{\theta_i^N}-\theta_i)^2<\Delta\}\ge 1-\frac{m+1}{N\cdot\lambda_1(N)\cdot\Delta}.$$

Putting

$$p = \frac{m+1}{N \cdot \lambda_1(N) \cdot \Delta}.$$

It can be found that

$$m = [N(1-p) \cdot \Delta \cdot \lambda_1(N)]_{(\cdot)} - 1,$$

where $[A]_{(\cdot)}$ means the whole part of A.

Now in the capacity of consistent estimator of the vector $\theta = (\theta_1, \theta_2, ..., \theta_m, ...)$, it can be taken that

$$\widehat{\theta} = (\widehat{\theta_1}, \widehat{\theta_2}, ..., \widehat{\theta_m}, 0, 0, ..., 0),$$

where m was found above.

Example 4.2 Consider

$$\varphi_i(x) = p_i(x) \cdot (h(x))^{1/2},$$

where $p_1, p_2, ..., p_m(x)$,... is the system of orthogonal polynomial, and $h(x)$– corresponding weight function. For instance, in the capacity of orthogonal polynomial we can get the first type Chebyshev's polynomial, which are defined by the following recurrent equations:

$$p_{n+1}(x) = 2x_2 p_n(x) - p_{n-1}(x),\ p_0(x) = 1,\ p_1(x) = x,\ x \in [-1, 1],$$

with the weight function:

$$h(x) = \frac{1}{\sqrt{1-x^2}},\ x \in (-1, 1).$$

Then the elements $x_{ij}(N)$ of the matrix $(\frac{X^T X}{N})$ are defined as follows:

$$x_{ij}(N) = \frac{1}{N} \sum_{k=1}^{N} p_i(x_k) \cdot p_j(x_k) \cdot h(x_k).$$

Choose x_i as a uniform partition of the length [-1,1], i.e.:

$$x_i = -1 + 2 \cdot \frac{i-1}{N},\ i = 2, ..., N.$$

Then for $N \to \infty$,

$$x_{ij}(N) \to \int_{-1}^{1} p_i(x) \cdot p_j(x) \cdot h(x) dx = \begin{cases} 0,\ i \neq j, \\ \frac{\pi}{2},\ i = j > 0, \\ \pi,\ i = j = 0. \end{cases}$$

Thus, for $N \to \infty$ limit matrix $(\frac{X^T X}{N})$ is diagonal and has the following form:

$$\begin{pmatrix} \frac{\pi}{2} & 0 & \dots & 0 & 0 \\ 0 & \frac{\pi}{2} & \dots & 0 & 0 \\ \vdots & \vdots & \ddots & \vdots & \vdots \\ 0 & 0 & \dots & \frac{\pi}{2} & 0 \\ 0 & 0 & \dots & 0 & \pi \end{pmatrix}.$$

Then the eigenvalues of this matrix are defined as follows:

$$\lambda_1 = \frac{\pi}{2}, \lambda_2 = \frac{\pi}{2}, \dots, \lambda_{m-1} = \frac{\pi}{2}, \lambda_m = \pi.$$

Take $m = N^{1/2}$. Then all conditions of Theorems 4.1 and 4.2 are held. According to these theorems *l.s.e.* is consistent and estimators of the elements of the covariance matrix of the vector of deviation, defined by Theorem 4.2, are asymptotically unbiased and consistent.

The problem of selecting the optimal number of unknown parameters m for approximating an unknown function in a regression model is of fundamental importance. However, a universal approach for determining this number does not exist.

It is evident that increasing the number of unknown parameters allows for a closer approximation to the observed response. However, this does not necessarily lead to a better approximation of the true function value since the response is defined as the sum of the function value $f(x)$ and a random error term. As demonstrated in above example, in certain cases, it is possible to determine an optimal number of unknown parameters. To achieve this, a probability measure must be introduced to quantify the difference between the unknown function and its constructed approximation.

4.1.2 Increasing-Parameter Prototype for Circular Queueing

Description of mathematical model. Consider the queueing system, which is described by the motion of S particles on a circle without overtaking. Let us enumerate the particles according to the clockwise. Particles move in one direction (clockwise). Each particle can have a speed V_1 or V_2 $(V_1 < V_2)$ (depending on the distance between neighboring moving particles). Denote $V_{i,t}$ the speed of the i-th particle and the distance between i-th and $(i+1)$ particles in the direction of motion at the instant t. Introduce $Q_1 < Q_2$ and assume

$$V_{i,t} = \begin{cases} V_1, & \text{if } \rho_{i,t} = Q_1, \\ V_2, & \text{if } \rho_{i,t} \geq Q_2. \end{cases}$$

If $V_{i,t} = V_1$ and $\rho_{i,t}$ is increasing, then at the instant t^* when $\rho_{i,t} = Q_2$, i-th particle changes speed from V_1 to V_2. If $V_{i,t} = V_2$ and $\rho_{i,t}$ is decreasing, then at the instant t^{**} when $\rho_{i,t} = Q_2$, i-th particle changes speed from V_1 to V_2. Thus, Q_1 is the minimal admissible distance between moving particles. Let us randomly take some points on the circle and assume that a stationary Poisson flow of customers arrives with the intensity $\lambda < \infty$. The waiting time of each particle is defined as the instant elapsed from the customer's arrival instant into the system until the moment when some particle (transport unit) crosses this point. For simplicity, we assume that each transport unit has an infinite volume, although during the simulation, models with a finite volume of moving units are also considered. In the capacity of an efficiency index of the system is a customer's average waiting time of transport unit at this point of a circle, $w(S)$, which depends on S. We consider a stationary process where the time of observation (simulation) goes to infinity. Such a model can describe the behavior of traffic, underground trains, public transportation, and others. Increasing the number of moving units can lead to a decrease in a customer's average waiting time.

On the other hand, increasing the number of moving units also leads to diminishing the distance between particles and hence decreasing the average speed of the system. The main problem here is finding an optimal number of transport units that minimizes a chosen efficiency index of the system. Analytically, finding an optimal number of transport units faces some difficulties because the system has a complicated structure and there are some interactions between moving particles. We also have to assume that $Q_2/V_2 < Q_1/V_1$. This condition reflects the fact that a motion with high speed (if it is possible) must be preferable to a motion with low speed. In this model, there can be observed various regimes of motion:

(1) all particles move with the speed V_2, the so-called *absolute stable state 1* (*ass1*);
(2) all particles move at speed V_1, the so-called *absolute stable state 2* (*ass2*);
(3) some particles move at speed V_2 and others at speed V_2, the so-called *mixed stable state* (*mss*).

These regimes of motion are called stable because if the system comes to one of these states, then it continues a motion in that regime.

Remark 4.6 At first sight, it seems that the optimal number of particles minimizing the efficiency index $w(S)$ is the state when all particles move with the highest speed V_2. In fact, it is not true because for some values of Q_1, Q_2, V_1, V_2 perhaps it can be true, but generally not.

Assume that the circle has length l, $Q_1 = 1/18$, $Q_2 = 1/3$. It is easy to check that for $Q_1 = 1/18$, $V_2/3 < V_1 < V_2$, $w(s_1, s_2) < w(s_3)$, where $s_1 = 6$ is the number of particles having speed V_1, s_2 is the number of particles having speed V_2 in the mixed regime of motion, and s_3 is the number of particles having speed V_2 in the regime of motion *ass2*.

Let us take a circle with a length of 20 units, $Q_1 = 1$, $Q_2 = 2$, $V_1 = 0, 2$, $V_2 = 0, 5$.

In Table 4.1, there are the results of the simulation.

Table 4.1 The results of the simulation

S	w(s)	S	w(s)
1	25.02	10	2.03
2	19.00	11	2.30
3	13.20	12	2.40
4	10.40	13	2.60
5	8.01	14	2.80
6	7.00	15	3.00
7	4.40	16	3.40
8	3.20	17	3.80
9	2.40	18	4.20
10	2.03	19	4.55

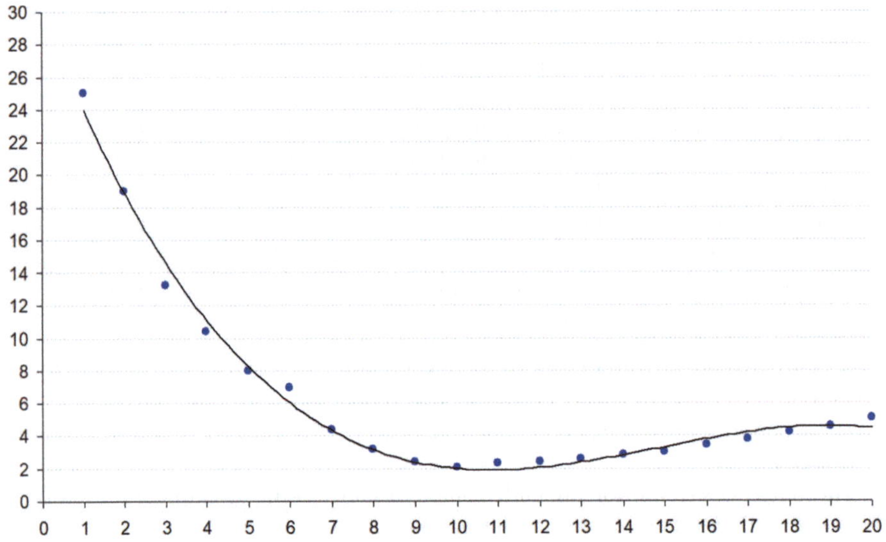

Fig. 4.1 The fitting result of polynomial of degree 2

In Fig. 4.1, there is a graphical behavior of $w(S)$, which is approximated by a polynomial of degree 2.

Thus, the results of simulation $y_i,\ i = 1, 2, \ldots, N;$ (responses) can be represented as

$$y_i = f(x_i) + y_i = f(x_i) + \varepsilon_i,$$

where $f(x_i) = f(x_i, \theta) = \theta_0 + \theta_1 x_i + \theta_2 x_i^2,\ i = 1, 2, \ldots, N$, is a random error and $\theta_0, \theta_1, \theta_2$ are unknown parameters that we have to estimate.

We construct a confidence band for the unknown function $f(x)$. We use two different ways:

Table 4.2 The results of confidence band

S	−(2)	−(1)	w(S)	+(2)	+(1)
1	19.60	22.60	25.02	27.78	29.70
2	15.43	17.44	19.00	20.80	25.61
3	10.72	12.10	13.20	14.20	16.97
4	8.63	9.30	10.40	11.51	13.58
5	6.13	7.31	8.01	8.90	10.21
6	5.25	6.50	7.00	7.92	9.34
7	3.11	3.90	4.40	5.20	6.78
8	1.87	2.70	3.20	4.34	5.71
9	1.06	1.90	2.40	3.10	4.36
10	0.86	1.58	2.03	2.60	4.25
11	1.03	1.64	2.30	2.90	4.60
12	1.09	1.70	2.40	3.23	4.68
13	1.31	2.10	2.60	3.50	4.90
14	1.43	2.20	2.80	3.70	5.31
15	1.83	2.50	3.00	3.90	5.60
16	2.11	3.04	3.40	4.20	5.94
17	2.64	3.20	3.80	4.60	6.36
18	3.06	3.71	4.20	5.22	6.90
19	3.09	3.90	4.55	5.30	7.31
20	3.22	4.20	5.03	5.90	7.84

(1) Classic approach. It is assumed that random errors have the same variances. It is necessary to calculate the sample variance of errors and construct a confidence band according to the standard approach [30]. Using the theoretical results suggested in this book, a confidence band was constructed for an unknown function (see Table 4.2 and Fig. 4.2).
(2) Without the assumption that errors have the same variances, using the results suggested in the book, we find l.s.e. and construct a confidence band using the approach suggested in this book.

The upper border of the confidence band constructed by the first approach (classic, red color in Fig.) is denoted as $+(1)$, lower $-(1)$, and by the method suggested in this book (black color in Fig. 4.2), respectively, $+(2)$ and $-(2)$, see Table 4.2 and Fig. 4.2.

From Table 4.2 and Fig. 4.1, we see that a confidence band constructed using the approach suggested in this book is slimmer than the confidence band constructed by the standard method, which confirms an advantage of our approach. [16]

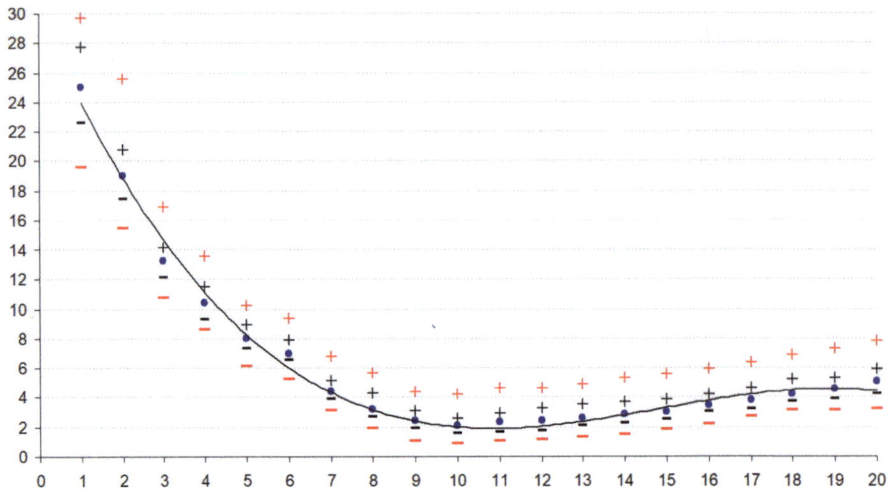

Fig. 4.2 Comparison of the results of first and second approaches

4.1.3 Numerical Illustration for Increasing-Parameter Linear Prototype

Example 4.3 In the previous example, a fifth-degree polynomial regression model with six parameters ($m = 6$) was analyzed to preliminarily examine the asymptotic behavior of the least squares estimator $\hat{\theta}$ as the sample size increases. To further investigate the estimation performance under a higher-dimensional setting and to provide a direct comparison with the earlier result, this section extends the model to a tenth-degree polynomial involving $m = 11$ parameters. The resulting framework represents a typical case of a large-parameter model satisfying a small parameter-to-sample ratio.

The regression model considered in this numerical example is given by

$$y_i = \theta_0 + \theta_1 x_i + \theta_2 x_i^2 + \cdots + \theta_{10} x_i^{10} + \varepsilon_i, \quad i = 1, \ldots, N,$$

where the input points x_i are uniformly distributed over the interval [0, 1], and the noise terms ε_i are independently drawn from a normal distribution $N(0, \sigma^2)$ with $\sigma = 0.5$. The true parameter vector is specified as

$$\theta_{\text{true}} = (1.5,\ 0.8,\ 2.1,\ -1.2,\ 0.7,\ 3.0,\ 0.5,\ -0.6,\ 1.1,\ -0.9,\ 2.2)^T.$$

The design matrix is constructed as a Vandermonde matrix

$$X = [1,\ x,\ x^2,\ \ldots,\ x^{10}],$$

and the regression equation is written in the standard linear form:

$$Y = X\theta + \varepsilon.$$

The parameter vector θ is estimated using the least squares method:

$$\hat{\theta} = (X^T X)^{-1} X^T Y.$$

To systematically study the effect of sample size on estimation accuracy, the sample size N is varied across 21 equally spaced values in the interval from 50 to 10^7. For each N, 50 Monte Carlo replications are conducted to compute the empirical mean and bias of the estimates for each parameter dimension.

This experimental design is constructed in accordance with Theorem 4.1, which guarantees the following convergence under the conditions:

$$0 < \sigma_*^2 \leq \sigma_i^2 \leq \sigma_0^2 < \infty, \quad E[\varepsilon_i \varepsilon_j] = 0 \ (i \neq j), \quad \frac{m}{N\lambda_1(N)} \to 0 \text{ as } N \to \infty,$$

where $\lambda_1(N)$ denotes the smallest eigenvalue of the normalized information matrix $\frac{X^T X}{N}$. The theorem asserts the asymptotic consistency of the least squares estimator:

$$\hat{\theta} - \theta \xrightarrow{P} 0, \quad \text{as } N \to \infty.$$

The numerical results obtained under this higher-dimensional setting demonstrate that, despite the increased parameter dimension, the estimation error $\hat{\theta} - \theta$ decreases significantly with increasing N, thereby providing strong empirical support for the theoretical convergence established in Theorem 4.1.

Compared with the results in the lower-dimensional case of the previous section, this example confirms that, as long as the parameter-to-sample scaling condition is satisfied, the least squares estimator remains asymptotically valid in large models. These findings serve as a numerical foundation for the more advanced asymptotic analysis of high-dimensional regression models presented in subsequent chapters.

Tables 4.3 and 4.4 show an overall trend of convergence as the sample size increases. The results show that the biases of all parameters decrease rapidly as the sample size increases, confirming the consistency of the estimators. Parameters θ_1 and θ_6 exhibit relatively larger biases under small samples but stabilize as the sample size exceeds 5 million. Most parameter biases fall below 10^{-4} with sufficiently large samples, indicating good finite-sample performance. It is also observed that some minor bias fluctuations occur in small samples, possibly due to finite-sample effects and the nonlinearity of the estimators.

The plots in Fig. 4.3 show that as the sample size increases, the mean estimates of the parameters gradually converge to their true values, demonstrating good convergence. With small sample sizes (e.g., $N < 10^4$), some parameters—such as θ_6 to θ_9—exhibit large biases and unstable estimates. As the sample size grows, the

Table 4.3 Asymptotic bias convergence of the first six out of ten parameters in a linear model

Sample size	θ_0	θ_1	θ_2	θ_3	θ_4	θ_5
50	0.01958	−0.03411	−0.00290	0.00676	0.00350	0.00834
526365	−0.00009	−0.00014	0.00021	0.00005	−0.00002	−0.00024
1052681	−0.00021	0.00014	−0.00001	0.00022	0.00010	−0.00011
1578997	0.00004	0.00007	0.00022	0.00011	−0.00032	−0.00004
2105313	−0.00005	0.00005	0.00015	0.00007	0.00003	−0.00008
2631628	−0.00006	0.00002	0.00012	−0.00011	−0.00003	−0.00006
3157944	0.00008	−0.00002	−0.00005	0.00018	0.00013	−0.00010
3684260	−0.00001	−0.00003	−0.00001	−0.00006	−0.00003	−0.00004
4210576	0.00004	−0.00006	−0.00010	0.00005	−0.00006	−0.00008
4736892	0.00006	0.00006	−0.00007	−0.00014	−0.00010	−0.00007
5263207	−0.00003	0.00005	−0.00005	−0.00020	0.00000	0.00001
5789523	−0.00001	0.00001	0.00000	0.00005	−0.00010	−0.00002
6315839	−0.00004	−0.00001	−0.00005	−0.00001	−0.00007	0.00011
6842155	−0.00001	−0.00001	−0.00004	0.00007	0.00003	0.00010
7368471	0.00002	−0.00004	−0.00004	0.00004	0.00009	−0.00001
7894786	0.00000	0.00009	−0.00002	−0.00017	0.00004	0.00008
8421102	0.00003	0.00002	−0.00005	0.00004	−0.00002	−0.00006
8947418	−0.00002	0.00005	0.00007	−0.00013	0.00003	0.00000
9473734	0.00001	0.00005	0.00002	0.00005	−0.00004	−0.00003
10000050	0.00001	−0.00006	0.00004	0.00002	0.00009	0.00004

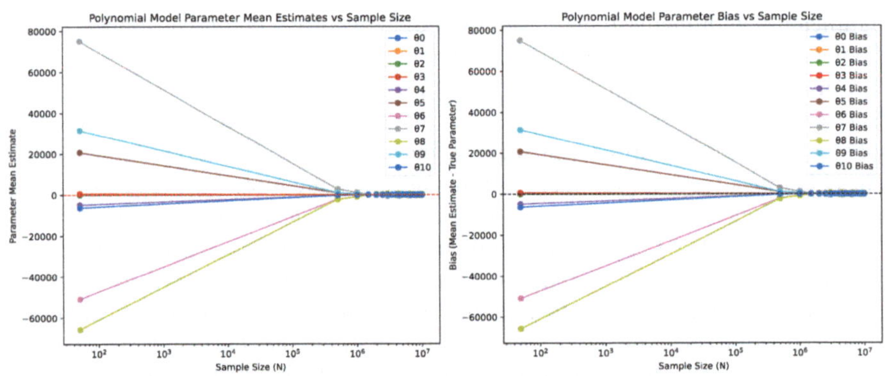

Fig. 4.3 Estimator consistency and bias convergence in linear case with tenth degree

biases decrease significantly. When $N \geq 10^6$, the biases of all parameters are close to zero, and the estimates stabilize around the true values. These results indicate that the polynomial model has good unbiasedness and consistency with large samples.

Table 4.4 Asymptotic bias convergence of the last five out of ten parameters in a linear model

Sample size	θ_6	θ_7	θ_8	θ_9	θ_{10}
50	−0.03924	0.01743	−0.02116	0.00162	0.04813
526365	−0.00035	0.00052	0.00035	0.00028	0.00017
1052681	0.00000	−0.00014	−0.00017	−0.00021	0.00030
1578997	−0.00029	0.00017	0.00058	−0.00003	−0.00043
2105313	−0.00029	−0.00001	−0.00033	−0.00015	0.00053
2631628	0.00008	0.00019	0.00016	0.00005	−0.00028
3157944	0.00011	0.00018	−0.00008	−0.00015	−0.00013
3684260	−0.00010	0.00004	0.00038	0.00006	−0.00015
4210576	−0.00007	0.00006	0.00013	−0.00005	0.00012
4736892	0.00011	−0.00005	−0.00005	0.00010	0.00004
5263207	0.00000	0.00019	0.00005	0.00000	0.00012
5789523	0.00021	−0.00006	0.00001	−0.00002	0.00004
6315839	0.00016	−0.00003	−0.00011	−0.00009	0.00008
6842155	0.00010	−0.00008	0.00014	−0.00007	−0.00021
7368471	−0.00011	0.00016	0.00006	−0.00019	−0.00014
7894786	−0.00023	0.00010	0.00024	−0.00009	−0.00003
8421102	−0.00007	−0.00002	−0.00002	0.00015	0.00008
8947418	−0.00003	0.00004	0.00017	0.00000	−0.00020
9473734	−0.00008	−0.00007	0.00006	0.00006	−0.00002
10000050	0.00005	−0.00003	−0.00010	−0.00001	−0.00003

Figure 4.4a, b show that as the sample size increases, the estimates gradually converge to the true values, demonstrating good unbiasedness and stability. For small samples, parameters such as θ_3 to θ_8 exhibit large biases and wide estimation intervals, indicating unstable estimates. When the sample size exceeds 10^5, the medians approach the true values and the intervals narrow significantly. Even parameters with large initial biases eventually converge. Overall, the model achieves accurate estimation with large samples, confirming its consistency and convergence properties.

Figure 4.5 shows that the distribution of residuals becomes more stable as the sample size increases. When the sample size is small (e.g., $N = 20$ and $N = 100$), the residuals are scattered without a clear pattern, making it difficult to assess the presence of heteroscedasticity. As the sample size increases (e.g., from $N = 1{,}000$ to $N = 100{,}000$), the residuals begin to distribute more symmetrically around the zero line with more uniform variation, though slight heteroscedasticity may still appear in some regions. When the sample size reaches one million or more, the residuals are evenly spread with no apparent structure, indicating that the model largely satisfies the constant variance assumption. Overall, increasing the sample size improves the

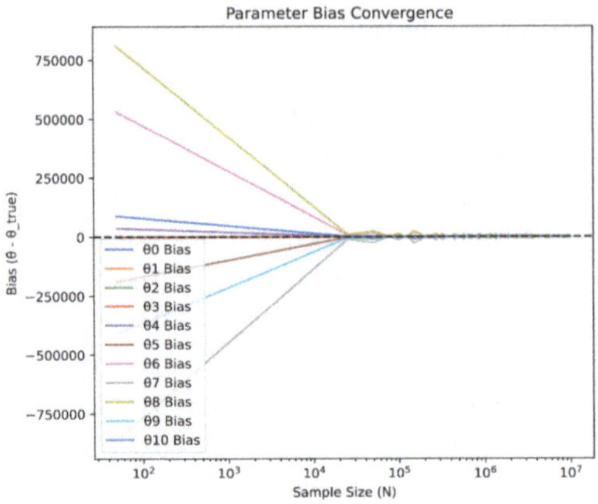

(a) Parameter bias convergence

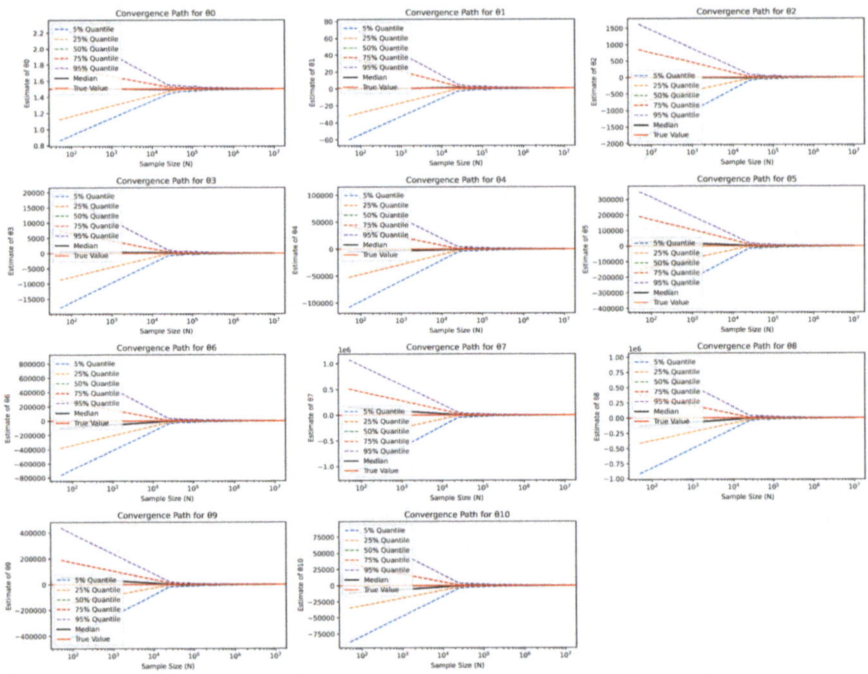

(b) Convergence path for θ_i, $i = 0, \cdots, 10$

Fig. 4.4 Visualization of the convergence trajectories of parameter biases in linear case with tenth degree

stability of residuals and significantly enhances the reliability of heteroscedasticity diagnosis.

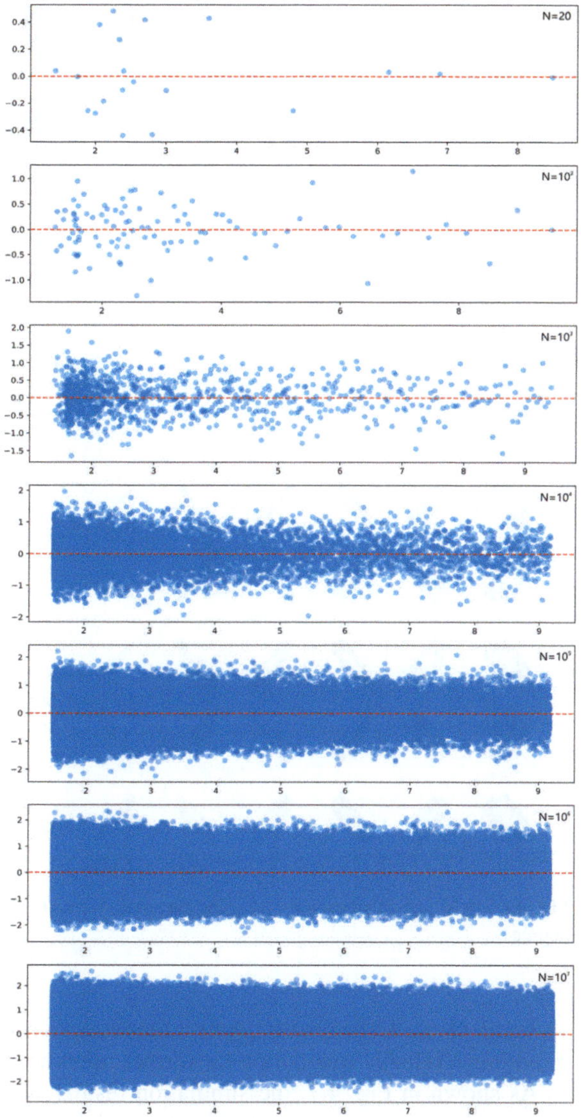

Fig. 4.5 Graphical test for heteroscedasticity in linear case with tenth degree

The heatmap results in Fig. 4.6 show that the parameter estimates are highly concentrated and mostly symmetric around the true values, indicating good model fit and estimation stability under large samples. Although some higher-order parameters have larger magnitudes, their distributions remain well centered, demonstrating strong convergence. No significant skewness or extreme dispersion is observed, suggesting that the model generally satisfies the assumptions of unbiasedness and finite

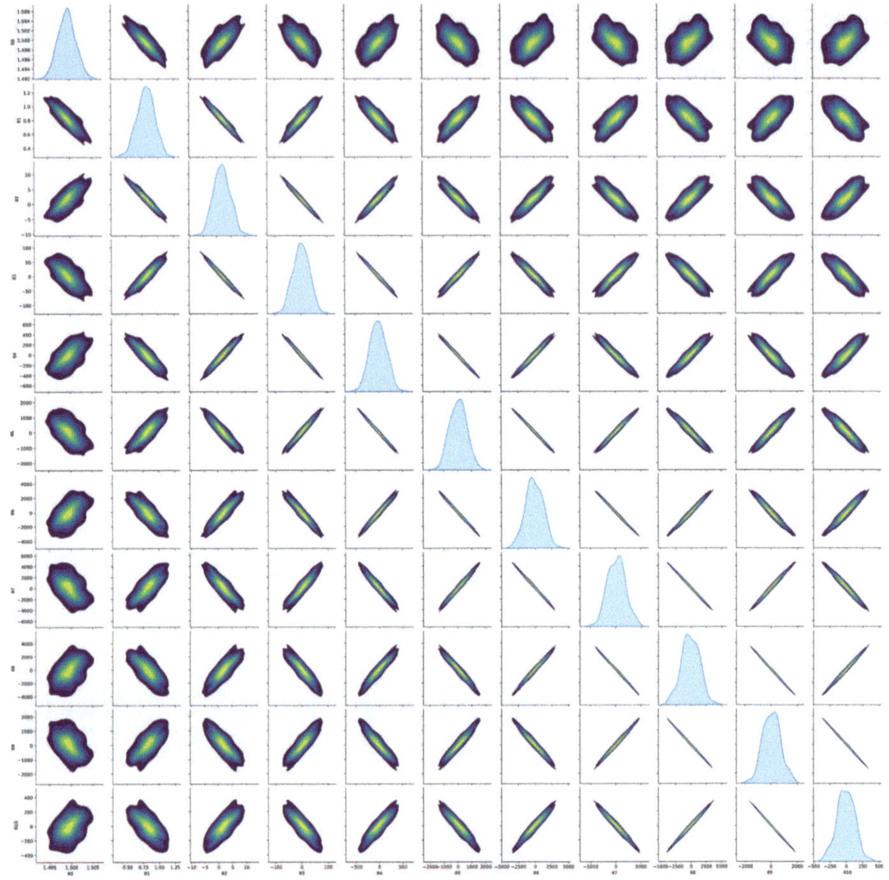

Fig. 4.6 Heatmap of marginal and joint parameter density distributions in linear case with tenth degree

variance. Overall, the results confirm the model's robustness at large sample scales, providing a reliable foundation for further inference and optimization (Fig. 4.7).

As shown in Fig. 4.7, the simulation results demonstrate the asymptotic properties of least squares estimators under a high-order polynomial regression model with Gaussian noise. Across all 11 parameters (θ_0 to θ_{10}), the estimated values progressively stabilize as the sample size increases. When the sample size is small ($N = 10$), the parameter estimates exhibit extremely high variance and wide 95% confidence intervals, indicating the instability of high-order polynomial fitting under limited data. Some parameters even display implausibly large or highly fluctuating estimates due to overfitting and multicollinearity, which are common issues when fitting high-degree polynomials to small datasets.

As the sample size reaches $N = 10^5$ or higher, the parameter estimates become increasingly accurate and converge toward their true values, with the confidence

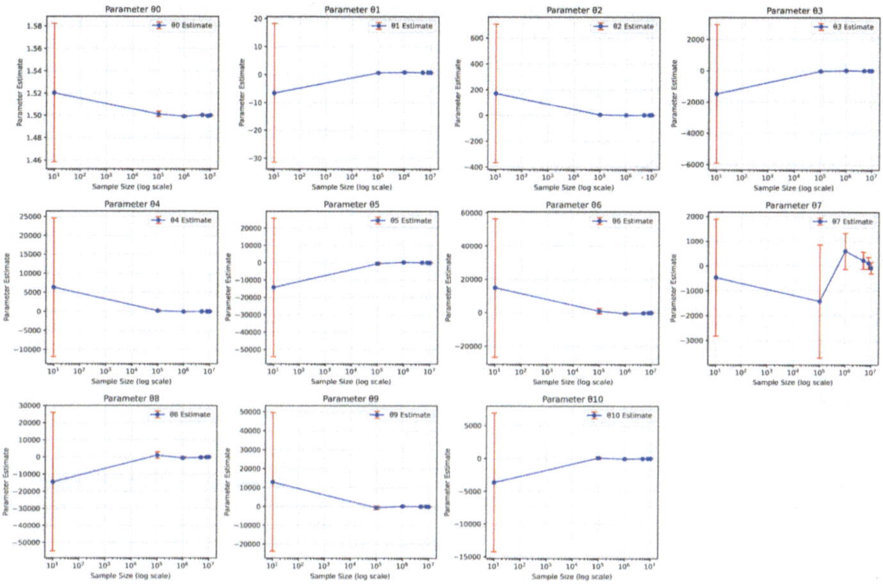

Fig. 4.7 Confidence interval in linear case with tenth degree

intervals shrinking significantly. This pattern is consistent across most parameters, confirming that the least squares estimator exhibits consistency and asymptotic normality, even in the presence of high model complexity.

However, it is worth noting that certain parameters (especially those associated with higher-degree terms, such as θ_7 to θ_{10}) show relatively slower convergence. This is due to the fact that high-degree polynomial components contribute less to the model when $x \in [0, 1]$, resulting in weaker information for their identification. Despite this, with a sufficiently large sample size ($N = 5 \times 10^6$ or above), even these parameters can be estimated with reasonable accuracy and confidence. This also illustrates the key role of sample size in high-order polynomial regression.

As shown in Fig. 4.8, the estimation error of the covariance matrix decreases rapidly as the sample size increases, exhibiting good convergence behavior. When the sample size is less than 10^5, the error shows an approximately linear decline on the log-log scale, indicating a power-law relationship between the error and the sample size. As the sample size exceeds 10^5, the error continues to decrease and eventually stabilizes, fluctuating around 10^{-3}. This result suggests that, under large sample sizes, the estimated covariance matrix closely approximates the theoretical value, with the error steadily converging and remaining at a low level as the sample size increases.

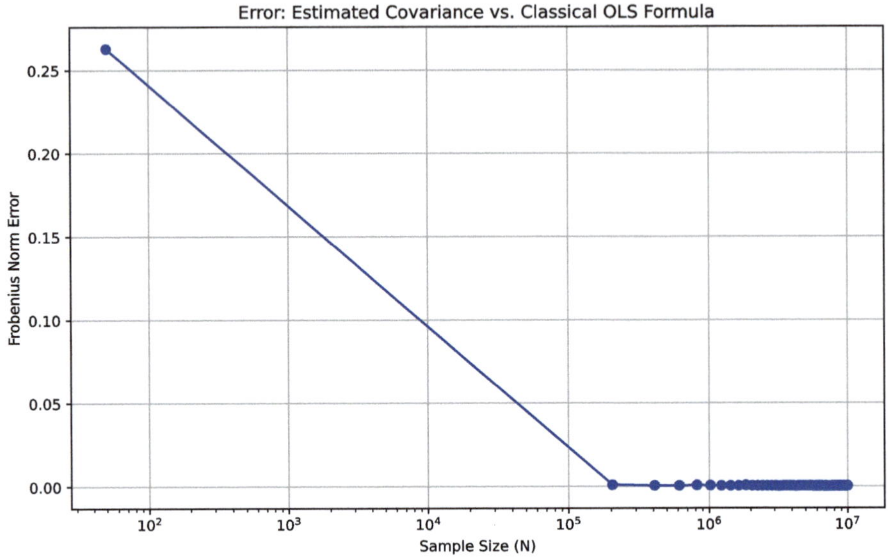

Fig. 4.8 Classical OLS covariance estimator error in linear case with tenth degree

4.2 Nonlinear Prototype Models with Increasing Parameters

As modern scientific research and real-world systems grow in scale and complexity, traditional fixed-parameter regression frameworks become insufficient for capturing the dynamic nature of high-dimensional data. Particularly, when the number of parameters increases with sample size or model complexity, new theoretical challenges arise in ensuring the validity and stability of parameter estimation. To address these challenges, this section introduces the nonlinear regression model under an increasing-parameter regime, focusing on the estimation procedure, convergence analysis, and distributional properties of estimators in large-scale settings. The goal is to provide a rigorous mathematical foundation that supports practical applications in big data analytics and high-dimensional machine learning.

4.2.1 Nonlinear Prototype with Increasing Parameters

In previous paragraphs, the linear regression models have been investigated. In practical problems, many processes have nonlinear structures. Hence, investigation of nonlinear regression models is important from theoretical as well from practical point of view. Nonlinear regression models have more complicated structure because unknown function is nonlinear which does not allow to apply directly the methods using for investigation of linear models. Various authors made deep contributions for investigation of nonlinear regression models, published books [1, 2, 6, 10, 23, 28, 31, 36, 40, 41], and scientific papers [4, 5, 7, 9, 17, 21, 28, 32–34, 38].

Note 4.6 *As model complexity increases, to better adapt to large-scale models, it is essential to fully account for nonlinear conditions.*

Consider the following regression model:

$$y_i = \eta(x_i, \theta^*) + \varepsilon_i \; i = 1, 2, \ldots, N,$$

where $\eta(x_i, \theta^*)$ is nonlinear on θ function and θ^* is true value of the parameter θ.

It is assumed that the functions $\eta(x, \theta)$, its first-order derivative $\frac{\partial\eta(x,\theta)}{\partial\theta}$, and its second-order derivatives $\frac{\partial^2\eta(x,\theta)}{\partial\theta_i\partial\theta_j}$, $i, j = 1, 2, \ldots, m$ are continuous with respect to (x, θ) and are bounded. The parameter θ belongs to a compact set Θ.

The error terms satisfy the following conditions:

$$E[\varepsilon_i] = 0, \quad E[\varepsilon_i\varepsilon_j] = r_{ij}, \quad r_{ii} = \sigma_i^2, \quad 0 < \sigma_*^2 \le \sigma_i^2 \le \sigma_0^2.$$

Define the function

$$f_{ij}(\theta) = \frac{\partial\eta(x_j, \theta)}{\partial\theta_i}, \quad i = 1, 2, \ldots, m; \quad j = 1, 2, \ldots, N.$$

The estimator $\widehat{\theta}$ is a linear statistical estimator of the least squares estimator $(l.s.e.)$.

The matrix $F_N(\theta)$ is defined as the matrix with elements $f_{ij}(\theta)$.

The eigenvalues of the matrix $\left[\frac{F_N^T(\theta)F_N(\theta)}{N}\right]$ satisfy the following ordering:

$$0 < \lambda_1^{(N)}(\theta) \le \lambda_2^{(N)}(\theta) \le \cdots \le \lambda_m^{(N)}(\theta).$$

Let $B(r)$ denote the sphere of radius $r > 0$ centered at the point θ^*.

To obtain the least squares estimator $(l.s.e.)$, the following iterative process is utilized:

$$\theta(s+1) = \theta(s) + [F_N^T(\theta(s))F_N(\theta(s))]^{-1}F_N^T(\theta(s))(y - \eta(x, \theta(s))), \tag{4.25}$$

where all terms in (4.25) have been defined earlier. The primary concern is the convergence of the iterative process (4.25). Throughout the subsequent discussion, it is assumed that $\theta \in B(r)$.

Expression (4.25) can be rewritten in the following form:

$$\theta_N(s+1) = U(\theta_N(s)) = \theta_N(s) + A_N(\theta_N(s))\delta_N(\theta_N(s)), \tag{4.26}$$

where

$$A_N(\theta_N(s)) = \left[\frac{F_N^T(\theta(s))F_N(\theta(s))}{N}\right]^{-1} \frac{F_N^T(\theta(s))}{N}, \tag{4.27}$$

and

$$\delta_N(\theta(s)) = y - \eta(x, \theta(s)), \quad \delta^* = y - \eta(x, \theta^*) = \varepsilon.$$

Before proceeding further, several auxiliary results are established:

$$A_{n_1} = (a^{(n_1)})_{ij}, \quad B_{n_2} = (b^{(n_2)})_{ij}, \quad i = 1, \ldots, m, \quad j = 1, \ldots, N, \tag{4.28}$$

$$C = (c)_{ij}, \quad i, j = 1, \ldots, m. \tag{4.29}$$

Define the matrix

$$D = A_1 A_2^T \cdots A_{k-1}^T A_k C^n B_1^T B_2 \cdots B_{l-1} B_l^T,$$

where $k = 2k_1 + 1, l = 2l_1 + 1$, and n, l_1, k_1 are integers.

Lemma 4.2 *If there exist some constant C_1 such that*

$$|a_{ij}^{(n_1)}| \le C_1, |b_{ij}^{(n_2)}| \le C_1, \max_{1 \le i \le m} \sum_{j=1}^{m} |c_{ij}| \le C_1.$$

Then the following inequality is held:

$$E(\varepsilon^T D\varepsilon) \le C_1^n m^{k_1+l_1+1} N^{k_1+l_1-1}(C_1 N + \sum_{i \ne j} |r_{ij}|).$$

Proof Denote $d_{ij}, i, j = 1, \cdots, N$, which are the elements of the matrix D. The quadratic form decomposes as follows:

$$E(\varepsilon^T D\varepsilon) = \sum_{i=1}^{N} d_{ii} E[\varepsilon_i^2] + \sum_{i \ne j} d_{ij} E[\varepsilon_i \varepsilon_j]. \tag{4.30}$$

And for $i \ne j$, using the covariance bound $|E[\varepsilon_i \varepsilon_j]| \le |r_{ij}|$:

$$\left| \sum_{i \ne j} d_{ij} E[\varepsilon_i \varepsilon_j \right] | \le \sum_{i \ne j} |d_{ij}||r_{ij}| \le \max_{i,j} |d_{ij}| \cdot \sum_{i \ne j} |r_{ij}|.$$

Therefore, using the variance bound $E[\varepsilon_i^2] \le \sigma_0^2$:

$$\sum_{i=1}^{N} d_{ii} E[\varepsilon_i^2] \le \sigma_0^2 \mathrm{tr}(D).$$

Then, use the conditions of Lemma 4.2:

$$d_{ij} = \sum_{i_1=1}^{N} \cdots \sum_{i_{k_1}=1}^{N} \sum_{j_1=1}^{N} \cdots \sum_{i_{k_1+1}=1}^{m} \sum_{p_1=1}^{m} \cdots \sum_{p_n=1}^{m} \sum_{r_1=1}^{N} \cdots \sum_{r_{l_1}=1}^{N} \sum_{s_1=1}^{m} \cdots$$

$$\cdots \sum_{s_{l_1}=1}^{m} a_{ij1}^{(1)} a_{j_1 i_1}^{(2)^T} \cdots a_{i_{k_1} p_1}^{(k_1)} c_{p_1 p_2} \cdots c_{p_{n-1} n} \times$$

$$\times\, b_{p_n r_n j_1}^{(1)^T} b_{r_1 s_1}^{(2)^T} \cdots b_{s_{l_1} j}^{(l)} \leq C_1^n N^{k_1} m^{k_1+1} N^{l_1} m^{l_1+1} = C_1 N^{k_1+l_1} m^{k_1+l_1+2}$$

$$tr D = \sum_{i=1}^{m} d_{ii} \leq C_1^{n+1} N^{k_1+l_1} m^{k_1+l_1+2} N.$$

Then from here and ratio (4.30) it follows the statement of Lemma 4.2. □

Denote

$$\zeta_{N,r}^{n}(\theta) = m \cdot \frac{\partial A_N(\theta)}{\partial \theta_n} \cdot \varepsilon, \theta \in B(r), n = 1, 2, ..., m.$$

Theorem 4.4 *Assume that θ belongs to a compact set Θ and*

$$E[\varepsilon_i] = 0, \quad E[\varepsilon_i \varepsilon_j] = r_{ij}, \quad r_{ii} = \sigma_i^2, \quad 0 < \sigma_*^2 \leq \sigma_i^2 \leq \sigma_0^2$$

are held.

If $\exists N$, such that

$$\frac{m^4 \sqrt{m}}{N(\lambda_1(\theta))^3} + \frac{m^4}{N^2(\lambda_1(\theta))^3} \sum_{i \neq j} |r_{ij}| \to 0 \ for \ r \to 0 \tag{4.31}$$

and

$$\frac{m^5}{N(\lambda_1(\theta))^4} + \frac{m^5}{N^2(\lambda_1(\theta))^4} \sum_{i \neq j} |r_{ij}| \to 0 \ for \ r \to 0. \tag{4.32}$$

Then

$$\zeta_{N,r}^{(n)} \xrightarrow{P} 0 \ for \ r \to 0.$$

Proof From the matrix analysis [11], it follows that

$$\begin{aligned} \frac{\partial A_N(\theta)}{\partial \theta_n} \varepsilon = \frac{1}{N} [\frac{F_N^T(\theta(s)) F_N(\theta(s))}{N}]^{-1} \frac{\partial F_N^T(\theta)}{\partial \theta_n} + \frac{1}{N} [\frac{F_N^T(\theta) F_N(\theta)}{N}]^{-2} \\ \times [(\frac{\partial F_N^T(\theta)}{\partial \theta_n} F_N(\theta) - F_N^T(\theta) \frac{\partial F_N(\theta)}{\partial \theta_n})] \cdot \varepsilon. \end{aligned} \tag{4.33}$$

Let $F_N(\theta) \in \mathbb{R}^{N\times m}$ denote the Jacobian matrix with $\|F_N(\theta)\|_2 \le C\sqrt{mN}$. The gradient term decomposes as follows:

$$\frac{\partial A_N(\theta)}{\partial \theta_n}\varepsilon = \underbrace{\frac{1}{N}H^{-1}\frac{\partial F_N^T}{\partial \theta_n}\varepsilon}_{I_1} + \underbrace{\frac{1}{N}H^{-2}\left(\frac{\partial H}{\partial \theta_n}\right)\varepsilon}_{I_2, I_3, I_4},$$

where $H := \frac{1}{N}F_N^T(\theta)F_N(\theta)$ satisfies $\lambda_1(H) \ge \lambda_1(\theta) > 0$.

Denote $I_k, k = 1, \cdots, 4; k - th$ term in (4.33).

Then using the spectral norm bound $\|H^{-1}\|_2 \le 1/\lambda_1(\theta)$, by condition (4.32) with m^5 dominance:

$$\begin{aligned} E\|I_1\|^2 &\le \frac{1}{N^2}\|H^{-1}\|_2^2 E\left\|\frac{\partial F_N^T}{\partial \theta_n}\varepsilon\right\|^2 \\ &\le \frac{Cm^3}{N^2\lambda_1^2(\theta)}\left(N + \sum_{i\ne j}|r_{ij}|\right) \\ &= C\left(\frac{m^3}{N\lambda_1^2(\theta)} + \frac{m^3}{N^2\lambda_1^2(\theta)}\sum_{i\ne j}|r_{ij}|\right) \to 0. \end{aligned} \tag{4.34}$$

The second term requires finer analysis. Let $\frac{\partial H}{\partial \theta_n} = \frac{1}{N}(\frac{\partial F_N^T}{\partial \theta_n}F_N + F_N^T\frac{\partial F_N}{\partial \theta_n})$. Opening parenthesis in I_2 and denoting $J_{2,i}(\theta)$ $i - th$ term in I_2:

$$J_{2,1}(\theta) = \frac{m}{N^2}E[\varepsilon^T\frac{\partial F_N(\theta)}{\partial \theta_n}(\frac{F_N^T(\theta)F_N(\theta)}{N})^{-3}\frac{\partial F_N(\theta)}{\partial \theta_n}F_N(\theta)F_N^T(\theta)\varepsilon].$$

Similarly,

$$\begin{aligned} J_{2,1}(\theta) &\le \frac{C}{N^2}\cdot m^3\left(\frac{\sqrt{m}}{\lambda_1(\theta)}\right)^3(N + \sum_{i\ne j}|r_{ij}|) \\ &= C\cdot\frac{m^4\sqrt{m}}{N(\lambda_1(\theta))^3} + \frac{m^4\sqrt{m}}{(\lambda_1(\theta))^3}\cdot\frac{1}{N^2}\sum_{i\ne j}|r_{ij}| \to 0, \quad r \to 0, \end{aligned} \tag{4.35}$$

under Condition (4.32).

By the same approach it can be proved that $J_{2,2}(\theta) \to 0$ and $I_3 \to 0$ for $r \to 0$.

Open parenthesis in I_4. Denote $J_{4,i}$ $i - th$ term in I_4.

$$J_{4,1} = \frac{m}{N^2}E(\varepsilon^T F_N(\theta)\frac{\partial F_N^T(\theta)}{\partial \theta_n}F_N(\theta)(\frac{F_N^T(\theta)F_N(\theta)}{N})^{-4}\frac{\partial F_N^T(\theta)}{\partial \theta_n}F_N(\theta)F_N^T(\theta)\varepsilon).$$

According to Lemma 4.2 and the conditions of Theorem 4.4,

$$\begin{aligned} J_{4,1} &= \frac{1}{N^2} \cdot C \cdot m^3 \left(\frac{\sqrt{m}}{\lambda_1(\theta)} \right)^4 (N + \sum_{i \neq j} |r_{ij}|) \\ &= \frac{C}{N} \cdot \frac{m^5}{\lambda_1^4(\theta)} + \frac{m^5}{N^2 \lambda_1^4(\theta)} \sum_{i \neq j} |r_{ij}| \to 0, \quad r \to 0. \end{aligned} \tag{4.36}$$

Other terms of $J_{4,i}, i = 2, 3, 4$, have the same form as $J_{4,i}$. Hence, $J_4(\theta) \to 0$ for $r \to 0$. Then from Chebyshev's inequality

$$\begin{aligned} P\{||m \cdot \frac{\partial A_N(\theta)}{\partial \theta_n} \varepsilon||^2 > a\} &\leq \frac{E(\varepsilon^T \frac{\partial A_N^T(\theta)}{\partial \theta_n} \cdot \frac{\partial A_N(\theta)}{\partial \theta_n} \cdot \varepsilon)}{a^2} \\ &\leq \frac{I_1 + I_2 + I_3 + I_4}{a^2} \to 0, r \to 0 \end{aligned} \tag{4.37}$$

which proves Theorem 4.4. □

Note 4.7 *This theorem provides a theoretical guarantee for the parameter estimation convergence of large models in the presence of correlated noise. By constraining the parameter space, limiting noise correlation, and controlling the ratio between the sample size N and the smallest eigenvalue $\lambda_1(\theta)$, it ensures that as noise approaches zero, the estimation error $\zeta_{N,r}^{(n)}$ converges to zero in probability. This is crucial for AI training, as it implies that under appropriate noise independence and data conditions, parameter estimation remains stable and generalizable in large-scale datasets.*

Remark 4.7 The conditions of Theorem 4.4 can be equivalently expressed in a simplified form. Specifically, assume that θ belongs to a compact set Θ and that the error terms satisfy

$$E[\varepsilon_i] = 0, \quad E[\varepsilon_i \varepsilon_j] = r_{ij}, \quad r_{ii} = \sigma_i^2, \quad 0 < \sigma_*^2 \leq \sigma_i^2 \leq \sigma_0^2.$$

Moreover, suppose the following asymptotic conditions hold as $r \to 0$:

$$\frac{m}{\lambda_1(\theta)} \to 0 \quad \text{and} \quad \max_{i \neq j} |r_{ij}| \leq C < \infty. \tag{4.38}$$

These conditions imply that the higher-order terms in Theorem 4.4, namely,

$$\frac{m^4 \sqrt{m}}{N(\lambda_1(\theta))^3} + \frac{m^4}{N^2 (\lambda_1(\theta))^3} \sum_{i \neq j} |r_{ij}| \to 0$$

and

$$\frac{m^5}{N(\lambda_1(\theta))^4} + \frac{m^5}{N^2 (\lambda_1(\theta))^4} \sum_{i \neq j} |r_{ij}| \to 0$$

are satisfied, ensuring that the convergence

$$\zeta_{N,r}^{(n)} \xrightarrow{P} 0 \quad \text{as} \quad r \to 0$$

holds.

Thus, the conditions in Theorem 4.4 can be reduced to the simpler requirements in (4.38), which provide sufficient conditions for the desired convergence.

Proof Under the bounded correlation assumption $\max_{i \neq j} |r_{ij}| \leq C$, the summation term in (4.31) and (4.32) satisfies

$$\frac{1}{N^2} \sum_{i \neq j} |r_{ij}| \leq \frac{C(N-1)}{N^2} \leq \frac{C}{N} \to 0.$$

This converts the original conditions to

$$\frac{m^{9/2}}{N\lambda_1^3(\theta)} + \frac{Cm^4}{N} \to 0$$
$$\frac{m^5}{N\lambda_1^4(\theta)} + \frac{Cm^5}{N} \to 0.$$

From the key ratio $\frac{m}{\lambda_1(\theta)} \to 0$, it can be derived that the asymptotic bounds:

$$\frac{m^{9/2}}{N\lambda_1^3(\theta)} = \left(\frac{m}{\lambda_1(\theta)}\right)^3 \cdot \frac{m^{3/2}}{N} \to 0$$
$$\frac{m^5}{N\lambda_1^4(\theta)} = \left(\frac{m}{\lambda_1(\theta)}\right)^4 \cdot \frac{m}{N} \to 0,$$

where the convergence follows from $\frac{m}{\lambda_1(\theta)} \to 0$ combined with polynomial growth rates.

The residual terms in (4.2.1) and (4.2.1) satisfy

$$\frac{Cm^4}{N} \leq C\left(\frac{m}{\lambda_1(\theta)}\right)^4 \cdot \frac{\lambda_1^4(\theta)}{N} \to 0$$
$$\frac{Cm^5}{N} \leq C\left(\frac{m}{\lambda_1(\theta)}\right)^5 \cdot \frac{\lambda_1^5(\theta)}{N} \to 0$$

through the boundedness $\lambda_1(\theta) \geq c > 0$ from Theorem 4.4's assumptions.

The original formulation contains an implicit requirement that needs explicit statement, the eigenvalue lower bound $\lambda_1(\theta) \geq c > 0$ must persist through the limit process $r \to 0$, without which the ratio $\frac{m}{\lambda_1(\theta)}$ could become undefined.

Therefore, under the simplified Conditions (4.38) with persistent eigenvalue bound, Theorem 4.4's conclusion remains valid through the demonstrated asymptotic relationships. □

Denote

$$L_n = \frac{\partial U_N(\theta)}{\partial \theta_n} \tau_N(r) = \max_{p=1,\dots,m} sup||l_p||.$$

Theorem 4.5 *Suppose that θ belongs to a compact set Θ and the error terms satisfy*

$$E[\varepsilon_i] = 0, \quad E[\varepsilon_i \varepsilon_j] = r_{ij}, \quad r_{ii} = \sigma_i^2, \quad 0 < \sigma_*^2 \leq \sigma_i^2 \leq \sigma_0^2.$$

Furthermore, assume that for some sufficiently large N, the following asymptotic conditions hold as $r \to 0$:

$$\frac{m^4\sqrt{m}}{N(\lambda_1(\theta))^3} + \frac{m^4}{N^2(\lambda_1(\theta))^3}\sum_{i \neq j}|r_{ij}| \to 0,$$

$$\frac{m^5}{N(\lambda_1(\theta))^4} + \frac{m^5}{N^2(\lambda_1(\theta))^4}\sum_{i \neq j}|r_{ij}| \to 0.$$

Then, the following convergence result holds:

$$m \cdot \tau_N(r) \xrightarrow{P} 0 \quad as \quad r \to 0.$$

Proof Consider

$\bar{f}_{ij}(\theta)$, which are the elements of the matrix $(\frac{F_N^T(\theta)F_N(\theta)}{N})^{-4}$, $i, j = 1, 2, \dots, m$;

$\partial f_{kl}(\theta)$, which are the elements of the matrix $(\frac{\partial F_N(\theta)}{\partial \theta_n})$, $k = 1, 2, \dots, m$; $l = 1, 2, \dots, N$.

Then from expression (4.27) it follows:

$$\begin{aligned}
&E||m \cdot \frac{\partial A_N(\theta)}{\partial \theta_n} \cdot \varepsilon||^2 = m \cdot E(\varepsilon^T \frac{\partial A_N^T(\theta)}{\partial \theta_n} \cdot \frac{\partial A_N(\theta)}{\partial \theta_n} \cdot \varepsilon) \\
&= \frac{m^2}{N} E[\varepsilon^T \frac{\partial F_N(\theta)}{\partial \theta_n} (\frac{F_N^T(\theta)F_N(\theta)}{N})^{-1} (\frac{F_N^T(\theta)F_N(\theta)}{N})^{-1} \frac{\partial F_N(\theta)}{\partial \theta_n} \varepsilon] \\
&+ \frac{m}{N^2} E[\varepsilon^T \frac{\partial F_N(\theta)}{\partial \theta_n} (\frac{F_N^T(\theta)F_N(\theta)}{N})^{-3} (\frac{\partial F_N(\theta)}{\partial \theta_n} \cdot F_N(\theta) + F_N^T(\theta) \frac{\partial F_N(\theta)}{\partial \theta_n}) F_N^T(\theta) \cdot \varepsilon] \qquad (4.39) \\
&+ \frac{m^2}{N} E[\varepsilon^T F_N(\theta) (\frac{\partial F_N^T(\theta)}{\partial \theta_n} \cdot F_N(\theta) + F_N^T(\theta) \frac{\partial F_N(\theta)}{\partial \theta_n}) (\frac{\partial F_N(\theta) F_N(\theta)}{N})^{-2} \\
&(\frac{\partial F_N(\theta) F_N(\theta)}{N})^{-2} (\frac{\partial F_N^T(\theta)}{\partial \theta_n} F_N(\theta) + F_N(\theta) \frac{\partial F_N(\theta)}{\partial \theta_n})] F_N^T(\theta) \cdot \varepsilon.
\end{aligned}$$

Denote $I_i, i = 1, 2, 3, 4$; $i - th$ term in (4.39). Using Condition (4.27) and the statement of Lemma 4.2 for elements of the matrix $\frac{F_N^T(\theta)F_N(\theta)}{N}$, it can be shown that

$$\frac{\partial \delta_N(\theta)}{\partial \theta_n} = -\frac{\partial \eta(x,\theta)}{\partial \theta_n} = -F_N(\theta).$$

Then from (4.31) it follows:

$$\begin{aligned} m \cdot \frac{\partial U_N(\theta)}{\partial \theta_n} &= [I_n + \frac{\partial A_N(\theta)}{\partial \theta_n}\delta_N(\theta) + A_N(\theta) \cdot \frac{\partial \delta_N(\theta)}{\partial \theta_n}]m \\ &= m \cdot [\frac{\partial A_N(\theta)}{\partial \theta_n}\delta(\theta) + \frac{\partial A_N(\theta)}{\partial \theta_n}(\delta_N(\theta) - \delta(\theta^*))] \\ &= \frac{\partial A_N(\theta)}{\partial \theta_n} \cdot \varepsilon \cdot m + \frac{\partial A_N(\theta)}{\partial \theta_n} \cdot \Delta\eta(x,\theta,\theta^*) \end{aligned} \quad (4.40)$$

where $\Delta\eta(x,\theta,\theta^*) = \eta(x,\theta^*) - \eta(x,\theta)$.

According to Theorem 4.4 the first term in (4.40) goes to zero for $r \to 0$.

The second term goes to zero because $\Delta\eta(x,\theta,\theta) \to 0$ for $r \to 0$ uniformly on x. Hence the statement of Theorem 4.5 has proved. □

Note 4.8 *This is the convergence law of parameter quantity of nonlinear large model under big data analysis. The theorem provides a rigorous mathematical guarantee for parameter estimation convergence in large-scale models (with parameters in the compact set* Θ*) under correlated noise. By bounding the noise correlation* $\sum_{i\neq j}|r_{ij}|$ *and maintaining a proper ratio between the sample size N and the smallest eigenvalue* $\lambda_1(\theta)$*, it ensures that as noise approaches zero, the estimation error term (measured by* $m \cdot \tau_N(r)$*) converges to zero in probability. In the context of AI, this means that for high-dimensional large models, as long as the relationship between model dimensionality m and data volume N is appropriately controlled, and the noise correlation remains constrained, parameter estimation remains stable and generalizable. This provides a solid theoretical foundation for large-scale AI models, including deep learning.*

Denote

$$\rho_N(\theta) = U_N(\theta) - \theta, \rho^* = \rho(\theta^*).$$

Theorem 4.6 *Suppose that there exists a constant A such that*

$$\frac{m}{\lambda_1(\theta)} + \frac{m}{N(\lambda_1(\theta))^2}\sum_{i\neq j}|r_{ij}| \leq A$$

Then

$$\lim_{N\to\infty}\lim_{k\to\infty} P\{\sqrt{N}\cdot\rho^* > k\} = 0.$$

Proof The goal is to show that the probability $P\left\{\sqrt{N}\cdot\rho^* > k\right\}$ approaches 0 as N grows large and k becomes large. To begin, consider the quantity $\rho_N(\theta) = U_N(\theta) - \theta$, and compute the expectation of its squared norm:

$$E[\|\rho_N(\theta)\|^2] = E\left[(U_N(\theta) - \theta)^T (U_N(\theta) - \theta)\right].$$

This can be rewritten as:

$$E[\|\rho_N(\theta)\|^2] = E\left[\varepsilon^T A_N^T(\theta) A_N(\theta) \varepsilon\right],$$

where ε represents a random error vector, and $A_N(\theta)$ is a matrix associated with the estimation of $U_N(\theta)$.

Next, the elements of $A_N(\theta)$ are bounded, and Lemma 4.2 provides a useful bound. Thus, the expectation can be further expressed as follows:

$$E[\|\rho_N(\theta)\|^2] = E\left[\varepsilon^T F_N(\theta)\left(\frac{F_N^T(\theta) F_N(\theta)}{N}\right)^{-1} \frac{F_N^T(\theta)}{N}\varepsilon\right],$$

where $F_N(\theta)$ is a matrix derived from $U_N(\theta)$.

Since $F_N(\theta)$ is bounded, it is possible to bound the expectation using the trace of the inverse matrix. This leads to the inequality:

$$N E[\|\rho_N(\theta)\|^2] \leq \sigma_0^2 \operatorname{tr}\left(\left(\frac{F_N^T(\theta) F_N(\theta)}{N}\right)^{-1}\right) + \frac{m}{N(\lambda_1(\theta))^2}\sum_{i \neq j} |r_{ij}|.$$

By applying properties of the trace, this simplifies to

$$N E[\|\rho_N(\theta)\|^2] \leq \frac{c_1 \cdot m}{\lambda_1(\theta)} + \frac{m}{N(\lambda_1(\theta))^2}\sum_{i \neq j} |r_{ij}|.$$

Thus, it follows that

$$N E[\|\rho_N(\theta)\|^2] \leq A, \quad \forall \theta \in B(r),$$

where A is the constant given in the assumption of the theorem and $B(r)$ denotes a ball of radius r centered at θ^*.

Now, since $\rho^*(\theta) = \rho(\theta^*)$ is bounded by the constant A, Chebyshev's inequality can be applied. It yields

$$P\left\{\sqrt{N} \cdot \rho^* > k\right\} \leq \frac{E[\|\rho^*\|^2]}{k^2}.$$

Using the inequality $N E[\|\rho^*\|^2] \leq A$, it is obtained that

$$P\left\{\sqrt{N} \cdot \rho^* > k\right\} \leq \frac{A}{k^2}.$$

Therefore, as $k \to \infty$, the probability tends to zero, which completes the proof.

Thus, the result is proven:

$$\lim_{N\to\infty}\lim_{k\to\infty} P\left\{\sqrt{N}\cdot\rho^* > k\right\} = 0.$$

In conclusion, the proof of Theorem 4.6 is now complete. □

Note 4.9 *This theorem is parameter domain radius collapse theorem under nonlinear large model big data analysis.*

Theorem 4.7 *Suppose that $\theta(0) \in B(r)$ and that the following inequality holds:*

$$\tau_N(r) + \frac{||\rho^*||}{r} < 1.$$

Furthermore, assume that there exists a constant A such that

$$\frac{m}{\lambda_1(\theta)} + \frac{m}{N(\lambda_1(\theta))^2}\sum_{i\neq j}|r_{ij}| \leq A.$$

Then, there exists a random variable $\widehat{\theta}_N$ such that

$$\widehat{\theta}_N(s) \xrightarrow{P} \widehat{\theta}_N, \quad as \quad s\to\infty.$$

Proof Since the partial derivative $\frac{\partial U_N(\theta)}{\partial\theta}$ exists by Condition (4.26), it follows from Lagrange's formula that

$$||U_N(\theta) - U_N(\theta^*)|| \leq ||\frac{\partial U_N(\widetilde{\theta})}{\partial\theta}|| \cdot ||\theta - \theta^*||, \quad \theta \in B(r),\ \widetilde{\theta} \in B(r).$$

By applying the statement of Theorem 4.5 for all $\widetilde{\theta} \in B(r)$, it holds that

$$||\frac{\partial U_N(\widetilde{\theta})}{\partial\theta}|| = \left|\left|\sum_{p=1}^{m}\frac{\partial U_N(\widetilde{\theta})}{\partial\theta_p}\right|\right| \leq m\cdot\sup_{1\leq p\leq m}||\frac{\partial U_N(\widetilde{\theta})}{\partial\theta_p}|| \to 0 \text{ as } r\to 0.$$

Thus, if $\theta(0) \in B(r)$, there exists some N such that, starting from some r, the mapping $U_N(\theta)$ becomes compressive.

On the other hand, using the following inequality:

$$||U_N(\theta) - \theta^*|| \leq ||U_N(\theta) - U_N(\theta^*)|| + ||U_N(\theta^*) - \theta^*|| \leq \tau_N(r)\cdot r + ||\rho^*|| \leq r,$$

according to Theorem 4.5. Therefore, the mapping $U_N(\theta)$ maps the ball $B(r)$ into itself.

Since $B(r)$ is a complete metric space, it follows from the Banach fixed-point theorem (or the principle of contractive mappings [26]) that for all $\theta(0) \in B(r)$, as $r \to 0$ and $s \to \infty$, the equation

$$U_N(\theta) = \theta$$

has a unique solution, denoted as $\widehat{\theta}$.

Furthermore, since $U_N(\theta(s-1)) = \theta(s)$ and

$$||U_N(\theta(s-1)) - U_N(\widehat{\theta})|| < r \quad \text{as} \quad r \to 0,$$

it follows that

$$||\theta_N(s) - \widehat{\theta}_N|| \xrightarrow{P} 0, \quad \text{in probability as} \quad s \to \infty, \quad r \to 0.$$

Thus, the desired result is established. □

Note 4.10 *In this theorem, the convergence condition of parameter analysis is proposed for the large model with nonlinear parameter structure, and the mathematical basis of Newton iterative method for solving parameters is established. The necessary condition of convergence and the final conclusion are clarified.*

Remark 4.8 Actually, Theorem 4.7 can be interpreted as follows. The least squares estimator ($l.s.e.$) is obtained as the solution of an iterative process that depends on the initial approximation. If, from a practical perspective, the initial value θ_0 is chosen such that $\theta_0 \in B(r)$, then as $s \to \infty$ (where s represents the number of iterations), the iterative process converges to a certain value θ_N. Furthermore, as $N \to \infty$, the limiting value $\theta(\infty)$ can be regarded as the true value of the unknown parameters in the sense of the $l.s.e.$

Theorem 4.8 *Suppose that $\theta(0) \in B(r)$ and that the following inequality holds:*

$$\tau_N(r) + \frac{||\rho^*||}{r} < 1.$$

Furthermore, assume that there exists a constant A such that

$$\frac{m}{\lambda_1(\theta)} + \frac{m}{N(\lambda_1(\theta))^2} \sum_{i \neq j} |r_{ij}| \leq A.$$

Additionally, let $\widehat{\theta}_N$ be a random variable satisfying

$$\widehat{\theta}_N(s) \xrightarrow{P} \widehat{\theta}_N, \quad as \quad s \to \infty.$$

Then, the following bound holds:

$$\sup_{\theta(0) \in B(r)} ||\theta_N(s) - \widehat{\theta}_N|| \leq \frac{(\tau_N(r))^s}{1 - \tau_N(r)} \cdot r.$$

Proof Let $s < t$ be arbitrary. According to the principle of contractive mappings (as detailed in [26]), it holds that

$$\sup_{\theta(0)\in B(r)} ||\theta(s) - \theta(t)|| \leq \frac{(\tau_N(r))^s}{1 - \tau_N(r)} ||\theta(0) - \theta(1)||.$$

This result follows from the fact that the mapping $U_N(\theta)$ is a contraction, and hence the distance between successive iterates decays exponentially with the number of iterations. The rate of this decay is controlled by $\tau_N(r)$, which is less than 1 by assumption.

Since $U_N(\theta)$ maps $B(r)$ into itself, the above inequality implies

$$\sup_{\theta(0)\in B(r)} ||\theta(s) - \theta(t)|| \leq \frac{(\tau_N(r))^2}{1 - \tau_N(r)} \cdot r.$$

This establishes that the distance between $\theta(s)$ and $\theta(t)$ is bounded by a term involving $\tau_N(r)$ raised to a power and the radius r.

Furthermore, since $\theta(t) \in B(r)$, the operator $U_N(\theta)$ is applied to $\theta(t)$, leading to the convergence of the sequence to the limit point $\widehat{\theta}_N$. By Theorem 4.7, this sequence converges to $\widehat{\theta}_N$. Consequently, it follows that

$$\sup_{\theta(0)\in B(r)} ||\theta_N(s) - \widehat{\theta}_N|| \leq \frac{(\tau_N(r))^s}{1 - \tau_N(r)} \cdot r.$$

This completes the proof, as it establishes the desired bound on the distance between $\theta_N(s)$ and the fixed point $\widehat{\theta}_N$, showing that this distance decays exponentially with s. □

Note 4.11 *The theorem establishes a theoretical framework for the convergence rate of parameter iteration for the large model under nonlinear parameter structure, and clarifies its convergence characteristics and influencing factors.*

Theorem 4.9 *Suppose that $\theta(0) \in B(r)$ and that the following inequality holds:*

$$\tau_N(r) + \frac{||\rho^*||}{r} < 1.$$

Furthermore, assume that there exists a constant A such that

$$\frac{m}{\lambda_1(\theta)} + \frac{m}{N(\lambda_1(\theta))^2} \sum_{i \neq j} |r_{ij}| \leq A.$$

Additionally, let $\widehat{\theta}_N$ be a random variable satisfying

$$\widehat{\theta}_N(s) \xrightarrow{P} \widehat{\theta}_N, \quad as \quad s \to \infty.$$

Then, the random variable

$$\sqrt{N}(\widehat{\theta}_N - \theta^*)$$

is bounded in probability as $N \to \infty$*, i.e., the estimator* $\widehat{\theta}_N$ *is* $\sqrt{N}$*-consistent.*

Proof Let $r = \frac{2k}{\sqrt{N}}$ be chosen. From Theorem 4.5, it follows that

$$\tau(r) \to 0 \quad \text{as} \quad N \to \infty \text{ in probability.}$$

According to Theorem 4.8, the random variable $\sqrt{N} \cdot \rho^*$ is bounded in probability. To apply this result, consider the choice of parameters:

$$\tau(r) = \frac{1}{2}, \quad r = \frac{2k}{\sqrt{N}}, \quad ||\rho^*|| < \frac{k}{\sqrt{N}}.$$

Using these values, the following inequality holds:

$$\tau_N(r) + \frac{||\rho^*||}{r} < \frac{1}{2} + \frac{||\rho^*|| \cdot \sqrt{N}}{2k}.$$

This inequality holds for sufficiently large N, ensuring that the term $\tau_N(r) + \frac{||\rho^*||}{r}$ remains less than $\frac{1}{2}$.

Next, consider the bound:

$$||\widehat{\theta}_N - \theta^*|| \leq \sup_{\theta(0) \in B(r)} ||\widehat{\theta} - \theta(s)||.$$

Using the conditions from Theorem 4.8, it follows that

$$\lim_{k \to \infty} \lim_{N \to \infty} P\left(||\widehat{\theta}_N - \theta^*|| > \frac{k}{\sqrt{N}}\right) = 0.$$

This result establishes that as N increases, the probability that $||\widehat{\theta}_N - \theta^*||$ exceeds $\frac{k}{\sqrt{N}}$ vanishes, implying that the estimator $\widehat{\theta}_N$ converges to θ^* at the rate of $\frac{1}{\sqrt{N}}$.

Thus, $\widehat{\theta}_N$ is $\sqrt{N}$-consistent, and the proof is complete. □

Note 4.12 *The theorem establishes a theoretical framework for the boundedness and consistency of parameters for large-scale model analysis under nonlinear parameter framework, and clarifies the necessary conditions for its establishment.*

Theorem 4.10 *Let* $\varepsilon_i, i = 1, 2, ...,$ *be the sequence of independent random variables. If*

$$\frac{m^3}{N \cdot \lambda_1^2(\theta)} \to 0 \tag{4.41}$$

for $N \to \infty$, $\forall \theta \in B(r)$, *then* $\sqrt{N} \cdot \rho^* \to N\{0, \sum(\theta^*)\}$ *for* $N \to \infty$
where

$$\sum(\theta^*) = \left(\frac{F_N^T(\theta^*)F_N(\theta^*)}{N}\right)^{-1} \frac{F_N^T(\theta^*)I(\sigma^2)F_N(\theta^*)}{N}(\frac{F_N^T(\theta^*)F_N(\theta^*)}{N})^{-1} \tag{4.42}$$

$$I(\sigma^2) = E[\varepsilon\varepsilon^T] = \begin{pmatrix} \sigma_1^2 & 0 & \dots & 0 \\ 0 & \sigma_2^2 & \dots & 0 \\ \vdots & \vdots & \ddots & \vdots \\ 0 & 0 & \dots & \sigma_N^2 \end{pmatrix}. \tag{4.43}$$

Proof Consider the nonzero vector $l = (l_1, l_2, \dots, l_m)^T$ of size m. Define

$$\sqrt{N}l^r\rho^* = \sqrt{N}l^T\left(\frac{F_N^T(\theta^*)F_N(\theta^*)}{N}\right)^{-1}\frac{F_N^T(\theta^*)}{N}(y - \eta(x, \theta^*)),$$

which can be expanded as

$$\sqrt{N}l^T\rho^* = \frac{1}{N}\sum_{k=1}^{N}\sum_{i=1}^{m}\sum_{j=1}^{m}\sqrt{N}l_i\bar{f}_{ij}f_{jk}(\theta^*)\varepsilon_k,$$

where $f_{jk}(\theta)$ represents the elements of the matrix $F_N^T(\theta^*)$, with $j = 1, 2, \dots, N$ and $k = 1, 2, \dots, m$, and $\left(\frac{F_N^T(\theta^*)F_N(\theta^*)}{N}\right)^{-1}$ contains elements denoted by x_{ij} for $i, j = 1, 2, \dots, m$.

Next, take the expectation of both sides:

$$E[\sqrt{N}l^T\rho^*] = 0,$$

which follows directly from Theorem 4.6.

Now, consider the second moment:

$$E[(\sqrt{N}l^T\rho^*)^2] = \frac{1}{N}\sum_{i=1}^{N}\left(\sum_{i=1}^{m}\sum_{j=1}^{m}l_i x_{ij}f_{jk}\right)^2\sigma_i^2.$$

This expression is bounded as follows:

$$E[(\sqrt{N}l^T\rho^*)^2] \le \frac{\sigma_0^2}{N}\left(\frac{\sqrt{m}}{\lambda_1(\theta)}\right)^2 C_1^2 \cdot N \cdot \left(\sum_{i=1}^{m} l_i\right)^2 \le C_1^2 \cdot \sigma_0^2 \cdot \frac{m^3}{\lambda_1^2(\theta)} \le C_2 \cdot \frac{m^3}{\lambda_1^2(\theta)},$$

which is finite for all $\theta \in B(r)$, according to the assumption of Theorem 4.9.

By the central limit theorem [12], the random variable $\sqrt{N}l^T\rho^*$ converges in distribution to a Gaussian random variable with mean 0 and covariance matrix $\sum(\theta^*)$:

$$E||\sqrt{N}l^T\rho^*|| = N \cdot El^T\rho^*\rho^{*T}l = l^T\left(\frac{F_N^T(\theta^*)F_N(\theta^*)}{N}\right)^{-1}$$
$$\times \frac{F_N^T(\theta^*)I(\sigma^2)F_N(\theta^*)}{N}\left(\frac{F_1^T(\theta^*)F_N(\theta^*)}{N}\right)^{-1} \cdot l = l^T\sum(\theta^*) \cdot l$$

which simplifies to

$$l^T\sum(\theta^*)l.$$

Thus, by the criterion for normality [53], the random variable $\sqrt{N}\rho^*$ converges in distribution to a normal distribution:

$$\sqrt{N}\rho^* \to N(0, \sum(\theta^*)),$$

as required.

To establish the minimal number of steps for asymptotic normality, assume the condition

$$\tau_N(r) + \frac{||\rho^*||}{r} < a < 1 \tag{4.44}$$

and

$$\frac{r \cdot a^b}{1-a} < \frac{k}{\sqrt{N}},$$

where k exists because the random variable $\sqrt{N}||\widehat{\theta}_N - \theta^*||$ is bounded, as shown in Theorem 4.9. □

Note 4.13 *The theorem establishes the theoretical framework of convergence distribution for large model analysis under nonlinear parameter framework, and clarifies the conditions and limitations of its establishment.*

Lemma 4.3 *Suppose that the following inequality holds:*

$$\tau_N(r) + \frac{||\rho^*||}{r} < a < 1.$$

Furthermore, assume that for some sufficiently large s:

$$s \geq \frac{\ln[(1-a)k] - \ln(r - \sqrt{N})}{\ln a}. \tag{4.45}$$

Additionally, let ε_i, $i = 1, 2, \ldots$, be a sequence of independent random variables, and suppose that

$$\frac{m^3}{N \cdot \lambda_1^2(\theta)} \to 0, \quad as \quad N \to \infty, \quad \forall \theta \in B(r).$$

Then, the following convergence result holds:

$$(\theta(s) - \theta^*) \xrightarrow{P} N(0, \Sigma(\theta^*)) \quad as \quad N \to \infty, \tag{4.46}$$

where

$$\Sigma(\theta^*) = \left(\frac{F_N^T(\theta^*)F_N(\theta^*)}{N}\right)^{-1} \frac{F_N^T(\theta^*)I(\sigma^2)F_N(\theta^*)}{N} \left(\frac{F_N^T(\theta^*)F_N(\theta^*)}{N}\right)^{-1},$$

and the matrix $I(\sigma^2)$ is given by

$$I(\sigma^2) = E[\varepsilon\varepsilon^T] = \begin{pmatrix} \sigma_1^2 & 0 & \dots & 0 \\ 0 & \sigma_2^2 & \dots & 0 \\ \vdots & \vdots & \ddots & \vdots \\ 0 & 0 & \dots & \sigma_N^2 \end{pmatrix}.$$

Proof From Theorem 4.8 under Condition (4.44), the iteration error satisfies

$$\|\theta_N(s) - \widehat{\theta}_N\| \le \frac{(\tau_N(r))^s}{1 - \tau_N(r)} \cdot r.$$

Substituting the lower bound of s from (4.45):

$$\begin{aligned}(\tau_N(r))^s &\le \exp(s \ln a) \quad (\text{since } \tau_N(r) < a) \\ &\le \exp\left(\frac{\ln[(1-a)k] - \ln(r - \sqrt{N})}{\ln a} \cdot \ln a\right). \\ &= \frac{(1-a)k}{r - \sqrt{N}}\end{aligned}$$

Thus the error bound becomes

$$\|\theta_N(s) - \widehat{\theta}_N\| \le \frac{(1-a)k}{(r - \sqrt{N})(1 - \tau_N(r))} \cdot r$$

which tends to 0 as $N \to \infty$ by Theorem 4.6.

Theorem 4.9 establishes that

$$\sqrt{N}(\widehat{\theta}_N - \theta^*) = O_P(1).$$

Combining through triangle inequality:

$$\sqrt{N}(\theta(s)-\theta^*)=\sqrt{N}(\theta(s)-\widehat{\theta}_N)+\sqrt{N}(\widehat{\theta}_N-\theta^*).$$

The first term vanishes in probability since

$$\sqrt{N}\|\theta(s)-\widehat{\theta}_N\|\le\sqrt{N}\cdot O_P(1)=O_P(1).$$

The remaining term $\sqrt{N}(\widehat{\theta}_N-\theta^*)$ converges in distribution to $N(0,\cdot)$ by Theorem 4.9's $\sqrt{N}$-consistency and the central limit theorem conditions embedded in Theorem 4.6.

$$\sqrt{N}(\theta(s)-\theta^*)=o_P(1)+O_P(1)\xrightarrow{d}N(0,\cdot).$$

Dividing both sides by $\sqrt{N}$ gives the final result (4.46). □

Theorem 4.11 *Suppose that the random variable*

$$\sqrt{N}(\widehat{\theta}-\theta^*)$$

is bounded in probability as $N\to\infty$. Furthermore, assume that

(1) There exists a constant C_1 such that

$$|a_{ij}^{(n_1)}|\le C_1,\quad |b_{ij}^{(n_2)}|\le C_1,\quad \max_{1\le i\le m}\sum_{j=1}^{m}|c_{ij}|\le C_1.$$

(2) The sequence of independent random variables ε_i, $i=1,2,\ldots,$ satisfies

$$\tau_N(r)+\frac{||\rho^*||}{r}<a<1.$$

Additionally, for some sufficiently large s

$$s\ge\frac{\ln[(1-a)k]-\ln(r-\sqrt{N})}{\ln a}.$$

Moreover, the asymptotic condition

$$\frac{m^3}{N\cdot\lambda_1^2(\theta)}\to 0,\quad as\quad N\to\infty,\quad \forall\theta\in B(r)$$

holds.

(3) The parameter θ belongs to a compact set Θ, and the error terms satisfy

$$E[\varepsilon_i]=0,\quad E[\varepsilon_i\varepsilon_j]=r_{ij},\quad r_{ii}=\sigma_i^2,\quad 0<\sigma_*^2\le\sigma_i^2\le\sigma_0^2.$$

Moreover, for some sufficiently large N, the following asymptotic conditions hold as $r\to 0$:

$$\frac{m^4\sqrt{m}}{N(\lambda_1(\theta))^3}+\frac{m^4}{N^2(\lambda_1(\theta))^3}\sum_{i\neq j}|r_{ij}|\to 0,$$

$$\frac{m^5}{N(\lambda_1(\theta))^4}+\frac{m^5}{N^2(\lambda_1(\theta))^4}\sum_{i\neq j}|r_{ij}|\to 0.$$

(4) There exists a constant A such that

$$\frac{m}{\lambda_1(\theta)}+\frac{m}{N(\lambda_1(\theta))^2}\sum_{i\neq j}|r_{ij}|\leq A.$$

Then, the following asymptotic normality result holds:

$$\sqrt{N}(\widehat{\theta}-\theta^*)\xrightarrow{D}N(0,\Sigma(\theta^*))\quad as\quad N\to\infty,$$

where

$$\Sigma(\theta^*)=\left(\frac{F_N^T(\theta^*)F_N(\theta^*)}{N}\right)^{-1}\frac{F_N^T(\theta^*)I(\sigma^2)F_N(\theta^*)}{N}\left(\frac{F_N^T(\theta^*)F_N(\theta^*)}{N}\right)^{-1}.$$

Proof ([14]) The estimating equation satisfies

$$A_N(\widehat{\theta})\cdot\delta_N(\widehat{\theta})=0. \tag{4.47}$$

Expanding the terms in Eq. (4.47), the following expression is obtained:

$$\begin{aligned}A_N(\hat{\theta})\cdot\delta_N(\hat{\theta})&=[A_N(\theta^*)+A_N(\hat{\theta})-A_N(\theta^*)][y-\eta(x,\theta^*)+\eta(x,\theta^*)-\eta(x,\hat{\theta})]\\&=A_N(\theta^*)(y-\eta(x,\theta^*))+A_N(\theta^*)(\eta(x,\theta^*)-\eta(x,\hat{\theta}))\\&+[A_N(\hat{\theta})-A_N(\theta^*)](y-\eta(x,\theta^*))+[A_N(\hat{\theta})-A_N(\theta^*)](\eta(x,\theta^*)-\eta(x,\hat{\theta}))\\&=\rho^*+[A_N(\hat{\theta})-A_N(\theta^*)]\cdot\delta^*+A_N(\hat{\theta})\cdot\Delta\eta(x,\theta^*,\hat{\theta})=0.\end{aligned}$$

For the term $A_N(\widehat{\theta})-A_N(\theta^*)$ and $\Delta\eta(x,\theta^*,\widehat{\theta})$, the following identities hold:

$$A_N(\widehat{\theta})-A_N(\theta^*)=\left[\frac{\partial A_N(\theta^*)}{\partial\theta}+\int_0^1\left(\frac{\partial A_N(\theta^*+z(\widehat{\theta}-\theta^*))}{\partial\theta}-\frac{\partial A_N(\theta^*)}{\partial\theta}\right)dz\right]\cdot(\widehat{\theta}-\theta^*), \tag{4.48}$$

$$\eta(\widehat{\theta})-\eta(\theta^*)=\left[\frac{\partial\eta(\theta^*)}{\partial\theta}+\int_0^1\left(\frac{\partial\eta(\theta^*+z(\widehat{\theta}-\theta^*))}{\partial\theta}-\frac{\partial\eta(\theta^*)}{\partial\theta}\right)dz\right]\cdot(\widehat{\theta}-\theta^*). \tag{4.49}$$

Substituting Eq. (4.49) into (4.48), the following expression is obtained:

$$\rho^* + \frac{\partial A_N(\theta^*)}{\partial \theta}\delta_N^*(\widehat{\theta} - \theta^*) \int_0^1 \frac{\partial A_N(\theta^* + z(\widehat{\theta} - \theta^*))}{\partial \theta}\delta_N^*(\widehat{\theta} - \theta^*) - \frac{\partial A_N(\theta^*)}{\partial \theta} \cdot$$
$$\delta_N^*(\widehat{\theta} - \theta^*)dz + A_N(\widehat{\theta})\frac{\partial \eta(\theta^*)}{\partial \theta}(\widehat{\theta} - \theta^*) + A_N(\widehat{\theta}) \int_0^1 \frac{\partial \eta}{\partial \theta}(\theta^* + z(\widehat{\theta} - \theta^*))(\widehat{\theta} - \theta^*)$$
$$- A_N(\widehat{\theta})\frac{\partial \eta(\widehat{\theta})}{\partial \theta}(\widehat{\theta} - \theta^*) - A_N(\widehat{\theta})\frac{\partial \eta(\theta^*)}{\partial \theta}(\widehat{\theta} - \theta^*) = 0. \tag{4.50}$$

From the assumption that

$$A_N(\widehat{\theta}) \cdot \frac{\partial \eta(\widehat{\theta})}{\partial \theta} = I_m,$$

it follows from Eq. (4.50) that

$$\sqrt{N}(\widehat{\theta} - \theta^*) = \sqrt{N} \cdot \rho^* + \sqrt{N} \cdot M_1 + \sqrt{N} \cdot M_2,$$

where

$$M_1 = \int_0^1 \frac{\partial A_N(\theta^* + z(\widehat{\theta} - \theta^*))}{\partial \theta}\delta_N^* dz(\widehat{\theta} - \theta^*)$$

and

$$M_2 = A_N(\widehat{\theta}) \int_0^1 \frac{\partial \eta(\theta^* + z(\widehat{\theta} - \theta^*))}{\partial \theta} dz(\widehat{\theta} - \theta^*).$$

Since the random variable $\sqrt{N}(\widehat{\theta} - \theta^*)$ is bounded in probability, and by Theorem 4.4, it follows that

$$\frac{\partial A_N(\theta)}{\partial \theta} \cdot \delta_N^* \xrightarrow{P} 0 \quad \text{as} \quad N \to \infty.$$

Thus, $\sqrt{N} \cdot M_1 \xrightarrow{P} 0$ as $N \to \infty$. Similarly, $\sqrt{N} \cdot M_2 \xrightarrow{P} 0$ as $N \to \infty$. Finally, using Theorem 4.10, it follows that

$$\sqrt{N}(\widehat{\theta} - \theta^*) \xrightarrow{d} N(0, \sum(\theta^*)).$$

This completes the proof. □

Note 4.14 *The theorem constructs the mathematical theory of parameter convergence distribution for the large model analysis under the nonlinear parameter framework, and clarifies the necessary conditions and constraints for its establishment.*

4.2.2 Confidence Interval for Nonlinear Prototype

In the case of nonlinear regression models, the least squares estimator ($l.s.e.$) can be obtained through an iterative process. Similar to the procedure described in the previous chapter, the elements of the covariance matrix of the deviation vector $\sqrt{N}(\widehat{\theta}-\theta)$ can be estimated. Given these estimators, a confidence interval for the function $f(\theta, x)$ at each observation point x_i can be constructed.

It has been established that, under certain conditions, the least squares estimator remains consistent for the model. Furthermore, the vector $\sqrt{N}(\widehat{\theta}-\theta)$ follows an asymptotically normal distribution as $N \to \infty$.

Consider the vector $\sqrt{Nm}(\widehat{\theta}-\theta)$ and its associated covariance matrix. The covariance matrix of $\sqrt{Nm}(\widehat{\theta}-\theta)$ satisfies

$$C_N^* = E(\sqrt{Nm}(\widehat{\theta}-\theta)^T \sqrt{Nm}(\widehat{\theta}-\theta)) = NmE(\widehat{\theta}-\theta)^T(\widehat{\theta}-\theta) = NmC_N.$$

According to Theorem 4.10, the vector $\sqrt{N}(\widehat{\theta}-\theta)$ follows a Gaussian distribution. Consequently, the vector $\sqrt{Nm}(\widehat{\theta}-\theta)$ also exhibits a Gaussian distribution. Consider the quadratic form:

$$\frac{1}{N}(\widehat{\theta}-\theta)^T C_N^{*\,-1}(\widehat{\theta}-\theta) = \frac{1}{Nm}(\widehat{\theta}-\theta)^T C_N^{-1}(\widehat{\theta}-\theta). \tag{4.51}$$

By Theorem 4.11, the quadratic form in (4.51) follows an asymptotic chi-square distribution as $N \to \infty$ with m degrees of freedom, since the vector $\sqrt{N}(\widehat{\theta}-\theta)$ is asymptotically Gaussian [8]. Consequently, with probability $\gamma > 0$, the following inequality holds:

$$(\widehat{\theta}-\theta)^T C_N^{-1}(\widehat{\theta}-\theta) \le \frac{\chi_\gamma^2(m)}{Nm}, \tag{4.52}$$

where $\chi_\gamma^2(m)$ represents the quantile of level $\gamma > 0$ for the chi-square distribution with m degrees of freedom.

As noted in Chapter 3, for sufficiently large m, the chi-square distribution can be approximated by a normal distribution with mean m and variance $2\,m$. Therefore, in practical applications, the normal approximation is often preferable. In this case, the chi-square distribution in (4.52) can be replaced by a normal distribution with mean m and variance $2\,m$, denoted as $N(m, 2\,m)$.

In the ratio (4.52), the matrix C_N^{-1} is unknown. Following the approach in Theorem 4.2, asymptotically unbiased and consistent estimators of the elements of the covariance matrix C_N have been constructed. Therefore, the estimator matrix $\widehat{C}_N^{-1}$ can be used in place of C_N^{-1}. Solving (4.52) for θ, the confidence ellipsoid is obtained:

$$\varepsilon_\gamma(\theta) = \{\theta : (\widehat{\theta}-\theta)^T \widehat{C}_N^{-1}(\widehat{\theta}-\theta) \le \frac{x_\gamma^2(m)}{N}\}. \tag{4.53}$$

Alternatively, instead of $\frac{x_\gamma^2(m)}{N}$, the quantile of level $\gamma > 0$ of the normal distribution $N(1, 2)$ can be used:

$$\varepsilon_\gamma(\theta) = \left\{\theta : (\widehat{\theta} - \theta)^T \widehat{C}_N^{-1} (\widehat{\theta} - \theta) < \frac{N_\gamma(1, 2)}{N}\right\},$$

where $N(1, 2)$ is the quantile of level $\gamma > 0$ for the normal distribution with mean 1 and variance 2.

Remark 4.9 The vector $\sqrt{Nm}(\widehat{\theta} - \theta)$ is introduced because, in (4.53), replacing the chi-square distribution with the normal distribution and dividing the right-hand side by m results in a normal distribution with parameters 1 and 2.

Remark 4.10 It should be emphasized that, in principle, the elements of the matrix $\widehat{C}_N$ are not necessarily required to converge as $N \to \infty$. The essential property is that the difference between c_{kl} and c_{kl}^* (elements of the deviation vector) converges to zero in probability, implying that the difference between the two vectors also vanishes.

To determine the lower and upper bounds of the confidence band, the following expressions must be evaluated:

$$\inf_{\varepsilon_\gamma(\theta)} f(x, \theta) \quad \text{and} \quad \sup_{\varepsilon_\gamma(\theta)} f(x, \theta),$$

which correspond to the lower and upper boundaries, respectively.

For convenience, the confidence ellipsoid $\varepsilon_\gamma(\theta)$ can be represented in the following form:

$$\sum_{i=1}^{m} \frac{(\widehat{\theta}_i - \theta_i)^2}{\chi_\gamma^2(m) \cdot \lambda_{m-i}} = \frac{1}{\sqrt{N}},$$

where $\chi_\gamma^2(m)$ denotes the quantile of level $\gamma > 0$ of the chi-square distribution with m degrees of freedom, and $\lambda_1, \lambda_2, \ldots, \lambda_m$ represent the eigenvalues of the matrix $\widehat{C}_N^{-1}$.

The next step involves determining the minimum and maximum of the function $f(x, \theta)$, subject to the constraint that θ belongs to the confidence ellipsoid. To achieve this, Lagrange's method of undetermined multipliers is employed [25]. The corresponding Lagrange function is given by

$$Z = \sum_{i=1}^{m} \varphi_i(x) \cdot \theta_i + \lambda^* \cdot \sum_{i=1}^{m} \left[\frac{(\widehat{\theta}_i - \theta_i)^2}{\chi_\gamma^2(m) \cdot \lambda_{m-i}} - \frac{1}{\sqrt{N}}\right],$$

where λ^* is the Lagrange multiplier and θ_i satisfies the following equation:

$$\frac{d}{d\theta_i} Z = 0, \quad i = 1, 2, \ldots, m.$$

These equations determine the required points.

Note 4.15 *In the above steps, the analytical conclusions of the statistical properties of the parameters are given for the large-scale model analysis under the nonlinear parameter architecture, and the corresponding algorithm steps are described in detail.*

4.2.3 Numerical Illustration for Increasing-Parameter Nonlinear Prototype

Example 4.4 In the previous example, a three-parameter nonlinear regression model was considered, consisting of an exponential function combined with a constant term. That setting served to verify the asymptotic behavior of the estimator in a relatively low-dimensional case. To further investigate how the estimation properties evolve as the number of parameters increases, this section introduces a more complex nonlinear model with six parameters, thereby facilitating a direct comparison with the previous example. The proposed model takes the following general form:

$$y = \alpha_1 \exp(\alpha_2 x) + \alpha_3 x^{\alpha_4} + \alpha_5 \sin(\alpha_6 x) + \varepsilon,$$

which combines exponential, power, and trigonometric components. This hybrid structure offers increased representational flexibility and allows for a more comprehensive examination of the estimator's performance in a higher-dimensional setting.

In this numerical experiment, the true parameter vector is specified as

$$(\alpha_1, \alpha_2, \alpha_3, \alpha_4, \alpha_5, \alpha_6) = (2.0,\ 0.3,\ 1.2,\ 0.5,\ 0.8,\ 1.2),$$

and the observation errors are generated from an autoregressive process of order one (AR(1)):

$$\varepsilon_i = \rho\varepsilon_{i-1} + \eta_i, \quad \eta_i \sim N(0, \sigma_{\text{error}}^2), \quad i = 1, 2, \ldots, N,$$

with $\rho = 0.5$ and $\sigma_{\text{error}} = 0.5$. The sample size N ranges from small ($N = 50$) to extremely large ($N = 10^7$), with 100 evenly spaced points in logarithmic scale. For each sample size, 50 Monte Carlo replications are performed to analyze the mean behavior and variability of the estimated parameters.

This experimental design is constructed based on the theoretical framework provided by Theorem 4.7, which ensures the asymptotic convergence of the estimator under appropriate conditions. Specifically, assuming that the initial value $\theta(0)$ lies within the ball $B(r)$, and the following inequality holds:

$$\tau_N(r) + \frac{\|\rho^*\|}{r} < 1,$$

Table 4.5 Parameter bias convergence under large-sample conditions in nonlinear case with six-mixed parameters

Sample size	α_1	α_2	α_3	α_4	α_5	α_6
50	−1.98332	0.70960	3.16630	−0.13995	−0.03673	0.33555
526363	−0.27371	0.01685	0.34025	−0.04624	−0.01721	0.00673
1052676	−0.45995	0.03130	0.58467	−0.07773	−0.03844	0.01080
1578989	0.02262	−0.00059	−0.02353	−0.00493	−0.00208	−0.00139
2105302	0.12443	−0.00717	−0.15353	0.03408	0.00956	−0.00194
2631615	0.08622	−0.00506	−0.10566	0.02243	0.00576	−0.00066
3157928	0.01541	−0.00083	−0.01523	−0.00011	−0.00240	0.00073
3684242	0.05206	−0.00260	−0.05983	0.00541	−0.00088	−0.00101
4210555	0.13423	−0.00737	−0.16265	0.03156	0.00645	−0.00277
4736868	0.15711	−0.00882	−0.19216	0.04209	0.00942	−0.00313
5263181	0.12686	−0.00716	−0.15548	0.03284	0.00749	−0.00258
5789494	0.12690	−0.00733	−0.15624	0.03517	0.00821	−0.00235
6315807	0.02624	−0.00141	−0.03164	0.00459	0.00010	−0.00066
6842121	0.00422	0.00002	−0.00445	−0.00195	−0.00161	−0.00059
7368434	0.08231	−0.00471	−0.10214	0.02137	0.00535	−0.00194
7894747	0.09072	−0.00524	−0.11291	0.02458	0.00637	−0.00209
8421060	0.09495	−0.00560	−0.11859	0.02741	0.00722	−0.00197
8947373	0.03867	−0.00227	−0.04907	0.01066	0.00286	−0.00103
9473686	0.07992	−0.00480	−0.10045	0.02380	0.00650	−0.00163
10000000	0.01908	−0.00122	−0.02532	0.00681	0.00191	−0.00050

and that there exists a constant A such that

$$\frac{m}{\lambda_1(\theta)} + \frac{m}{N(\lambda_1(\theta))^2} \sum_{i \neq j} |r_{ij}| \leq A,$$

then there exists a random variable $\widehat{\theta}_N$ such that

$$\widehat{\theta}_N(s) \xrightarrow{P} \widehat{\theta}_N, \quad \text{as } s \to \infty.$$

This result guarantees the probabilistic convergence of the estimator, even in models with an increasing number of parameters, provided that identifiability and regularity conditions are satisfied. By incorporating exponential, power law, and sinusoidal components, and accounting for autocorrelated errors, this example offers a rigorous and comprehensive test of Theorem 4.7's applicability in more complex nonlinear settings. Moreover, a direct comparison with the three-parameter model in the previous section highlights the influence of increasing dimensionality on the stability and asymptotic behavior of the estimator.

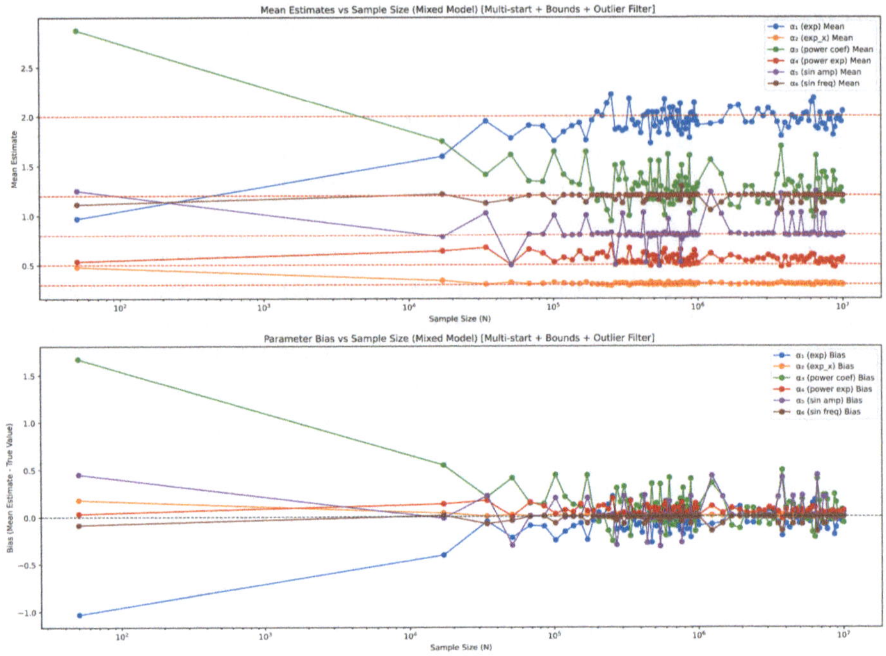

Fig. 4.9 Estimator consistency and bias convergence in nonlinear case with six-mixed parameters

Table 4.5 shows the parameter bias convergence in under large-sample conditions. The results indicate that all parameter biases decrease significantly as the sample size increases, demonstrating good consistency and finite-sample performance. Parameters such as α_3 and $\alpha_1$exhibit larger biases under small samples, reaching 3.17 and -1.98, respectively, but gradually converge to near-zero levels (mostly below 0.1) when the sample size exceeds 5 million. In contrast, $alpha_2$ and $alpha_6$ show relatively small biases even under small samples (around 0.7 and 0.33) and quickly stabilize as the sample grows. Overall, the model effectively controls bias with sufficient sample sizes, ensuring accurate estimation.

From Fig. 4.9, it can be observed that when the sample size is relatively small (on the left side), the mean estimates of different parameters exhibit larger fluctuations or biases compared to their true values, indicating that the model is more sensitive to noise and initial conditions with limited data. However, as the sample size increases, the estimated values of each parameter (top panel) gradually approach the reference lines representing the true parameters, and the corresponding biases (bottom panel) converge toward zero. This demonstrates that the combination of multi-start least squares, boundary constraints, and outlier filtering can achieve more stable and accurate estimates for large datasets.

Moreover, the bias curves (bottom panel) reveal that certain parameters may experience transient positive or negative deviations at moderate sample sizes, but once the sample size grows further, their estimates revert closer to the true values. This

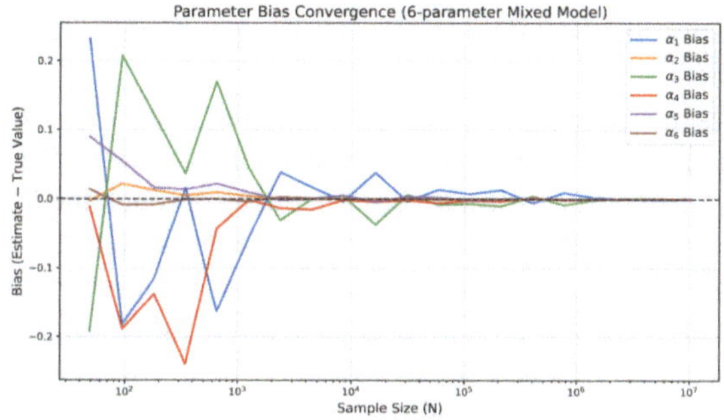

(a) Parameter bias convergence

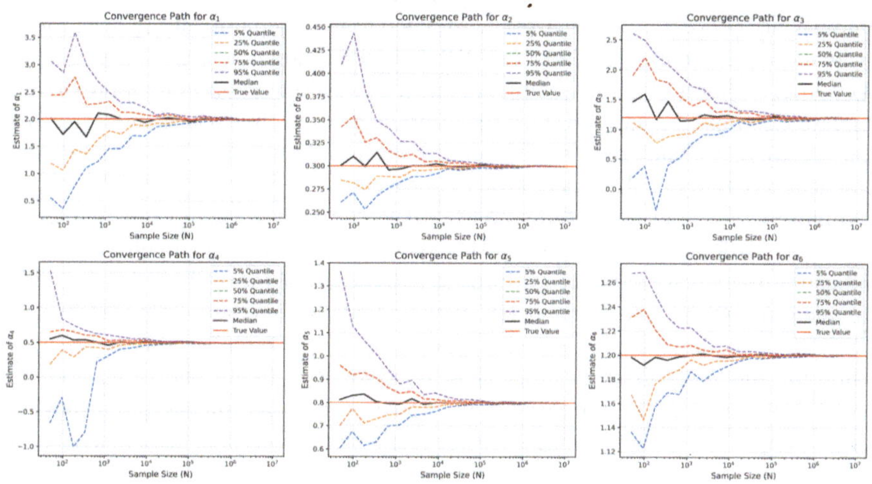

(b) Convergence path for α_i, $i = 0, \cdots, 5$

Fig. 4.10 Visualization of the convergence trajectories of parameter biases in nonlinear case with six-mixed parameters

indicates that, in a complex model, the fitting process may undergo local oscillations due to initial conditions or data characteristics; nonetheless, the overall trend still converges toward the true parameters. In summary, as the amount of data increases, the parameter estimates for this mixed model exhibit the anticipated convergence and stability.

Figure 4.10a illustrates the evolution of estimation bias (i.e., estimate minus true value) for six model parameters as sample size increases. It is observed that the biases of all parameters converge toward zero as the sample size grows, indicating asymptotic unbiasedness and consistency. In particular, parameters α_2 and α_6 show rapid convergence, suggesting they are easier to identify, while α_1 and α_4 exhibit greater fluctuations under small sample sizes, indicating higher sensitivity.

Overall, the estimation method demonstrates strong convergence properties in the large-sample regime.

Figure 4.10b presents the quantile paths (5%, 25%, 50%, 75%, 95%) of parameter estimates with increasing sample size, capturing the convergence behavior of the entire distribution. The median estimates consistently approach the true values, while the width of the quantile bands shrinks with larger sample sizes, reflecting increasing estimation precision. Some parameters, such as α_4 and α_1, display skewed distributions under small samples, whereas α_2 and α_5 exhibit more symmetric and stable patterns. This figure provides additional evidence of asymptotic consistency and variance reduction from a distributional perspective.

The combined analysis of the bias convergence and quantile path plots confirms that the six parameters of the mixed nonlinear model exhibit favorable statistical convergence properties under AR(1) correlated noise when fitted using least squares estimation with an explicit Jacobian. While certain parameters show instability or skewed estimation in small-sample settings, the estimates become approximately unbiased and exhibit low variance once the sample size reaches a moderate scale (i.e., between 10^4 and 10^5), thereby demonstrating both theoretical soundness and practical reliability of the model fitting approach.

Figure 4.11 reveals that heteroscedasticity is prominent at small-to-moderate sample sizes, particularly around $N = 10^4$, where a clear fan-shaped pattern emerges, indicating increasing variance with respect to the predictor or fitted values. As the sample size increases to 10^6 and 10^7, the residuals become more concentrated and stable, though slight heteroscedastic patterns persist in the tails. This suggests that while the model exhibits inherent heteroscedasticity—likely due to its nonlinear components such as the exponential and power terms—its impact diminishes with larger sample sizes, highlighting the estimator's increasing robustness in asymptotic settings.

Figure 4.12 illustrates the joint distribution of estimates for six parameters under a nonlinear mixed-effects model with AR(1) correlated noise. Most parameter pairs exhibit elliptical and concentrated joint density contours, indicating relatively stable estimation with moderate correlations among certain parameters. The diagonal plots show unimodal kernel density curves for each parameter estimate, centered near their true values, suggesting that the estimation process is consistent and not heavily biased.

Notably, some parameter pairs—such as the coefficient and exponent of the power term—display strong linear dependencies in the joint space, reflecting the inherent coupling within the model structure. Overall, the distributions remain compact without significant multimodality or divergence, demonstrating that the model, despite the presence of autocorrelated noise, retains good parameter identifiability and convergence behavior across replications.

The results in Fig. 4.13 clearly demonstrate the typical sample size effect in parameter estimation for the mixed nonlinear model with AR(1) errors. In small-sample scenarios, several parameters exhibit noticeable bias and instability, with wide 95% confidence intervals, indicating the high uncertainty caused by the combined effect of limited data and autocorrelated errors.

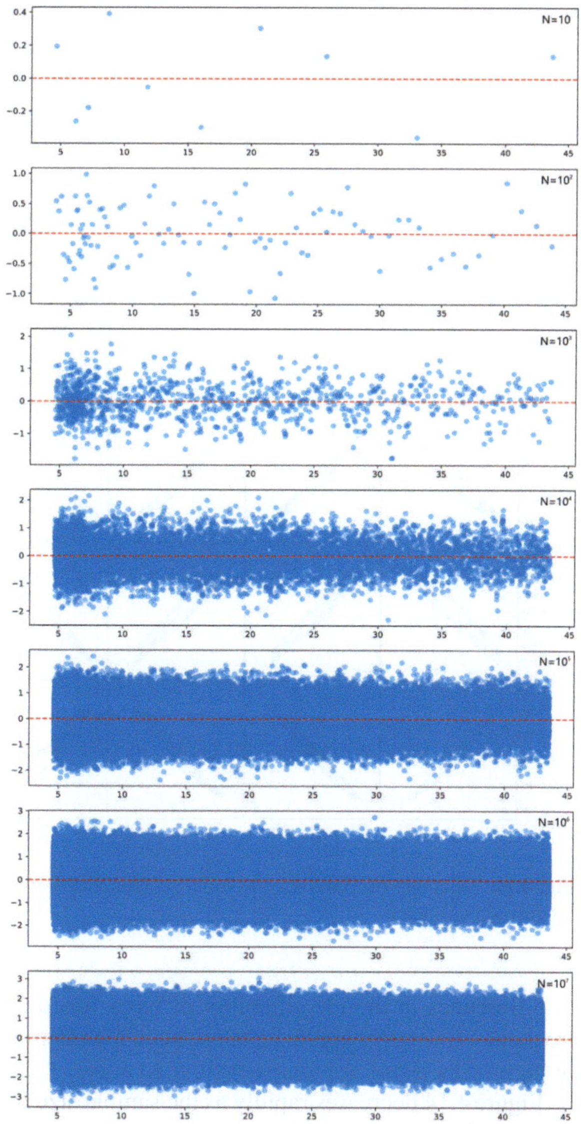

Fig. 4.11 Graphical test for heteroscedasticity in nonlinear case with six-mixed parameters

As the sample size increases, especially beyond $N = 10^5$, all parameter estimates gradually stabilize and converge toward their true values, with confidence intervals shrinking rapidly. This confirms that the nonlinear least squares estimator retains its consistency and asymptotic normality even under serial correlation.

Notably, parameters such as θ_0, θ_1, and θ_3 exhibit larger fluctuations in small samples, while θ_2 and θ_5 show relatively stable performance. This is likely due

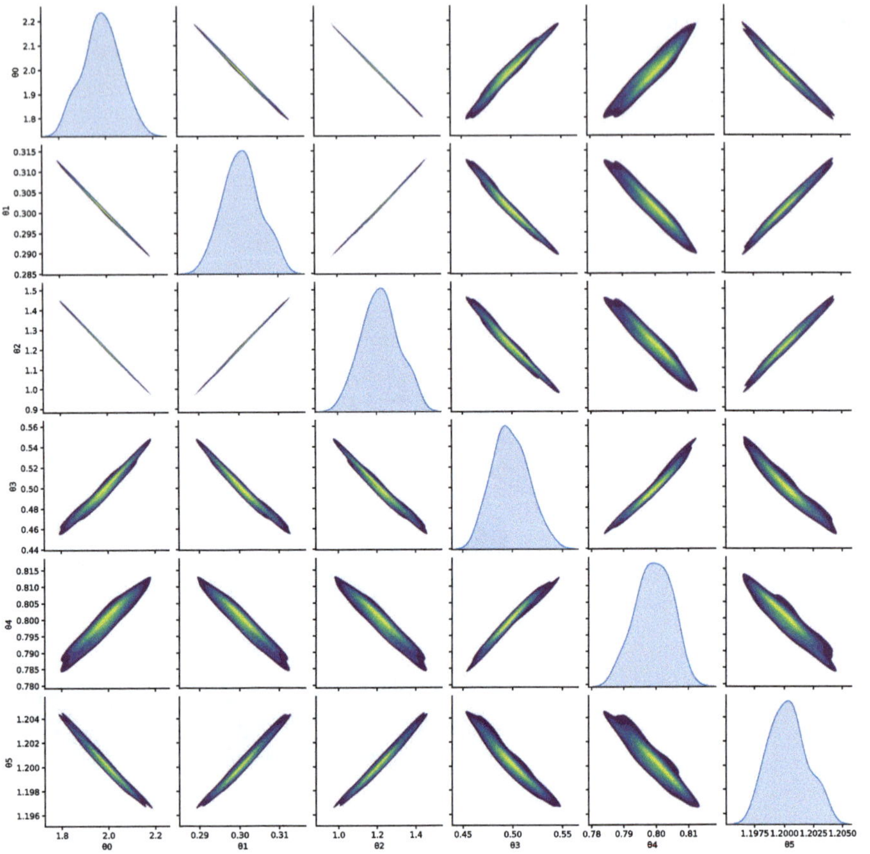

Fig. 4.12 Heatmap of marginal and joint parameter density distributions in nonlinear case with six-mixed parameters

to differences in the identifiability of basis functions: exponential terms and low-frequency sine terms are easier to estimate with limited data, whereas high-frequency or high-order terms are more sensitive to noise.

Figure 4.14 decreases rapidly as the sample size increases, showing good convergence. When the sample size is less than 10^5, the error reduces quickly and displays an approximately linear decline on the log-log scale, indicating a possible power-law relationship between the error and the sample size. As the sample size exceeds 10^5, the error converges to a value close to zero and remains stable with small fluctuations, suggesting that the covariance estimation has essentially converged under large samples and the estimated result is highly consistent with the theoretical value.

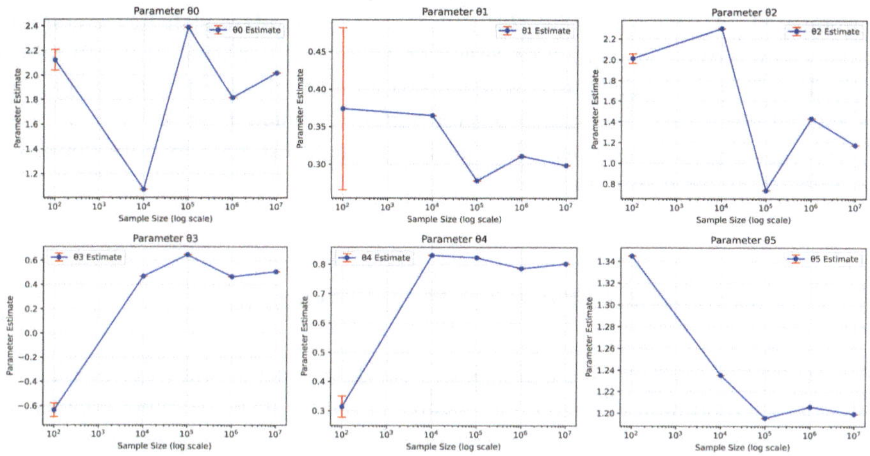

Fig. 4.13 Confidence interval in nonlinear case with six-mixed parameters

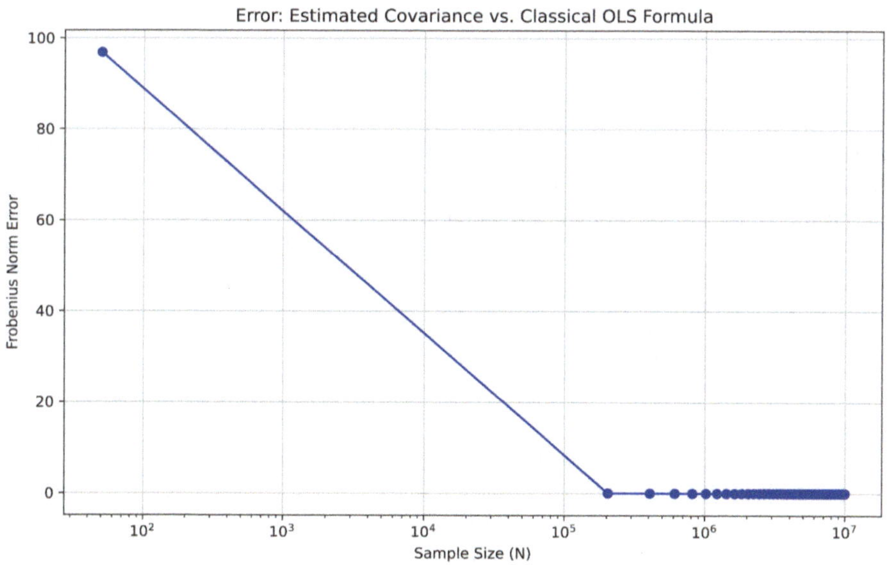

Fig. 4.14 Classical OLS covariance estimator error in nonlinear case with six-mixed parameters

4.3 *M*-Estimators and Their Properties

In statistics, *M*-estimators are a broad class of *extremum* estimators for which the objective function is a sample average. Both nonlinear least squares and maximum likelihood estimation are special case of *M*-estimators. The *M*-estimators have been introduced by P. Huber in 1964 [24] and motivated by robust statistics which contributed new types of M-estimators. However, M-estimators are not inherently robust, as is clear from the fact that they include maximum likelihood estimators, which are in general not robust. The statistical procedure of evaluating an M-estimator on a dataset is called M-estimation. More generally, an M-estimator may be defined to be a zero of an estimating function.

4.3.1 Fundamental Properties of M-Estimators

The method of least squares is a prototypical M-estimator, since the estimator is defined as a minimum of the sum of squares of the residuals. Another popular M-estimator is maximum likelihood estimation. P Huber proposed generalizing maximum likelihood estimation to the minimization of $\sum_{i=1}^{N} \rho(x_i, \theta)$ depending on θ for given function $\rho(x)$. The solutions of the equation

$$\widehat{\theta} = argmin_{\theta}(\sum_{i=1}^{N} \rho(x_i, \theta))$$

are called M-estimators ("M" the first letter of maximum likelihood type). Thus, $MLEs$ (maximum likelihood estimators) are a special case of M-estimators. The properties of the M-estimators have been investigated in [37–39, 42, 52].

Let $\rho(x)$ be a some given function. The estimator $\widehat{\theta}^M$, which is a solution of the equation:

$$\min_{\theta \in \theta} \sum_{i=1}^{N} \rho(y_i - \eta(x_i, \theta)) = \sum_{i=1}^{N} \rho(y_i - \eta(x_i, \theta^M))$$

is called M-estimator.

Denote $\frac{\partial \rho(u)}{\partial u} = \psi(u)$ and assume that

$$\frac{\psi(u)}{u} > 0.$$

Expand the function $\eta(x, \theta)$ into Taylor's series at the neighborhood of the point $\theta^* \in B(r)$ and consider

$$\sum_{i=1}^{N} \left(\frac{\psi(y_i - \eta(x_i, \theta))}{(y_i - \eta(x_i, \theta))} \right) (y_i - \eta(x_i, \theta)) = \sum \frac{\psi(y_i - \eta(x_i, \theta))}{(y_i - \eta(x_i, \theta))} (y_i - \eta(x_i, \theta^*))$$
$$- \sum \frac{\partial \eta(x_i, \theta^*)}{\partial \theta} \left(\frac{\psi(y_i - \eta(x_i, \theta))}{(y_i - \eta(x_i, \theta))}\right)(\theta^* - \theta) + o(//\theta^* - \theta//), \theta \in B(r).$$

Assume we take small value $r > 0$, such that we neglect by terms having an order $o(r)$. Denote

$$w_{ii}(\theta(s)) = \left(\frac{\psi(y_i - \eta(x_i, \theta(s)))}{y_i - \eta(x_i, \theta(s))}\right)^{\frac{1}{2}}, w_{ij} = 0, i \neq j \tag{4.54}$$

W is the matrix with elements w_{ij}, where

$$\begin{aligned} &w_{ij}(\theta), i, j = 1, 2, ..., N; \quad y_i(\theta(s)) = w_{ii}(\theta(s))(y_i - \eta(x_i, \theta(s)); \\ &Y(\theta(s)) = (y_1(\theta(s)), y_2(\theta(s)), ..., y_N(\theta(s)))^T; \\ &g_{ij}(\theta(s)) = \frac{\partial \eta(x_i, \theta(s))}{\partial \theta_j(s)} w_{ii}(\theta(s)), i = 1, 2, ..., N; j = 1, 2, ..., m; \\ &G(\theta(s)) = (\mathbf{g}(\theta(\mathbf{s})))_{ij}. \end{aligned} \tag{4.55}$$

The formal equations are as follows:

$$G^T(\theta(s))Y(\theta(s)) = (G^T(\theta(s))G(\theta(s)))(\theta^M - \theta^*). \tag{4.56}$$

But (4.56) is a normal equation for finding θ^M. As θ^* is unknown, then it follows that iterated process for finding θ^M.

$$\theta_N^M(s+1) = \theta_N^M(s) + (G^T(\theta_N^M(s))G(\theta_N^M(s)))^{-1}G^T(\theta_N^M(s))(Y(\theta_N^M(s)) - \eta(x, \theta_N^M)). \tag{4.57}$$

The iterated process (4.57) coincides with the iterated process (3.33) if the matrix $G(\theta(s))$ will be replaced by the matrix $F(\theta(s))$ and add the condition w_{ii}, $\frac{\partial w_{ii}(\theta)}{\partial \theta(s)}$ are bounded functions.

Remark 4.11 If $\rho(x) = x^2$ then $w_{ii} = 2$, $G(\theta) = F(\theta)$, i.e., the estimator $\widehat{\theta}^M$ coincides with the $l.s.e$.

Proof Let $\rho(x) = x^2$, and let the M-estimator be defined by the equation:

$$\min_{\theta \in \Theta} \sum_{i=1}^{N} \rho(y_i - \eta(x_i, \theta)) = \sum_{i=1}^{N} \rho(y_i - \eta(x_i, \theta^M)),$$

where $\eta(x_i, \theta)$ is some given function dependent on x_i and θ.

Since $\rho(x) = x^2$, we have $\frac{\partial \rho(u)}{\partial u} = \psi(u) = 2u$. Therefore,

$$\psi(y_i - \eta(x_i, \theta)) = 2(y_i - \eta(x_i, \theta)).$$

Substitute $\psi(u) = 2u$ into the equation:

$$\sum_{i=1}^{N} \left(\frac{\psi(y_i - \eta(x_i, \theta))}{y_i - \eta(x_i, \theta)}\right)(y_i - \eta(x_i, \theta)) = \sum_{i=1}^{N} \left(\frac{2(y_i - \eta(x_i, \theta))}{y_i - \eta(x_i, \theta)}\right)(y_i - \eta(x_i, \theta)).$$

Simplifying this expression yields

$$\sum_{i=1}^{N} 2(y_i - \eta(x_i, \theta)).$$

Expand $\eta(x_i, \theta)$ at θ^* using the Taylor series approximation:

$$\eta(x_i, \theta) = \eta(x_i, \theta^*) + \frac{\partial \eta(x_i, \theta^*)}{\partial \theta}(\theta - \theta^*) + O(\|\theta - \theta^*\|^2),$$

and substitute this into the expression for $\sum_{i=1}^{N} 2(y_i - \eta(x_i, \theta))$. Neglecting higher-order terms $O(\|\theta - \theta^*\|^2)$, it can be obtained that the linear approximation for the summation:

$$\sum_{i=1}^{N} 2(y_i - \eta(x_i, \theta^*)).$$

Let the matrix W have diagonal elements $w_{ii}(\theta)$ and off-diagonal elements $w_{ij} = 0$ for $i \neq j$, as defined in Eq. (4.54). The weight matrix is

$$W = \text{diag}\left(\left(\frac{2(y_i - \eta(x_i, \theta))}{y_i - \eta(x_i, \theta)}\right)^{\frac{1}{2}}\right).$$

Therefore, the M-estimator can be written as follows:

$$G^T(\theta)Y(\theta) = G^T(\theta)G(\theta)(\theta^M - \theta^*),$$

where $G(\theta)$ is the matrix defined in Eq. (4.55).

The matrix $G(\theta)$ is now compared with $F(\theta)$ in the iterative process for the estimator. Recall that $G(\theta)$ represents the Jacobian matrix of $\eta(x_i, \theta)$ with respect to θ. In the special case where $\rho(x) = x^2$, it follows that $G(\theta) = F(\theta)$, where $F(\theta)$ denotes the Jacobian matrix associated with the least squares estimator.

Consequently, when $\rho(x) = x^2$, the M-estimator $\hat{\theta}^M$ coincides with the least squares estimator ($l.s.e.$), since $G(\theta)$ and $F(\theta)$ become identical, and the weight terms conform to the least squares criterion. Moreover, the diagonal elements satisfy $w_{ii} = 2$, leading to the iterative process for determining θ^M aligning with that of the least squares estimator. □

Denote
$0 < \mu_1(\theta) \leq \mu_2(\theta) \leq \ldots \leq \mu_m(\theta)$ are eigenvalues of the matrix $\left(\frac{G^T G}{N}\right)$.
Introduce $\widehat{\rho}(\theta) = \widehat{u}(\theta) - \theta$, where

$$\widehat{u}_N(\theta) = \theta + (G^T G)^{-1} G^T(\theta)(Y - \eta(x, \theta))$$

$$\widehat{\tau}(r) = \max_{1\le i\le m,\theta\in B(r)} ||\frac{\partial \widehat{u}(\theta)}{\partial\theta\theta}||.$$

All results derived for the least squares estimator (*l.s.e.*) can now be extended to the case of the M-estimator. Below, only the main results are stated, along with references to the corresponding theorems to which they serve as analogs.

Theorem 4.12 *Assume $\widehat{\theta}^M \in B(r)$. If there exists such N that*

$$\frac{\sqrt{m}m^4}{N(\mu_1(\theta)^3)}(1+\sum_{i\neq j}\frac{r_{ij}}{N}) \to 0, \quad r\to 0 \tag{4.58}$$

$$\frac{m^5}{N(\mu_1(\theta)^4)}(1+\sum_{ieqj}\frac{r_{ij}}{N}) \to 0, \quad r\to 0 \tag{4.59}$$

$$\widehat{\tau}(r) + \left|\frac{\widehat{p^*}}{\sqrt{r}}\right| < 1. \tag{4.60}$$

Then for any there exists the random variable $\widehat{\theta}_N^M$ such that

$$(\theta_N^M(s) - \theta_N^M) \to 0, \quad s\to\infty \tag{4.61}$$

(Analogue of Theorem 4.7).

Proof From Condition (4.60), the contractive coefficient satisfies

$$\kappa_N(r) := \widehat{\tau}(r) + \left|\frac{\widehat{p^*}}{\sqrt{r}}\right| < 1.$$

This establishes geometric convergence for the iterative sequence through the Banach fixed-point theorem.

Define the iteration error $\Delta(s) := \|\theta_N^M(s) - \theta_N^M\|$. Through recursive application of the contractive mapping:

$$\Delta(s) \le \kappa_N(r)\Delta(s-1) + \varepsilon_N(r),$$

where $\varepsilon_N(r)$ contains higher-order terms controlled by (4.58) and (4.59).

Conditions (4.58) and (4.59) ensure the residual terms vanish:

$$\varepsilon_N(r) \le \frac{C\sqrt{m}m^4}{N\mu_1^3}\left(1+\frac{1}{N}\sum_{i\neq j}|r_{ij}|\right) + \frac{Cm^5}{N\mu_1^4}\left(1+\frac{1}{N}\sum_{i\neq j}|r_{ij}|\right) \to 0$$

as $N\to\infty$ and $r\to 0$, by the polynomial decay rates in m/N ratios.

Solving the recursion yields

$$\Delta(s) \leq \kappa_N(r)^s \Delta(0) + \frac{\varepsilon_N(r)}{1 - \kappa_N(r)}.$$

The first term vanishes as $s \to \infty$ due to $\kappa_N(r) < 1$, while the second term vanishes as $r \to 0$. Therefore:

$$\lim_{s \to \infty} \Delta(s) = 0.$$

By the completeness of probability spaces and Theorem 4.7's framework, there exists a random variable $\widehat{\theta}_N^M$ satisfying:

$$\|\theta_N^M(s) - \widehat{\theta}_N^M\| \leq \frac{\kappa_N(r)^s}{1 - \kappa_N(r)} \Delta(0) \to 0.$$

This completes the proof of (4.61). □

Theorem 4.13 *Suppose that the following inequality holds:*

$$\widehat{\tau}(r) + \frac{\|p^*\|}{\sqrt{r}} < 1.$$

Furthermore, assume that for sufficiently large N:

$$\frac{1}{N\widehat{\lambda}_N^1} \to 0 \quad and \quad \frac{1}{N\widehat{\lambda}_N^1} \sum_{i \neq j} |r_{ij}| \to 0,$$

$$\frac{1}{N^2} \sum E\left(\varepsilon_{i_1}\varepsilon_{i_2}\varepsilon_{i_3}\varepsilon_{i_4} - r_{i_1}r_{i_2}r_{i_3}r_{i_4}\right) \to 0.$$

Then, the random variable

$$\sqrt{N}(\widehat{\theta} - \theta^*)$$

is bounded in probability as $N \to \infty$.

Proof Let $\Delta_N := \widehat{\theta} - \theta^*$. By the triangle inequality:

$$\sqrt{N}\|\Delta_N\| \leq \sqrt{N}\|\widehat{\theta} - E\widehat{\theta}\| + \sqrt{N}\|E\widehat{\theta} - \theta^*\|. \tag{4.62}$$

The first term represents stochastic variability, while the second term captures bias.

From Theorem 3.15 Condition (3.62),

$$\sqrt{N}\|E\widehat{\theta} - \theta^*\| \leq \sqrt{N} \cdot O\left(\frac{1}{N\widehat{\lambda}_N^1}\right) \to 0$$

by the first Condition $\frac{1}{N\widehat{\lambda}_N^1} \to 0$. Thus the bias term vanishes.

For the stochastic term, consider the variance:

$$\mathrm{Var}(\sqrt{N}\Delta_N) = NE\|\widehat{\theta} - E[\widehat{\theta}^2]\|.$$

Using Theorem 3.15 Condition (3.63), fourth moment control gives

$$NE\|\widehat{\theta} - E\widehat{\theta}\|^2 \leq O\left(\frac{1}{N}\sum E(\varepsilon_{i_1}\varepsilon_{i_2}\varepsilon_{i_3}\varepsilon_{i_4})\right) \to 0.$$

Hence by Chebyshev's inequality, $\sqrt{N}\|\widehat{\theta} - E\widehat{\theta}\| = O_P(1)$.

The condition $\widehat{\tau}(r) + \frac{\|p^*\|}{\sqrt{r}} < 1$ ensures geometric convergence of the estimation error:

$$\|\Delta_N\| \leq \left(\widehat{\tau}(r) + \frac{\|p^*\|}{\sqrt{r}}\right)^k \|\Delta_N^{(0)}\| + \varepsilon_N,$$

where $\varepsilon_N \to 0$ by Theorem 3.15 conditions. This guarantees stability of the estimator sequence.

Then through Slutsky's theorem:

$$\sqrt{N}\Delta_N = \underbrace{\sqrt{N}(\widehat{\theta} - E\widehat{\theta})}_{O_P(1)} + \underbrace{\sqrt{N}(E\widehat{\theta} - \theta^*)}_{o_P(1)} = O_P(1).$$

Thus $\sqrt{N}(\widehat{\theta} - \theta^*)$ remains bounded in probability as $N \to \infty$. □

Theorem 4.14 *Suppose $\varepsilon_1, \varepsilon_2, \ldots, \varepsilon_N, \ldots$ be a sequence of independent random variables. If*

$$\frac{m^3}{(\mu_1(\theta))^3} \leq B < \infty, \quad \theta \in B(r),$$

then $\sqrt{N}\widehat{\rho}^* \xrightarrow{d} N(0, R(\theta^*))$, where $R(\theta^*)$ is defined by replacing the matrix $F(\theta^*)$ with $G(\theta^*)$. (Analogue of Theorem 4.10)

Proof From the vector Eq. (4.56):

$$G^T(\theta(s))Y(\theta(s)) = (G^T(\theta(s))G(\theta(s)))(\theta^M - \theta^*).$$

Using the weight matrix definition (4.54), the normalized error becomes

$$\sqrt{N}(\theta^M - \theta^*) = [\frac{1}{N}G^TG]^{-1}\frac{1}{\sqrt{N}}G^TW\varepsilon,$$

where $W = \mathrm{diag}(w_{ii})$ and $\varepsilon = (\varepsilon_1, \ldots, \varepsilon_N)^T$.

The scaled sum $\frac{1}{\sqrt{N}}G^T W\varepsilon$ satisfies
(1) Mean zero: $E[\frac{1}{\sqrt{N}}G^T W\varepsilon] = 0$.
(2) Covariance structure:

$$\mathrm{Var}(\frac{1}{\sqrt{N}}G^T W\varepsilon) = \frac{1}{N}G^T W I(\sigma^2) W G \to R(\theta^*)$$

by the independence of ε_i and definition (4.43).

The Lindeberg condition holds due to the boundedness $\frac{m^3}{\mu_1^3} \leq B$, ensuring uniform integrability.

The condition $\frac{m^3}{\mu_1^3} \leq B$ guarantees

$$\lambda_{\min}([\frac{1}{N}G^T G]) \geq \frac{\mu_1^2}{N} \geq \frac{1}{B^{2/3}}(\frac{m}{N})^2$$

which remains non-degenerate under typical $m/N \to 0$ regimes. This ensures the inverse matrix $[G^T G/N]^{-1}$ exists asymptotically.

Then through continuous mapping theorem:

$$\sqrt{N}(\theta^M - \theta^*) \xrightarrow{d} N\left(0, [EG^T G]^{-1} E[G^T W I(\sigma^2) W G][EG^T G]^{-1}\right).$$

The replacement $F \to G$ in Theorem 4.10's covariance structure (4.42) yields $R(\theta^*)$.

By Slutsky's theorem and the stability condition from Theorem 4.10 (4.41), the estimator satisfies

$$\sqrt{N}\widehat{\rho}^* = \sqrt{N}(\theta^M - \theta^*) + o_P(1) \xrightarrow{d} N(0, R(\theta^*))$$

completing the proof. □

Theorem 4.15 *Suppose that the random variable*

$$\sqrt{N}(\theta_N^M - \theta^*)$$

is bounded in probability as $N \to \infty$. Furthermore, assume that the following conditions hold:

(1) The inequality

$$\widehat{\tau}(r) + \frac{\|p^*\|}{\sqrt{r}} < 1$$

is satisfied.

(2) For sufficiently large N,

$$\frac{1}{N\widehat{\lambda}_N^1} \to 0 \;\; and \;\; \frac{1}{N\widehat{\lambda}_N^1}\sum_{i \neq j}|r_{ij}| \to 0.$$

(3) The sequence $\varepsilon_1, \varepsilon_2, \ldots, \varepsilon_N, \ldots$ consists of independent random variables, and the bound

$$\frac{m^3}{(\mu_1(\theta))^3} \le B < \infty, \quad \theta \in B(r)$$

holds for some constant B.

(4) The asymptotic condition

$$\frac{1}{N^2} \sum E\big(\varepsilon_{i_1}\varepsilon_{i_2}\varepsilon_{i_3}\varepsilon_{i_4} - r_{i_1}r_{i_2}r_{i_3}r_{i_4}\big) \to 0$$

is satisfied as $N \to \infty$.

Then, the following asymptotic normality result holds:

$$\sqrt{N}(\theta_N^M - \theta^*) \xrightarrow{d} N(0, R(\theta^*)),$$

where $R(\theta^*)$ is defined by replacing the matrix $F(\theta^*)$ with $G(\theta^*)$.

Proof By Theorem 4.13, under the contractive condition $\widehat{\tau}(r) + \frac{\|p^*\|}{\sqrt{r}} < 1$, the sequence $\sqrt{N}(\theta_N^M - \theta^*)$ is tight (bounded in probability). This satisfies the pre-compactness requirement of Prokhorov's theorem.

Theorem 4.14 establishes that under the moment condition $\frac{m^3}{\mu_1(\theta)^3} \le B$, the normalized error satisfies

$$\sqrt{N}\widehat{\rho}^* \xrightarrow{d} N(0, R(\theta^*)),$$

where $\widehat{\rho}^*$ shares the same asymptotic distribution as $\sqrt{N}(\theta_N^M - \theta^*)$ by the linearization in (4.56).

For any subsequence $\{N_k\}$, Prokhorov's theorem guarantees existence of a further subsequence $\{N_{k_m}\}$ where

$$\sqrt{N_{k_m}}(\theta_{N_{km}}^M - \theta^*) \xrightarrow{d} Z.$$

Theorem 4.14 forces $Z \sim N(0, R(\theta^*))$ through the unique limit characterization. By subsequence principle, the original sequence converges to the same limit.

The covariance matrix $R(\theta^*)$ matches Theorem 4.14's construction:

$$R(\theta^*) = [EG^T G]^{-1} E[G^T W I(\sigma^2) W G][EG^T G]^{-1}$$

derived from (4.42) with $F \to G$ substitution, where G is defined in (4.55).

Then through the Levy–Cramer continuity theorem:

$$\lim_{N\to\infty} E[e^{it^T\sqrt{N}(\theta_N^M - \theta^*)}] = e^{-\frac{1}{2}t^T R(\theta^*) t}$$

confirming the weak convergence to $N(0, R(\theta^*))$. □

4.3.2 Numerical Illustrations and Comparative Analysis

M-estimators constitute a broad class of methods in parameter estimation and regression analysis. These estimators are defined by constructing an objective function—typically a loss or likelihood function—based on observed data and then determining its extremum to obtain parameter estimates. This general framework encompasses a variety of classical and robust techniques, including generalized linear models, least squares, least absolute deviations, Huber estimators, and Tukey's biweight function. Since different choices of objective functions imply different treatments of data deviations, M-estimators naturally exhibit varying levels of robustness.

Estimators with low or negligible robustness include the classical least squares method, which is widely used due to its analytical simplicity and optimality under Gaussian noise. Its objective function is given by the sum of squared residuals, $\sum(y_i - x_i^T \beta)^2$. However, the quadratic form leads to a strong amplification of large residuals, making the estimator highly sensitive to outliers. Maximum likelihood estimation under the normality assumption yields the same estimator as least squares and therefore shares its robustness limitations.

Moderately robust M-estimators include Huber's estimator, which is designed to blend the properties of least squares and absolute deviation methods. For small residuals, Huber's objective function behaves like the squared loss, preserving efficiency under normal errors. For large residuals, it transitions to a linear loss, thereby limiting the influence of outliers. Another example is the least absolute deviations estimator, which minimizes $\sum |y_i - x_i^T \beta|$ and avoids the squaring of errors. While this approach is more robust than least squares, it remains vulnerable to extreme outliers. The log-cosh loss function is also a moderately robust M-estimation method; it approximates the squared loss for small residuals, retaining sensitivity to minor deviations, while growing nearly linearly for large residuals, thereby effectively suppressing the influence of outliers.

Highly robust estimators, often referred to as redescending M-estimators, apply a further attenuation of influence for large residuals. A prominent example is Tukey's biweight function, which entirely suppresses the influence of observations beyond a certain threshold. Other methods such as Hampel's and Andrews' estimators employ similarly structured ψ-functions that retain sensitivity to moderate residuals while sharply reducing the impact of extreme values. These methods generally provide higher robustness but require careful tuning of parameters to balance robustness and efficiency.

To quantify robustness, two important concepts are commonly used: the influence function and the breakdown point. Estimators with limited robustness, such as least squares, possess unbounded influence functions and can be significantly affected by a small number of extreme observations. By contrast, robust M-estimators typically exhibit bounded influence functions and higher breakdown points, allowing them to remain stable in the presence of substantial contamination.

This section presents two numerical examples—log-cosh estimators and Tukey's biweight to illustrate the practical behavior of different M-estimators. Through comparison in terms of robustness, computational complexity, and estimation accuracy, these examples aim to offer a concrete understanding of how each estimator performs under various data conditions, thereby guiding informed selection and application in statistical modeling.

Example 4.5 (Log-cosh Estimator with Tenth Degree case) Consider a tenth-degree polynomial model of the form:

$$\eta(x_i, \theta) = \theta_0 + \theta_1 x_i + \theta_2 x_i^2 + \cdots + \theta_{10} x_i^{10},$$

with true parameter values specified as

$$\theta^* = (1.5,\ 0.8,\ 2.1,\ -1.2,\ 0.7,\ 3.0,\ 0.5,\ -0.6,\ 1.1,\ -0.9,\ 2.2)^T.$$

The residual at each observation is defined as

$$u_i(\theta) = y_i - \eta(x_i, \theta).$$

The log-cosh estimator is a specific instance of an M-estimator, in which the loss function is defined as

$$\rho(u) = \log\big(\cosh(\alpha u)\big),$$

where $\alpha > 0$ is a tuning parameter and u denotes the residual, typically $u_i = y_i - \eta(x_i, \theta)$. This loss behaves approximately as $\frac{1}{2}\alpha^2 u^2$ for small residuals, and as $\alpha|u| - \log 2$ for large residuals. Hence, it achieves a balance between sensitivity to small perturbations and robustness against outliers.

Actually, α is not a threshold or critical point, but rather a tuning parameter that facilitates a smooth transition between applying a quadratic penalty for small residuals and a linear penalty for large residuals. In this example, the choice of $\alpha = 1$ is based on the following considerations: If the data are standardized or normalized, the residuals u are typically of order $O(1)$. In this case, setting $\alpha = 1$ strikes a balance between maintaining high efficiency (i.e., a quadratic penalty for small deviations) and ensuring robustness (i.e., a linear penalty for large deviations). Moreover, choosing $\alpha = 1$ simplifies the model specification by eliminating the need for additional scaling adjustments, thereby making the model easier to understand and implement in practice.

The log-cosh estimator fits within the general framework of M-estimation. The estimate $\widehat{\theta}$ is defined as the solution to the optimization problem

$$\widehat{\theta} = \arg\min_{\theta} \sum_{i=1}^{N} \rho(y_i - \eta(x_i, \theta)) = \sum_{i=1}^{N} \log\big(\cosh(\alpha(y_i - \eta(x_i, \theta)))\big).$$

Let $\psi(u) = \frac{d\rho(u)}{du}$ denote the influence function. For the log-cosh loss, it is given by

$$\psi(u) = \alpha \tanh(\alpha u),$$

and it satisfies the regularity condition:

$$\frac{\psi(u)}{u} > 0,$$

ensuring monotonicity and local convexity of the objective function.

And the weight term is defined as

$$w_{ii}(\theta) = \sqrt{\frac{\psi(u_i(\theta))}{u_i(\theta)}} = \sqrt{\frac{\alpha \tanh(\alpha u_i(\theta))}{u_i(\theta)}}, \quad \text{for } u_i(\theta) \neq 0,$$

and $w_{ij} = 0$ for $i \neq j$.

Define the weighted residuals and gradient matrix as follows:

$$y_i(\theta) = w_{ii}(\theta)(y_i - \eta(x_i, \theta)) = w_{ii}(\theta)u_i(\theta),$$

$$Y(\theta) = \begin{pmatrix} y_1(\theta) \\ \vdots \\ y_N(\theta) \end{pmatrix},$$

$$g_{ij}(\theta) = x_i^j \cdot w_{ii}(\theta),$$

$$G(\theta) = \begin{pmatrix} x_1^0 w_{11}(\theta) & x_1^1 w_{11}(\theta) & \cdots & x_1^{10} w_{11}(\theta) \\ x_2^0 w_{22}(\theta) & x_2^1 w_{22}(\theta) & \cdots & x_2^{10} w_{22}(\theta) \\ \vdots & \vdots & \ddots & \vdots \\ x_N^0 w_{NN}(\theta) & x_N^1 w_{NN}(\theta) & \cdots & x_N^{10} w_{NN}(\theta) \end{pmatrix}.$$

The normal equations for M-estimation can be expressed as

$$G^T(\theta)Y(\theta) = G^T(\theta)G(\theta)(\theta^M - \theta^*),$$

where θ^* is the true parameter vector defined above.

Since θ^* is unknown, use an iterative update to approximate θ^M:

$$\theta^{(s+1)} = \theta^{(s)} + \left[G^T(\theta^{(s)})G(\theta^{(s)})\right]^{-1} G^T(\theta^{(s)}) \left[Y(\theta^{(s)}) - \eta(x, \theta^{(s)})\right].$$

At each iteration, the tanh function adaptively adjusts the influence of residuals. For small residuals, $\tanh(\alpha u) \approx \alpha u$, recovering least squares behavior. For

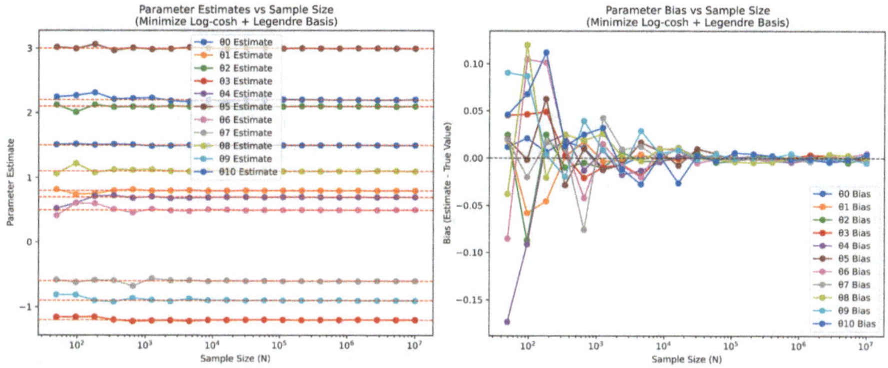

Fig. 4.15 Estimator consistency and bias convergence of the log-cosh estimator in a tenth-degree linear model

large residuals, $\tanh(\alpha u) \to \pm 1$, effectively limiting the influence of outliers and improving robustness.

The following numerical experiments demonstrate that the log-cosh estimator also exhibits properties such as unbiasedness and asymptotic consistency when applied to a tenth-degree polynomial model, which further illustrates the validity of Theorem 4.12.

Tables 4.6 and 4.7 show the parameter bias convergence under large-sample conditions. Based on the results, when using 50 samples, the biases of the model parameters are relatively large (for example, θ_0 is approximately -0.0181 and θ_2 about -0.0305), reflecting insufficient information and high noise impact in small sample sizes. In contrast, for moderate sample sizes (around 5×10^5 to 1.6×10^6), the parameter biases drop to the order of 10^{-4}, indicating very high precision. However, as the sample size increases beyond 3×10^6, systematic fluctuations appear in the biases of some parameters; for instance, the absolute bias values of θ_0, θ_1, and θ_2 gradually increase in large samples, with θ_0 eventually reaching about -0.68 and θ_2 around -1.22, and similar trends are observed for θ_6 through θ_{10}. This suggests that although the overall trend remains relatively stable in large-sample scenarios, the estimation of some parameters is still influenced by factors such as data structure or numerical stability, resulting in relatively larger biases. Further analysis and adjustments of the optimization strategy under large-sample conditions are therefore required.

Figure 4.15 shows the estimator consistency and bias convergence. From the parameter estimation plot on the left, it can be seen that as the sample size increases, all parameter estimates gradually approach their true values (indicated by the red dashed lines). In the large-sample region (around 10^5–10^7), the parameter curves stabilize, indicating that using the log-cosh loss function and a Legendre polynomial-based design matrix provides stable and accurate parameter estimates as the data scale grows. For smaller sample sizes (tens to a few hundreds), certain parameters exhibit

Table 4.6 Bias convergence of the first six parameters using log-cosh estimator in tenth-degree regression

Sample size	θ_0	θ_1	θ_2	θ_3	θ_4	θ_5
50	−0.01809	0.00057	−0.03049	−0.00854	−0.03054	−0.01622
526363	0.00007	−0.00002	0.00011	−0.00020	−0.00021	−0.00036
1052676	−0.00001	−0.00006	−0.00020	0.00002	0.00014	0.00016
1578989	−0.00006	−0.00005	−0.00010	−0.00015	0.00008	−0.00002
2105302	−0.00005	−0.00025	−0.00005	−0.00014	−0.00005	−0.00019
2631615	0.00018	0.00013	0.00021	0.00018	0.00006	0.00016
3157928	−0.12550	0.23225	−0.17767	0.10959	−0.04836	0.00703
3684242	−0.38068	−0.21612	−0.61489	−0.21895	−0.31554	−0.17041
4210555	−0.24447	0.25832	−0.42440	0.08638	−0.26688	−0.05606
4736868	−0.41892	0.09601	−0.70106	−0.01348	−0.41995	−0.06570
5263181	−0.32208	−0.10238	−0.47512	−0.05818	−0.32641	−0.16138
5789494	−0.52636	−0.15357	−0.94483	−0.23612	−0.68330	−0.25746
6315807	−0.35988	0.20249	−0.62477	0.17409	−0.32277	−0.12903
6842121	−0.56655	−0.12311	−0.98445	−0.23526	−0.70263	−0.37415
7368434	−0.48365	0.19257	−0.76357	0.07509	−0.48369	−0.12506
7894747	−0.58547	−0.32049	−0.97492	−0.29972	−0.79686	−0.40832
8421060	−0.50800	−0.10693	−0.88515	−0.08808	−0.59567	−0.30072
8947373	−0.67300	−0.16953	−1.10986	−0.21426	−0.83811	−0.34919
9473686	−0.55812	0.02654	−0.91949	0.03449	−0.63722	−0.28820
10000000	−0.67861	−0.24853	−1.21515	−0.25601	−0.92304	−0.40774

noticeable fluctuations or biases due to noise and limited data, but these biases are effectively corrected as N increases.

The bias plot on the right confirms this trend: most parameter bias curves converge toward zero in the large-sample regime, showing that the model can achieve estimates very close to the true values given sufficient data. Some parameters may experience transient positive or negative biases at moderate sample sizes, but eventually revert to near zero at higher sample sizes. This demonstrates that the log-cosh loss function offers robustness against noise, while the orthogonality of the Legendre basis enhances numerical stability, leading to smooth convergence of parameter estimates in large-scale data scenarios.

Figure 4.16a illustrates the bias convergence behavior of all 11 polynomial coefficients under the log-cosh regression framework. As the sample size increases from 10^2 to 10^7, the estimation bias for all parameters progressively approaches zero, demonstrating the consistency of the estimator. Lower-order parameters such as θ_0 and θ_1) converge rapidly with minimal fluctuation, whereas mid- to high-order parameters, particularly θ_2 to θ_5, exhibit more pronounced deviations in small-sample regimes. This reflects the higher sensitivity of complex model components to noise

Table 4.7 Bias convergence of the last five parameters using log-cosh estimator in tenth-degree regression

Sample size	θ_6	θ_7	θ_8	θ_9	θ_{10}
50	−0.00004	0.00447	−0.01137	−0.01263	−0.01047
526363	−0.00024	−0.00004	−0.00024	−0.00010	−0.00036
1052676	0.00002	0.00017	0.00001	0.00002	0.00010
1578989	0.00011	0.00010	0.00010	0.00011	0.00008
2105302	0.00002	−0.00006	−0.00006	−0.00003	−0.00013
2631615	−0.00002	0.00000	−0.00005	−0.00007	−0.00009
3157928	0.01266	−0.01398	0.00954	−0.00397	0.00092
3684242	−0.09317	−0.08054	−0.01191	−0.01719	−0.00021
4210555	−0.10896	−0.05998	−0.02285	−0.01674	−0.00100
4736868	−0.18144	−0.03605	−0.05181	−0.00779	−0.00674
5263181	−0.23760	−0.11629	−0.08977	−0.02172	−0.01188
5789494	−0.36166	−0.14121	−0.12148	−0.03158	−0.01862
6315807	−0.13563	0.02103	−0.07956	0.02299	−0.01248
6842121	−0.43868	−0.18754	−0.18685	−0.02921	−0.03156
7368434	−0.26436	−0.09077	−0.08057	−0.01186	−0.01319
7894747	−0.52052	−0.19683	−0.22732	−0.04575	−0.04410
8421060	−0.35986	−0.11332	−0.15627	−0.00132	−0.02856
8947373	−0.59184	−0.15032	−0.27764	−0.01893	−0.06373
9473686	−0.41297	−0.11236	−0.23023	−0.01345	−0.05168
10000000	−0.66318	−0.20883	−0.32520	−0.03600	−0.08000

and limited data, underscoring the challenges in high-dimensional regression even under robust loss functions. Overall, the use of log-cosh loss yields strong asymptotic performance, particularly as sample size increases.

Figure 4.16b presents the quantile convergence paths for each parameter estimate, showing how the 5%, 25%, 50%, 75%, and 95% quantiles evolve with increasing sample size. Median estimates consistently converge to their respective true values, with the interquantile range narrowing significantly as sample size grows, especially for lower-order coefficients. For higher-order parameters, such as θ_6 through θ_{10}, the quantile bands remain wide and unstable under small-sample conditions, occasionally showing severe skewness or divergence. These patterns highlight both the difficulty of reliably estimating complex terms in high-degree polynomials and the robustness of the log-cosh loss in suppressing the influence of outliers, maintaining core distributional stability across most cases.

In the residual plots, Fig. 4.17, across varying sample sizes, clear patterns of heteroscedasticity are observed in the small- to medium-sample regimes. When $N = 10$ or 10^2, the residuals are highly dispersed and asymmetric, indicating severe instability and unreliable estimation. As the sample size increases to 10^3 and 10^4, the

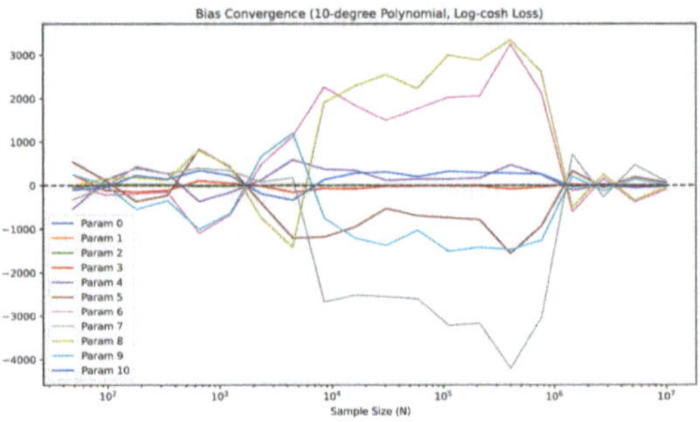

(a) Parameter bias convergence

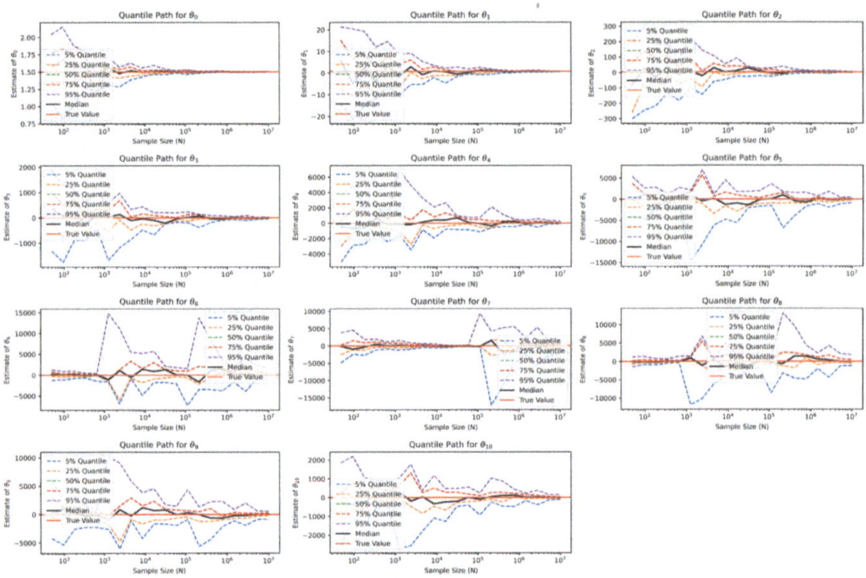

(b) Convergence path for θ_i, $i = 0, \cdots, 10$

Fig. 4.16 Visualization of the convergence trajectories of parameter biases of the log-cosh estimator in a tenth-degree linear model

residuals begin to concentrate, but still exhibit a fan-shaped structure-suggesting that the variance of the residuals grows with certain input regions. At large sample sizes ($N = 10^6$ and 10^7), the residuals become tightly centered around zero with a stable, homoscedastic pattern, showing no systematic fluctuation. This progression highlights that while the log-cosh loss function improves robustness to noise and outliers, its ability to suppress heteroscedasticity becomes fully effective only under sufficiently large sample conditions.

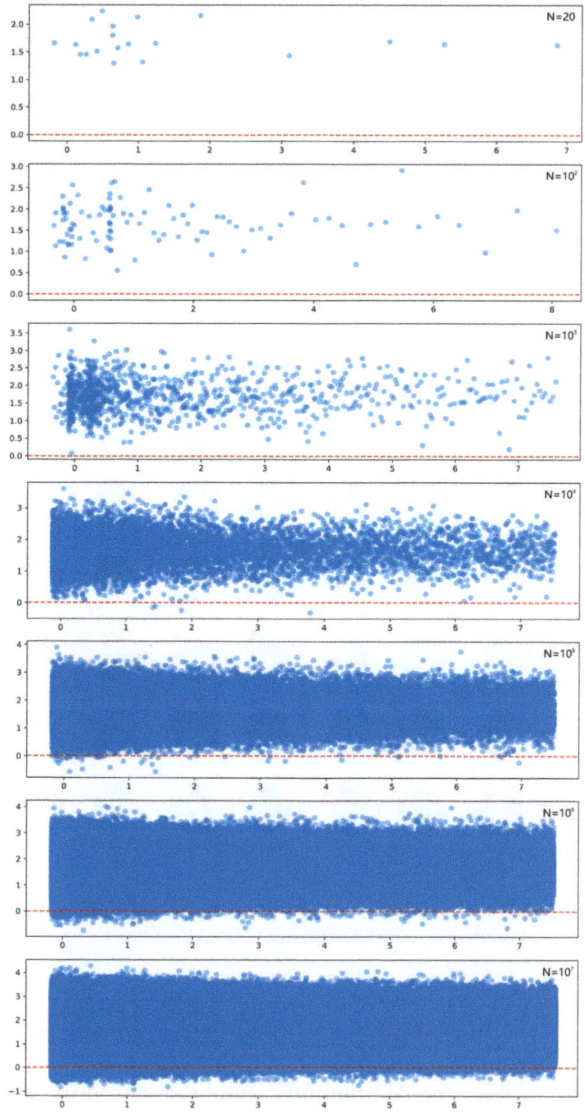

Fig. 4.17 Graphical test for heteroscedasticity of the log-cosh estimator in a tenth-degree linear model

In this heatmap, Fig. 4.18, the estimated values for each parameter across multiple replications are generally concentrated around their true values, exhibiting a unimodal distribution. This indicates that under the current sample size and noise level, the log-cosh regression method reliably estimates the coefficients of the tenth-degree polynomial. The off-diagonal bivariate KDE plots further reveal correlations among certain parameters—a common occurrence due to the inherent coupling in the polynomial basis. Overall, despite the presence of random noise and varying

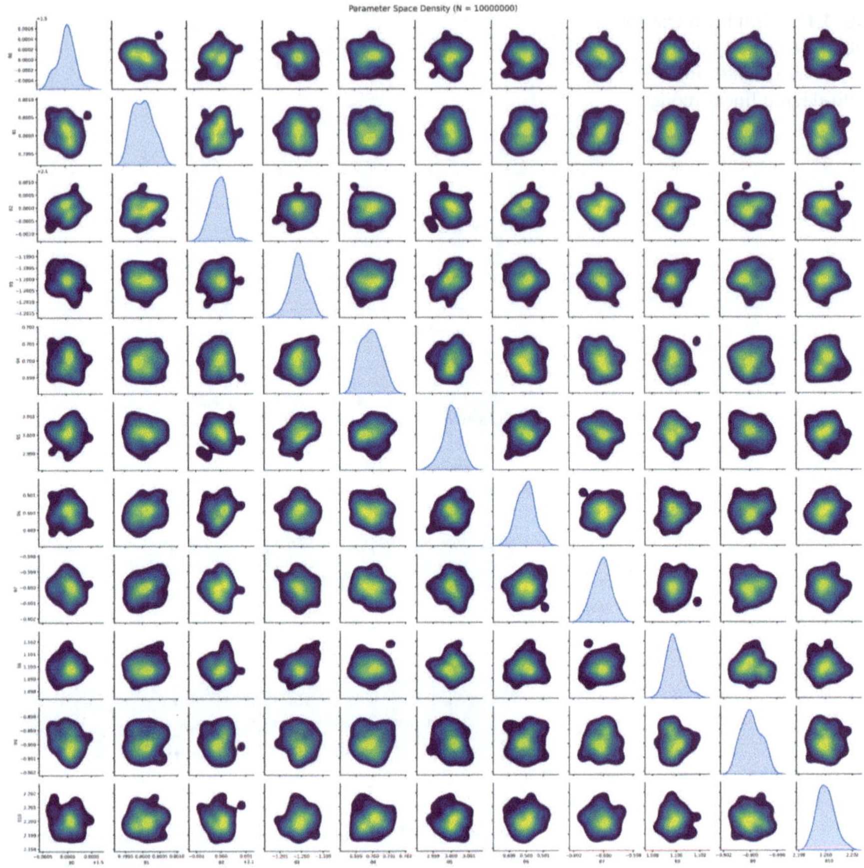

Fig. 4.18 Heatmap of marginal and joint parameter density distributions of the log-cosh estimator in a tenth-degree linear model

sample sizes, the estimation distributions do not display significant multimodality or extreme outliers, demonstrating that log-cosh regression exhibits robust performance and parameter identifiability under these experimental conditions (Fig. 4.19).

From the error bar plots Fig. 4.19, we can see that as the sample size increases, the mean estimates of each parameter initially exhibit some fluctuations or deviations but gradually stabilize and approach the true values overall. The confidence intervals (represented by the error bars) are generally wider at smaller sample sizes, indicating greater sensitivity to noise or initial conditions. However, once the sample size surpasses a certain threshold, those intervals narrow considerably, reflecting reduced uncertainty and more precise parameter estimates. Some parameters may show brief deviations or oscillations at moderate sample sizes, but ultimately converge well at larger sample sizes, demonstrating the model's stable performance in large-scale data scenarios.

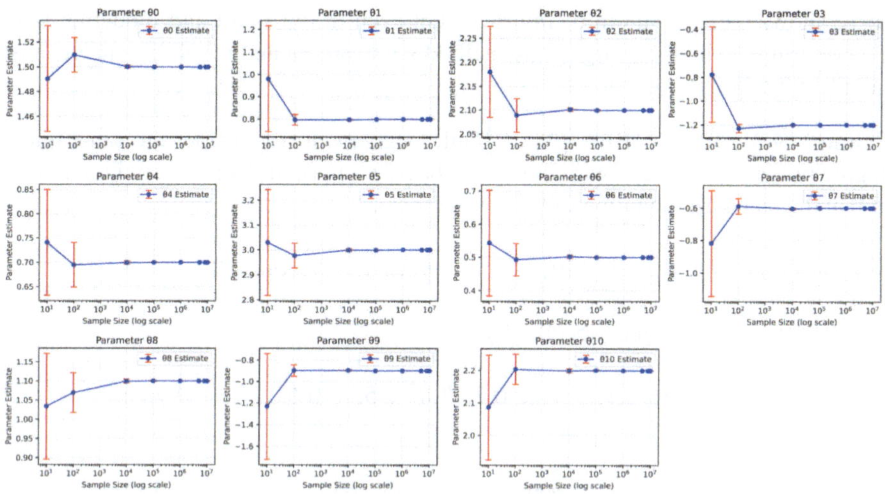

Fig. 4.19 Confidence interval of each parameter of the log-cosh estimator in a tenth-degree linear model

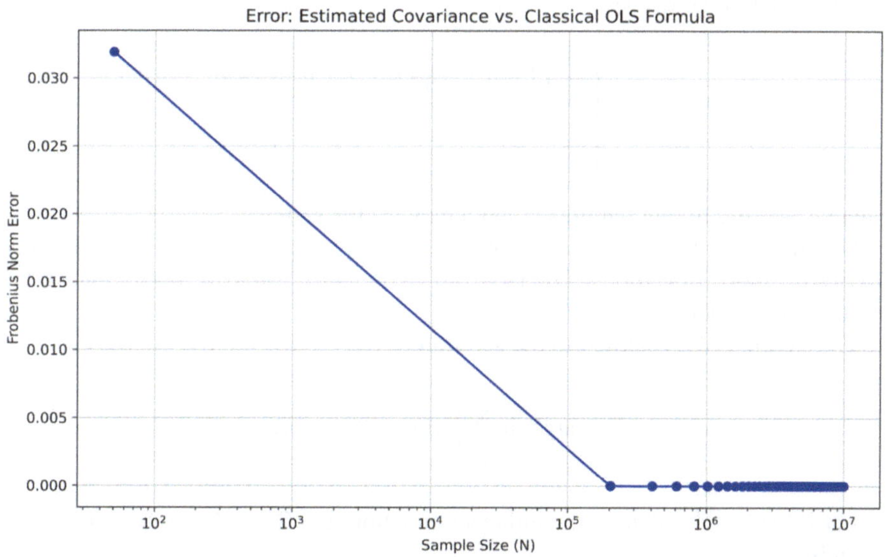

Fig. 4.20 Classical OLS covariance estimator error of the log-cosh estimator in a tenth-degree linear model

Figure 4.20 illustrates that the covariance estimation error, measured by the Frobenius norm, consistently decreases as the sample size grows. When the sample size exceeds approximately 10^5, the error quickly drops to a very low level and remains stable thereafter. This result suggests that the log-cosh regression achieves highly accurate covariance estimation when sufficient data is available. Furthermore, the curve shows a clear pattern of improvement even in the small and moderate sample size range, which highlights the method's robustness and reliability under finite sample conditions.

Example 4.6 (Tukey's Biweight in Tenth Degree Case) Consider the same tenth-degree polynomial model with the error obeying a normal distribution with a mean of 0 and a standard deviation of 0.5. of the form:

$$\eta(x_i, \theta) = \theta_0 + \theta_1 x_i + \theta_2 x_i^2 + \cdots + \theta_{10} x_i^{10},$$

with true parameter values specified as

$$\theta^* = (1.5,\ 0.8,\ 2.1,\ -1.2,\ 0.7,\ 3.0,\ 0.5,\ -0.6,\ 1.1,\ -0.9,\ 2.2)^T.$$

$$u_i(\theta) = y_i - \eta(x_i, \theta).$$

The Tukey's biweight estimator is a specific instance of an M-estimator, in which the loss function is defined as

$$\rho(u) = \begin{cases} \frac{c^2}{6}\left[1 - \left(1 - \left(\frac{u}{c}\right)^2\right)^3\right], & \text{if } |u| \le c, \\ \frac{c^2}{6}, & \text{if } |u| > c, \end{cases}$$

where $c > 0$ is a tuning parameter and u denotes the residual.

For small $|u|$, this loss behaves roughly like a quadratic function, retaining sensitivity to minor deviations. For large $|u|$, it remains bounded, thus limiting the influence of outliers.

The estimator $\widehat{\theta}$ is defined as the solution to

$$\widehat{\theta} = \arg\min_{\theta} \sum_{i=1}^{N} \rho\big(y_i - \eta(x_i, \theta)\big).$$

Let $\psi(u) = \frac{d\rho(u)}{du}$ denote the influence function. For Tukey's biweight loss, it is given by

$$\psi(u) = \begin{cases} u\left[1 - \left(\frac{u}{c}\right)^2\right]^2, & |u| < c, \\ 0, & |u| \ge c. \end{cases}$$

It satisfies the regularity condition $\frac{\psi(u)}{u} > 0$ for $|u| < c$, ensuring a smoothly decreasing influence as $|u|$ grows, and completely zero influence once $|u| \ge c$.

Accordingly, the weight term is defined as

$$w_{ii}(\theta) = \sqrt{\frac{\psi\left(u_i(\theta)\right)}{u_i(\theta)}} = \sqrt{\frac{u_i(\theta)\left[1 - \left(\frac{u_i(\theta)}{c}\right)^2\right]^2}{u_i(\theta)}}, \quad \text{if } |u_i(\theta)| < c,$$

and $w_{ii}(\theta) = 0$ otherwise.

Also set $w_{ij} = 0$ for $i \neq j$.

In Tukey's biweight loss function, the scale parameter c plays a key role in determining the boundary of the saturated region of the loss function. The parameter c controls the point beyond which the loss function no longer increases, thereby effectively truncating the influence of extreme outliers.

To ensure that this robust loss function achieves high statistical efficiency under normally distributed noise while adequately suppressing the effect of outliers, a common approach to determine c is as follows. First, one estimates the noise standard deviation $\hat{\sigma}$ using a robust scale estimator (e.g., the median absolute deviation, MAD):

$$\hat{\sigma} = \frac{\text{median}(|r_i|)}{0.6745}.$$

Then, one sets

$$c = k \cdot \hat{\sigma},$$

= where the constant $k = 4.685$ is typically used to ensure approximately 95% efficiency under normality. In the present example, given that the noise standard deviation is known to be $\sigma = 0.5$ (and taking $\hat{\sigma} \approx 0.5$), we obtain

$$c = 4.685 \times 0.5 \approx 2.3425.$$

For ease of numerical computation and implementation, we round c to 2.34. This value strikes a balance between suppressing extreme outliers and maintaining a good fit for regular observations.

Define the weighted residuals and gradient matrix as follows:

$$y_i(\theta) = w_{ii}(\theta)\left(y_i - \eta(x_i, \theta)\right) = w_{ii}(\theta)\, u_i(\theta),$$

$$Y(\theta) = \begin{pmatrix} y_1(\theta) \\ \vdots \\ y_N(\theta) \end{pmatrix},$$

$$g_{ij}(\theta) = x_i^j\, w_{ii}(\theta),$$

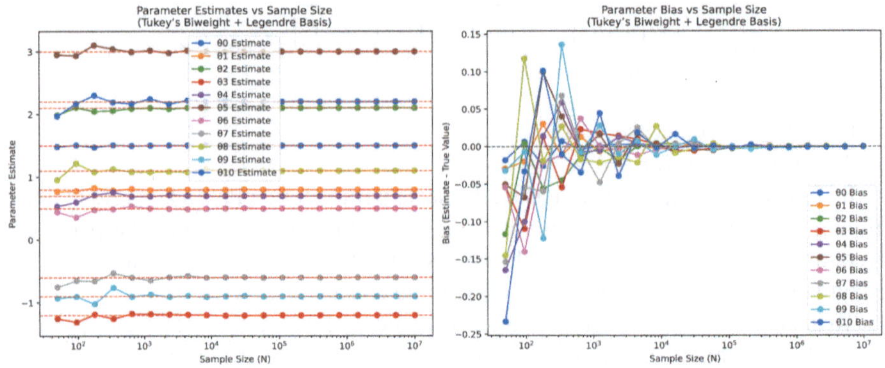

Fig. 4.21 Estimator consistency and bias convergence using Tukey's biweight in tenth-degree case

$$G(\theta) = \begin{pmatrix} x_1^0\, w_{11}(\theta) & x_1^1\, w_{11}(\theta) & \cdots & x_1^{10}\, w_{11}(\theta) \\ x_2^0\, w_{22}(\theta) & x_2^1\, w_{22}(\theta) & \cdots & x_2^{10}\, w_{22}(\theta) \\ \vdots & \vdots & \ddots & \vdots \\ x_N^0\, w_{NN}(\theta) & x_N^1\, w_{NN}(\theta) & \cdots & x_N^{10}\, w_{NN}(\theta) \end{pmatrix}.$$

The normal equations for M-estimation can be expressed as

$$G^T(\theta)\, Y(\theta) = G^T(\theta)\, G(\theta)\, \left(\theta^M - \theta^*\right),$$

where θ^* is the true parameter vector defined above.

Since θ^* is unknown, we use an iterative update to approximate θ^M:

$$\theta^{(s+1)} = \theta^{(s)} + \left[G^T(\theta^{(s)})\, G(\theta^{(s)})\right]^{-1} G^T(\theta^{(s)}) \left[Y(\theta^{(s)}) - \eta\left(x, \theta^{(s)}\right)\right].$$

At each iteration, the biweight function completely downweighs residuals exceeding $\pm c$. For small $|u|$, $\rho(u) \approx \frac{u^2}{2}$, recovering near-quadratic behavior, while for large $|u|$, $\rho(u)$ remains bounded.

Hence, Tukey's biweight effectively limits the influence of outliers and enhances robustness compared to non-robust estimators such as ordinary least squares. The following experimental results will fully illustrate the convergence of the estimator under large sample size.

Tables 4.8 and 4.9 show the parameter bias convergence under large-sample conditions. The results show that all parameter biases decrease significantly as the sample size increases, confirming the consistency of the estimators. Some parameters, such as θ_8 to θ_{10}, exhibit relatively large biases in small samples (0.0371, 0.0234, and 0.0582, respectively), but their biases reduce to the order of 10^{-3} or smaller when the sample size exceeds 5 million. Parameters like θ_0 to θ_3 maintain smaller biases throughout and exhibit faster convergence. Overall, the model demonstrates effective bias control and improved estimation accuracy with large samples.

Table 4.8 Bias convergence of the first six parameters using Tukey's biweight in tenth-degree regression

Sample size	θ_0	θ_1	θ_2	θ_3	θ_4	θ_5
50	−0.00428	0.00001	−0.02761	−0.00792	0.01515	0.01890
526365	0.00004	0.00029	0.00042	0.00004	−0.00192	0.00016
1052681	−0.00007	−0.00015	0.00019	0.00012	−0.00105	−0.00021
1578997	0.00002	−0.00013	0.00035	0. 00010	−0.00108	0.00019
2105313	0.00003	−0.00012	0.00013	0.00008	−0.00125	0.00003
2631628	0.00008	−0.00009	0.00019	0.00010	−0.00146	0.00047
3157944	−0.00001	−0.00007	0.00029	0.00012	−0.00138	0.00039
3684260	0.00003	−0.00015	0.00025	0.00004	−0.00129	0.00012
4210576	0.00005	−0.00008	0.00026	0.00009	−0.00133	0.00033
4736892	−0.00005	−0.00007	0.00022	0.00000	−0.00132	0.00019
5263207	0.00000	0.00001	0.00027	−0.00019	−0.00123	0.00034
5789523	0.00005	−0.00005	0.00024	−0.00015	−0.00147	0.00036
6315839	−0.00003	0.00000	0.00025	0.00001	−0.00176	0.00017
6842155	−0.00001	−0.00004	0.00022	−0.00001	−0.00138	0.00023
7368471	0.00000	−0.00002	0.00027	0.00011	−0.00143	0.00006
7894786	−0.00002	−0.00015	0.00028	−0.00001	−0.00138	0.00046
8421102	0.00004	−0.00008	0.00040	0.00001	−0.00137	0.00019
8947418	−0.00006	−0.00002	0.00021	0.00003	−0.00131	0.00049
9473734	0.00001	−0.00005	0.00031	−0.00018	−0.00153	0.00054
10000050	−0.00001	−0.00006	0.00022	0.00000	−0.00132	0.00020

Figure 4.21 shows the estimator consistency and bias convergence. From the parameter estimation plot on the left, we can see that as the sample size increases, most parameter estimates exhibit moderate fluctuations or shifts in the smaller-sample regime but generally converge toward their corresponding true parameters (indicated by the red dashed lines). By the time the sample size reaches the range of 10^5 to 10^7, these estimates stabilize. This suggests that, with the combination of Tukey's biweight loss and a Legendre polynomial basis, the model demonstrates robustness to noise and provides increasingly accurate estimates as more data become available.

The bias plot on the right further supports this observation: for most parameters, the bias curves converge toward zero at large sample sizes, indicating that discrepancies between the estimates and true values are effectively corrected. Although a few parameters may exhibit temporary positive or negative deviations at moderate sample sizes, these biases are ultimately mitigated as the sample size continues to grow. Overall, this approach shows stable convergence and an ability to suppress outlier effects when dealing with large-scale data.

Table 4.9 Bias convergence of the last five parameters using Tukey's biweight in tenth-degree regression

Sample size	θ_6	θ_7	θ_8	θ_9	θ_{10}
50	−0.02768	−0.06956	0.03710	0.02339	0.05816
526365	0.00208	0.00037	−0.00043	−0.00073	−0.00091
1052681	0.00203	0.00049	−0.00024	−0.00027	−0.00139
1578997	0.00197	−0.00007	0.00064	−0.00034	−0.00231
2105313	0.00219	0.00029	0.00026	−0.00064	−0.00132
2631628	0.00197	0.00013	−0.00001	−0.00041	−0.00110
3157944	0.00193	−0.00031	0.00002	−0.00024	−0.00081
3684260	0.00161	0.00013	−0.00006	−0.00004	−0.00102
4210576	0.00192	0.00004	0.00051	−0.00069	−0.00151
4736892	0.00199	0.00028	−0.00050	−0.00037	−0.00068
5263207	0.00191	0.00009	−0.00015	−0.00025	−0.00091
5789523	0.00214	0.00051	0.00040	−0.00075	−0.00158
6315839	0.00251	0.00007	0.00009	−0.00021	−0.00124
6842155	0.00192	0.00023	0.00059	−0.00043	−0.00155
7368471	0.00197	0.00030	0.00068	−0.00053	−0.00152
7894786	0.00190	0.00025	0.00079	−0.00071	−0.00168
8421102	0.00171	−0.00020	0.00029	−0.00017	−0.00111
8947418	0.00170	−0.00012	0.00080	−0.00069	−0.00162
9473734	0.00201	−0.00018	0.00028	−0.00053	−0.00112
10000050	0.00175	0.00018	0.00090	−0.00056	−0.00162

By conducting repeated simulation experiments across varying sample sizes, the parameter bias convergence plots and quantile path diagrams are generated to evaluate the asymptotic properties and robustness of the method in Fig. 4.22.

Figure 4.22a presents the bias convergence behavior of the 11 parameters estimated using Tukey's biweight loss in a tenth-degree polynomial regression model. As the sample size increases, the bias of all parameters gradually approaches zero, confirming the consistency of the estimator. Lower-order parameters such as θ_0, θ_1, and θ_2 converge rapidly with minimal deviation, reflecting the model's robustness in capturing dominant structural components. In contrast, mid- to high-order parameters, particularly θ_5 through θ_{10}, exhibit larger and more unstable biases in the small-sample regime, likely due to sensitivity to localized noise or numerical amplification in higher-degree terms. Nonetheless, as the sample size grows, all parameter estimates stabilize, indicating the effectiveness of Tukey's loss in mitigating the influence of extreme residuals and improving asymptotic reliability.

Figure 4.22b illustrates the quantile convergence paths of each parameter estimate, revealing how the 5%, 25%, 50%, 75%, and 95% quantiles evolve with increasing

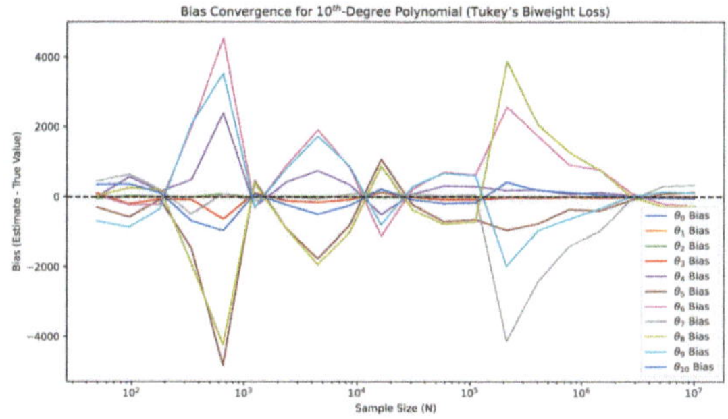

(a) Parameter bias convergence

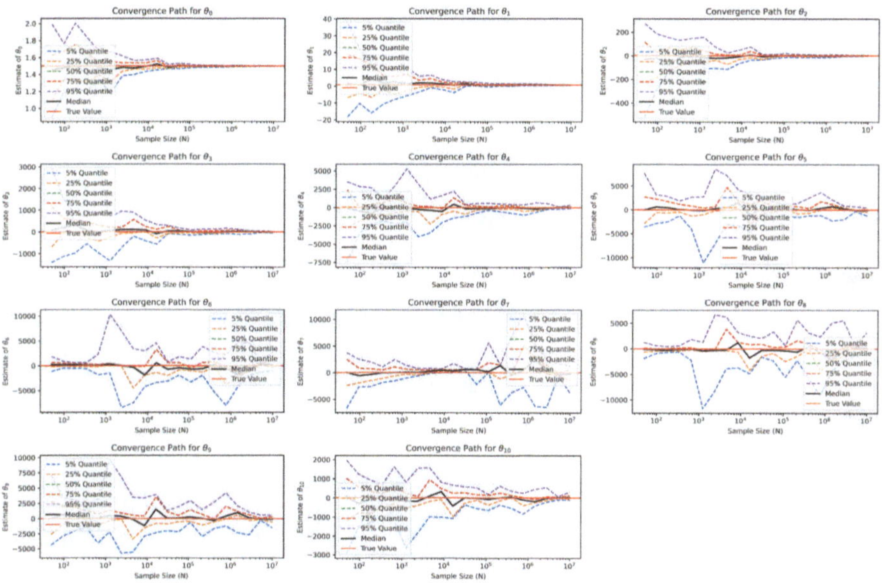

(b) Convergence path for θ_i, $i = 0, \cdots, 11$

Fig. 4.22 Visualization of the convergence trajectories of parameter biases using Tukey's biweight in tenth-degree case

sample size. The median estimates for all parameters steadily approach their true values, with the interquantile bands shrinking significantly as sample size increases—especially for lower-order terms like θ_0, θ_1, and θ_2, where convergence is both fast and stable. For higher-order parameters such as θ_8 through θ_{10}, the estimates display pronounced skewness and volatility under small-sample conditions, yet the use of Tukey's loss effectively suppresses extreme deviations and results in more concentrated and centered distributions.

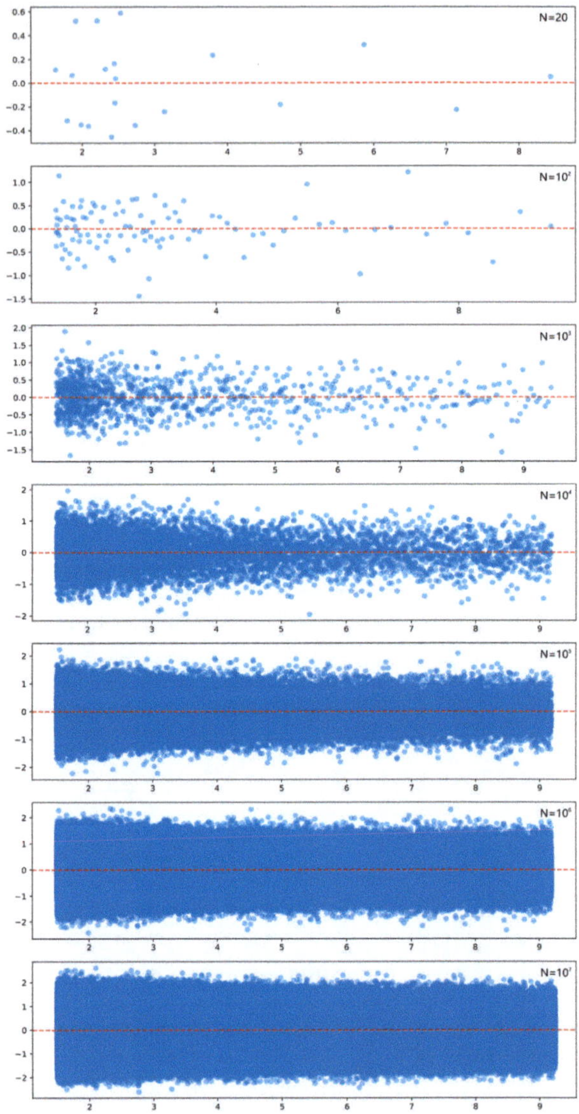

Fig. 4.23 Graphical test for heteroscedasticity using Tukey's biweight in tenth-degree case

Figure 4.23 shows the graphical test for heteroscedasticity. In the small-sample regime (e.g., $N = 20$ and $N = 10^2$), the residuals exhibit high variability and irregular dispersion, indicating unstable estimation and a strong presence of noise and outliers. Although Tukey's biweight loss provides robustness against extreme values, its ability to stabilize residuals is limited under such sparse data conditions. The residual plots at this stage show clear signs of heteroscedasticity, with no consistent variance structure and several residuals deviating significantly from the central mass.

As the sample size increases to moderate levels ($N = 10^3$ to $N = 10^5$), the residuals become more concentrated around zero, and their distribution becomes more structured. While fan-shaped patterns still emerge in certain input regions—suggesting localized heteroscedasticity—the overall variance begins to stabilize. Tukey's loss function starts to exert its robustness advantage here by mitigating the influence of outliers and reducing the residual spread, especially in regions where higher-order polynomial terms could otherwise amplify noise.

In the large-sample regime ($N = 10^6$ and $N = 10^7$), the residuals display a tightly clustered and symmetric distribution around zero, with no apparent heteroscedastic patterns. The residuals resemble white noise, exhibiting no trend or structure, which aligns with the homoscedasticity assumption in well-specified models. These results confirm that Tukey's biweight loss achieves excellent stability and convergence properties under sufficient data, making it particularly well suited for high-dimensional regression tasks involving noise and potential outliers.

From Fig. 4.24, we can see that, for different sample sizes, the polynomial parameter estimates obtained via Tukey's biweight loss (with regularization) generally exhibit relatively concentrated distributions after multiple simulations, and there is some correlation among the parameters. At smaller sample sizes, the estimates appear more scattered or biased, but as the data scale increases, most parameter distributions become more compact and closer to the true values, reflecting the effectiveness of the warm-start strategy and orthogonal polynomial basis in enhancing numerical stability. Overall, despite large differences in sample sizes, the model demonstrates good convergence and resilience to outliers at higher data scales, indicating that the regularized Tukey's biweight regression approach offers notable robustness and accuracy in handling noise and multi-parameter fitting.

From these error bar plots (as shown in Fig. 4.25), it is evident that when the sample size is small, each parameter's estimated value diverges somewhat from its true value, and the confidence intervals (error bars) are relatively wide, reflecting the model's sensitivity to noise and initial conditions in a limited data setting. However, as the sample size increases, most parameter estimates rapidly converge toward the true values, and the confidence intervals narrow considerably, indicating that the uncertainty in the parameter estimates gradually decreases. Ultimately, at larger sample sizes, the estimates reach a relatively stable and accurate level. This demonstrates that the chosen regression and regularization strategies effectively handle noise, suppress outliers, and improve numerical stability, allowing the model to exhibit strong convergence and robustness with large-scale data.

In Fig. 4.26, the Frobenius norm error decreases significantly as the sample size increases, and reaches a negligible level when the sample size exceeds approximately 10^5. After that, the error remains stable and close to zero even as the sample size further grows. This result suggests that the covariance estimation based on Tukey regression converges well to the classical OLS covariance formula in large-sample settings. Moreover, in the small and medium-sample regimes (e.g., $N < 10^5$), the error exhibits a smooth and consistent reduction, reflecting the progressive improvement in the estimation accuracy as more information becomes available.

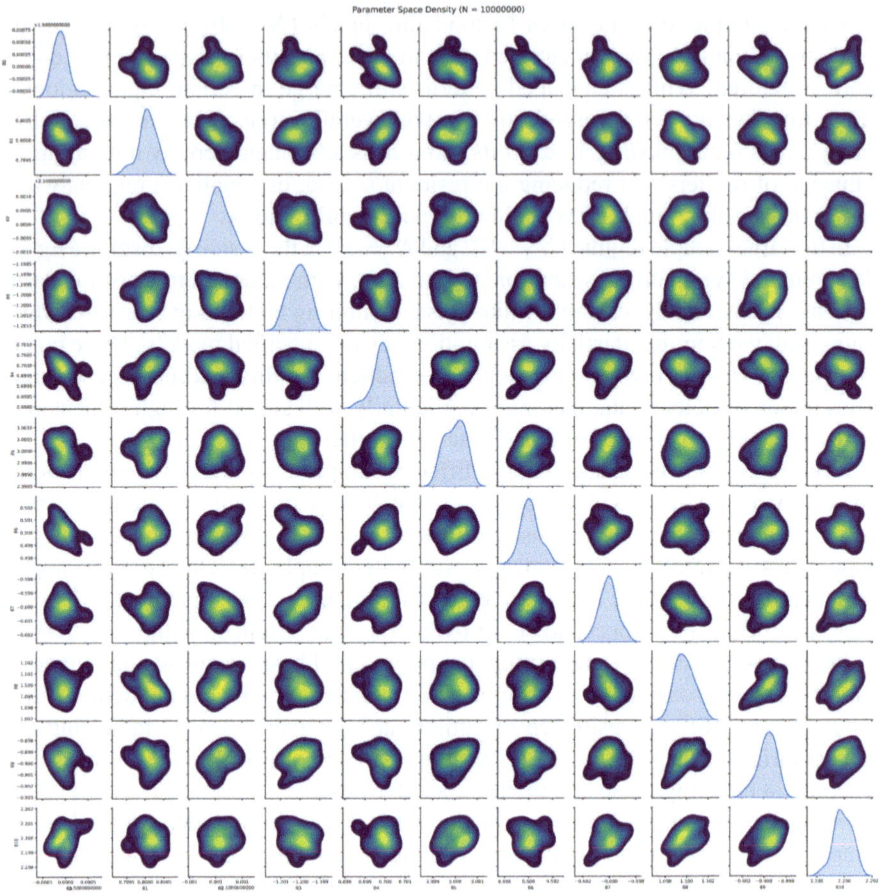

Fig. 4.24 Heatmap of marginal and joint parameter density distributions using Tukey's biweight in tenth-degree case

In this study, we focus on the following tenth-degree polynomial model:

$$y_i = \theta_0 + \theta_1 x_i + \theta_2 x_i^2 + \cdots + \theta_{10} x_i^{10} + \varepsilon_i, \quad i = 1, \ldots, N,$$

where the input points x_i are uniformly distributed over the interval [0, 1], and the noise terms ε_i are independently drawn from a normal distribution $N(0, \sigma^2)$ with $\sigma = 0.5$. The true parameter vector is given as

$$\theta_{\text{true}} = (1.5,\ 0.8,\ 2.1,\ -1.2,\ 0.7,\ 3.0,\ 0.5,\ -0.6,\ 1.1,\ -0.9,\ 2.2)^T.$$

In order to evaluate and compare different parameter estimation methods, we employed the least squares estimator, the log-cosh estimator, and Tukey's biweight estimator, and generated a series of visualization charts based on these methods. The

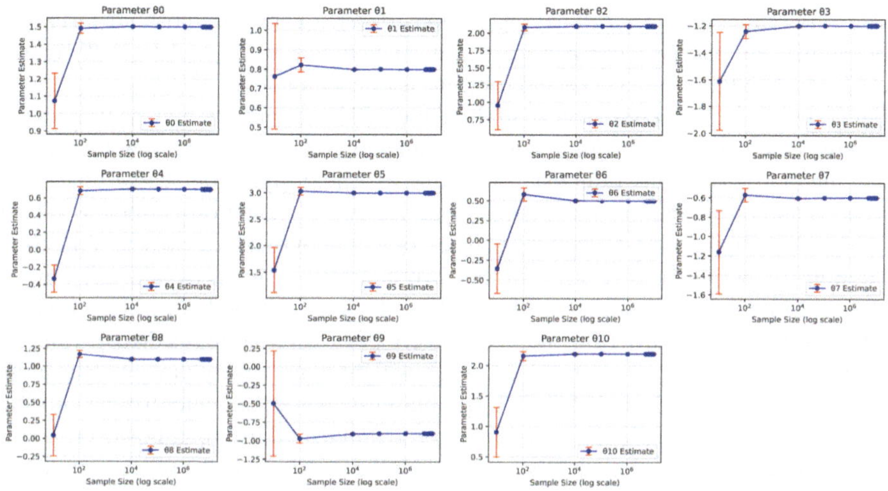

Fig. 4.25 Confidence interval of each parameter using Tukey's biweight in tenth-degree case

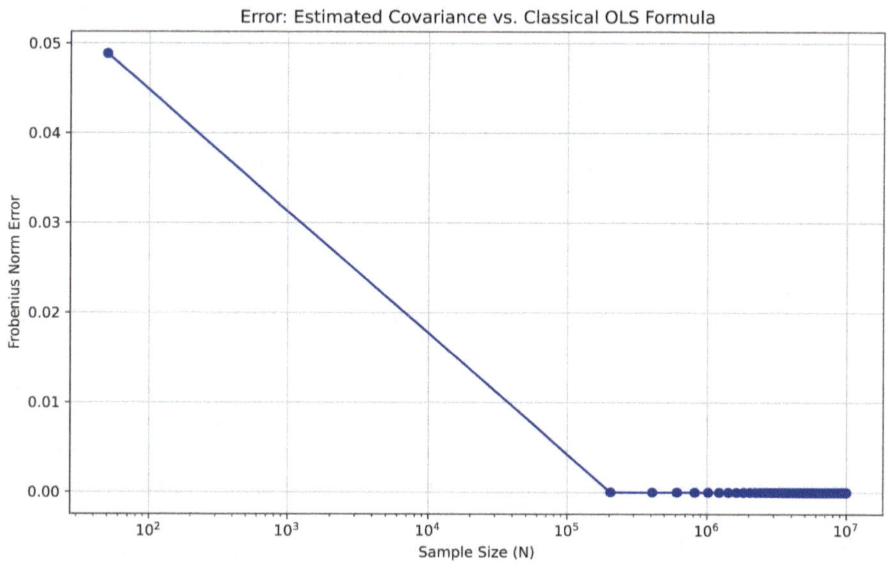

Fig. 4.26 Classical OLS covariance estimator error using Tukey's biweight in tenth-degree case

following sections provide an in-depth analysis of these charts, exploring differences in performance with respect to estimation accuracy, robustness, and other aspects, thereby offering both theoretical and practical insights for parameter estimation in complex models.

(1) Figures 4.3, 4.15, and 4.21 illustrate how, in a tenth-order polynomial model, the mean estimates and biases of the model parameters vary with increasing sample

size using three different estimation methods (least squares, log-cosh, and Tukey's biweight). By visually comparing the mean and bias curves, the charts reveal differences in the convergence and robustness of each method.

For convergence and consistency, all three methods demonstrate that, as the sample size increases, the mean estimates of the parameters converge toward their true values and the biases gradually diminish, exhibiting excellent asymptotic consistency. For robustness considerations, the log-cosh and Tukey's biweight estimators, as robust methods, offer significant advantages when potential outliers are present. Although the noise is normally distributed in the current scenario, in cases of small sample sizes (or in practical situations with outliers), the log-cosh method, which employs a relatively mild penalty (thus not excessively penalizing large residuals), exhibits a stable bias pattern. In contrast, Tukey's biweight aggressively downweighs large residuals, which may lead to slightly larger biases for some parameters (especially the higher-order terms), reflecting a more vigorous robust adjustment; however, it still converges to the true values in large samples. In addition, while the robust estimators may yield coefficients with slightly larger bias under ideal noise assumptions, they tend to avoid the extremely large magnitudes observed in high-order coefficients estimated by least squares, thereby improving numerical stability and interpretability.

Under purely normal noise conditions, the least squares method proves efficient (however, it cannot be ignored that the starting deviation is the largest and the deviation decreases faster), whereas the two robust methods can provide more reliable estimates in small-sample scenarios or when outliers are present, albeit possibly at the cost of some efficiency. In practical applications, if concerns arise about outliers or model imperfections, employing either the log-cosh or Tukey's biweight method may be more appropriate.

(2) Figures 4.4a, 4.16a, and 4.22a illustrate the bias convergence behavior of parameter estimates in a tenth-degree polynomial model, obtained using the ordinary least squares estimator, the log-cosh estimator, and Tukey's biweight estimator, respectively. Specifically, the plots visualize how the estimation bias—defined as the difference between the estimated and true parameter values—diminishes as the sample size increases. The figures not only reveal the trend of bias reduction across varying sample sizes for each estimation method, but also provide an intuitive comparison of their robustness in handling model complexity and noise disturbances in high-dimensional settings.

In the context of high-degree polynomial regression, the performance of different estimators varies significantly, particularly under limited sample sizes. The least squares (OLS) estimator demonstrates large-parameter biases when the sample size is small, primarily due to the numerical instability induced by the high-order polynomial design matrix. Although LS is theoretically optimal under Gaussian noise, it requires a substantially large sample size to achieve stable and unbiased parameter estimation. This observation is consistent with the experimental results, where biases only diminish significantly as the sample size reaches the scale of hundreds of thousands. On the other hand, robust estimators such as the log-cosh and Tukey's biweight methods exhibit faster bias convergence within a much smaller sample range (up to approximately 4000). The log-cosh estimator, due to its smooth and differentiable loss function, achieves gradual and consistent reduction in bias as the sample size increases. This smooth convergence behavior indicates its capability to mitigate the influence of moderate outliers or numerical irregularities without introducing abrupt changes in parameter estimates. In contrast, Tukey's biweight estimator employs a redescending loss function that heavily downweighs large residuals. While this leads to strong robustness against outliers, it may introduce relatively large initial bias fluctuations for certain parameters when the sample size is limited. Nevertheless, as the sample size grows, the bias converges rapidly, highlighting the method's effectiveness in stabilizing parameter estimation under noisy conditions.

The comparison reveals that the least squares estimator, despite its asymptotic optimality under Gaussian noise, suffers from instability in high-degree polynomial regression models when the sample size is small. It demands an exceedingly large number of observations to attain satisfactory bias reduction. In contrast, robust estimators, particularly log-cosh and Tukey's biweight, offer superior bias convergence properties under limited data scenarios. The log-cosh loss facilitates smooth and reliable parameter estimation, while the biweight loss provides stronger protection against large residuals at the cost of potential initial bias variability. These results underscore the practical advantage of robust estimators in complex modeling tasks where data availability is constrained and model complexity exacerbates numerical instability.

(3) Figures 4.4a, 4.16b, and 4.22b show the convergence path and quantile evolution of each parameter by using the least squares method, log-cosh estimator, and Tukey's biweight estimator in the tenth-degree polynomial model.

Under ideal conditions with normally distributed noise, the ordinary least squares estimator is theoretically the most efficient; however, it is highly sensitive to outliers and extreme samples. This sensitivity can lead to considerable variability in finite-sample settings, especially for the higher-order coefficients. In contrast, the log-cosh estimator employs a smooth loss function that reduces the impact of large residuals. As a result, it exhibits improved stability even with smaller sample sizes and a more concentrated distribution relative to the ordinary least squares method, although its overall robustness is slightly inferior to that of Tukey's biweight estimator. Tukey's biweight estimator, by virtue of its redescending loss function, effectively downweighs observations that deviate significantly from the center. This characteristic yields notably compact quantile paths and exceptionally stable median estimates across both low- and high-order parameters, demonstrating superior performance in the presence of potential outliers. Although the ordinary least squares estimator guarantees unbiasedness and high efficiency under ideal, normally distributed noise conditions, the analysis of the three sets of charts clearly shows that robust estimators, such as the log-cosh and Tukey's biweight methods, exhibit faster convergence and reduced variability in finite-sample scenarios or when the data deviates slightly from ideal assumptions. In particular, Tukey's biweight estimator not only converges rapidly for all parameters but also achieves a more concentrated distribution, making it a more robust choice in the presence of outliers. Therefore, in practical applications where outliers or deviations from normality are a concern, adopting the log-cosh or

Tukey's biweight estimator can provide more stable and reliable parameter estimates compared to the traditional least squares approach.

(4) Figures 4.5, 4.17, and 4.23 illustrate the fitting performance of three estimation methods—least squares, log-cosh, and Tukey's biweight—under heteroscedastic noise conditions for a tenth-degree polynomial model, across varying sample sizes (from $N = 20$ to $N = 10^7$).

In terms of sensitivity to outliers, the least squares (OLS) method is highly susceptible to the influence of large residuals, often resulting in significant fluctuations in the fitted curve. In contrast, the log-cosh estimator mitigates the impact of outliers by applying a smoothed penalty on large residuals, leading to a more stable fit. Tukey's biweight estimator goes one step further by nearly eliminating the influence of outliers through a truncation mechanism, achieving the strongest robustness among the three.

Regarding the effect of sample size variation, the differences between the three methods are most evident when the sample size is small ($N = 20$): OLS exhibits pronounced oscillations in the fitted curve, while both log-cosh and Tukey's biweight estimators are better at capturing the true underlying trend. As the sample size increases ($N = 10^2$ to $N = 10^7$), LS performance improves but may still be affected by extreme values in localized regions. In comparison, the results of log-cosh and Tukey's biweight become increasingly stable, with the latter maintaining high consistency across all sample sizes. When the data is relatively clean and normally distributed, LS can provide unbiased and efficient estimates. However, in real-world scenarios where outliers are common, log-cosh and Tukey's biweight offer greater resistance to anomalous disturbances. Among them, Tukey's biweight achieves the highest robustness due to its truncation strategy, while log-cosh provides a smoother transition and serves as a balanced alternative between sensitivity and robustness.

(5) Figures 4.6, 4.18, and 4.24 demonstrate the joint distribution of the parameter estimates under multiple repetitions of the experiments, from which the effects of different estimation methods on parameter uncertainty and correlation can be visualized. For the least squares method (left panel), the heatmaps show highly skewed and elongated elliptical contours, indicating strong linear correlations among the parameters, which are especially pronounced among the higher-order terms, reflecting a serious multicollinearity problem in the model. The middle panel shows the results of the log-cosh estimator, whose heat map profile is significantly more regular than that of OLS, with a reduced slope and a shape closer to an ellipse or a circle-like shape, which indicates that this method alleviates the covariance among parameters to a certain extent and improves the stability of the estimation. On the other hand, the right figure corresponds to Tukey's biweight estimator, whose heat map shows the most rounded and symmetric distribution, and the joint distributions of most parameters are approximately normal with weak correlation, which indicates that this method is the most effective in reducing the correlation of parameter estimation, improving the numerical stability, and suppressing the noise interference, which further verifies its strong robustness property.

The comparative analysis reveals that the Ordinary Least Squares (OLS) estimator is highly sensitive to noise and high-order polynomial terms, leading to wide parameter distributions, large variances, and strong correlations among high-order coefficients. These issues result in numerical instability and considerable estimation bias. In contrast, the log-cosh estimator significantly improves robustness compared to OLS, effectively narrowing the distribution of parameter estimates and reducing the impact of large residuals. However, it still exhibits certain instability and parameter correlation, especially for high-order coefficients. Tukey's biweight estimator demonstrates the best robustness among the three methods. It yields the most concentrated parameter distributions, substantially reduces the variance of high-order coefficients, and effectively weakens parameter correlations, achieving superior stability and resistance to noise and outliers. Overall, for scenarios involving high noise levels and high-dimensional settings (e.g., high-order polynomial regression), robust estimators such as Tukey's biweight offer a clear advantage over traditional OLS and simple robust loss functions like log-cosh, making them a more reliable choice for stable and accurate modeling.

Table 4.27 shows the comprehensive computational framework in our numerical examples. The objective of this study is to thoroughly investigate the asymptotic properties and numerical stability of parameter estimation in regression models. Specifically, we compare the performance of different regression models and estimation methods under various error structures (independent errors versus correlated errors) and varying numbers of parameters, providing theoretical support and methodological guidance for practical applications.

In "Problem Definition and Key Comparisons", we focus on four main areas of comparison:

(1) **Model Type**

we categorize the regression models into simple linear regression, general linear regression, and nonlinear regression, with a primary focus on comparing the advantages and limitations of different model types in parameter estimation.

(2) **Number of Parameters**

In Examples 3.1–3.4, we use 6 parameters for thc regression model, while in Examples 4.3–4.6, the number of parameters is expanded to 11 (through higher-degree polynomials). This change aims to assess the impact of model complexity on estimation results.

(3) **Error Structure**

We compare the effects of independent normal errors versus AR(1) correlated errors on parameter estimation. Notably, correlated errors introduce larger biases in small samples, which converge as the sample size increases.

(4) **Estimation Methods**

We compare traditional ordinary least squares (OLS) with more robust M-estimators. The M-estimators, including log-cosh and Tukey's biweight, provide superior robustness in handling outliers.

Then, we design different regression models based on the problem definition:

(1) **Linear Regression Models**

Example 3.1 uses a simple polynomial regression model with independent errors. Examples 3.2 and 3.3 employ the general model type but with AR(1) errors to analyze their impact on parameter estimation.

(2) **Nonlinear Regression Models**

Example 3.4 features a low-dimensional nonlinear model solved by Gauss–Newton iterations. Example 4.4 involves a more complex hybrid nonlinear model (combining exponential, power, and sine components) with AR(1) errors.

(3) **High-Dimensional Model Extension**

Example 4.3 is an extended linear model with a tenth-degree polynomial (11 parameters) and independent errors. Examples 4.5 and 4.6 are also tenth-degree polynomial models but use robust M-estimators: log-cosh loss and Tukey's biweight loss.

For experimental design and numerical validation, we perform extensive numerical testing for each model. Firstly, we vary the sample size from 50 to 10^7 to assess model stability and convergence properties under different conditions. Each example undergoes 50 repetitions to ensure the stability and reliability of the experimental results. Then, we compute and plot the following: convergence curves of parameter estimates and their means; parameter bias, quantile trajectories, and convergence paths; residual diagnostic plots (residuals versus fitted values) and covariance estimation error graphs and parameter density heatmaps.

Next, we conduct in-depth analysis and comparison of the results.

(1) **Compare Different Model Classes**

We compare simple linear regression, general linear regression, and nonlinear regression models in terms of parameter estimation performance, evaluating their behavior under different error structures.

(2) **Analyze the Effect of Parameter Count**

We compare Examples 3.1–3.4 (6 parameters) with Examples 4.3–4.6 to assess how increasing the number of parameters impacts model stability and convergence.

(3) **Discuss the Role of Error Structure**

We analyze the impact of independent errors and AR(1) errors on parameter estimation. The results show that AR(1) errors introduce larger bias in small samples, but as the sample size increases, the bias decreases.

(4) **Emphasize Robustness**

We highlight that the OLS method is sensitive to outliers, while robust M-estimators such as log-cosh and Tukey's biweight provide superior robustness, effectively reducing the impact of outliers on parameter estimates.

Finally, we draw the following conclusions:

(1) In large samples, all models achieve asymptotic unbiasedness and consistency, both linear and nonlinear regression models exhibit asymptotic unbiasedness and consistency under large sample sizes.

(2) High-Dimensional extensions perform well when the parameter-to-sample ratio condition is met, despite the increase in the number of parameters, as long as the appropriate parameter-to-sample size ratio is maintained, models still yield stable and consistent estimates.

(3) Robust M-estimators exhibit clear advantages under noisy or outlier-prone conditions, for data with noise or outliers, robust M-estimators (such as log-cosh and Tukey's biweight) significantly enhance model robustness and estimation accuracy.

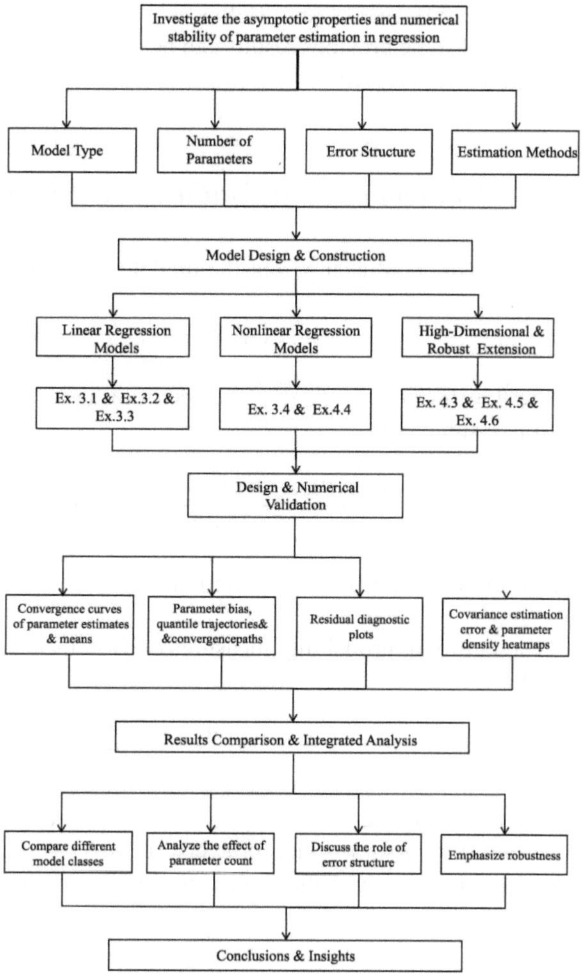

Fig. 4.27 Comprehensive computational framework for parameter estimation in regression models

(4) The overall workflow provides a systematic framework, this computational framework offers a systematic approach from model construction and algorithm implementation to experimental validation and integrated result analysis, bridging theoretical findings with practical applications.

4.4 Generalized Prototype Models

In the literature, various types of regression models were published which can be considered as a general form of regression model, for instance, functional regression, spline and quantile regression models, regression models with missing data [35,

44, 46–51, 54], and others. Below is considered the general form of regression models and some previous results can be considered as a special case of generalized regression models [15].

4.4.1 Foundations and Mathematical Framework

The following nonlinear regression model is considered:

$$y_i = \eta(x_i, \theta) + \varepsilon_i, \quad i = 1, 2, \ldots, N,$$

where $\theta = (\theta_1, \theta_2, \ldots, \theta_m)^T$ is the vector of unknown parameters.

It is assumed that

$$\eta(x, \theta), \quad \frac{\partial \eta(x, \theta)}{\partial \theta_i}, \quad \frac{\partial^2 \eta(x, \theta)}{\partial \theta_i \partial \theta_j}, \quad i = 1, 2, \ldots, m, \quad j = 1, 2, \ldots, m$$

are continuous and bounded functions, where $x \in X \subset \mathbb{R}$, $\theta \in \Theta \subset \mathbb{R}^p$, and $0 < p < N$.

The family of probability measures on Borel subsets A of the space X is defined as follows:

$$\mu_N(A) = \frac{1}{N} \sum_{i=1}^{N} I_A(x_i),$$

where $I_A(x_i)$ denotes the indicator function of the set A.

It is further assumed that for every A, as $N \to \infty$, the sequence $\mu_N(A)$ weakly converges to some probability measure $\mu(A)$, denoted as follows:

$$\mu_N \xrightarrow{w} \mu \quad \text{as} \quad N \to \infty. \tag{4.63}$$

Moreover, the error terms satisfy the following conditions:

$$E[\varepsilon_i] = 0, \quad E[\varepsilon_i^2] = \sigma^2(x_i) = \sigma_i \leq \sigma_0^2,$$

where $\sigma^2(x)$ is an unknown bounded function.

Define the matrix $F_N(\theta)$ with elements:

$$F_N(\theta) = \left[\frac{\partial \eta(x_i, \theta)}{\partial \theta_j} \right], \quad i = 1, 2, \ldots, N, \quad j = 1, 2, \ldots, m.$$

Explicitly, this matrix takes the form:

$$\frac{\partial \eta(x,\theta)}{\partial \theta} = \begin{bmatrix} \frac{\partial \eta(x_1,\theta)}{\partial \theta_1} & \cdots & \frac{\partial \eta(x_1,\theta)}{\partial \theta_m} \\ \vdots & \ddots & \vdots \\ \frac{\partial \eta(x_N,\theta)}{\partial \theta_1} & \cdots & \frac{\partial \eta(x_N,\theta)}{\partial \theta_m} \end{bmatrix}.$$

The normalized information matrix is given by

$$M_N(\theta) = \frac{1}{N} F_N^T(\theta) F_N(\theta) = \frac{1}{N} \int_X \frac{\partial \eta^T(x,\theta)}{\partial \theta} \cdot \frac{\partial \eta(x,\theta)}{\partial \theta} \mu_N(dx).$$

The eigenvalues of $M_N(\theta)$ are denoted by $\lambda_1^{(N)}(\theta), \ldots, \lambda_m^{(N)}(\theta)$, and satisfy

$$0 < \lambda_1^{(N)}(\theta) \leq \lambda_2^{(N)}(\theta) \leq \cdots \leq \lambda_m^{(N)}(\theta).$$

Following the approach in Sect. 4.3, an iterative process can be constructed to compute the least squares estimator ($l.s.e.$), similar to the procedure used for nonlinear regression models. This process is given by

$$\theta_N(s+1) = \theta_N(s) + (N \cdot M_N(\theta(s)))^{-1} F_N^T(\theta(s))(y - \eta(x, \theta(s))). \tag{4.64}$$

The convergence properties of the process (4.64) are now analyzed.

Before proceeding, the following theorem, established in [30], is recalled.

Theorem 4.16 *Assume that there is given the families of functions $G_N(x,\theta)$ which are continuous on $X \times \Theta$, where X, Θ are compact sets. $G_N(x,\theta) \to G(x,\theta)$ (uniformly) for $N \to \infty$.*

Then

$$\lim_{N\to\infty} \sup_{\theta} \left| \int_X G_N(x,\theta)\mu_N(dx) - \int_X G(x,\theta)\mu(dx) \right| = 0.$$

Proof By the assumption of uniform convergence, for any $\varepsilon > 0$, there exists a sufficiently large N_0 such that for all $N \geq N_0$:

$$|G_N(x,\theta) - G(x,\theta)| < \varepsilon \quad \text{for all } x \in X \text{ and } \theta \in \Theta.$$

This implies that for $N \geq N_0$:

$$\left| \int_X G_N(x,\theta)\, \mu_N(dx) - \int_X G(x,\theta)\, \mu(dx) \right| \leq \int_X |G_N(x,\theta) - G(x,\theta)|\, \mu_N(dx).$$

Using the fact that $|G_N(x,\theta) - G(x,\theta)| < \varepsilon$, it can be obtained that:

$$\left| \int_X G_N(x,\theta)\, \mu_N(dx) - \int_X G(x,\theta)\, \mu(dx) \right| \leq \varepsilon \int_X 1\, \mu_N(dx) = \varepsilon \mu_N(X).$$

Since μ_N is a probability measure on X, $\mu_N(X) = 1$. Hence, the difference between the integrals is bounded by ε, i.e.:

$$\left|\int_X G_N(x,\theta)\,\mu_N(dx) - \int_X G(x,\theta)\,\mu(dx)\right| \le \varepsilon.$$

Now, take the supremum over θ. Since the right-hand side does not depend on θ, for any $\varepsilon > 0$, there exists an N_0 such that for all $N \ge N_0$, it follows that

$$\sup_\theta \left|\int_X G_N(x,\theta)\,\mu_N(dx) - \int_X G(x,\theta)\,\mu(dx)\right| \le \varepsilon.$$

This implies that

$$\lim_{N\to\infty} \sup_\theta \left|\int_X G_N(x,\theta)\,\mu_N(dx) - \int_X G(x,\theta)\,\mu(dx)\right| = 0.$$

Thus, the proof of the theorem is complete. □

From Theorem 4.16.

Corollary 4.1 *Under the following conditions:*

$$\mu_N(A) = \frac{1}{N}\sum_{i=1}^{N} I_A(x_i), \tag{4.65}$$

where $I_A(x_i)$ denotes the indicator function of the set A.

$$\mu_N \xrightarrow{w} \mu \quad as \quad N \to \infty \tag{4.66}$$

uniformly on x the following expression is held: $M_N(\theta) \to M(\theta)$ for $N \to \infty$.

Proof Let $M_N(\theta) = \int_X G_N(x,\theta)\mu_N(dx)$ and $M(\theta) = \int_X G(x,\theta)\mu(dx)$. The matrix elements of $G_N(x,\theta)$ are given by

$$[G_N(x,\theta)]_{jk} = \frac{1}{N}\sum_{i=1}^{N} \frac{\partial\eta(x_i,\theta)}{\partial\theta_j}\frac{\partial\eta(x_i,\theta)}{\partial\theta_k}.$$

By Condition (4.65), the partial derivatives $\frac{\partial\eta(x,\theta)}{\partial\theta}$ are continuous and bounded on $X \times B(r)$. The Uniform Law of Large Numbers [12] implies

$$\sup_{\theta\in B(r)} |G_N(x,\theta) - G(x,\theta)| \xrightarrow{P} 0 \quad \text{as } N \to \infty,$$

where $G(x,\theta) = E\left[\frac{\partial\eta(x,\theta)}{\partial\theta}\frac{\partial\eta(x,\theta)}{\partial\theta}^T\right]$.

Condition (4.66) states $\mu_N \xrightarrow{w} \mu$. For any continuous bounded function $h(x, \theta)$, it follows that

$$\int_X h(x, \theta)\mu_N(dx) \xrightarrow{P} \int_X h(x, \theta)\mu(dx) \quad \text{uniformly in } \theta \in B(r).$$

Since $G_N(x, \theta)$ converges uniformly to $G(x, \theta)$ and $\mu_N \xrightarrow{w} \mu$, the Portmanteau theorem [53] gives

$$M_N(\theta) = \int_X G_N(x, \theta)\mu_N(dx) \to \int_X G(x, \theta)\mu(dx) = M(\theta),$$

uniformly in $\theta \in B(r)$. The uniformity follows from the equicontinuity of $\{G_N\}$ ensured by Condition (4.65).

By the Arzela–Ascoli Theorem [26], the uniform boundedness and equicontinuity of $\{G_N\}$ guarantee uniform convergence on compact $B(r)$. Thus,

$$\sup_{\theta \in B(r)} |M_N(\theta) - M(\theta)| \xrightarrow{P} 0 \quad \text{as } N \to \infty.$$

Thus, the proof of the theorem is complete. □

Corollary 4.2 *Suppose that the sequence of probability measures $\mu_N(A)$ is given by*

$$\mu_N(A) = \frac{1}{N} \sum_{i=1}^{N} I_A(x_i), \tag{4.67}$$

where $I_A(x_i)$ denotes the indicator function of the set A.

Furthermore, assume that $\mu_N(A)$ weakly converges to a probability measure $\mu(A)$ as $N \to \infty$, i.e.:

$$\mu_N \xrightarrow{w} \mu \quad as \quad N \to \infty. \tag{4.68}$$

If the sequence of normalized information matrices satisfies the uniform convergence condition:

$$M_N(\theta) \to M(\theta) \quad \textit{uniformly on } \theta \textit{ as } N \to \infty,$$

then the following relation holds uniformly on θ:

$$\frac{\partial M_N(\theta)}{\partial \theta} \to \frac{\partial M(\theta)}{\partial \theta} \quad as\ N \to \infty.$$

Proof Let $F_N(\theta) = \left(\frac{\partial \eta(x_i, \theta)}{\partial \theta}\right)_{N \times m}$ be the Jacobian matrix. The derivative of $M_N(\theta) = \frac{1}{N} F_N^T(\theta) F_N(\theta)$ is

$$\frac{\partial M_N(\theta)}{\partial \theta_k} = \frac{1}{N}\left(\frac{\partial F_N^T(\theta)}{\partial \theta_k} F_N(\theta) + F_N^T(\theta)\frac{\partial F_N(\theta)}{\partial \theta_k}\right).$$

By Condition 4.67, the second derivatives $\frac{\partial^2 \eta(x,\theta)}{\partial \theta_k \partial \theta_j}$ exist and are continuous on $X \times B(r)$. The Uniform Law of Large Numbers [12] implies

$$\sup_{\theta \in B(r)} \left| \frac{1}{N}\frac{\partial F_N^T(\theta)}{\partial \theta} F_N(\theta) - E\left[\frac{\partial}{\partial \theta_k}\left(\frac{\partial \eta}{\partial \theta}\right)^T \frac{\partial \eta}{\partial \theta}\right]\right| \xrightarrow{P} 0.$$

For the empirical measure μ_N, the weak convergence $\mu_N \xrightarrow{w} \mu$ (Condition 4.68) and continuity of $x \mapsto \frac{\partial}{\partial k}\left(\frac{\partial}{\partial \theta}\right)^T \frac{\partial}{\partial \theta}$ imply

$$\frac{1}{N} F_N^T(\theta)\frac{\partial F_N(\theta)}{\partial \theta_k}\left(\frac{\partial \eta}{\partial \theta}\right)^T \frac{\partial \eta}{\partial \theta}\mu_N(dx) \xrightarrow{P} \int_X \frac{\partial}{\partial \theta_k}\left(\frac{\partial \eta}{\partial \theta}\right)^T \frac{\partial \eta}{\partial \theta}\mu(dx).$$

The family $\left\{\frac{\partial M_N(\theta)}{\partial \theta_k}\right\}$ is equicontinuous by

Boundedness of third derivatives: $\sup_{x,\theta}\left|\frac{\partial^3 \eta}{\partial \theta_k \partial \theta_j \partial \theta_i}\right| \le C$, Arzela–Ascoli theorem for compact $B(r)$.

With Theorem 4.16 (uniform convergence of $M_N \to M$) and the Moore–Osgood theorem for exchanging limits:

$$\sup_{\theta \in B(r)}\left|\frac{\partial M_N(\theta)}{\partial \theta} - \frac{\partial M(\theta)}{\partial \theta}\right| \xrightarrow{P} 0 \quad \text{as } N \to \infty.$$

The theorem is thus proven. □

Remark 4.12 In Corollaries 4.1 and 4.2, there have been proved only existence of the limits $M_N(\theta)$ and $\frac{\partial M_N(\theta)}{\partial \theta}$, although the values of these limits are unknown. Consider measurable families:

$$g_{N,y}(x,\theta) \in C(X \times \Theta).$$

Introduce the following conditions:

$$\lim_{N \to \infty} \sup_{X \times \theta} |g_N(x,\theta) - g(x,\theta)| = 0$$

$$\sup_{X \times \theta} |g_{N,y}(x,\theta)| \le C < \infty$$

$$G_N(x,\theta) = E g_{N,y}(x,\theta)$$

which is continuous function on $X \times \theta$.

$$\lim_{N\to\infty} \sup_{\Theta} \frac{1}{N} D(\sum_{i=1}^{N} g_{N,y_i}(x_i, \theta^*)) = 0.$$

Denote

$$\begin{aligned} S_N(\theta) &= \frac{1}{N}\sum_{i=1}^{N} g_{N,y_i}(x_i,\theta), \theta \in B(r); \\ G(x,\theta) &= \lim_{N\to\infty} G_N(x,\theta) = \lim_{N\to\infty} E g_{N,y}(x,\theta); \\ S(\theta) &= \int_X^{\theta} G(x,\theta)\mu(dx). \end{aligned} \tag{4.69}$$

Theorem 4.17 *Assume $\theta \in B(r)$ and for*

$$y_i = \eta(x_i,\theta) + \varepsilon_i, \quad i = 1, 2, \ldots, N,$$

the conditions

$$\theta_N(s+1) = \theta_N(s) + (N \cdot M_N(\theta(s)))^{-1} F_N^T(\theta(s))(y - \eta(x,\theta(s))),$$

$$[G_N(x,\theta)]_{jk} = \frac{1}{N}\sum_{i=1}^{N} \frac{\partial\eta(x_i,\theta)}{\partial\theta_j}\frac{\partial\eta(x_i,\theta)}{\partial\theta_k}, \tag{4.70}$$

$$\lim_{N\to\infty} \sup_{X\times\theta} |g_N(x,\theta) - g(x,\theta)| = 0 \tag{4.71}$$

$$\sup_{X\times\theta} |g_{N,y}(x,\theta)| \le C < \infty \tag{4.72}$$

are held. Then

$$(S_N(\theta) - S(\theta)) \xrightarrow{P} 0, \, for N \to \infty, r \to 0.$$

Proof Since $\mu_N(\cdot)$ is the discrete measure, the ratio in Eq. (4.69) can be rewritten as follows:

$$S_N(\theta) = \frac{1}{N}\sum_{i=1}^{N} g_{N,y_i}(x_i,\theta) = \int_X^{\theta} g_{N,y}(x,\theta)\mu_N(dx).$$

Using Condition (4.72), it follows that

$$E[S_N(\theta)] = \int_X^{\theta} E[g_{N,y}(x,\theta)]\mu_N(dx) = \int_X^{\theta} G_N(x,\theta)\mu_N(dx).$$

Applying Theorem 4.16, it is obtained that

$$E[S_N(\theta)] = \int_X^\theta G_N(x,\theta)\mu_N(dx) \to \int_X^\theta G(x,\theta)\mu(dx), \quad \theta \in B(r), \quad \text{as} \quad N \to \infty.$$

Thus, for $N \to \infty$, it follows that

$$E[S_N(\theta)] \to \int_X^\theta G(x,\theta)\mu(dx) = S(\theta), \quad \theta \in B(r).$$

For $\theta \in B(r)$, considering Eq. (4.69), the following expression is obtained:

$$||S_N(\theta) - E[S_N(\theta)]|| = \left|\left|\int_X^\theta g_{N,y}(x,\theta)\mu_N(dx) - \int_X^\theta G_N(x,\theta)\mu_N(dx)\right|\right|$$
$$= \left|\left|\int_X^\theta \left(g_{N,y}(x,\theta) - G_N(x,\theta) - [g_{N,y}(x,\theta^*) - G_N(x,\theta^*)]\right)\mu_N(dx)\right|\right|$$
$$+ \left|\left|\int_X^\theta [g_{N,y}(x,\theta^*) - G_N(x,\theta^*)]\mu_N(dx)\right|\right|.$$

Let $\Delta\varphi(x,\theta,\theta^*) = g_{N,y}(x,\theta) - G_N(x,\theta) - g_{N,y}(x,\theta^*) + G_N(x,\theta^*)$. Substituting this into Eq. (4.4.1), the following expression is obtained:

$$P\left(||S_N(\theta) - E[S_N(\theta)]|| > a\right) - P\left(||\int_X^\theta \Delta\varphi(x,\theta,\theta^*)\mu_N(dx)|| > a\right)$$
$$\le P\left(||S_N(\theta^*) - E[S_N(\theta^*)]|| > a\right).$$

Applying Chebyshev's inequality to the first term in Eq. (4.4.1), the following result is obtained:

$$P\left(||S_N(\theta^*) - E[S_N(\theta^*)]|| > a\right) \le \frac{DS_N(\theta^*)}{a^2},$$

where

$$DS_N(\theta^*) = \frac{1}{N^2}\sum_{i=1}^N g_{N,y_i}(x_i,\theta^*),$$

which tends to zero as $N \to \infty$.

Denote $\omega_{N,y}(r)$ as the modulus of continuity of $\varphi_N(x,\theta)$. By Condition (4.64), it follows that $\omega_{N,y}(r) \le C$.

Since $\varphi_N(x,\theta)$ is defined as

$$\varphi_N(x,\theta) = g_{N,y}(x,\theta) - G_N(x,\theta),$$

and given the conditions in Eq. (4.71) and $G_N(x,\theta) = Eg_{N,y}(x,\theta)$, the family of functions $\varphi_N(x,\theta)$ is compact in $X \times \theta$. Therefore:

$$\lim_{N\to\infty}\lim_{r\to 0} E[\omega_{N,y}(r)] = 0,$$

almost surely for all y.

Thus, all conditions of Lebesgue's dominated convergence theorem [26] are satisfied, and it follows that

$$\lim_{N\to\infty}\lim_{r\to 0} E[\omega_{N,y}(r)] = 0.$$

Taking into account Eq. (4.70), it follows that

$$(S_N(\theta) - S(\theta)) \xrightarrow{P} 0 \quad \text{as} \quad N \to \infty.$$

Theorem 4.17 is therefore proved. □

Note 4.16 *For the large-scale model analysis under the generalized nonlinear parameter framework, the theorem establishes the mathematical conditions of parameter iterative convergence, and gives the corresponding theoretical proof and convergence conclusion.*

Denote

$$\delta_N^* = \delta_N(\theta^*) = y - \eta(x, \theta^*),$$

$$A_N(\theta) = \frac{1}{N}[\frac{F_N^T(\theta)\cdot F_N(\theta)}{N}]^{-1} \cdot F_N^T(\theta),$$

where $a_{ij}(\theta)$;$i = 1, \cdots, N$; $j = 1, \cdots, m$, which are the elements of the matrix $A_N(\theta)$;

$\frac{\partial A_N(\theta)}{\partial\theta}$ is the matrix for elements of which $\partial a_{ij}(\theta)$ are held:

$$\partial\alpha_{ij}(\theta) = \sum_{k=1}^{m}\frac{\partial\alpha_{ij}(\theta)}{\partial\theta_k},$$

where

$$\theta = (\theta_1, \theta_2, ..., \theta_m).$$

Theorem 4.18 *Suppose $\theta \in B(r)$. If the following condition*

$$\frac{\sqrt{m}}{N\cdot\lambda_1^{(N)}(\theta)} \le C < \infty, \;\; for\ N \to \infty, r \to 0$$

is held then

$$\frac{\partial A_N(\theta)}{\partial\theta}\cdot\delta_N(\theta^*) \xrightarrow{P} 0\ for\ N \to \infty, r \to 0.$$

Proof The proof uses Theorem 4.17. For this, let $g_{N,y}(x, \theta)$ be taken as the family of functions:

$$g_{N,y}(x, \theta) = a_{iN}(x, \varepsilon) \cdot \delta_N(\theta^*), \quad i = 1, \cdot, m,$$

where $a_{iN}(x, \varepsilon)$ and (i, N) represent the elements of the matrix $\frac{\partial A_N(\theta)}{\partial \theta_p}$.

Denote $x_{ij}^{(N)}(\theta), i, j = 1, 2, \ldots, m$, as the elements of the matrix $(M_N(\theta))^{-1}$. For $a_{iN}(x, \varepsilon)$, it holds that

$$a_{iN}(\theta) = \sum_{k=1}^{m} x_{ij}^{(N)}(\theta) \cdot f_{KN}(\theta) \leq C_1 \cdot \frac{\sqrt{m}}{\lambda_1^{(N)}(\theta)}, \quad \theta \in B(r).$$

Thus, from equation $(S_N(\theta) - S(\theta)) \xrightarrow{P} 0, for N \to \infty, r \to 0.$, it follows that the conditions of Theorem 4.17 are satisfied. For $S_N(\theta)$, it is obtained that

$$E[S_N(\theta)] = \frac{1}{N} \sum_{i=1}^{N} E[g_{N,y_i}(x_i, \theta)] = \frac{1}{N} \sum_{i=1}^{N} a_{iN} \cdot \delta_N^* = 0.$$

By applying Theorem 4.16, it follows that for $\theta \in B(r)$,

$$\frac{\partial A_N(\theta)}{\partial \theta_p} \cdot \delta_N(\theta^*) \xrightarrow{P} 0 \quad \text{as} \quad N \to \infty, r \to 0.$$

Thus, Theorem 4.18 is proved. □

Denote

$$U_N(\theta) = \theta + A_N(\theta) \cdot \delta_N(\theta);$$

$$L_i^{(N)}(\theta) = \frac{\partial U_N(\theta)}{\partial \theta_i}, \theta \in B(r)$$

$$\tau_N(r) = \max_{1 \leq i \leq m} \sup_{\theta \in B(r)} ||l_i^{(N)}(\theta)||.$$

Theorem 4.19 *Suppose that $\theta \in B(r)$ and that the following condition holds:*

$$\frac{\sqrt{m}}{N \cdot \lambda_1^{(N)}(\theta)} \leq C < \infty, \quad as\ N \to \infty, r \to 0.$$

Furthermore, assume that

$$\frac{\partial A_N(\theta)}{\partial \theta} \cdot \delta_N(\theta^*) \xrightarrow{P} 0 \quad as\ N \to \infty, r \to 0.$$

Then, the following convergence result holds:

$$\tau_N(r) \xrightarrow{P} 0 \quad as\ N \to \infty, r \to 0.$$

Proof Consider the relation $(F_N^T(\theta) \cdot F_N(\theta))^{-1} F_N^T(\theta) \cdot \frac{\partial \delta_N(\theta)}{\partial \theta} = -I_N$, where I_N is the identity matrix. Therefore, for each $L_i^{(N)}(\theta)$, the following expression holds:

$$L_i^{(N)}(\theta) = \left[I_i + \frac{\partial A_N(\theta)}{\partial \theta_i} \cdot \delta_N(\theta) + A_N(\theta) \cdot \frac{\partial \delta_N(\theta)}{\partial \theta_i} \right].$$

This can be rewritten as follows:

$$L_i^{(N)}(\theta) = \left[\frac{\partial A_N(\theta)}{\partial \theta_i} \cdot \delta_N(\theta^*) + \frac{\partial A_N(\theta)}{\partial \theta_i} \cdot \Delta\delta_N(\theta^*, \theta) \right],$$

where $\Delta\delta_N(\theta^*, \theta) = \delta_N(\theta) - \delta_N(\theta^*)$.

Since the relation $A_N(\theta) \cdot \frac{\partial \eta_N(x,\theta)}{\partial \theta_i} = -I_i$ holds, we consider the following inequality:

$$P(\tau_N(r) > a) \le P\left(\left\| \frac{\partial A_N(\theta)}{\partial \theta_p} \cdot \delta_N(\theta^*) \right\| > a \right) + P\left(\left\| \frac{\partial A_N(\theta)}{\partial \theta_p} \cdot \Delta\delta_N(\theta, \theta^*) \right\| > a \right). \quad (4.73)$$

The first term on the right-hand side of Inequality (4.73) tends to zero as $N \to \infty$ and $r \to 0$ by Theorem 4.18. Similarly, the second term also tends to zero as $N \to \infty$ and $r \to 0$ by Theorem 4.17.

Therefore, it follows that

$$\tau_N(r) \xrightarrow{P} 0 \quad \text{as} \quad N \to \infty, r \to 0.$$

This completes the proof of Theorem 4.19. □

Note 4.17 *In this theorem, the mathematical theory of radius collapse of parameter domain is proposed for the large model analysis under the generalized nonlinear parameter framework, and its convergence condition and strict proof are given.*

Below, convergence of random variables ς_1, ς_2 will be denoted as follows:

$$\varsigma_1 \to \varsigma_2$$

and it means weak convergence [43].

Theorem 4.20 *Suppose that for some $\gamma > 0$, the following moment condition holds:*

$$E|\varepsilon_i|^{2+\gamma} \le C < \infty,$$

where $\varepsilon_1, \varepsilon_2, \ldots, \varepsilon_N$ are independent random variables. Furthermore, assume that $\theta \in B(r)$ and that the following asymptotic condition is satisfied:

$$\frac{\sqrt{m}}{N \cdot \lambda_1^{(N)}(\theta)} \leq C < \infty, \quad as\ N \to \infty, r \to 0. \tag{4.74}$$

Additionally, suppose that

$$\frac{\partial A_N(\theta)}{\partial \theta} \cdot \delta_N(\theta^*) \xrightarrow{P} 0 \quad as\ N \to \infty, r \to 0.$$

Then, the following asymptotic normality result holds:

$$\sqrt{N} \cdot (\widehat{\theta} - \theta^*) \xrightarrow{d} N(0, \widehat{\sum}(\theta^*)),$$

where

$$\widehat{\sum}(\theta^*) = (M(\theta^*))^{-1} \cdot \widetilde{M}(\theta^*) \cdot (M(\theta^*))^{-1},$$

and

$$\widetilde{M}(\theta^*) = \lim_{N \to \infty} \widetilde{M}_N(\theta^*) = \lim_{N \to \infty} \frac{F_N^T(\theta^*) \cdot I(\sigma^2) \cdot F_N(\theta^*)}{N}.$$

Here, the matrix $I(\sigma^2)$ is given by

$$I(\sigma^2) = \begin{pmatrix} \sigma_1^2 & 0 & \dots & 0 \\ 0 & \sigma_2^2 & \dots & 0 \\ \vdots & \vdots & \ddots & \vdots \\ 0 & 0 & \dots & \sigma_N^2 \end{pmatrix}.$$

The existence of the limit $\widetilde{M}_N(\theta^*)$ follows from Corollary 4.1.

Proof In a manner similar to the proof of Theorem 4.2, let a nonzero vector $l = (l_1, l_2, \dots, l_m)^T$ of size m be introduced.

Then, the following expression for $\sqrt{N} \cdot l^T \rho^*$ can be shown:

$$\sqrt{N} \cdot l^T \rho^* = \sqrt{N} \cdot l^T A_N(\theta^*) \cdot \delta_N^* = \frac{1}{\sqrt{N}} \sum_{i=1}^{N} N l^T a_i(\theta^*)(y_i - \eta(x_i, \theta^*)),$$

where $a_i(\theta^*)$ represents the i-th row of the matrix $A_N(\theta^*)$.

According to the condition (4.63), it follows that

$$E[N \cdot l^T a_i(\theta^*)(y_i - \eta(x_i, \theta^*))] = l^T N \cdot E[a_i(\theta^*)] \cdot \varepsilon_i = 0.$$

Thus, $E[\sqrt{N} \cdot l^T A_N(\theta^*) \cdot \delta_N^*] = 0$.

Next, consider:

$$E(\sqrt{N} \cdot l^T A_N(\theta^*) \cdot \delta_N^*)^2 = E\sqrt{N} \cdot l^T A_N(\theta^*) \cdot \varepsilon \cdot \varepsilon^T A_N^T(\theta^*) l \sqrt{N}$$
$$= N \cdot l^T \left[\frac{F_N^T(\theta^*) \cdot F_N(\theta^*)}{N} \right]^{-1} \frac{F_N^T(\theta^*)}{N} \cdot I(\sigma^2) \cdot \frac{F_N(\theta^*)}{N} \left[\frac{F_N^T(\theta^*) \cdot F_N(\theta^*)}{N} \right]^{-1}$$
$$= l^T \cdot \widehat{\sum}(\theta^*) \cdot l.$$

Since $E[\varepsilon_i] = E[(y_i - \eta(x_i, \theta^*))^{2+\gamma}] \le C < \infty$, the following inequality holds: $\frac{C}{\varepsilon} > \frac{1}{\varepsilon \cdot N} \sum_{i=1}^{N} \int |y_i - \eta(x_i, \theta^*)|^{2+\gamma} dF_{y_i}(x_i)$, where $\varepsilon > 0$ and $F_{y_i}(x)$ is the distribution function of y_i.

Next, consider Lindeberg's fraction:

$$\Lambda_N(\varepsilon) = \frac{1}{N} \sum_{i=1}^{N} \int_{|y_i - \eta(x_i, \theta^*)| > \sqrt{N} \cdot \varepsilon} (y_i - \eta(x_i, \theta^*))^2 dF_{y_i}(x_i).$$

It follows that

$$\frac{C}{\varepsilon} > \frac{1}{\varepsilon \cdot N} \sum_{i=1}^{N} \int |y_i - \eta(x_i, \theta^*)|^{2+\gamma} dF_{y_i}(x_i)$$
$$\ge \frac{1}{\varepsilon \cdot N} \sum_{i=1}^{N} \int_{|y_i - \eta(x_i, \theta^*)| \ge \sqrt{N} \cdot \varepsilon} |y_i - \eta(x_i, \theta^*)|^2 \cdot (\varepsilon \cdot \sqrt{N})^\gamma dF_{y_i}(x_i)$$
$$\ge \frac{(\varepsilon \sqrt{N})^\gamma}{\varepsilon} \cdot \Lambda_N(\varepsilon).$$

From Eq. (4.74), it follows that

$$0 \le \Lambda_N(\varepsilon) \le \frac{C}{\varepsilon^\gamma \cdot N^{\frac{\gamma}{2}}} \quad \text{as} \quad N \to \infty.$$

By the central limit theorem, it follows that

$$\sqrt{N} \cdot l^T A_N(\theta^*) \cdot \delta_N^* \xrightarrow{d} N(0, l^T \cdot \widehat{\sum}(\theta^*) \cdot l).$$

Using the criterion of normality [53], it is obtained that

$$\sqrt{N} \cdot A_N(\theta^*) \cdot \delta_N^* \xrightarrow{d} N(0, \widehat{\sum}(\theta^*)).$$

Thus, Theorem 4.20 is proved. □

Note 4.18 *Aiming at the large-scale model analysis under the generalized nonlinear parameter framework, the theorem establishes the mathematical framework of parameter convergence distribution, and clarifies the conditions and theoretical basis for its establishment.*

In the statement of Theorem 4.20, the matrix $\widehat{\sum}(\theta^*)$ is unknown. Hence, it is necessary to estimate the elements of this matrix.

Suppose $r_B > 0$ and $r_B \to 0$ for $s \to \infty$.

Theorem 4.21 *If for some N and τ_B the following inequality is held:*

$$\tau_N(r_B) + \frac{||\rho^*||}{r_B} < 1.$$

Then there exists the random variable $\widehat{\theta}_N$, such that

$$(\theta_N(s) - \widehat{\theta}_N) \xrightarrow{P} 0 \ for \ s \to \infty$$

and $\sqrt{N}(\widehat{\theta}_N - \theta^*) \to N(0, \widehat{\sum}(\theta^*))$ *for* $N \to \infty$.

Proof Theorem 4.21 can be considered as a corollary of Theorems 4.20 and 4.21 because all conditions of these theorems are held [15]. Consider the condition $\tau_N(r_B) + \frac{\|\rho^*\|}{r_B} < 1$. This condition ensures that the mapping $U_N(\theta)$ is contractive on the set $B(r_B)$, and hence there exists a unique fixed point $\widehat{\theta}_N$ such that

$$U_N(\widehat{\theta}_N) = \widehat{\theta}_N.$$

From Theorem 4.18, it follows that

$$\theta_N(s) \xrightarrow{P} \widehat{\theta}_N \quad \text{as} \quad s \to \infty.$$

Thus, the desired result can be shown as follows:

$$(\theta_N(s) - \widehat{\theta}_N) \xrightarrow{P} 0 \quad \text{as} \quad s \to \infty.$$

Next, consider the asymptotic distribution of $\widehat{\theta}_N$. By Theorem 4.20, it follows that

$$\sqrt{N}(\widehat{\theta}_N - \theta^*) \xrightarrow{d} N(0, \widehat{\sum}(\theta^*)) \quad \text{as} \quad N \to \infty,$$

where

$$\widehat{\sum}(\theta^*) = (M(\theta^*))^{-1} \cdot \widetilde{M}(\theta^*) \cdot (M(\theta^*))^{-1}.$$

Here, $\widetilde{M}(\theta^*)$ converges to the matrix:

$$\lim_{N\to\infty} \frac{F_N^T(\theta^*) \cdot I(\sigma^2) \cdot F_N(\theta^*)}{N}.$$

Thus, the result of Theorem 4.21 follows directly from Theorems 4.20 and 4.18. Therefore, $\widehat{\theta}_N$ converges in probability to θ^*, and the distribution of the estimator is asymptotically normal with covariance matrix $\widehat{\sum}(\theta^*)$. □

Note 4.19 *The theorem constructs the mathematical theory of parameter convergence and its distribution for the large model analysis under the generalized nonlinear parameter framework, and clarifies the corresponding convergence conditions and theoretical basis.*

4.4.2 Estimating Covariance Matrix Elements

In Theorem 4.20, the matrix $\widehat{\sum}(\theta^*)$ is unknown. Below, using the $\sqrt{N}$-consistent estimator of the unknown parameter θ^*, an estimation of the elements of the matrix $\widehat{\sum}(\theta^*)$ is provided.

Recall that, according to Theorem 4.20, the following relation holds:

$$\widehat{\sum}(\theta^*) = (M(\theta^*))^{-1} \cdot \widetilde{M}(\theta^*) \cdot (M(\theta^*))^{-1},$$

where

$$\widetilde{M}(\theta^*) = \lim_{N\to\infty} \widetilde{M}_N(\theta^*) = \lim_{N\to\infty} \frac{F_N^T(\theta^*) \cdot I(\sigma^2) \cdot F_N(\theta^*)}{N}.$$

Denote $\widetilde{m}_{k,l}(\theta)$ as the elements of the matrix $\widetilde{M}(\theta^*)$, and let $\widetilde{m}_{k,l}^{(N)}(\theta)$ be the elements of the matrix $\widetilde{M}_N(\theta)$, which is given by

$$\widetilde{m}_{k,l}^{(N)} = \frac{1}{N}(y - \eta(x, \widetilde{\theta}))^T \cdot I_{k,l}^2(\widetilde{\theta}) \cdot (y - \eta(x, \hat{\theta})).$$

The matrix $I_{k,l}^2(\widetilde{\theta})$ is defined as

$$I_{k,l}^2(\widetilde{\theta}) = \begin{bmatrix} \varphi_k(x_1, \hat{\theta}) \cdot \varphi_l(x_1, \hat{\theta}) & \dots & 0 \\ 0 & \varphi_k(x_2, \hat{\theta}) \cdot \varphi_l(x_2, \hat{\theta}) & 0 \\ 0 & 0 & \varphi_k(x_N, \hat{\theta}) \cdot \varphi_l(x_N, \hat{\theta}) \end{bmatrix}.$$

Finally, the eigenvalues of the matrix $M_N(\hat{\theta})$ are denoted as follows:

$$0 < \hat{\lambda}_1(\theta) \leq \dots \leq \hat{\lambda}_m(\theta).$$

Theorem 4.22 *Assume that* $\frac{m}{N(\hat{\lambda}_1(\theta))^2}$ *is bounded for all* $\theta \in B(r)$. *Then:*

$$E[\widetilde{m}_{kl}^{(N)} - \widetilde{m}_{kl}^{(N)}(\theta^*)] \to 0$$

and

$$(\widetilde{m}_{kl}^{(N)} - \widetilde{m}_{kl}^{(N)}(\theta^*)) \xrightarrow{P} 0 \quad as \quad N \to \infty.$$

Proof Consider the following expression:

$$\begin{aligned} y - \eta(x, \widehat{\theta}) &= \eta(x, \theta^*) + \varepsilon - \eta(x, \widehat{\theta}) \\ &= \frac{\partial \eta(x, \widehat{\theta})}{\partial \theta}(\theta^* - \widehat{\theta}) + \varepsilon + o(r) \\ &= -F_N(\widehat{\theta})(\widehat{\theta} - \theta^*) + \varepsilon + o(r). \end{aligned} \tag{4.75}$$

For $\theta(s+1) \in B(r)$, using Eq. (4.75), the following expression is obtained:

$$\begin{aligned} \theta(s+1) - \theta^* &= \theta(s) + \left[F_N^T(\theta(s)) \cdot F_N(\theta(s))\right]^{-1} F_N^T(\theta(s)) \cdot (y - \eta(x, \theta(s))) - \theta^* \\ &= \theta(s) + \left[F_N^T(\theta(s)) \cdot F_N(\theta(s))\right]^{-1} F_N^T(\theta(s)) \cdot \left[(\theta^* - \theta(s)) + \varepsilon + o(r)\right] - \theta^* \\ &= \theta(s) + \left[F_N^T(\theta(s)) \cdot F_N(\theta(s))\right]^{-1} F_N^T(\theta(s)) \cdot \varepsilon + o(r). \end{aligned}$$

Thus, the update for $\theta(s+1)$ involves a linear term in ε and a remainder term of order $o(r)$.

Since the elements of the matrix $F_N(\theta)$ are continuous, as $s \to \infty$ and $r \to 0$, Theorem 4.21 implies

$$[(\widehat{\theta} - \theta^*) - \left(F_N^T(\widehat{\theta}) \cdot F_N(\widehat{\theta})\right)^{-1} F_N^T(\widehat{\theta}) \cdot \varepsilon]^P \to 0.$$

From Eq. (4.75), it follows that

$$\begin{aligned} y - \eta(x, \widehat{\theta}) &= \varepsilon - F_N(\widehat{\theta}) \left(F_N^T(\widehat{\theta}) \cdot F_N(\widehat{\theta})\right)^{-1} F_N^T(\widehat{\theta}) \cdot \varepsilon \\ &= \left[I - F_N(\widehat{\theta}) \left(F_N^T(\widehat{\theta}) \cdot F_N(\widehat{\theta})\right)^{-1} F_N^T(\widehat{\theta})\right] \cdot \varepsilon. \end{aligned} \tag{4.76}$$

Substituting Eq. (4.76) into (4.75) and expanding the terms, the following expression is obtained:

$$\begin{aligned} &\frac{1}{N}(y - \eta(x, \widehat{\theta}))^T \cdot I_{k,l}^{(2)}(\widehat{\theta}) \cdot (y - \eta(x, \widehat{\theta})) \\ &= \frac{\varepsilon^T \cdot I_{k,l}^{(2)}(\widehat{\theta}) \cdot \varepsilon}{N} - \frac{\varepsilon^T I_{kl}^{(2)}(\widehat{\theta}) \cdot F_N(\widehat{\theta})(F_N^T(\widehat{\theta}) \cdot F_N(\widehat{\theta}))\varepsilon}{N} \\ &- \varepsilon^T \cdot F_N(\widehat{\theta}) \times \frac{(F_N^T(\widehat{\theta}) \cdot F_N(\widehat{\theta}))^{-1} \cdot F_N^T(\widehat{\theta}) \cdot I_{kl}(\widehat{\theta}) \cdot \varepsilon}{N} + \varepsilon^T \cdot F_N(\widehat{\theta}) \\ &\times \frac{(F_N^T(\widehat{\theta}) \cdot F_N(\widehat{\theta}))^{-1} \cdot F_N^T(\widehat{\theta}) \cdot I_{k,1}(\widehat{\theta}) \cdot F_N(\widehat{\theta})(F_N^T(\widehat{\theta}) \cdot F_N(\widehat{\theta}))^{-1} \cdot F_N^T(\widehat{\theta}) \cdot \varepsilon}{N} \end{aligned} \tag{4.77}$$

By applying the same method as in Theorem 4.22, it can be shown that the terms in Eq. (4.77) tend to zero in probability as $N \to \infty$, uniformly in x. This implies that $\widehat{m}_{kl}^{(N)}(\widehat{\theta})$ is an asymptotically unbiased estimator of $\widetilde{m}_{kl}(\theta^*)$, and, by Theorem 4.22, it is also a consistent estimator.

By applying Corollary 4.1, it follows that for the matrix $\widetilde{M}_N(\widehat{\theta})$, the limit matrix $\widetilde{M}$ exists, and the elements of $\widetilde{M}_N(\widehat{\theta})$ are asymptotically unbiased and consistent estimators of the corresponding elements of the matrix $M(\theta^*)$.

Next, consider the matrix:

$$\bar{B}_N = \frac{1}{N}\left[\frac{F_N^T(\widehat{\theta}) \cdot F_N(\widehat{\theta})}{N}\right]^{-1} \cdot \widehat{B} \cdot \left[\frac{F_N^T(\widehat{\theta}) \cdot F_N(\widehat{\theta})}{N}\right]^{-1}.$$

Denote $\bar{b}_{kl}$ as the elements of the matrix $\bar{B}_N$ and c_{kl} as the elements of the matrix $\widehat{\sum}(\theta^*)$. Then

$$\bar{b}_{kl} = \frac{1}{N}\sum_{i_l=1}^{m}\sum_{i_2=1}^{m} x_{kl_1} \cdot \bar{b}_{i_l i_2} \cdot x_{i_2 l},$$

$$\widetilde{c}_{kl} = \frac{1}{N}\sum_{i_l=1}^{m}\sum_{i_2=1}^{m} x_{kl_1} \cdot \widetilde{b}_{i_l i_2} \cdot x_{i_2 l}.$$

Under the conditions of Theorem 4.21, it follows that

$$|\bar{b}_{kl} - \bar{c}_{kl}| = \frac{1}{N}\left|\sum_{i_1=1}^{m}\sum_{i_2=1}^{m} x_{kl_1} \cdot (\bar{b}_{i_1 i_2} - \widetilde{b}_{i_1 i_2}) \cdot x_{i_2 l}\right| \le \frac{m}{(\widehat{\lambda_1}(N))^2} \max |\bar{b}_{i_1 i_2} - \widetilde{b}_{i_1 i_2}| \to 0.$$

Thus, Theorem 4.22 is proved. □

Note 4.20 *This theorem establishes the unbiased theory of covariance matrix estimation for the large model analysis under the generalized nonlinear parameter framework, and clarifies the conditions and mathematical basis for its establishment.*

For the construction of a confidence band for the function $\eta(x, \theta)$, Theorems 4.20 and 4.21, along with the approach suggested in the previous paragraph, can be utilized.

Note 4.21 *In the large-scale model analysis under the generalized nonlinear parameter framework, the analysis algorithm of parameter statistical properties is similar to the previous method, and it is easy to construct and promote.*

Remark 4.13 The results established above are also valid for M-estimators, and their proofs follow the same reasoning.

The advantages and limitations of the proposed method for constructing a confidence band for unknown functions in regression models should be noted. One of the main advantages of this approach is that it enables the analysis of regression models where the errors of random variables have different and unknown variances. Moreover, it applies to cases where only a single response is available at each observation point. Such models are commonly encountered in practical applications but have not been extensively studied in the literature.

4.5 Optimization of High-Dimensional Multi-parameter Regression

In multi-parameter polynomial fitting, increasing the number of parameters introduces several challenges, including reduced fitting accuracy, decreased numerical stability, and lower computational efficiency. First, as the number of parameters grows, the model tends to suffer from underfitting or overfitting, especially when the sample size is insufficient or the data is noisy, which reduces the fitting accuracy. Second, the increase in parameters makes the matrix operations involved in the fitting process highly sensitive to small data perturbations, leading to numerical instability and unreliable fitting results. Third, the enlarged parameter scale significantly increases the complexity and computational cost of matrix operations, such as solving and decomposing matrices, which in turn reduces the overall efficiency of the fitting process. Therefore, under high-dimensional parameter settings, it is necessary to improve the model's fitting accuracy, numerical stability, and computational efficiency to achieve reliable polynomial fitting. Accordingly, this subsection is organized into two parts. The first part addresses the improvement of fitting accuracy under multi-parameter settings, and the second part focuses on computational efficiency enhancement.

4.5.1 Methods for Improving Fitting Accuracy

To enhance the accuracy and robustness of multi-parameter polynomial fitting, improvements are made from four aspects: basis function design, parameter estimation, numerical optimization, and experimental strategy, as detailed below.
(1) Basis Function Design and Sampling Strategy

Orthogonal Polynomial Basis. In multi-parameter polynomial fitting, orthogonal polynomials (e.g., Legendre or Chebyshev polynomials) are used to replace the traditional Vandermonde basis (constructed with x^k to form the design matrix. Due to their integral orthogonality over the interval, orthogonal polynomials effectively reduce the correlation among basis functions, alleviate the multicollinearity of high-order terms, and thus significantly improve the numerical stability and fitting accuracy of parameter estimation.

Chebyshev Node Sampling. Instead of uniform sampling, Chebyshev nodes are employed to generate sampling points for model fitting. These nodes are denser near the interval boundaries, which helps suppress the oscillations of high-order polynomials near the edges and improves the numerical stability of the design matrix. This sampling strategy also provides a solid foundation for the application of orthogonal polynomials and regularization techniques, further enhancing the fitting performance.

Logarithmically Spaced Sampling. By employing dense sampling in lower numerical ranges to capture subtle variations and sparse sampling in higher ranges to reduce redundant computations, this approach achieves an efficient representation and balanced sampling of data with multiscale characteristics. Consequently, computational efficiency and resource utilization are significantly enhanced, establishing it as an effective and precise technique for cross-order-of-magnitude data analysis.

(2) Robust Parameter Estimation

Regularization Techniques. Regularization techniques, such as ridge regression, are introduced into the least squares objective function. By penalizing the magnitude of high-order coefficients, regularization effectively suppresses their volatility, mitigating the amplification of noise-induced fluctuations. This approach achieves a better bias-variance trade-off, thereby improving the generalization ability and predictive accuracy of the model, especially under multi-parameter settings.

Loss Function Optimization. To improve robustness against outliers, the Huber loss is employed instead of the commonly used L2 loss (least squares). The Huber loss behaves like L2 when residuals are small and switches to L1 when residuals are large, thereby reducing the adverse impact of outliers on the model. This is particularly effective in small-sample or noisy data scenarios.

(3) Tolerance Control for Optimizers

By tightening the optimizer's tolerance parameters (e.g., reducing `gtol` or `ftol`, or increasing `maxiter`, the optimization process can achieve better convergence and parameter estimation accuracy, particularly in large-sample or high-dimensional problems.

(4) Replications for Improving Fitting Robustness

Increasing the number of replications in Monte Carlo experiments effectively reduces simulation errors. This is particularly beneficial when the original number of replications is insufficient, as it helps alleviate random fluctuations in the fitting results and significantly improves the accuracy and robustness of the fitting process.

4.5.2 Methods for Enhancing Computational Efficiency

To address the computational challenges caused by the increase in parameters and sample sizes, this section presents several techniques to improve the computational efficiency of multi-parameter polynomial fitting. The methods are organized into three categories: data structure optimization, acceleration of stochastic simulation, and improvements in numerical optimization and convergence.

(1) Data Structure Optimization

Pre-computation and Reuse of Design Matrices. In repeated experiments, the design matrix X and the Chebyshev sampling nodes x are reused instead of being recomputed for every repetition. By pre-computing these matrices outside the simulation loop, redundant computations are avoided, leading to a significant reduction in runtime.

Vectorized Resampling for i.i.d. Noise. To enhance the efficiency of generating independent resampled data, NumPy's vectorized operations are used to perform noise generation and sample reconstruction in batch, avoiding explicit loops. This technique is specifically applied to the generation of *i.i.d.* (independent and identically distributed) random errors, which are added to the model output for constructing resampled datasets. The vectorization significantly accelerates the process of generating independent noise for each repetition without affecting the statistical properties of the generated samples.

Elimination of Redundant Operations. Certain redundant operations, such as sorting Chebyshev nodes that are inherently ordered, are removed to avoid unnecessary computations and further improve computational efficiency.

Block-Wise Updates for Large-Scale Experiments. For large-scale experiments involving massive repetitions, a block-wise updating strategy is adopted. Parameters are updated based on aggregated information from multiple replications, reducing the computational burden while maintaining consistency in parameter estimation.

(2) Acceleration of Stochastic Simulation

Vectorized Generation of AR(1) Errors. The generation of AR(1) correlated errors is vectorized using efficient NumPy or SciPy functions, such as `scipy.signal.lfilter`, replacing inefficient Python loops. Unlike the independent noise used for simple resampling, the AR(1) process generates serially correlated errors with predefined correlation structures. Vectorization significantly accelerates the error generation process, particularly when the sample size and the number of repetitions are large.

Parallel Execution of Fitting Tasks. Since different samples and repetitions are independent, parallel computation is employed using multiprocessing tools (e.g., `multiprocessing` or `joblib`). This fully utilizes the available multi-core CPUs to reduce total computation time.

(3) Numerical Stability and Convergence Control

L-BFGS-B Optimizer. The L-BFGS-B algorithm, a quasi-Newton method with limited-memory and box-constraint capabilities, is adopted for parameter estimation. It approximates the Hessian information without explicitly storing large matrices and efficiently performs iterative updates, making it suitable for large-scale or high-dimensional fitting problems.

Warm-start Strategy. An effective warm-start strategy is adopted to accelerate convergence. In large-scale optimization or when the problem exhibits numerical instability, initializing the optimizer from a zero vector or random values may lead to local optima or unstable optimization trajectories. By using the solution from the previous optimization as the initial value for the next run, the optimizer can quickly identify a more stable or promising search direction. This approach reduces the number of iterations required, enhances the convergence stability, and significantly saves computational time and resources in large-scale data analysis and model training.

Standardization of Input Variables. Input variables are standardized (zero mean and unit variance) before constructing the polynomial design matrix. This step mitigates the scaling problem of high-order terms, improving the numerical stability of the optimization process and accelerating convergence.

Standardization of Output Variables. When necessary, output variables are standardized or centered to match the scale of the loss function. This adjustment avoids gradient calculation anomalies due to scale mismatch and ensures a stable optimization process.

4.5.3 Comparative Analysis of Pre- and Post-optimization

To validate the effectiveness of the proposed methods in improving the fitting accuracy and computational efficiency for multi-parameter polynomial fitting, this subsection selects a simulation experiment from Sects. 4.1 and 4.4, respectively, as a case study.

(1) **Case 1**

The experimental setup in Sect. 4.1 involves estimating an 11-parameter model under varying sample sizes (ranging from 50 to 10,000,050) with 50 replications for each sample size. Comparison before and after optimization is shown in Table 4.10.

1. Improvement in Fitting Accuracy

As shown in the bias estimation results, significant improvements in parameter estimation accuracy are observed after the optimization:

Before optimization, severe biases were found, especially for high-order parameters when the sample size is small or moderate. For example, under a small sample size of $N = 50$, the biases for parameters θ_2 to θ_{10} are extremely large, exceeding hundreds or even thousands in magnitude. Even when the sample size increases to several millions, the biases for high-order parameters such as θ_4 to θ_{10} remain noticeably large and fluctuating.

After optimization, the biases across all parameters are drastically reduced to near-zero levels, even for small sample sizes. For instance, when $N = 50$, the maximum absolute bias drops to less than 0.05, and for larger sample sizes (e.g., $N > 10^6$), all parameter biases are stably maintained within 10^{-3} to 10^{-4} levels.

This improvement is primarily attributed to the use of orthogonal polynomial bases, regularization techniques, and optimized sampling strategies, which effectively reduce the ill-conditioning of the design matrix and control the estimation variance of high-order coefficients.

2. Improvement in Computational Efficiency

In addition to accuracy improvement, the proposed methods significantly accelerate the computational process:

Table 4.10 Comparison of parameter bias convergence under large-sample conditions

(1) Pre-optimization				(2) Post-optimization			
Sample size	θ_0	θ_1	θ_2	Sample size	θ_0	θ_1	θ_2
50	0.0063	0.4350	−28.7557	50	0.0196	−0.0341	−0.0029
2631628	−0.0007	0.0137	−0.3580	2631628	−0.0001	−0.0001	0.0002
5263207	0.0001	0.0065	−0.2265	5263207	−0.0002	0.0001	0.0000
7894786	0.0000	0.0003	−0.0367	7894786	0.0000	0.0001	0.0002
10526365	0.0000	0.0013	0.1230	10526365	−0.0001	0.0000	0.0001
13157944	0.0000	−0.0074	0.2371	13157944	−0.0001	0.0000	0.0001
15789523	0.0001	−0.0090	0.3077	15789523	0.0001	0.0000	−0.0001
18421102	0.0000	0.0012	−0.0503	18421102	0.0000	0.0000	0.0000
21052681	−0.0006	0.0289	−0.4543	21052681	0.0000	−0.0001	−0.0001
23684260	0.0001	−0.0051	0.1150	23684260	0.0001	0.0001	−0.0001
26315839	0.0001	0.0031	−0.1480	26315839	0.0000	0.0001	0.0000
28947418	−0.0001	0.0092	−0.2267	28947418	0.0000	0.0000	0.0000
31578997	−0.0002	0.0086	−0.1109	31578997	0.0000	0.0000	0.0000
34210576	−0.0001	−0.0004	0.0851	34210576	0.0000	0.0000	0.0000
36842155	−0.0001	0.0061	−0.0919	36842155	0.0000	0.0000	0.0000
39473734	0.0001	−0.0074	0.1571	39473734	0.0000	0.0001	0.0000
42105313	0.0001	−0.0019	−0.0034	42105313	0.0000	0.0000	−0.0001
44736892	0.0000	−0.0060	0.1244	44736892	0.0000	0.0000	0.0001
47368471	0.0003	−0.0139	0.2379	47368471	0.0000	0.0001	0.0000
50000050	0.0001	−0.0042	0.1277	50000050	0.0000	−0.0001	0.0000
Sample size	θ_3	θ_4	θ_5	Sample size	θ_3	θ_4	θ_5
50	594.0912	−4879.9412	20727.2699	50	0.0068	0.0035	0.0083
2631628	4.1894	−25.1701	89.0739	2631628	0.0001	0.0000	−0.0002
5263207	2.5921	−15.4405	54.7579	5263207	0.0002	0.0001	−0.0001
7894786	0.7960	−6.4940	27.0478	7894786	0.0001	−0.0003	0.0000

(continued)

Table 4.10 (continued)

(1) Pre-optimization				(2) Post-optimization			
Sample size	θ_0	θ_1	θ_2	Sample size	θ_0	θ_1	θ_2
10526365	−2.8807	24.4263	−106.3858	10526365	0.0001	0.0000	−0.0001
13157944	−3.1897	21.6973	−83.8239	13157944	−0.0001	0.0000	−0.0001
15789523	−4.0852	27.1372	−102.8333	15789523	0.0002	0.0001	−0.0001
18421102	0.7820	−5.3969	20.2789	18421102	−0.0001	0.0000	0.0000
21052681	3.4358	−14.8722	39.9071	21052681	0.0001	−0.0001	−0.0001
23684260	−1.1513	5.9057	−16.7286	23684260	−0.0001	−0.0001	−0.0001
26315839	2.0340	−13.7812	53.6789	26315839	−0.0002	0.0000	0.0000
28947418	2.4787	−14.4993	50.1669	28947418	0.0001	−0.0001	0.0000
31578997	0.5734	−1.0108	−1.7393	31578997	0.0000	−0.0001	0.0001
34210576	−1.3633	9.4067	−35.2383	34210576	0.0001	0.0000	0.0001
36842155	0.6542	−3.0152	10.0429	36842155	0.0000	0.0001	0.0000
39473734	−1.4717	7.4369	−22.7072	39473734	−0.0002	0.0000	0.0001
42105313	0.4494	−4.4152	19.6385	42105313	0.0000	0.0000	−0.0001
44736892	−1.1026	5.3582	−15.9252	44736892	−0.0001	0.0000	0.0000
47368471	−1.8345	7.4810	−17.3357	47368471	0.0000	0.0000	0.0000
50000050	−1.5889	9.9731	−35.6040	50000050	0.0000	0.0001	0.0000
Sample size	θ_6	θ_7	θ_8	Sample size	θ_6	θ_7	θ_8
50	−50873.7792	75046.2150	−65697.5653	50	−0.0392	0.0174	−0.0212
2631628	−197.3130	276.7316	−238.2753	2631628	−0.0004	0.0005	0.0004
5263207	−121.3210	169.1687	−144.1510	5263207	0.0000	−0.0001	−0.0002
7894786	−65.0872	94.4625	−81.6754	7894786	−0.0003	0.0002	0.0006
10526365	267.5545	−404.4182	362.8762	10526365	−0.0003	0.0000	−0.0003
13157944	195.4170	−280.1234	241.4265	13157944	0.0001	0.0002	0.0002
15789523	236.3566	−335.3610	286.9355	15789523	0.0001	0.0002	−0.0001
18421102	−45.3929	62.2120	−51.1147	18421102	−0.0001	0.0000	0.0004

(continued)

Table 4.10 (continued)

(1) Pre-optimization				(2) Post-optimization			
Sample size	θ_0	θ_1	θ_2	Sample size	θ_0	θ_1	θ_2
21052681	−68.5965	75.4801	−51.2484	21052681	−0.0001	0.0001	0.0001
23684260	26.4860	−21.5418	5.3354	23684260	0.0001	−0.0001	0.0000
26315839	−127.6460	188.0872	−167.4488	26315839	0.0000	0.0002	0.0000
28947418	−107.6540	144.7918	−118.7394	28947418	0.0002	−0.0001	0.0000
31578997	11.2674	−22.3187	22.7258	31578997	0.0002	0.0000	−0.0001
34210576	78.1459	−105.3344	84.5954	34210576	0.0001	−0.0001	0.0001
36842155	−23.6059	36.7480	−35.3379	36842155	−0.0001	0.0002	0.0001
39473734	44.3274	−56.1471	44.8537	39473734	−0.0002	0.0001	0.0002
42105313	−48.5978	71.3823	−62.0923	42105313	−0.0001	0.0000	0.0000
44736892	30.2766	−37.0059	28.1383	44736892	0.0000	0.0000	0.0002
47368471	22.6044	−14.0335	−0.1788	47368471	−0.0001	−0.0001	0.0001
50000050	76.9543	−102.7736	83.0369	50000050	0.0000	0.0000	−0.0001
Sample size	θ_9	θ_{10}		Sample size	θ_9	θ_{10}	
50	31462.2757	−6350.2277		50	0.0016	0.0481	
2631628	114.7012	−23.5930		2631628	0.0003	0.0002	
5263207	68.5659	−13.9526		5263207	−0.0002	0.0003	
7894786	38.7551	−7.7684		7894786	0.0000	−0.0004	
10526365	−178.2019	36.9057		10526365	−0.0002	0.0005	
13157944	−114.7225	23.0888		13157944	0.0000	−0.0003	
15789523	−135.6614	27.2127		15789523	−0.0001	−0.0001	
18421102	23.0602	−4.3791		18421102	0.0001	−0.0001	
21052681	19.4604	−3.1403		21052681	−0.0001	0.0001	
23684260	3.4179	−1.8333		23684260	0.0001	0.0000	
26315839	82.4501	−17.2295		26315839	0.0000	0.0001	
28947418	54.2638	−10.5908		28947418	0.0000	0.0000	

(continued)

Table 4.10 (continued)

(1) Pre-optimization				(2) Post-optimization			
Sample size	θ_0	θ_1	θ_2	Sample size	θ_0	θ_1	θ_2
31578997	−12.0074	2.6122		31578997	−0.0001	0.0001	
34210576	−37.1458	6.8490		34210576	−0.0001	−0.0002	
36842155	18.8458	−4.2461		36842155	−0.0002	−0.0001	
39473734	−20.5563	4.1144		39473734	−0.0001	0.0000	
42105313	29.6337	−5.9936		42105313	0.0001	0.0001	
44736892	−12.1150	2.2572		44736892	0.0000	−0.0002	
47368471	4.9046	−1.8321		47368471	0.0001	0.0000	
50000050	−37.2541	7.1328		50000050	0.0000	0.0000	

Pre-computation of the design matrix, vectorized resampling, and block-wise updating effectively reduce redundant computations and improve data preparation speed. The vectorized solution of ridge regression eliminates repetitive loops and leverages efficient matrix operations to simultaneously compute estimates for multiple replications. The combination of pre-computation, vectorization, and matrix-based solution strategies significantly improves the numerical stability and computational efficiency of the fitting process.

As a result, the total runtime for processing all sample sizes and replications is substantially reduced compared to the unoptimized version, making the proposed approach more scalable and efficient for large-scale applications.

(2) **Case 2**

The simulation experiment in Sect. 4.4 was chosen as the case study for this subsection. The experimental setup consists of estimating an 11-parameter model at different sample sizes (ranging from 50 to 10,000,000) with 50 replications per sample size. Comparison before and after optimization is shown in Fig. 4.28.

Improvement in Fitting Accuracy

As illustrated in the bias comparison plots, the proposed optimization method significantly improves the accuracy of parameter estimation:

Before optimization, the parameter estimates suffer from severe biases, especially for higher-order parameters under moderate or small sample sizes. For instance, when $N = 10^6$, biases for parameters θ_2 to $\theta_1 0$ frequently exceed hundreds or even thousands, showing extremely unstable and fluctuating patterns. Even as N grows to over 10^7, these biases for high-order parameters still do not converge well and remain volatile.

After optimization, all parameter biases are dramatically reduced to near-zero levels across the entire range of sample sizes. Even in small-sample scenarios (e.g., $N = 50$), the maximum absolute bias is less than 0.1, and when N reaches 10^6 or higher, the biases consistently stabilize within the 10^{-3} level. This substantial improvement is mainly attributed to the combined use of the aforementioned optimization methods, such as orthogonal polynomial basis, warm-start strategy, and parallel execution, which effectively alleviate the ill-conditioning issue of the model and reduce the variance of high-order parameter estimates.

2. Improvement in Numerical Stability and Computational Efficiency

Beyond improving estimation accuracy, the proposed approach also enhances numerical stability and computational efficiency:

The combination of the aforementioned optimization methods resulted in a more stable estimation process for all sample sizes. The previously observed large fluctuations, oscillations, and outliers in parameter estimates are no longer present after optimization. This stability not only improves the interpretability and reliability of the results but also contributes to faster convergence and reduced computational overhead during optimization.

Furthermore, by leveraging basis orthogonality, the optimization problem becomes better conditioned, allowing matrix inversion and gradient-based optimization to be executed more efficiently. As a result, the optimized approach is

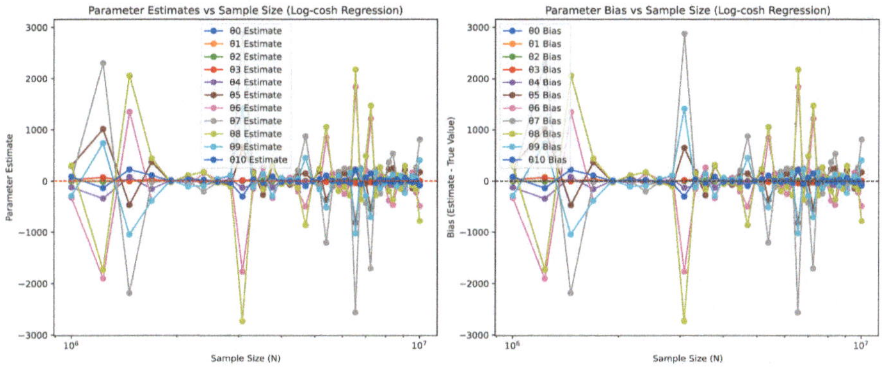

(a) Pre-optimization

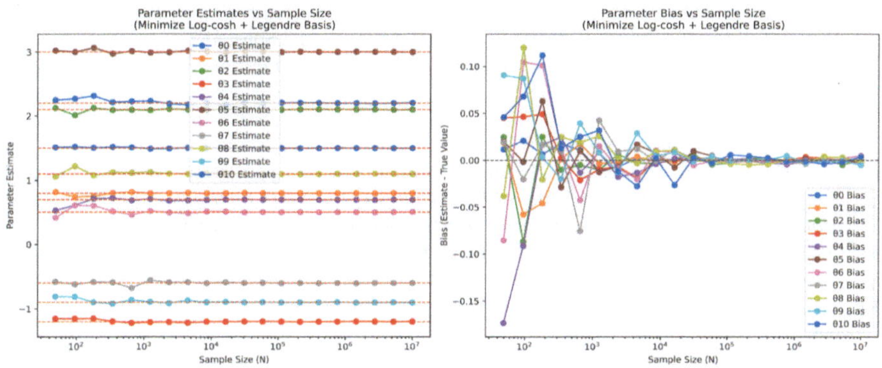

(b) Post-optimization

Fig. 4.28 Visualization of the convergence trajectories of parameter biases in nonlinear case

highly scalable, enabling reliable parameter estimation even for ultra-large sample sizes (up to 10^7) without numerical instability or excessive computation time.

(3) **Overall Findings**

The case study confirms that the combined optimization techniques not only alleviate the challenges of bias and instability in multi-parameter polynomial fitting but also greatly enhance computational efficiency. The results demonstrate the practical applicability of the proposed methods in real-world scenarios where both accuracy and efficiency are essential.

References

1. John Aldrich. Fisher and regression. *Statistical Science*, pages 401–417, 2005.
2. Robert Bartels. Re-interpreting r2, regression through the origin, and weighted least squares. *Univ. Sydney Bus. Sch*, 2015.
3. author Billingsley, Patrick. *Probability and measure*. Wiley series in probability and statistics. Anniversary edition.. edition, 2019.
4. David Birkes and Yadolah Dodge. *Alternative methods of regression*. John Wiley & Sons, 2011.
5. Åke Björck. *Numerical methods for least squares problems*. SIAM, 2024.
6. Matias D Cattaneo, Nicolás Idrobo, and Rocio Titiunik. A practical introduction to regression discontinuity designs: Extensions. *arXiv preprint* arXiv:2301.08958, 2023.
7. Samprit Chatterjee and Jeffrey S Simonoff. *Handbook of regression analysis*. John Wiley & Sons, 2013.
8. Kuang-Fu Cheng and Pi-Erh Lin. Nonparametric estimation of a regression function: Limiting distribution 2. *Australian Journal of Statistics*, 23(2):186–195, 1981.
9. P Diaconis and DA Freedman. On inconsistent m-estimators. *Ann. Statist*, 10:454–461, 1982.
10. Francis Galton. *Typical laws of heredity*. William Clowes and Sons, 1877.
11. Vi͡acheslav Leonidovich Girko. *An introduction to statistical analysis of random arrays*. Walter de Gruyter GmbH & Co KG, 2018.
12. Boris Vladimirovich Gnedenko and Andreĭ Kolmogorov. *Limit distributions for sums of independent random variables*.
13. Asaf Hajiyev and Abdullayeva Narmina. Optimal choice of parameters for unknown function in big data analysis problems. In *Proceedings of the Ninth International Conference on Management Science and Engineering Management*, pages 399–408. Springer, 2015.
14. Asaf Hajiyev and Jiuping Xu. Gauss-newton method for nonlinear regression models with an increasing number of unknown parameters. *Azerbaijan J. Mathematics*, v.15, No.2, 2025.
15. Asaf Hajiyev and Jiuping Xu. Generalized regression models with an increasing number of unknown parameters. *Proceedings of the Institute of Mathematics and Mechanics, Ministry of Science and Education*, No.1-2, 2025.
16. Asaf Hajiyev, Jiuping Xu, and Tiannan Deng. Application of regression models for statistical analysis of simulation data. Research Report, 2025.
17. Asaf H Hajiyev. Estimators in regression models with dependent errors. *Proc. Instit. Math. Mech. Baku*, 5:41–55, 1999.
18. Asaf H Hajiyev. Regression models with increasing numbers of unknown parameters. *International Encyclopedia. Statist. Sciences, Springer*, pages 1208–1211, 2011.
19. Asaf H Hajiyev and Singpurwalla N. Risk analysis in oil exploration. *Intern. J. Appl. Comput.Mathem*, 1:103–118, 2013.
20. Moritz Hanke, Louis Dijkstra, Ronja Foraita, and Vanessa Didelez. Variable selection in linear regression models: Choosing the best subset is not always the best choice. *Biometrical Journal*, 66(1):2200209, 2024.
21. Xuming He and Qi-Man Shao. On parameters of increasing dimensions. *Journal of multivariate analysis*, 73(1):120–135, 2000.
22. Ronald R Hocking. A biometrics invited paper. the analysis and selection of variables in linear regression. *Biometrics*, pages 1–49, 1976.
23. Peter J Huber. Robust regression: asymptotics, conjectures and monte carlo. *The annals of statistics*, pages 799–821, 1973.
24. Peter J Huber. *Robust statistical procedures*. SIAM, 1996.
25. PJ Huber and EM Ronchetti. Robust statistics, wiley & sons. *Inc.*, 2009.
26. Andreĭ Kolmogorov. *Introductory real analysis*.
27. Ingrid Kreuzmair and Helge Toutenburg. Sensitivity of the restricted least squares estimator under mis specified restrictions and multicollinearity. *Advances and Applications in Statistics*, 8(2):153–175, 2008.

28. William La Cava, Bogdan Burlacu, Marco Virgolin, Michael Kommenda, Patryk Orzechowski, Fabrício Olivetti de França, Ying Jin, and Jason H Moore. Contemporary symbolic regression methods and their relative performance. *Advances in neural information processing systems*, 2021(DB1):1, 2021.
29. Weihao Lu, Haobo Zhang, Yicheng Li, Manyun Xu, and Qian Lin. Optimal rate of kernel regression in large dimensions. *arXiv preprint* arXiv:2309.04268, 2023.
30. M.B Malyutov. Lower boarders for average duration sequentially design of experiment. *Izvestiya vuzo*, 11:19–41, 1983. (in Russian).
31. Dastan Maulud and Adnan M Abdulazeez. A review on linear regression comprehensive in machine learning. *Journal of Applied Science and Technology Trends*, 1(2):140–147, 2020.
32. Daniel Q Naiman. Conservative confidence bands in curvilinear regression. *The Annals of Statistics*, pages 896–906, 1986.
33. Sahand N Negahban, Pradeep Ravikumar, Martin J Wainwright, and Bin Yu. A unified framework for high-dimensional analysis of m-estimators with decomposable regularizers. 2012.
34. AA Novikov. Consistency of least squares estimators in regression models with martingales. *Proc. Steklov Mathematical Institute: Statistical Control of Random Processes. Optimization Software*, 1985.
35. Raymond J O'Connor and Tansy L Wagner. A test of a regression-tree model of species distribution. *The Auk*, 121(2):604–609, 2004.
36. Iain Pardoe. *Applied regression modeling*. John Wiley & Sons, 2020.
37. Stephen Portnoy. Asymptotic behavior of m-estimators of p regression parameters when p 2/n is large. i. consistency. *The Annals of Statistics*, pages 1298–1309, 1984.
38. Stephen Portnoy. Asymptotic behavior of m estimators of p regression parameters when $\frac{p^2}{n}$ is large; ii. normal approximation. *The Annals of Statistics*, 13(4):1403–1417, 1985.
39. Enayetur Raheem, S Ejaz Ahmed, and Shuangzhe Liu. Stein-rule m-estimation in sparse partially linear models. *Japanese Journal of Statistics and Data Science*, 7(1):507–535, 2024.
40. DA Ratkowsky. Handbook of nonlinear regression models, 1990.
41. William JJ Rey and William JJ Rey. Bootstrap methods, sampling distributions. *Introduction to Robust and Quasi-Robust Statistical Methods*, pages 78–88, 1983.
42. VS Sampath Kumar. Some aspects of general m-estimation theory for dependent random variables. *Statistics: A Journal of Theoretical and Applied Statistics*, 20(1):95–107, 1989.
43. A. N Shiryaev. *Probability*. Springer, 2013.
44. Qiang Sun, Wen-Xin Zhou, and Jianqing Fan. Adaptive huber regression. *Journal of the American Statistical Association*, 115(529):254–265, 2020.
45. Charles C Thigpen. A sample-size problem in simple linear regression. *The American Statistician*, 41(3):214–215, 1987.
46. Vladimir Vapnik. *Estimation of dependences based on empirical data*. Springer Science & Business Media, 2006.
47. I.N Vuchkov, B.B Boyadjieva, and E Molakov. *Applied linear regression analysis*. Finance and Statist, 1987. (in Russian).
48. Matt P Wand. A comparison of regression spline smoothing procedures. *Computational Statistics*, 15:443–462, 2000.
49. Lan Wang, Yichao Wu, and Runze Li. Quantile regression for analyzing heterogeneity in ultra-high dimension. *Journal of the American Statistical Association*, 107(497):214–222, 2012.
50. Richard H Williams. A new method for teaching multiple regression to behavioral science students. *Teaching of Psychology*, 2(2):76–78, 1975.
51. Fang Yao, Hans-Georg Müller, and Jane-Ling Wang. Functional linear regression analysis for longitudinal data. 2005.
52. Victor J Yohai and Ricardo A Maronna. Asymptotic behavior of m-estimators for the linear model. *The Annals of Statistics*, pages 258–268, 1979.
53. Sh Zaks. *Theory of statistical inference*. John Wiley & Sons, 1972.
54. Antonia Zapf, Christian Wiessner, and Inke Regina König. Regression analyses and their particularities in observational studies: Part 32 of a series on evaluation of scientific publications. *Deutsches Ärzteblatt International*, 121(4):128, 2024.

Chapter 5
Case Studies in Machine Learning Interpretation with Confidence Regions

I propose to consider the question, "Can machines think?"

— Turing (1912–1954)

This chapter proposes a comprehensive framework for evaluating the confidence of interpretable models. The primary objective is to quantify the uncertainty associated with surrogate models used for explanation by constructing statistically grounded confidence regions. To this end, a novel Confidence Interval Comprehensive Evaluation Index is developed, which jointly considers both parameter uncertainty and predictive uncertainty of the fitted local models. The framework is validated through three real-world case studies, each involving predictive modeling using machine learning algorithms such as Support Vector Machines (SVM), XGBoost, and neural networks. For each selected explanation point, Gaussian-based perturbations are generated to simulate local neighborhoods, and linear or nonlinear regression models are fitted to approximate the local decision boundary. The uncertainty of these local approximations is then visualized through confidence ellipsoids for model parameters and confidence bands for fitted functions, offering a more intuitive and quantitative understanding of the "Black-box" interpretability. The proposed evaluation index is further used for comparative and robustness analysis across different models and settings. Collectively, this framework supports the development of an interpretable platform prototype for assessing and visualizing the uncertainty of local model explanations.

A. Hajiyev and J. Xu, *High-Dimensional Regression Analysis and Artificial Intelligence*,
https://doi.org/10.1007/978-981-95-2514-0_5

5.1 Confidence Interval Credibility Evaluation Index

In interpretable machine learning, local surrogate models—such as linear or nonlinear regressors fitted within the neighborhood of a target instance—are commonly used to approximate the decision behavior of complex black-box models. While such local approximations have been widely adopted (e.g., in methods like LIME), relatively little attention has been paid to systematically quantifying their uncertainty. This section introduces a novel framework that addresses this gap by constructing confidence ellipsoids and confidence bands for local surrogate models, providing a principled way to assess both parameter-level and function-level uncertainty.

To enhance the interpretability and practical utility of these confidence structures, we further propose a unified Confidence Interval Comprehensive Evaluation Index (CICE). This index is designed to rigorously evaluate the quality, reliability, and informativeness of the constructed intervals. It accounts for (i) the width of the confidence band for each sample point; (ii) the deviation of the center of the confidence band interval from the predicted value for each sample point, and (iii) the prediction accuracy across the chosen dataset. Together, these metrics offer a holistic view of explanations' stability, credibility, and accuracy, enabling meaningful comparisons across models, instances, and settings.

5.1.1 Constructing the Confidence Interval Evaluation Framework

The three evaluation metrics are defined as follows:

Definition 5.1 Given a prediction point $x_i, i = 1, 2, \ldots, N$, let $[L(x_i),\ U(x_i)]$ denote the associated confidence interval at a confidence level γ (e.g., 95%), where $L(x_i)$ and $U(x_i)$ are the lower and upper bounds, respectively. The confidence interval width at x_i is defined as:

$$\text{Width}(x_i) = U(x_i) - L(x_i)$$

The confidence interval width quantifies the uncertainty of the prediction at a given point. A larger width indicates higher predictive uncertainty, whereas a narrower width suggests that the model yields more precise estimates at the point in question. It is directly influenced by local data density, model smoothness, and variance estimates. This metric is a **interpretability indicator**, as it is defined pointwise for each input x_i.

In the context of model interpretability, a narrower confidence interval implies that the surrogate explanation is more certain and stable in approximating the local behavior of the black-box model. Conversely, wider intervals may signal model instability or high local variance, potentially reducing the user's trust in the generated explanation.

Definition 5.2 Let $\hat{f}(x_i)$ be the center of the confidence interval at point x_i, which may be taken as either the midpoint $\frac{L(x_i)+U(x_i)}{2}$ or a point estimate obtained from the model.

And let $f(x_i)^*$ denote the predicted value. The center deviation at x_i is defined as:

$$\text{Center Deviation}(x_i) = \left|\hat{f}(x_i) - f(x_i)^*\right|$$

The center deviation captures the discrepancy between the estimated center of the confidence interval and the predicted value. Smaller deviations correspond to higher local accuracy in prediction. This metric is particularly sensitive to systematic bias or local model misspecification. This is a **interpretability indicator**, since it measures the estimation accuracy at each point x_i.

In locally interpretable models, the center of the interval often corresponds to the surrogate model's prediction. A smaller center deviation indicates a higher fidelity between the explanation and the actual output of the black-box model, which is essential for maintaining the credibility of the explanation.

Definition 5.3 Prediction accuracy (PA) is defined as a weighted index that combines goodness of fit and the inverse of mean squared error. This is a **accuracy indicator** because it measures the accuracy of the model. This composite evaluation approach aims to reflect the model's performance from different angles and balances the influence of both metrics through weighting:

$$PA = \alpha \times R^2 + (1 - \alpha) \times \frac{1}{1 + MSE}$$

where R^2 measures the model's ability to explain the variance in the data, while $\frac{1}{1+MSE}$ provides a reverse measure of the error size, with smaller MSE indicating higher prediction accuracy. By combining these two, the model's performance across different samples is evaluated more comprehensively, ensuring that the prediction results have both high goodness of fit and low prediction error.

In practice, R^2 measures the model's explanatory power over the training data, with values closer to 1 indicating a better fit and a greater proportion of the variance being explained. However, relying solely on R^2 may overlook the model's performance on individual samples, particularly when outliers or large errors are present. To address this limitation, we introduce $\frac{1}{1+MSE}$, which emphasizes the accuracy of predictions by penalizing higher errors. The smaller the MSE, the higher the accuracy, and taking its inverse strengthens the evaluation of samples with smaller errors, reflecting the model's precision and stability.

The weighted combination of prediction accuracy can comprehensively consider the performance of the model in different aspects. Different weights ($\alpha \in [0, 1]$) can be assigned to adjust the focus on fit and accuracy according to the specific requirements of the task. This approach not only improves the comprehensiveness of the evaluation, but also provides more refined guidance in the model optimization process, adjusting and improving the model according to the accuracy requirements, thus improving the actual predictive ability of the model while ensuring the interpretability.

Taken together, these three metrics provide a multifaceted assessment of the uncertainty, accuracy, and reliability of local surrogate models, thus serving as key indicators for evaluating the quality of model interpretability in a statistically grounded manner. We designed a two-stage evaluation framework that separately addresses interpretability indicators and accuracy indicators, and then combines them through suitable transformations into a comprehensive evaluation index.

Step 1. Interpretability Indicator Evaluation.

At this stage, we define an interpretable index that incorporates the width of the confidence interval and the central deviation, aiming to capture pointwise uncertainty and accuracy. Specifically, we define the interpretability metric as:

$$\text{Interpretability Index}(x_i) = \beta \times \frac{1}{1 + \text{Width}(x_i)} + (1 - \beta) \times \frac{1}{1 + \text{Center Deviation}(x_i)}$$

where $\text{Width}(x_i)$ denotes the confidence interval width at point x_i, and Center Deviation(x_i) measures the absolute difference between the interval's center estimate and the predicted parameter. The coefficients β is user-defined weights to reflect the relative importance of interval width versus center deviation, $\beta \in [0, 1]$.

From an interpretability perspective, this local index reflects the fidelity and stability of the explanation at each prediction point. A higher local score implies that the surrogate model provides both a narrow interval (greater confidence) and a center estimate close to the true value (higher fidelity), both of which are essential for trustworthy local explanations.

Step 2. Combining Interpretability and Accuracy Indicators.

To appropriately combine the interpretability and accuracy indicators, we first take the average of the interpretability evaluation indicator across all sample points. Then, using a weighted average, we combine this with the accuracy indicator (PA) to obtain the final Confidence Interval Comprehensive Evaluation Index (CICE). The formula for CICE is given as follows:

$$\text{CICE} = \omega \cdot \left(\frac{1}{n} \sum_{x=1}^{n} \text{Local Index}(x_i) \right) + (1 - \omega) \cdot \text{PA},$$

where n is the total number of sample points, ω is a weight parameter that determines the relative importance of the interpretability index and PA, $\omega \in [0, 1]$. The interpretability index assesses the stability and uncertainty across samples, while the predictive accuracy measures the combined predictive power of the model. The value range of the CICE composite index is [0, 1], with a larger index representing a better combination of model interpretability and accuracy.

Thus, CICE provides an overall score that reflects both the solvability of the "black-box" model and its predictive accuracy over the entire dataset. It allows for robust comparisons of the explanatory quality of different models, explanation points, and uncertainty estimation methods.

5.1.2 *Embedding the Framework in Interpretable AI*

While many existing interpretability methods—such as LIME and SHAP—focus on generating local surrogate models to approximate the behavior of complex predictors, they often overlook the uncertainty associated with these explanations. In practice, a locally fitted model may yield drastically different explanations depending on neighborhood sampling, noise sensitivity, or model variance. Therefore, quantifying the uncertainty and reliability of such local approximations is essential to establish trust in the interpretability process.

The proposed confidence evaluation framework fills this gap by integrating statistically grounded confidence ellipsoids and bands into the explanation workflow. More importantly, the Confidence Interval Comprehensive Evaluation Index (CICE) enables direct comparison of explanation quality across different surrogate models, explanation points, or neighborhood configurations. This positions our framework as a crucial complement to existing interpretability techniques, offering both diagnostic power and quantitative evaluation capabilities.

5.1.3 *Case Study Workflow for Confidence Evaluation*

To validate the proposed confidence evaluation framework and demonstrate its applicability in real-world settings, this chapter conducts three case studies involving predictive modeling and explanation analysis. The overall case workflow is divided into five stages, each designed to assess the reliability and interpretability of local surrogate models under both linear and nonlinear fitting schemes.

First, for each case, we introduce the real-world dataset and describe the corresponding prediction task. This includes defining the input features, target variable, and the practical significance of accurate predictions within the application domain.

Second, we present the baseline predictive models, including machine learning algorithms such as SVM, XGBoost, and neural networks. We then outline the overall methodological pipeline, including the construction of local neighborhoods, surrogate model fitting, and uncertainty quantification. A schematic diagram is provided to illustrate the complete process.

Third, we select representative explanation points and generate 500 Gaussian-perturbed samples around each one to simulate the local decision region. We apply linear regression models as local surrogates, and analyze the parameter estimates, the associated covariance matrices, and the resulting confidence ellipsoids and confidence bands.

Fourth, we extend the analysis by adopting nonlinear polynomial regressors to serve as more flexible local surrogates. Motivated by limitations observed in the linear models, this section investigates whether nonlinear fitting yields more faithful or stable explanations. The same evaluation procedure is repeated, including covariance matrix analysis, confidence ellipsoid visualization, and function-level uncertainty bands.

Finally, a comparative analysis is conducted to evaluate the difference in interpretability quality between linear and nonlinear local approximations. This includes both quantitative comparisons based on the proposed CICE metric, and visual comparisons of confidence structures.

To validate the proposed confidence evaluation framework, three real-world case studies are conducted. Each study focuses on a different application domain and predictive model, offering a diverse testbed for assessing local interpretability. In each case, machine learning models such as Support Vector Machines (SVM) and XGBoost are first trained. A representative sample is selected as the focal explanation point, and local perturbation samples are generated to simulate the neighborhood of the decision surface. Linear and nonlinear surrogate models are then fitted to approximate the model's local behavior. To quantify the reliability of these local approximations, we construct a confidence interval comprehensive evaluation index for the fitted functions. This index jointly supports the development of an interpretable platform prototype, offering both visualization and quantitative assessment of local explanation uncertainty.

5.2 Airfoil Self-Noise Prediction Case

In the fields of aerospace and wind energy, airfoil self-noise is a critical factor limiting system performance and comfort, primarily caused by phenomena such as boundary layer separation, vortex shedding at the trailing edge, and flow separation as the airfoil moves through the airflow. With increasingly stringent noise emission standards, accurately predicting airfoil self-noise and understanding its mechanisms are crucial for noise reduction and aerodynamic optimization in both aircraft and wind turbine design. In this study, we employ Support Vector Regression (SVR) for predicting airfoil self-noise, as SVR effectively handles nonlinear regression problems and provides strong predictive performance. However, the "black-box" nature of SVR limits the interpretability of its decision-making process. Therefore, this section selects a representative airfoil samples, generates perturbation data, and applies both local linear and nonlinear regression models to approximate the response behavior of SVR in the local region. Furthermore, we provide visualization support by constructing confidence ellipses and confidence bands and computing CICE metrics to quantify the uncertainty and interpretability of local predictions, helping engineers understand and trust model predictions to inform design decisions for noise reduction and aerodynamic optimization [6].

5.2.1 Airfoil Problem Context

In the field of aerospace engineering, particularly in the study of airfoil self-noise, it is widely acknowledged that noise is generated due to the turbulent boundary layer around the surface and trailing edge of the airfoil. This noise, considered as endogenous in nature, is caused by the airfoil itself as it interacts with smoother airflow environments, typically characterized by low inlet turbulence. Based on classical studies, airfoil self-noise can be categorized into several noise mechanisms, each with distinct origins and characteristics:

Turbulent boundary layer-trailing edge noise (TBL-TE Noise). Eddy structures in the turbulent boundary layer interact strongly at the trailing edge, causing dipole radiation at the trailing edge, which is one of the main sources of self-noise.

Separation-stall noise. When the increased attack angle leads to localized flow separation on the airfoil or into the deep stall region, large-scale vortex structures are formed and low-frequency noise is radiated.

Laminar boundary layer-vortex shedding noise (Laminar BL-Vortex Shedding Noise). When the wing surface remains laminar, periodic vortex shedding occurs at the trailing edge of the wing, creating a relatively narrow band or quasi-tuned noise peak.

Tip Vortex Formation Noise. When used for three-dimensional wing or rotor blades, strong vortices are generated in the tip region due to pressure differences, and the turbulence of the vortex nuclei causes localized high-intensity noise.

Trailing-Edge-Bluntness-Vortex-Shedding Noise. If the wing or blade trailing edge has a certain thickness or blunt edge, high-frequency vortex shedding will occur and cause noise.

These noise mechanisms have been extensively studied through theoretical and wind tunnel experiments conducted by NASA [3] and have led to the development of a comprehensive set of airfoil self-noise analysis methods based on the NACA series airfoils. These studies have significantly influenced subsequent noise reduction designs and noise prediction models for airfoils and wind turbine blades. However, the aerodynamic noise characteristics under varying conditions such as different frequencies, angles of attack, and free-stream velocities show significant differences. In real-world applications, it is often necessary to comprehensively consider the noise intensity (measured in decibels, dB) across multiple operating states. Therefore, the ability to predict airfoil self-noise more comprehensively and with greater precision holds significant engineering value.

In light of these complex noise mechanisms, the objective of this study is to develop a prediction model for airfoil self-noise that can efficiently predict noise levels under various conditions. By employing a robust machine learning framework, such as Support Vector Regression (SVR), we aim to capture the intricate relationships between input features like frequency, angle of attack, and chord length, which significantly influence noise generation. Furthermore, the study emphasizes enhancing the interpretability and robustness of the model by introducing local linear and nonlinear regression techniques, enabling more reliable predictions and insights into the factors contributing to self-noise. This approach is crucial for further advancing noise reduction strategies and optimizing the design of quieter and more efficient airfoils or blades in aerospace and wind energy applications.

5.2.2 Airfoil Dataset Overview

The data used in this study are primarily drawn from a dataset obtained from a series of aerodynamic and acoustic tests of 2D and 3D wing blade sections in an anechoic wind tunnel at NASA Langley Research Center (source: UCI Machine Learning Repository [2] and the research paper [3]). This dataset includes different sizes of NACA 0012 airfoils at different wind tunnel speeds and angles of attack, and the span of the airfoils and the position of the observer are the same in all experiments. It provides rich and systematic measurements of airfoil self-noise and boundary layer characteristics and lays a solid experimental foundation for the study of quantitative prediction of the five mechanisms of airfoil self-noise. The dataset is described in Table 5.1.

Characteristics of the dataset include:

Wide angle of attack range. From 0° to exceeding the critical stall angle, covering laminar flow, turbulence, separation, and stall.

Different wing sizes. The experiments include chord length levels from 2.54 to 30.48 cm, both 2D airfoils and 3D airfoils with tips.

Table 5.1 Variables description of airfoil self-noise dataset

Variable name	Role	Type	Units	Missing values
Frequency	Feature	Integer	Hz	No
Attack angle	Feature	Binary	deg	No
Chord length	Feature	Continuous	m	No
Free-stream-velocity	Feature	Continuous	m/s	No
Suction side displacement thickness	Feature	Continuous	m	No
Scaled sound pressure	Target	Continuous	dB	No

Table 5.2 Airfoil self-noise dataset

Frequency (Hz)	Angle of attack (deg)	Chord length (m)	Free-stream velocity (m/s)	Suction side displacement thickness (m)	Scaled sound pressure (dB)
800	0	0.3048	71.3	0.00266337	126.201
1000	0	0.3048	71.3	0.00266337	125.201
1250	0	0.3048	71.3	0.00266337	125.951
1600	0	0.3048	71.3	0.00266337	127.591
4000	3	0.3048	55.5	0.00452492	118.689
5000	3	0.3048	55.5	0.00452492	117.309
2500	3	0.3048	39.6	0.00495741	120.162
8000	3	0.3048	39.6	0.00495741	110.652
1600	3	0.3048	31.7	0.00529514	122.088
2000	3	0.3048	31.7	0.00529514	120.088
1250	4	0.2286	55.5	0.0042862	131.264
1000	5.3	0.2286	71.3	0.0051942	134
250	7.3	0.2286	71.3	0.0104404	139.918

Wide speed coverage. The incoming Mach numbers range from low speeds to subsonic medium and high end, ensuring the typical applicability of the experiment in engineering.

Rich measurement parameters. In addition to the overall noise spectrum (Scaled Sound Pressure in dB), Suction Side Displacement Thickness, Vortex Shedding Frequency, Chord Length, etc. were also measured.

Appropriate sample size. The dataset contains a total of 1503 samples, each consisting of 5 feature variables and 1 target variable (scaled sound pressure), and none of them have missing values. The dataset samples are displayed in Table 5.2.

Through the wing noise dataset provided by NASA and other research organizations, combined with various machine learning models (including support vector machine, random forest, deep neural network, etc.), it is expected to obtain the complex mapping relationship between these variables and the noise level (dB) by using the data of angle of attack, velocity, chord length, and the thickness of the boundary layer in the high-dimensional space, to realize the noise prediction of higher precision and a larger range. Given the special circumstances of the application scenario of adjusting the airfoil, it is important to know not only the noise prediction results but also the prediction causation.

5.2.3 *Airfoil Forecasting Model Development*

This section will focus on the prediction needs of the airfoil self-noise (unit dB), introduce the selected regression model and the corresponding method flow in detail, and show the overall idea with the framework diagram. Firstly, based on the Support Vector Regression (SVR) method in machine learning, a prediction model is constructed to complete the prediction of the overall noise intensity. Since the black box is so complex that it is impossible to accurately find the explanatory reason through the global fitting model, a local explanatory model is introduced to enhance the interpretability of the behavior of the black-box model. The overall approach not only focuses on the prediction accuracy, but also emphasizes the explanatory power of the model on the response of the perturbed samples in the local region, further combines linear and nonlinear local fitting to analyze the parameter uncertainty, and finally achieves a credible explanation of the predicted behavior of the model with the help of confidence intervals (including ellipsoids and confidence bands) and the CICE index.

5.2.3.1 Airfoil Modeling Background

The importance of airfoil self-noise in fields such as aerospace and wind energy has been illustrated in the previous data background. If the noise level (in dB) generated by an airfoil can be accurately predicted based on the given environmental conditions and airfoil characteristics (e.g., angle of attack, chord length, surface smoothness, etc.), it will be of great value for noise control and optimal design of airfoils/blades. However, the following challenges often exist in practical engineering:

Complex noise mechanisms. The airfoil self-noise is generated by the multi-factor coupling of boundary layer turbulence, vortex shedding, and aerodynamic disturbances, resulting in the obvious nonlinear characteristics of both noise intensity and spectral characteristics.

Data availability and measurement error. Although a certain number of experimental measurements are available, factors such as experimental conditions, instrumental accuracy, etc. may introduce noise.

Interpretability and uncertainty measures. While pursuing the prediction accuracy, an interpretability analysis of the model is also needed, especially in the local input space (e.g., a specific wing operating condition) where confidence intervals or confidence ellipsoids need to be evaluated for the reliability of the model.

Based on these needs and challenges, the research focus of this chapter can be summarized as follows: constructing high-precision regression prediction models for airfoil self-noise and achieving better generalization performance through suitable algorithms and parameter tuning. Linear and nonlinear proxy models (e.g., power function form) are used in local regions to enhance the interpretability, and the stability and uncertainty of the prediction results in the neighborhood range are evaluated with the tools of parameter covariance estimation and confidence intervals (including confidence ellipsoids and confidence bands). A complete procedure for "sound pressure level (dB) prediction" is proposed, based on which the local confidence intervals and explanatory significance of the model are quantified.

5.2.3.2 Airfoil Model Selection Principles

In terms of model selection, we mainly adopt the Support Vector Regression (SVR) as the overall prediction model, and locally use linear regression (OLS) and nonlinear regression (represented by the power function) for the refinement analysis.

The reasons and rationale for selecting SVR as the overall prediction model are summarized below:

Higher resistance to overfitting. SVR introduces relaxation variables and kernel function techniques, which can find a suitable "maximum spacing regression plane" in the high-dimensional feature space, and it is relatively easy to obtain a better generalization on a limited dataset.

Nonlinearly tractable. With appropriate kernel functions (e.g., RBF kernel, etc.), SVR can deal with the significant nonlinear relationships in the wing noise; meanwhile, it can utilize grid search and other methods to tune the parameters and improve the model accuracy.

Mature theoretical foundation. Under statistical learning theory, SVR has good theoretical support for small samples with high-dimensional data. This advantage of SVR fits well with the small amount of noise prediction data in this study.

The theoretical and practical value of using local linear and nonlinear regression as a proxy model is as follows:

Interpretability enhancement. In contrast to the more "black-box" SVR, after global fitting we select local points of interest and approximate the local structure of the SVR in its neighborhood using linear regression (OLS) or nonlinear regression (e.g., power function form). This provides an intuitive explanation of the relationship between the noise and the input features under that local region.

Confidence intervals and covariance estimation. Linear (or simple forms of nonlinear) regression can be used to estimate parameter variances with the help of classical statistical methods, and then construct local confidence intervals and joint confidence ellipsoids or confidence bands to measure prediction uncertainty and determine parameter significance.

Mature theoretical foundation. Physical motivation for the power function: In the study of wing self-noise, noise often shows power-law variation with certain flow variables (e.g., flow velocity, chord length, etc.), so local nonlinear regression in the form of a power function is a more reasonable physical approximation.

With this two-tier structure—global SVR for overall prediction and local linear/nonlinear regression for explanatory analysis—we can provide engineering-oriented interpretability and quantitative measures of uncertainty while maintaining prediction accuracy.

5.2.3.3 Airfoil Methodology Overview

Combining the realistic background of the problem and the characteristics of the test dataset, the methodological process of this paper can be divided into the following stages:

(1) Data splitting (Training and Testing)

Split the available dataset into training and test sets, randomly dividing them according to 80%:20% while keeping the data distribution representative.

(2) SVR model training and parameter tuning (Grid Search)

Grid Search combined with Cross Validation is used to find the optimal hyper-parameters within a given range of candidate kernel parameters; the performance of different hyperparameters is evaluated by MSE, R^2, etc. in the test/validation set. The best set of parameters with the best prediction index is finally selected to train the global SVR regressor.

(3) Select typical sample points to generate neighborhood data (±10% perturbation, 500 samples total)

From the test set or global dataset, select a target sample point (or points) to be interpreted and focused on; perform a random perturbation (Gaussian noise × 10%) in the vicinity of that target point, generating a neighborhood of 500 sample points to approximate the distribution of features in the local space (make sure that the perturbation is appropriately sized to stay within the physically feasible range).

(4) Construct local linear and nonlinear regression models based on neighborhood data

Local linear regression (OLS): fit a multivariate linear model to the neighborhood data $y = \beta_0 + \beta_1 x_1 + \cdots + \beta_n x_n$; estimate each coefficient β and compute the covariance matrix $\mathrm{Cov}(\beta)$;

Local nonlinear regression (power function): assume the form $y = a_1 \times (x_1^{k_1}) + a_2 \times (x_2^{k_2}) + \cdots$ or simplify to a single power function relationship $y = a \times x^k$ (depending on specific needs), find the parameters by least squares or least squares + canonical, etc., and estimate the covariance matrix as well.

(5) Parameter estimation, covariance matrix calculation, and confidence interval (ellipsoid and confidence band) construction

First, for the above regression model (linear/power function), after estimating the parameter vector $\widehat{\theta}$, the parameter covariance matrix $\widehat{\sum}$ can be obtained statistically (e.g., classical results of least squares or Gauss–Newton methods). Second, a parameter confidence ellipsoid (e.g., an ellipsoid or a high-dimensional ellipsoid) can be constructed based on $\widehat{\sum}$ and further confidence bands can be computed for $f(x, \theta)$ within a neighborhood, which can be used to assess the uncertainty of the model's function values in that localization. In addition, it is possible to observe whether the parameter intervals contain 0 to assess the significant contribution of each feature to the noise (i.e., to determine whether the feature coefficient is statistically significant or not).

(6) Calculation of the CICE index

The CICE Index (Comprehensive Interval Confidence Evaluation Index) comprehensively evaluates the model's interpretability and prediction accuracy by the weighted average of the interpretability index and prediction accuracy index. In the calculation process, the Interpretability Index is first calculated based on the width of the confidence interval and the center deviation of each sample point, which can reflect the credibility and stability of the model at that point. Next, the local indices of all samples are averaged to assess the global interpretability effect. Subsequently, the model's predictive accuracy (PA), derived from the weighted values of R^2 and $\frac{1}{1+MSE}$, is incorporated to obtain the final CICE value. By adjusting the weight parameter (ω) between the interpretability and accuracy indicators, the relative importance of each can be flexibly balanced, providing a comprehensive and quantifiable evaluation result. This approach not only enhances the credibility of local model explanations but also provides a standardized basis for comparing different models, explanation points, and uncertainty estimation methods through CICE.

(7) Comparative analysis of linear and nonlinear local model effects and differences

First, the local linear model is innovatively compared with the power function model in terms of goodness of fit, residual distribution, and confidence interval range (confidence ellipsoid and confidence band) on the data in the same neighborhood. It helps to choose a more suitable local proxy model in the final engineering, and to understand the complex correlation between the noise and the feature parameters more comprehensively, so as to enhance the interpretability and robustness of the local region of the "black box".

5.2.3.4 Airfoil Process Flowchart

Figure 5.1 illustrates the main flow from raw data to final local model interpretability analysis.

5.2.4 Airfoil SVR Training for Local Interpretation

After completing the feature preprocessing and grid search (searching for the optimal C, γ, ε, and other hyperparameters) on the original dataset (or its training set portion),

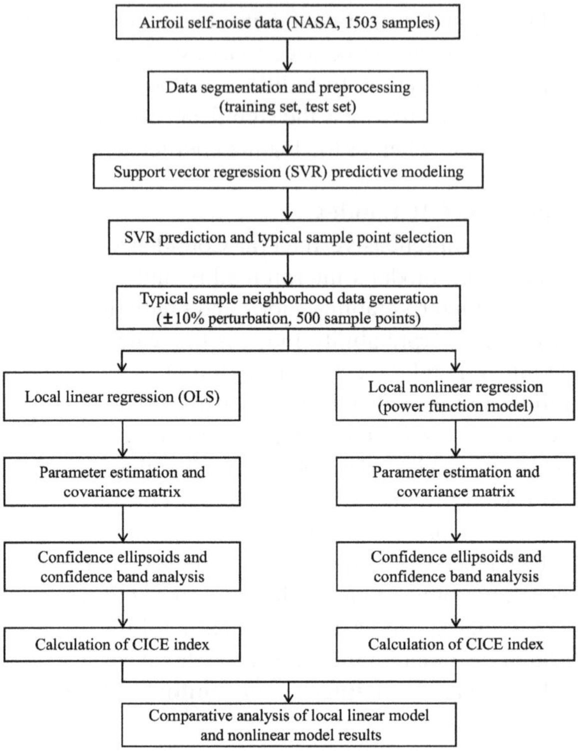

Fig. 5.1 Methodological framework diagram

Table 5.3 Global prediction results

Indicator	Value
Test MSE	6.861
Test R2	0.863

a SVR regressor that performs well on a global scale can be obtained. The prediction results for the airfoil self-noise (dB) are shown in Table 5.3.

The results of the global fitting using Support Vector Regression (SVR) in Table 5.3 show that the optimal hyperparameters are $C = 100$, $\varepsilon = 1$, and $gamma = scale$. Here, C controls the regularization strength of the model, with a higher C value indicating a greater emphasis on fitting the training data well. ε represents the maximum allowable prediction error, where a larger ε value smooths the model and reduces sensitivity to noise. $Gamma = scale$ means that gamma is automatically adjusted to $1/(n_{features} * X.var())$, which is suitable for most problems. A fivefold cross-validation was used for hyperparameter tuning, resulting in a total of 135 fits, ensuring the model's generalizability.

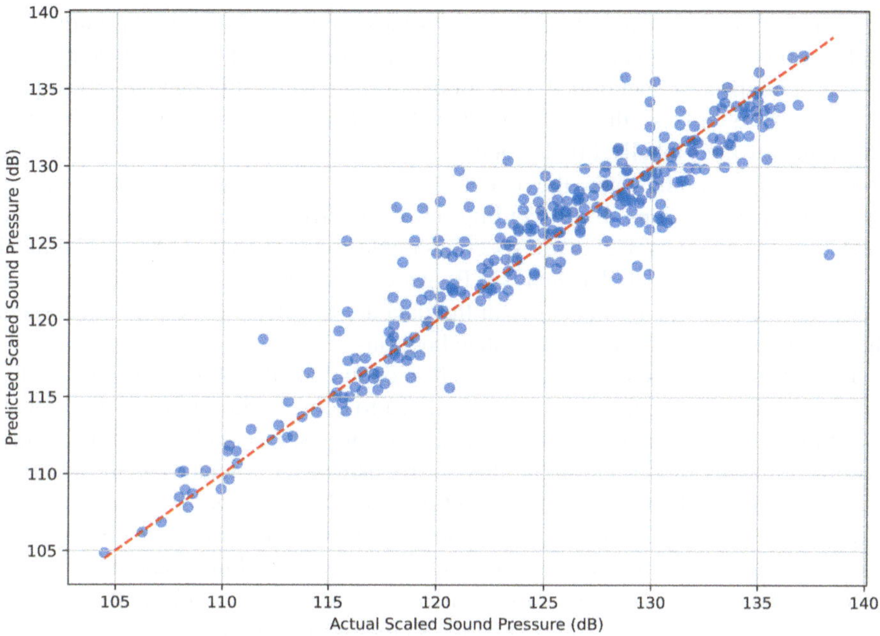

Fig. 5.2 SVR prediction versus actual

The evaluation results on the test set show that the model's Mean Squared Error (MSE) is 6.8614, indicating a relatively low prediction error on the test set, though there is still room for improvement. The R^2 value is 0.8630, meaning the model can explain approximately 86.3% of the variance in the noise data, indicating a good fit. Overall, the SVR model demonstrates strong fitting capability on this dataset and effectively predicts the variation in airfoil self-noise. However, there is still potential for improvement, particularly in terms of error control and model accuracy.

Figure 5.2 visualizes the performance of the Support Vector Regression (SVR) model by comparing the predicted and actual values of the scaled sound pressure (dB) in the test set. The x-axis represents the actual values of the sound pressure, while the y-axis represents the predicted values from the SVR model.

As observed, there is a strong positive correlation between the predicted and actual values, with most data points lying close to the red dashed diagonal line, which represents the ideal "perfect prediction" line where the predicted values match the actual values exactly. The scatter plot shows that the SVR model performs well, with a high level of accuracy in predicting the sound pressure levels.

In comparison with the previously discussed results, where the model achieved an MSE of 6.8614 and an R^2 value of 0.8630, the plot further emphasizes that the model effectively captures the underlying relationship between the features and the target variable. However, some deviations from the diagonal line are still visible, indicating minor prediction errors that could potentially be improved with further fine tuning or using different models. Overall, this visualization complements the numerical evaluation by providing an intuitive sense of the model's predictive performance.

Based on the aforementioned SVR global fitting results, although the model predictions are better (e.g., the MSE and R^2 values indicate more accuracy), the interpretability of the globally fitted model is poor, especially in the vicinity of some specific sample points, where the model is unable to accurately reflect the complexity of the local variations. To further improve the interpretability and accuracy of the model, especially in the vicinity of specific sample points, we selected several typical test sample points for detailed analysis. By generating 500 perturbation samples near these sample points, small changes in the data are simulated to further understand the performance of the model in localized regions.

Local fitting methods (local linear regression and local high-dimensional multi-parameter nonlinear regression) are introduced to compensate for the lack of interpretability of the global model. In practical applications, especially in complex engineering problems, model interpretability and robustness are crucial. With local fitting, we are not only able to adjust the prediction of each sample point more finely, but also provide confidence intervals (confidence ellipsoids and confidence bands) for each prediction of the model, which enhances the trustworthiness and adaptability of the model in practical applications. Therefore, we employ two local proxy models: local linear regression (OLS) and local nonlinear regression (power function) in order to comprehensively assess the prediction accuracy and uncertainty of the sample points.

5.2.5 Airfoil Local Linear Regression Modeling with OLS

The local linear regression (OLS) model $y = \beta_0 + \beta_1 x_1 + \beta_2 x_2 + \cdots + \beta_n x_n$ calculates the estimates of each parameter by the least squares fitting to the neighborhood data.

5.2.5.1 Airfoil Localized Linear Regression Model Construction and Results

Here, the "neighborhood data" is obtained by perturbing selected sample points; in this study we chose 500 perturbed sample points. These sample points are generated by applying $\pm 10\%$ random noise to the original values of each feature, resulting in a new set of data points that are used as inputs to the local regression.

By fitting these perturbed data, the OLS model gives regression coefficients for each input feature (e.g., frequency, angle of attack, chord length, etc.). These coefficients are key measures of how much each feature affects the target variable (noise dB). The goal of the regression model is to minimize the prediction error so that a best-fit model can be found.

Table 5.4 OLS localized regression results

	Coef	Std err	t	$P > \mid t \mid$	[0.025,0.975]
Const	121.075	0.122	990.978	0.000	[120.835, 121.315]
Frequency (Hz)	0.006	6.1e-05	91.865	0.000	[0.005,0.006]
Angle of attack (deg)	−1.26e-12	7.58e-15	−166.15 7	0.000	[−1.27e-12, −1.25e-12]
Chord length (m)	81.141	1.268	63.969	0.000	[78.649, 83.633]
Free-stream velocity (m/s)	−0.100	0.002	−54.559	0.000	[−0.104, −0.097]
Suction side displacement thickness (m)	326.531	75.114	4.347	0.000	[178.950, 474.112]
MSE	0.021				
R^2	0.971				

The OLS regression results, as shown in Table 5.4, indicate that the model's R^2 value is 0.971, meaning that approximately 97.1% of the variation in noise can be explained by the regression model, which suggests a good fit. The p-values for all regression coefficients are significantly less than 0.05, implying that the influence of each feature is statistically significant, and these features play an important role in predicting noise. Specifically, the coefficients for Frequency (Hz), Chord Length (m), and Suction Side Displacement Thickness (m) are 0.0056, 81.1409, and 326.5309, respectively, indicating that these factors have a significant impact on noise levels. Among them, the coefficient for Chord Length (m) demonstrates its strong influence on noise variation, suggesting that for each unit increase in chord length, the noise level increases by approximately 81.14 dB. Conversely, the coefficient for Free-stream Velocity (m/s) is −0.1003, indicating a relatively small but the negative effect of velocity on noise.

Additionally, the regression analysis provides the standard errors and confidence intervals for each parameter, allowing for further evaluation of the precision of these estimates. For example, the coefficient for Frequency (Hz) is estimated at 0.0056, with a standard error of 6.1e-05, demonstrating that the estimate is highly precise, and its 95% confidence interval is [0.005, 0.006], further validating the reliability of this estimate. The parameter covariance matrix from the local linear regression model also serves as a basis for subsequent uncertainty analysis. By visualizing the confidence ellipsoids, we can observe the uncertainty of different parameter combinations. For instance, the confidence ellipsoid for Frequency (Hz) and Chord Length (m) highlights the correlation and estimation errors between these variables in predicting noise. The size and shape of the ellipsoids reflect the variability of the regression coefficients, which is critical for understanding the relationships and uncertainties between the parameters. These results provide a solid foundation for further model optimization and uncertainty quantification.

5.2.5.2 Airfoil Parametric Covariance Matrix Calculation and Analysis

The covariance matrix is a very important element in regression analysis, which reflects the reliability of parameter estimates and the correlation between different characteristics. In this study, the covariance matrix was computed from the results of the OLS model showing the correlation between each parameter and their standard errors. The results in Table 5.5 illustrate that in this model, the covariance of the constant term with the other variables is small, which indicates that the correlation between the constant term and the individual features is low. On the other hand, the covariance between Chord Length (m) and Free-stream Velocity (m/s) is 0.0840, indicating some positive correlation between these two parameters. This phenomenon may reflect that the effects of these two variables on the noise level show a similar trend in the model, i.e., when one variable changes, the other variable may also change accordingly. The diagonal elements in the covariance matrix give the standard errors of the individual regression coefficients, indicating the uncertainty in the estimates of these parameters. If the covariance between certain characteristics is large, it may affect the stability of the model, and therefore the effect of these correlations on the model needs to be considered in practical applications.

Table 5.5 Covariance matrix

	Const	Frequency (Hz)	Angle of attack (deg)	Chord length (m)	Free-stream Velocity (m/s)	Suction side displacement thickness (m)
Const	0.0149	−0.0000	−0.0000	−0.0840	−0.0001	−4.4067
Frequency (Hz)	−0.0000	0.0000	0.0000	−0.0000	0.0000	−0.0003
Angle of attack (deg)	−0.0000	0.0000	0.0000	0.0000	0.0000	0.0000
Chord length (m)	−0.0840	−0.0000	0.0000	1.6090	0.0001	2.0425
Free-stream velocity (m/s)	−0.0001	0.0000	0.0000	0.0001	0.0000	0.0011
Suction side displacement thickness (m)	−4.4067	−0.0003	0.0000	2.0425	0.0011	5,642.0794

In addition, the covariance matrix indicates a weaker relationship between certain features. For example, the covariance of Angle of Attack (deg) with the other variables is very close to zero, indicating that the correlation between the Angle of Attack variable and the other variables is almost non-existent, which to some extent suggests that the feature is more independent of the noise prediction. Meanwhile, Suction Side Displacement Thickness (m), an important variable affecting noise, has a large covariance with the other variables, especially the negative correlation with the constant term (−4.4067). This may imply that Suction Side Displacement Thickness (m) has a more significant effect on the model under specific operating conditions. Overall, the calculation of the covariance matrix provides an important quantitative basis for understanding the interactions between the model parameters and their contribution to the noise prediction results, as well as the basis for subsequent uncertainty analysis and the construction of confidence intervals.

5.2.5.3 Airfoil Confidence Ellipsoid Analysis

The confidence ellipsoid is derived by calculating the covariance matrix, which demonstrates the range of possible variations in the parameter estimates of the model for a given level of confidence. In local regression, the size and shape of the confidence ellipsoid reflect the estimation error and uncertainty of the regression coefficients and reveal the interaction effects between different features. The confidence ellipsoid is defined by the quadratic form:

$$(\boldsymbol{\theta} - \hat{\boldsymbol{\theta}})^T \mathbf{C}^{-1} (\boldsymbol{\theta} - \hat{\boldsymbol{\theta}}) \leq \frac{\chi^2_{1-\alpha}(m)}{N} \tag{5.1}$$

where $\hat{\boldsymbol{\theta}}$ denotes the estimated coefficients, $\mathbf{C}$ the covariance matrix, and $m = 5$ the number of parameters. At a 95% confidence level, the chi-square critical value is 11.0705. Therefore, the resulting ellipsoid is given by:

$$\left\{\boldsymbol{\theta} \in \mathbb{R}^5 \parallel Q(\boldsymbol{\theta}) \leq 11.0705\right\} \tag{5.2}$$

To visualize the uncertainty structure of parameter estimates in the local linear model, six 3D parameter groups were selected, each composed of three variables from the full set. For each group, the corresponding 3×3 submatrix was extracted from the overall covariance matrix and decomposed into eigenvalues and eigenvectors. The semi-axis lengths of each ellipsoid were determined using the 95% chi-square quantile for three degrees of freedom ($\chi^2_{0.95}(3) = 7.8147$), allowing for the construction of confidence ellipsoids that represent the joint variability of parameters in each subspace.

Each ellipsoid was rendered by transforming a unit sphere by scaling and rotation according to the covariance structure. The estimated parameter vector $\hat{\boldsymbol{\theta}}$ was plotted at the center of each ellipsoid to visualize its relative position within the uncertainty region. In Fig. 5.3, we see that the confidence ellipsoids for different

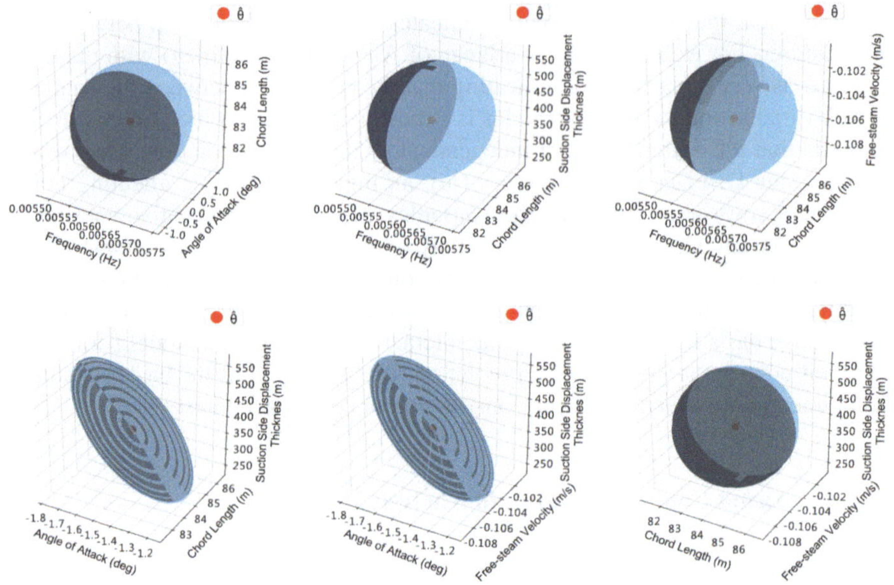

Fig. 5.3 95% confidence ellipsoid of the linear local model

parameter combinations demonstrate the correlation and error size of these parameters. The larger the confidence ellipsoid, the greater the estimation uncertainty for that parameter combination; conversely, the smaller the ellipsoid, the more stable the model's prediction for that combination.

The first plot shows the combination of "Frequency (Hz)" with "Angle of Attack (deg)" and "Chord Length (m)". The larger confidence ellipsoid suggests that this combination of variables exhibits higher uncertainty in the model. This larger ellipsoid also indicates a strong interaction between "Frequency (Hz)" and "Chord Length (m)", which may impact the prediction stability of the model. The second plot displays the combination of "Frequency (Hz)" with "Chord Length (m)" and "Suction Side Displacement Thickness (m)". In this plot, the shape of the ellipsoid shows a strong correlation between "Chord Length (m)" and "Suction Side Displacement Thickness (m)", implying that these two variables have a significant effect on noise prediction.

Other plots further reveal the varying contributions of features to the noise prediction. For example, the combination of "Angle of Attack (deg)" and "Free-stream Velocity (m/s)" shows a relatively smooth ellipsoid, suggesting that these two variables have a more stable influence on noise prediction, and the uncertainty of the model is smaller for these features. In contrast, the combination of "Free-stream Velocity (m/s)" and "Suction Side Displacement Thickness (m)" displays a more elongated ellipsoid, indicating that the interaction between these two variables may introduce higher error and uncertainty in the model.

The results from the confidence ellipsoids not only provide a quantitative understanding of the regression coefficients but also reveal the potential interactions between features. This is crucial for further optimizing the model, reducing uncertainty, and understanding the model's dependence on specific parameters. Next, we will analyze the results of the confidence band to understand the confidence intervals of the predicted values and assess the model's robustness.

5.2.5.4 Airfoil Functional Confidence Band Analysis

In the previous analysis, we examined the uncertainty of the parameters in the local linear regression model through the results of the confidence ellipsoid and illustrated the relationships and interactions between different feature combinations. The confidence ellipsoid, by displaying the uncertainty regions of the regression coefficients, helps us understand the dependencies between parameters and the model's sensitivity to parameter estimates. Next, we will further explore the construction and analysis of the function confidence band, which aims to assess the uncertainty in the predicted values and provide support for the model's robustness.

The function confidence band visualizes the confidence intervals of the predicted values on a given sample set. The blue curve in the figure represents the predicted outcomes of the local linear regression model, while the light blue shaded area indicates the 95% confidence band for these predictions. The wider the confidence band, the greater the uncertainty in the model's predictions, indicating a broader range of variation in the prediction results. For this dataset, the variation trend of the confidence band shows that around certain sample points, the prediction confidence is relatively high, while in other areas, there is considerable uncertainty in the prediction, which is typically caused by a large variation in input features or sparse sample distribution.

It can be observed in Fig. 5.4 that as the sample index changes, the predicted values fluctuate significantly, indicating that the model's predictions are not consistent throughout but are influenced by specific sample points and data features. In local regions, the fluctuation of the prediction values is small, indicating that predictions in these regions are more stable and reliable. In contrast, in other areas, especially where the sample points are more dispersed, the width of the confidence band increases, reflecting the model's poorer adaptability in these regions, necessitating further optimization.

The width and shape of the confidence band reflect the robustness of the model, i.e., its responsiveness to different input data. In certain feature combinations, such as "Frequency (Hz)" and "Chord Length (m)", the confidence band is relatively narrow, suggesting that these features contribute more consistently to noise prediction, and the model performs robustly in these regions. In other feature combinations, such as "Angle of Attack (deg)" and "Suction Side Displacement Thickness (m)", the confidence band is wider, indicating greater uncertainty in the model's predictions in these areas and the need for further improvement.

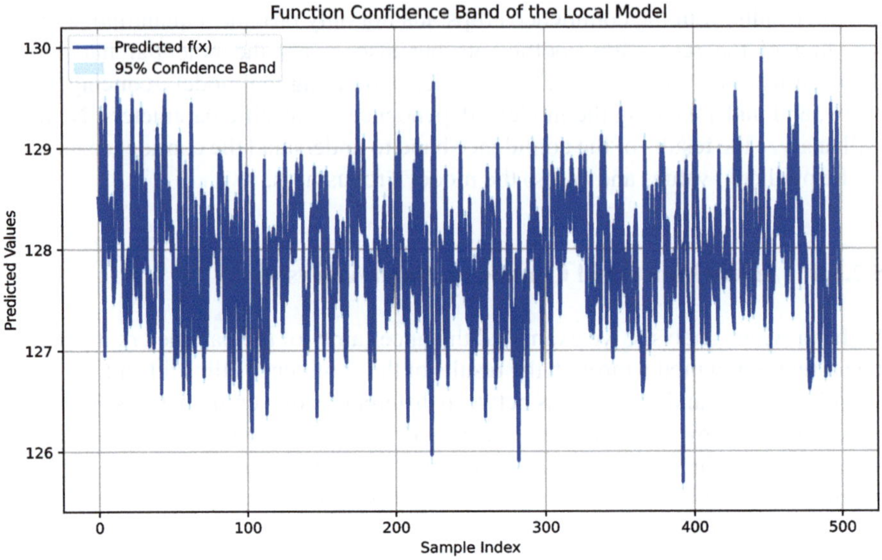

Fig. 5.4 Functional confidence band of the linear local model

Through the analysis of the function confidence band, we gain a more intuitive understanding of the model's performance and uncertainty in different sample regions. This analytical method enhances the interpretability of the model, helping to identify regions where the predictions are more reliable and where further optimization is needed. Combined with the analysis of the confidence ellipsoid, the function confidence band provides a complete evaluation of the error range, further enhancing our understanding of the model's predicted results.

5.2.5.5 Airfoil Analysis of CICE Calculations

In this study, we construct a local surrogate model based on local linear regression to explain the behavior of a black-box model and introduce a Confidence Interval Comprehensive Evaluation Index (CICE) to simultaneously assess both interpretability and predictive accuracy. The CICE metric integrates the width of the confidence intervals, the deviation between the confidence interval centers and the predicted values, and the predictive accuracy (PA), thus providing a quantitative and comprehensive framework for model interpretability evaluation. In the calculation process, a weighted average approach is employed—assigning weights of **0.8** to the confidence interval width and **0.2** to the center deviation for interpretability, while predictive accuracy is computed based on a weighted combination **(0.5, 0.5)** of R^2 and mean squared error (MSE). At the same time, in order to better explain the explainability and accuracy of the model, the weight combination of the two indicators is also set as **(0.5, 0.5)**.

Regarding interpretability, we measure the stability and reliability of the local linear surrogate model by computing the width of each sample's confidence interval and the deviation between the confidence interval center and the predicted value. When the confidence interval width is small and the center deviation is near zero, it indicates that the model's explanation in that local region is both stable and trustworthy. Our results reveal that, for most samples, the deviation between the confidence interval center and the predicted value is negligible, indicating that the midpoint of the confidence interval aligns almost perfectly with the prediction. This high interpretability score enhances the transparency of the model's explanations, thereby increasing user confidence in the interpretability process.

In terms of predictive accuracy, a high PA value (approximately 0.985) was obtained through a weighted combination of the R^2 value and the mean squared error (MSE), reflecting the model's strong performance in prediction tasks. When integrating both interpretability and predictive accuracy, the final CICE value is calculated to be **0.9303**, demonstrating that the local surrogate model achieves a favorable balance between transparency and accuracy. This result not only attests to the model's excellent fit on local data but also validates the effectiveness of incorporating confidence interval-based metrics in constructing a comprehensive evaluation method. Such findings provide both theoretical support and practical guidance for further optimization and interpretation of more complex models.

In summary, this book innovatively combines the analysis methods of the confidence ellipsoid and confidence band, providing visual insights into the uncertainty of model parameters and predicted values. The use of this approach in this section not only proves helpful in understanding the behavior and performance of the model but also provides strong support for future model optimization and uncertainty measures.

5.2.6 Airfoil Local Nonlinear Fitting with High-Dimensional Multiparameter Model

In traditional regression analysis, linear models are often used to describe the relationship between the dependent and independent variables. However, when dealing with complex real-world problems, linear models often struggle to capture the nonlinear features in the data. In this study, the prediction of airfoil self-noise is a complex system involving multiple factors, where the relationships between these factors may exhibit significant nonlinear characteristics. Therefore, it is necessary to choose a local nonlinear regression model as a supplement and optimization.

5.2.6.1 Airfoil Localized Nonlinear Fitting Model Construct

Local nonlinear fitting not only better adapts to the nonlinear relationships in the data but also provides an effective quantification of the model's local uncertainty.

Table 5.6 Local nonlinear fitting results

Indicator	Value
MSE	0.029
R2	0.955
RMSE	0.172

Specifically, this study adopts a power function model to describe the relationship between input features and noise levels. The model structure is as follows:

$$f(x) = a_1 x_1^{k_1} + a_2 x_2^{k_2} + a_3 x_3^{k_3} + a_4 x_4^{k_4} + a_5 x_5^{k_5} = \sum_{j=1}^{5} a_j x_j^{k_j}$$

where x_j is the input feature, a_j and k_j are the coefficients and exponents to be estimated. The advantage of this model is that it can fit most nonlinear behaviors commonly observed in physical phenomena. The power function model not only adapts to the changes in the exponents of the input features but also reflects the complex nonlinear coupling relationships between each feature and the output result. Through local nonlinear fitting, we can more accurately characterize the impact of each feature on the noise level, thereby further improving the predictive accuracy of the model.

In order to provide a more rigorous comparison with the results of the local linear fitting and better illustrate the interpretability and robustness issues innovatively proposed in this book, we selected the same domain sample points as in the previous section to construct the local nonlinear regression model. During the nonlinear fitting process, we first calculated the estimated values of the parameters (coefficients $\widehat{\theta}$) and evaluated the model's predictive ability.

Based on the fitting results of the nonlinear model in Table 5.6, the obtained parameter estimates indicate a complex nonlinear relationship between each input feature and the noise level. For example, Frequency (Hz) and Angle of Attack (deg) have a significant effect on the noise level, while Chord Length (m) has a lesser effect on the noise. The R^2 value of the model is 0.955, which indicates that it can explain about 95.5% of the noise variability, and the mean square error (MSE) is 0.029, and the root mean square error (RMSE) is 0.172, which indicates that the model has a good fit and low prediction error. Taken together, the nonlinear regression model effectively captures the nonlinear coupling effects between input features and accurately reflects the changes in noise level.

5.2.6.2 Airfoil Parametric Covariance Matrix Calculation and Analysis

In nonlinear regression models, the covariance matrix of parameters is essential for assessing both the precision of parameter estimates and their interdependencies.

This matrix quantifies the uncertainty associated with each parameter and reveals correlations between them. By calculating the covariance matrix, we gain insight into how parameter estimates may vary together, providing a foundation for uncertainty analysis such as confidence intervals and ellipsoids. In our model, the covariance matrix is calculated after fitting the nonlinear model using least squares, allowing us to examine how parameters like a_1 and a_5 relate to one another, and how the model's uncertainty is distributed across its parameters.

The covariance matrix in Table 5.7 reveals high variances for certain parameters, such as a_1, indicating substantial uncertainty in its estimate. This suggests that the model is particularly sensitive to changes in this parameter. On the other hand, smaller variances for parameters like a_2 and k_5 indicate more stable estimates. Furthermore, off-diagonal covariances, such as between a_1 and a_4, demonstrate significant interdependencies among parameters, highlighting nonlinear relationships within the model. These results suggest that certain parameters are likely to influence one another, requiring careful interpretation in further model refinement.

Ultimately, the covariance matrix plays a critical role in understanding the model's robustness and reliability. The large variances and covariances highlight areas of uncertainty, guiding where the model may need improvement or further refinement. These findings will be instrumental in the subsequent analysis of confidence ellipsoids, which will visually capture the uncertainty in the parameters. By integrating the covariance matrix with confidence ellipsoid analysis, we aim to enhance the model's interpretability and robustness, contributing to a more reliable model for real-world applications.

5.2.6.3 Airfoil Confidence Ellipsoid Analysis

In nonlinear regression models, confidence ellipsoids are an important tool for assessing the uncertainty of parameter estimates and help us understand the robustness and interpretability of the model by visualizing the range of possible values for different parameter combinations. In this study, we select six representative parameter combinations and demonstrate the estimation ranges of these parameters by plotting confidence ellipsoids for the nonlinear parameters. The size and shape of the confidence ellipsoids reflect the estimation accuracy and interdependence of different parameter combinations in the model. The confidence ellipsoids in the figure show the distribution of each parameter combination in three-dimensional space, which can help us to understand which parameters have a greater impact on the prediction effect of the model and which parameters contribute less to the stability of the model.

For example, some parameter combinations have larger confidence ellipsoids, indicating higher uncertainty in the parameter estimates. This may be due to the strong nonlinear coupling effects between these parameters. In Fig. 5.5, for instance, the confidence ellipsoid for the parameter combination a_3, a_4, a_5 is larger, indicating that the estimates for these parameters are less stable, reflecting the complex interactions between these features and the model's uncertainty. In contrast, some parameter combinations, such as k_3, k_4, k_5, show smaller confidence ellipsoids, sug-

Table 5.7 Covariance matrix for nonlinear model parameters

	a_1	a_2	a_3	a_4	a_5	k_1	k_2	k_3	k_4	k_5
a_1	6714.7755	−0.0000	−214.4290	−288.5793	−7209.7734	−12.4898	−0.0000	2.7387	−2.0840	−1.0442
a_2	−0.0012	0.0000	−0.0000	−0.0001	0.0016	0.0000	0.0000	0.0000	−0.0000	0.0000
a_3	−214.4300	−0.0000	653.7012	68.2696	−681.5429	0.3988	−0.0000	−8.2159	0.4825	−0.0937
a_4	−288.5796	−0.0000	68.2695	1733.6664	−1750.7219	0.5362	−0.0000	−0.8498	12.6734	−0.2487
a_5	−7209.7718	0.0000	−681.5441	−1750.7223	11306.0050	13.4110	0.0000	8.5106	−12.8088	1.6268
k_1	−12.4898	0.0000	0.3988	0.5362	13.4110	0.0232	0.0000	−0.0051	0.0039	0.0019
k_2	−0.0000	0.0000	−0.0000	−0.0000	0.0000	0.0000	0.0000	0.0000	−0.0000	0.0000
k_3	2.7387	0.0000	−8.2159	−0.8498	8.5106	−0.0051	0.0000	0.1033	−0.0060	0.0012
k_4	−2.0840	−0.0000	0.4825	12.6734	−12.8088	0.0039	−0.0000	−0.0060	0.0927	−0.0018
k_5	−1.0442	0.0000	−0.0937	−0.2487	1.6268	0.0019	0.0000	0.0012	−0.0018	0.0002

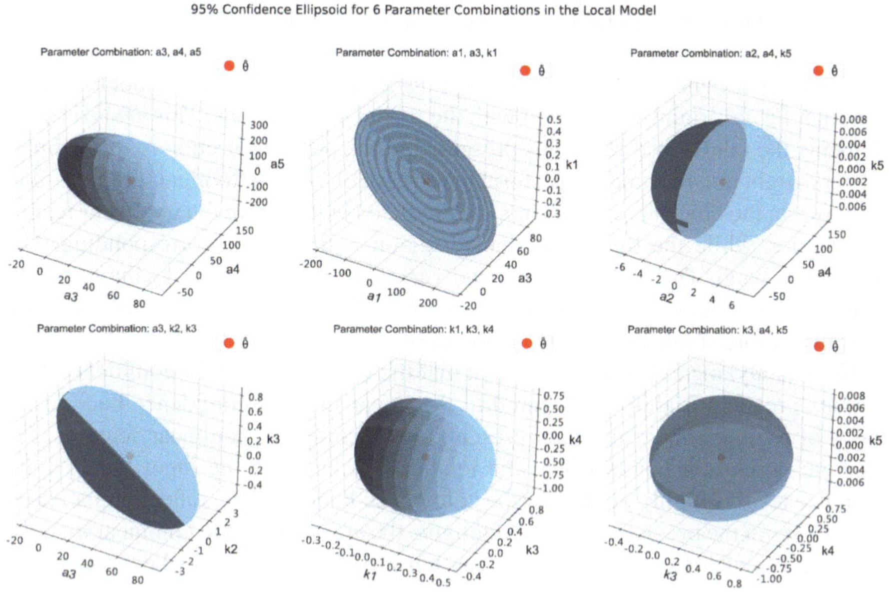

Fig. 5.5 95% confidence ellipsoid of the nonlinear local model

gesting higher estimation precision for these parameters and greater sensitivity of the model to these parameters. These observations allow us to further explore the model's performance under different feature conditions, identifying which features have a greater impact on noise prediction and which contribute less to the prediction.

In addition to parameter uncertainty, the confidence ellipsoids also illustrate the impact of nonlinear coupling effects on the model's predictive outcomes. When multiple parameters act in unison, their nonlinear relationships lead to more complex prediction results, and this interaction is reflected in the shape of the confidence ellipsoids. For instance, in the combination of a_3 and a_4, the evolving shape of the confidence ellipsoid showcases the complex coupling between these parameters. By further analyzing these ellipsoids, we gain a deeper understanding of the model's local behavior and its nonlinear responses in different regions. This provides valuable information for the subsequent optimization of the model. Overall, the visualization analysis of confidence ellipsoids enhances the interpretability of the model and offers a robust foundation for uncertainty quantification, aiding in the understanding of the model's robustness and its applicability in real-world scenarios.

This process provides a solid foundation for our subsequent analysis, especially in constructing and assessing model uncertainty. The combination of confidence ellipsoids and covariance matrices allows for a comprehensive evaluation of the interrelations between parameters and their contributions to the model's predictive results. Based on this, we can further analyze and optimize the model to achieve higher predictive accuracy and greater stability, ultimately providing more reliable support for real-world applications.

5.2.6.4 Airfoil Functional Confidence Band Analysis

The confidence band of the nonlinear function offers a valuable visual tool for understanding the uncertainty associated with the model's predictions. This analysis complements the confidence ellipsoids presented in the previous section by providing a deeper insight into the model's performance, particularly at individual sample points. In this case, the blue curve represents the predicted values from the nonlinear regression model, while the light blue shaded region indicates the corresponding 95% confidence band. The width of this band varies across the range of sample points, with broader regions signifying higher uncertainty in predictions and narrower areas indicating more stable and precise predictions.

Upon analyzing Fig. 5.6, it is evident that the nonlinear model's confidence band exhibits greater variability compared to the linear model's confidence band, especially in regions where the data presents more complex, nonlinear features. For example, at certain points where the relationship between the input features and the noise level is highly nonlinear, the confidence band widens, reflecting increased uncertainty in the predictions. This highlights the areas where the nonlinear model is particularly sensitive and where more caution should be exercised when interpreting predictions.

In contrast to the linear model, the nonlinear model's confidence band generally appears wider, particularly in regions where nonlinear characteristics are more prominent in the data. This suggests that while the nonlinear model provides a more accurate fit for complex data, it also exposes regions of higher uncertainty. The

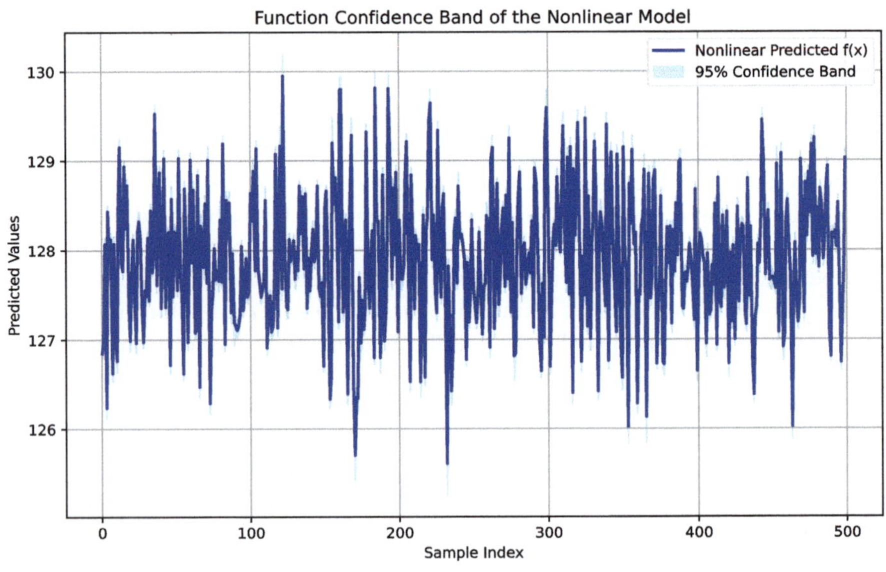

Fig. 5.6 Functional confidence band of the nonlinear local model

broader confidence band reflects the model's difficulty in making highly accurate predictions in certain areas, particularly where interactions between multiple features become more intricate. However, this also means the nonlinear model captures more of the data's complexity, improving prediction accuracy in other regions of the feature space.

By comparing the confidence bands of the linear and nonlinear models, we can better understand the performance of each model under varying conditions. The nonlinear model's broader confidence band in more complex regions indicates that it can handle intricate feature relationships but also shows areas of high uncertainty. This comparison allows us to evaluate the robustness and stability of both models and to determine where the nonlinear model significantly improves upon the linear model in terms of accuracy and interpretability. Moreover, these insights are crucial for refining the models and ensuring their robustness when applied to real-world data.

5.2.6.5 Airfoil Analysis of CICE Calculations

In this study, we utilize a local nonlinear regression surrogate model to approximate the behavior of a black-box model and introduce a Confidence Interval Comprehensive Evaluation (CICE) index to quantify both interpretability and predictive accuracy. The interpretability score for each sample is computed by integrating the width of its confidence interval with the deviation between the interval center and the predicted value. Smaller interval widths and lower deviations indicate that the model provides a more stable and reliable explanation within the local region. Our results demonstrate that the local nonlinear regression model achieves a high interpretability score, suggesting that for most samples, the model accurately captures the black-box behavior without relying excessively on complex nonlinear features. This approach enhances the transparency of the model's explanations, thereby increasing the user's confidence in the interpretability process. Moreover, each weight combination is consistent with the local linear agent model.

In terms of predictive accuracy, we evaluate model performance on the test set by combining the coefficient of determination (R^2) with the mean squared error (MSE) to derive a predictive accuracy (PA) metric, which in this case is approximately 0.985. When integrated with the interpretability metric, the final CICE value is computed to be **0.9244**. This high CICE value indicates that the model strikes an effective balance between explanation transparency and predictive precision. The result demonstrates that the local nonlinear regression surrogate not only approximates the local behavior of the black-box model with high fidelity but also provides consistently accurate predictions. These findings support the utility of such surrogate models in complex systems, particularly in real-world applications that require both a high degree of interpretability and exceptional prediction accuracy.

5.2.7 *Airfoil Integrated Interpretation and Comparative Analysis*

In this section, we present a systematic comparative analysis of the results of locally linear regression models (OLS) and locally nonlinear regression models (e.g., power function models). The central difference between the two models lies in the form of the relationship between the features they assume and the predictive target: linear models assume that the relationship is linear, whereas nonlinear models are able to reveal more complex interactions between features. Specifically, local linear regression models provide an intuitive and easily interpretable relationship between features and outputs through simple linear assumptions, whereas local nonlinear regression models, using forms such as power functions, are able to reveal complex nonlinear coupling effects between features, which are crucial for understanding the deeper mechanisms behind the data.

The local linear fit in this study provides a basic model structure whose linear interpretation of features is easy to understand, and due to its simplicity, it shows good stability in terms of prediction accuracy. However, when the data exhibit nonlinear relationships, the local linear model may not adequately capture the complex interactions between features, thus limiting its predictive effectiveness and robustness. In contrast, local nonlinear models demonstrate significant advantages in dealing with situations with complex feature relationships, especially in capturing the nonlinear coupling effects among features. By adjusting the power exponents of features, nonlinear regression is able to adapt more flexibly to the variability of data, and thus improve the predictive ability of the model. By comprehensively comparing the fitting results, the magnitude of feature effects, the model stability, and the adaptability to different contexts, we are able to deeply analyze the advantages and disadvantages of the two methods in different tasks and provide a theoretical basis for subsequent model selection.

Model Comparison. The local linear regression model (OLS) in Table 5.8 shows an R^2 of 0.971 and an MSE of 0.021. This suggests that the model fits the data well, explaining 97.1% of the variance in the target variable (noise level). The estimated coefficients for various input parameters, such as Frequency (Hz), Chord Length (m), and Suction Side Displacement Thickness (m), are all statistically significant (p-values < 0.05). In contrast, the local nonlinear regression model (Table 5.8) yields an R^2 of 0.955, indicating a slight reduction in goodness of fit compared to the linear

Table 5.8 Comparison of linear and nonlinear local surrogate models

Metric	Linear model	Nonlinear model
R^2	0.971	0.955
MSE	0.021	0.029
CICE	0.9303	0.9244

model, but it has a higher MSE (0.029), reflecting a relatively less accurate prediction performance compared to the linear model.

Feature Influence and Stability. The OLS regression analysis shows that Chord Length (m) and Suction Side Displacement Thickness (m) have a significant influence on the noise level, with large coefficients suggesting strong effects. On the other hand, the nonlinear model allows for more flexibility in capturing complex relationships, as seen in the confidence ellipsoids in Fig. 5.5, where the shape and size of these ellipsoids indicate the uncertainty and interactions between parameters. The nonlinear model introduces additional interactions, such as between Suction Side Displacement Thickness (m) and Free-stream Velocity (m/s), reflecting the more intricate coupling between features in the nonlinearity.

Predictive Certainty Assessment. Both models show a 95% confidence band around their predictions, with the nonlinear model's confidence band being wider and exhibiting more variability (as shown in Fig. 5.6). This suggests that the nonlinear model, while capturing the complex dynamics better, also exhibits higher predictive uncertainty in certain regions. Specifically, the nonlinear model's confidence band fluctuates more significantly, indicating greater sensitivity to input changes in complex regions, where the relationship between features and output is highly nonlinear.

CICE Index Assessment. Based on the comparison data in Table 5.8, the linear model outperforms the nonlinear model in terms of R^2 and MSE. The R^2 value of **0.971** is significantly higher than the **0.955** of the nonlinear model, indicating that the linear model better explains the variance in the data and is more suitable for scenarios requiring a strong fit to the data relationships. Additionally, the MSE of **0.021** for the linear model is superior to the **0.029** of the nonlinear model, demonstrating its advantage in predictive accuracy and lower prediction errors. However, despite the linear model's superior performance in R^2 and MSE, the CICE value of the nonlinear model is **0.9244**, slightly lower than the linear model's **.9303**, indicating that the nonlinear model also performs well in balancing interpretability and accuracy. The nonlinear model, with its flexible fitting approach, is better suited for handling more complex data patterns and exhibits higher robustness, although its predictive accuracy is slightly lower. In conclusion, the linear model is better suited for applications requiring higher interpretability, while the nonlinear model is more effective for handling complex data and provides reliable interpretability and prediction performance.

Conclusion. In summary, while the OLS linear model provides a high R^2 and a good fit for the data, it falls short in capturing the complex nonlinearities between the features. The nonlinear model, on the other hand, delivers a lower R^2 but demonstrates better prediction accuracy and more robustness by accommodating the nonlinear relationships between features. The confidence ellipsoid and confidence band analyses provide a richer understanding of model uncertainties, highlighting the importance of considering nonlinearity for more accurate predictions, especially in systems with complex, interdependent variables. These results suggest that by using high-dimensional multiparameter nonlinear models, we can obtain more accurate predictions, especially when confronted with data with complex, interdependent features.

5.3 Bank Marketing Decisions Case

While the XGBoost model demonstrates strong predictive performance in identifying potential long-term deposit subscribers, its lack of interpretability presents a challenge for practical deployment in bank marketing. To bridge this gap, this section conducts a local interpretability analysis by selecting representative customer samples and generating perturbed neighborhoods around each. Local linear or nonlinear regressions are fitted to approximate the XGBoost decision surface, and parameter covariance matrices are used to construct confidence ellipsoids and bands. These visual tools enable the quantification of prediction uncertainty and offer intuitive explanations of model behavior, thereby enhancing transparency and supporting informed customer targeting decisions under resource constraints [7].

5.3.1 Marketing Problem Statement

In the financial industry, particularly in the banking sector, telemarketing has become a commonly used strategy to promote long-term deposit products. However, due to the uncertainty of customer responses and the limitation of resources, efficiently identifying potential customers remains a significant challenge. Although previous studies have explored the potential of data mining in bank marketing, there is still a lack of systematic development and validation of predictive models based on real-world data—especially for forecasting whether a customer will successfully subscribe to a long-term deposit. Traditional models often rely on information obtained during or after the call (such as call duration), making them unsuitable for pre-call prediction and real-time decision-making.

To address this limitation, a predictive model based on XGBoost is first developed, followed by a focus on enhancing its interpretability. Given that the practical application of machine learning models in decision-making relies heavily on transparency and comprehensibility—particularly in commercially sensitive tasks such as customer targeting—merely providing a prediction is insufficient. The key question, therefore, is how to explain why a given customer is predicted to subscribe (or not subscribe) to a long-term deposit, and how stable and credible such explanations are under local perturbations. To this end, we construct local linear or nonlinear surrogate models and introduce confidence ellipsoids, bands and a new evaluation index—CICE—to visualize and assess the stability of local approximations, thereby enhancing the interpretability and practical utility of the model in banking marketing contexts.

The main contribution of this approach lies in providing individual-level explanations of model predictions by integrating local surrogate modeling, uncertainty visualization, and calculation. Unlike previous studies that primarily focus on predictive accuracy, this method places greater emphasis on the reliability and stability of explanations, which are essential for fostering managerial trust in model outputs during real-world decision-making. By fitting local surrogate models and visualizing confidence ellipsoids and bands, this method offers a concrete and actionable technique for improving model transparency in financial marketing settings, ultimately supporting more informed customer communication and resource allocation.

5.3.2 Marketing Data Description

The dataset used in this study originates from a Portuguese retail bank's telemarketing campaigns aimed at promoting long-term deposit subscriptions (source: research paper & UCI Machine Learning Repository [1, 4]). It contains approximately 4500 records collected between 2008 and 2013, each representing a single contact with a prospective customer. The dataset includes more than 20 variables encompassing client personal attributes, financial information, call-related features, and macroeconomic indicators. The prediction target is a binary variable y indicating whether the client subscribed to a term deposit (yes or no). The dataset is described in Table 5.9.

Several notable characteristics define this dataset. First, the class distribution is highly imbalanced, with only around 11% of the observations corresponding to successful subscriptions. This imbalance poses challenges for model training and evaluation. Second, the dataset features a variety of variable types, including categorical variables (e.g., job, marital status, contact method), numerical variables (e.g., account balance, call duration). Some features also exhibit temporal or cyclical patterns (e.g., month of contact, day of the week), which may influence customer decisions. Overall, the dataset is of moderate dimensionality and rich in information, making it well suited for predictive modeling and interpretability analysis. The dataset samples are displayed in Table 5.10.

5.3.3 Marketing Forecasting Models and Methodological Frameworks

To address the challenge of identifying potential long-term deposit subscribers in bank telemarketing, this section introduces the forecasting models and methodological framework adopted in this study. Building upon the data-driven motivation discussed earlier, the proposed approach aims not only to achieve high predictive accuracy, but also to ensure that the model outputs are interpretable and trustworthy for practical decision-making. The section is structured as follows: begin by presenting

Table 5.9 Variables table

Variable name	Role	Type	Missing values
Age	Feature	Continuous	No
Job	Feature	Categorical	No
Marital	Feature	Categorical	No
Education	Feature	Categorical	No
Default	Feature	Binary	No
Balance	Feature	Continuous	No
Housing	Feature	Binary	No
Loan	Feature	Binary	No
Contact	Feature	Categorical	No
Day	Feature	Integer	No
Month	Feature	Categorical	No
Duration	Feature	Continuous	No
Campaign	Feature	Integer	No
pdays	Feature	Integer	No
Previous	Feature	Integer	No
poutcome	Feature	Categorical	No
y	Target	Binary	No

the modeling background and formal problem definition; this is followed by an explanation of model selection and its theoretical foundations. Then describe the overall methodological process and conclude with a flowchart summarizing the complete framework. Together, these components constitute the technical core of the proposed prediction and interpretability system.

5.3.3.1 Marketing Modeling Background

The task of forecasting customer subscription to long-term deposits is formulated as a binary classification problem. Given a set of input features $x \in \mathbb{R}^d$, the objective is to learn a predictive function $f : \mathbb{R}^d \rightarrow \{0, 1\}$ that estimates the likelihood of a customer subscribing to the financial product. In addition to achieving accurate predictions, it is essential to ensure that the model outputs are interpretable and reliable, particularly in the context of financial decision-making. To this end, the modeling framework is designed to serve dual purposes: (1) constructing a robust classifier with strong generalization ability, and (2) enabling localized interpretability and uncertainty assessment for individual predictions. This consideration motivates the integration of ensemble-based learning with post-hoc local regression techniques and confidence visualization tools, which collectively support the subsequent methodological design.

Table 5.10 Sample dataset

Age	Job	Marital	Education	Default	Balance	Housing	Loan	
30	Unemployed	Married	Primary	No	1787	No	No	
33	Services	Married	Secondary	No	4789	Yes	Yes	
35	Management	Single	Tertiary	No	1350	Yes	No	
30	Management	Married	Tertiary	No	1476	Yes	No	
59	Blue-Collar	Married	Secondary	No	0	Yes	No	
35	Management	Single	Tertiary	No	747	No	No	
36	Self-Employed	Married	Tertiary	No	307	Yes	No	
39	Technician	Married	Secondary	No	147	Yes	No	
41	Entrepreneur	Married	Tertiary	No	221	Yes	No	
43	Services	Married	Primary	No	−88	Yes	Yes	
39	Services	Married	Secondary	No	9374	Yes	No	
43	Admin.	Married	Secondary	No	264	No	No	
36	Technician	Married	Tertiary	No	1109	No	No	
20	Student	Single	Secondary	No	502	No	No	
Contact	Day	Month	Duration	Campaign	pdays	Previous	poutcome	Y
Cellular	19	Oct	79	1	−1	0	Unknown	No
Cellular	11	May	220	1	339	4	Failure	No
Cellular	16	Apr	185	1	330	1	Failure	No
Unknown	3	Jun	199	4	−1	0	Unknown	No
Unknown	5	May	226	1	−1	0	Unknown	No
Cellular	23	Feb	141	2	176	3	Failure	No
Cellular	14	May	341	1	330	2	Other	No
Cellular	6	May	151	2	−1	0	Unknown	No
Unknown	14	May	57	2	−1	0	Unknown	No
Cellular	17	Apr	313	1	147	2	Failure	No
Unknown	20	May	273	1	−1	0	Unknown	No
Cellular	17	Apr	113	2	−1	0	Unknown	No
Cellular	13	Aug	328	2	−1	0	Unknown	No
Cellular	30	Apr	261	1	−1	0	Unknown	Yes

5.3.3.2 Marketing Model Selection and Basic Principles

XGBoost is selected as the predictive model in this study due to its strong performance in structured data scenarios and their proven effectiveness in handling complex, non-linear classification tasks. Compared to traditional statistical models such as logistic regression and decision trees, XGBoost provides enhanced flexibility through its ensemble architecture and robust regularization mechanisms. These characteristics are particularly suited for high-dimensional, heterogeneous data with potential class imbalance, as is typical in financial marketing applications.

XGBoost belongs to the family of gradient boosting decision tree algorithms and operates by sequentially fitting additive regression trees to minimize a specified loss function. At each iteration, a new tree is trained to predict the gradient of the loss function with respect to the current model's predictions, thereby improving overall accuracy. The algorithm incorporates both first- and second-order derivatives to enhance optimization efficiency and employs regularization terms in its objective function to control model complexity and reduce overfitting. The reasons and rationale for selecting XGBoost as the overall prediction model are summarized below:

High Predictive Performance on Structured Data. XGBoost has consistently demonstrated strong classification performance on structured, tabular datasets. Its ensemble architecture enables it to capture complex nonlinear relationships among input features, which is especially useful in financial decision-making contexts.

Robustness to Imbalanced and Noisy Data. Financial marketing data often exhibit class imbalance. XGBoost supports class weighting and loss customization, enabling it to handle skewed distributions and maintain robustness in the presence of noise.

Regularization for Overfitting Control. Unlike many standard boosting algorithms, XGBoost incorporates both L_1 and L_2 regularization terms into its objective function, effectively controlling model complexity and reducing the risk of overfitting.

Efficient Training and Scalability. XGBoost is optimized for speed and memory efficiency through advanced features such as histogram-based splitting, parallel computation, and out-of-core training, making it suitable for large-scale datasets.

Compatibility with Post-hoc Interpretability Tools. Despite its black-box nature, XGBoost can be effectively combined with model-agnostic interpretability techniques, such as local surrogate models, SHAP values, and partial dependence plots, thereby enabling explanation of individual predictions.

5.3.3.3 Marketing Overall Methodological Process

Combining the realistic background of the problem and the characteristics of the test dataset, the methodological process can be divided into the following stages:

(1) Model Training with XGBoost

The data is used to train an XGBoost classifier, selected for its high predictive accuracy and robustness. The model is fitted to the binary target variable indicating whether a customer subscribed to a long-term deposit. Hyperparameters are tuned using cross-validation to optimize performance on the minority (positive) class, addressing the class imbalance inherent in the dataset.

(2) Selection of Representative Instances for Interpretation

To analyze model behavior at the individual level, a set of representative instances is selected from the test set. These instances are chosen based on proximity to the classification threshold or coverage of diverse feature profiles. Each instance serves as a focal point for localized interpretability analysis.

(3) Local Neighborhood Generation and Surrogate Modeling

Around each selected instance, a local neighborhood is constructed by generating perturbed samples. These samples are created by slightly varying the original feature values within a controlled range. A local surrogate model, typically a linear or low-complexity regression, is then fitted to approximate the behavior of the original XGBoost model in the vicinity of the instance.

(4) Confidence Region Construction and Visualization

To assess the stability and reliability of the local surrogate model, confidence ellipsoids, confidence bands and CICE index are constructed based on the estimated regression parameters and their covariance structure. These visualizations provide an intuitive understanding of local uncertainty, enabling users to judge the trustworthiness of the explanation provided by the model.

(5) Integration of Predictive and Interpretive Insights

The outputs of the predictive model and the local interpretability framework are integrated to support decision-making. Accurate predictions are complemented by interpretable explanations and quantified uncertainty, allowing for more informed and transparent customer targeting strategies in bank marketing campaigns.

5.3.3.4 Marketing Method Flowchart

The overall methodological process is outlined in Fig. 5.7. The procedure begins with a bank telemarketing dataset containing 4,522 contact records. Following data segmentation and preprocessing, the dataset is split into training and testing subsets to facilitate model development.

XGBoost is utilized as the core predictive model to estimate the likelihood of customer subscription to long-term deposit products. Based on the model's output, a set of typical instances is selected to serve as focal points for local interpretability analysis. For each selected instance, a local neighborhood is constructed by perturbing feature values within a defined range ($\pm 10\%$), generating 500 synthetic samples per point.

Two types of surrogate models are fitted to the neighborhood data: a Tukey robust linear regression model and a local nonlinear regression model. Parameter estimation and covariance matrix computation are performed for both models to support uncertainty quantification. Confidence ellipsoids and confidence bands are subsequently derived to visualize the stability of the local approximations.

A comparative analysis of the two local modeling approaches is conducted to evaluate their consistency, interpretability, and robustness in explaining the behavior of the original predictive model.

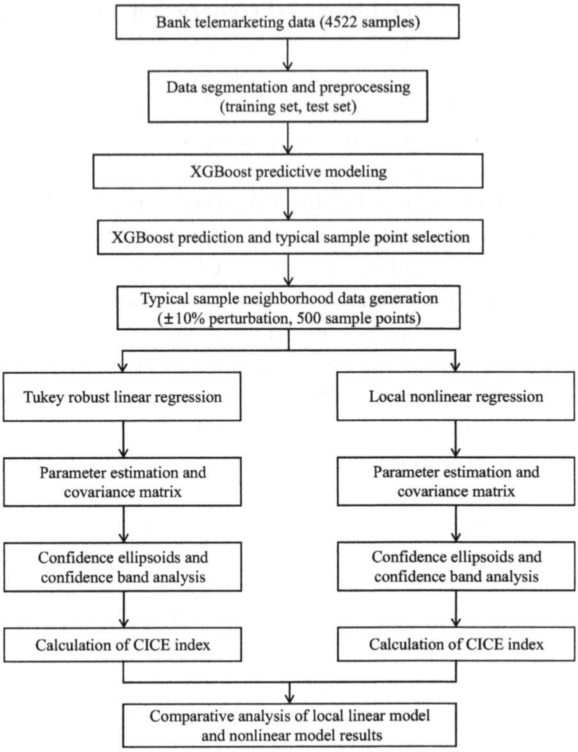

Fig. 5.7 Methodological framework diagram

5.3.4 Marketing XGBoost Training for Local Interpretation

After binary encoding of the target variable, nine major categorical features were transformed using smoothed target encoding to preserve predictive signals. The dataset was split into training and test subsets, and the core hyperparameters of the XGBoost classifier were tuned using randomized search with five-fold cross-validation. The parameter space included *max_depth*, *min_child_weight*, *colsample_bytree*, *subsample*, and *scale_pos_weight*, with 30 candidate configurations evaluated. The final model was trained using early stopping and subjected to evaluation and interpretability analysis on the held-out test set.

On the test set, the XGBoost model achieved an Area Under the Curve (AUC) of 0.889, as shown in Fig. 5.8, indicating strong global discriminatory capability. Using the automatically selected decision threshold of 0.497, the model yielded an accuracy of 88% and an F1 score of 0.566. The precision was 0.500, meaning that half of the instances predicted as subscribers were indeed true subscribers. The recall reached 0.651, reflecting the model's ability to identify 65.1% of all true subscribers. Considering the imbalanced nature of the data and the business importance of correctly identifying positive cases, these values of precision and recall are considered acceptable and practically useful.

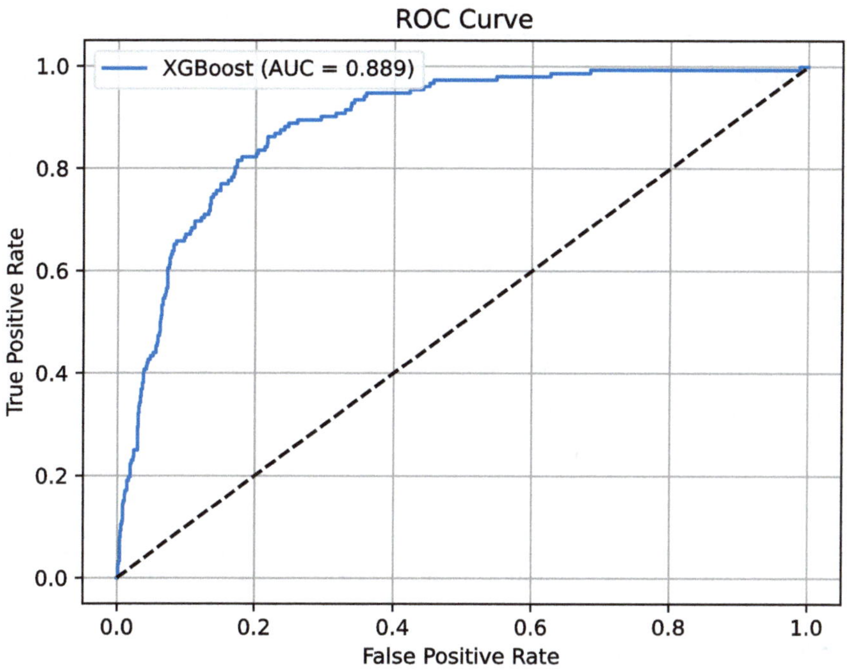

Fig. 5.8 Receiver Operating Characteristic (ROC) curve

According to feature importance analysis, as shown in Fig. 5.9, the most influential variables in the model include *balance*, *duration*, *age*, *day*, and *month*. These features were most frequently used as split nodes in the decision trees, indicating their strong contribution to the classification results. In particular, *balance* and *duration* reflect a customer's financial status and level of engagement, which may directly influence the likelihood of subscribing to financial products. This suggests that the model leverages both static attributes and interaction-based behaviors to enhance prediction accuracy and robustness.

To enable local interpretability, the test instance with the predicted probability closest to the optimal classification threshold (0.497) was selected. This sample, as shown in Table 5.11, indexed as 448, was chosen based on the smallest absolute difference between predicted probability and threshold value. The instance lies near the decision boundary where the model exhibits the highest uncertainty, making it an informative candidate for interpretability analysis. Notably, the instance has a true positive label but was predicted as negative, indicating a potentially critical misclassification.

To explore the model's behavior around this instance, the feature vector was first encoded and standardized in the same way as the training set. A local neighborhood of 500 samples was generated by adding Gaussian noise drawn from $N(0, 0.1^2)$ to each standardized feature. This perturbation process preserves the local structure

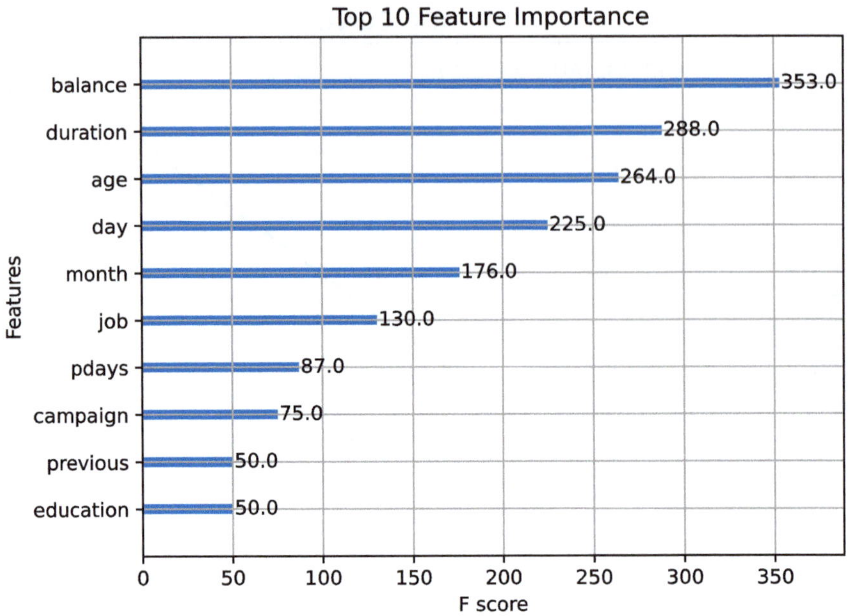

Fig. 5.9 Top 10 feature importance

while introducing controlled variability, allowing the construction of an interpretable neighborhood suitable for surrogate modeling.

Selecting a borderline instance enables the investigation of how the model responds in its most uncertain regions. The perturbation-based neighborhood facilitates local surrogate modeling, helping quantify sensitivity and interpretability at the individual prediction level. This process not only improves the model's transparency but also prepares the ground for subsequent linear and nonlinear approximation with confidence region analysis.

5.3.5 Marketing Local Linear Surrogate Modeling with Tukey Regression

To approximate the behavior of the XGBoost classifier in the neighborhood of the selected instance, a robust linear regression model was fitted using Tukey's Biweight M-estimator. The model was trained on 500 perturbed samples generated around the focal point, with standardized features and an intercept term included. The iterative reweighted least squares (IRLS) algorithm was applied to estimate coefficients, mitigating the influence of outliers in the local neighborhood.

Table 5.11 Feature values of the selected instance with index 448

Feature	Description	Value	Unit/type
Age	Client age	33.000	Years
Job	Job type (encoded)	0.134	Categorical
Marital	Marital status (encoded)	0.103	Categorical
Education	Education level (encoded)	0.146	Categorical
Default	Has credit in default (encoded)	0.116	Binary
Balance	Average yearly account balance	79.000	Euro
Housing	Has housing loan (encoded)	0.085	Binary
Loan	Has personal loan (encoded)	0.125	Binary
Contact	Contact communication type	0.144	Categorical
Day	Last contact day	22.000	Day (1–31)
Month	Last contact month (encoded)	0.410	Categorical
Duration	Duration of last contact	249.000	Seconds
Campaign	Number of contacts in this campaign	2.000	Count
pdays	Days since last contact	335.000	Days
Previous	Previous contacts	2.000	Count
poutcome	Outcome of previous campaign	0.125	Categorical

5.3.5.1 Marketing Estimated Coefficients of the Local Linear Model

The estimated coefficients from the Tukey robust linear regression model indicate that several features play a statistically significant role in shaping the local behavior of the predictive model, as shown in Table 5.12. Notably, variables such as *contact*, *duration*, *loan*, *pdays*, and *previous* exhibit significance at the 5% level. Their positive coefficients suggest that higher values are associated with an increased probability of successful prediction, which aligns with domain expectations in telemarketing contexts.

Table 5.12 Coefficient estimates of the Tukey robust linear model

Feature	Coef.	Std. Err	z	$P > \|z\|$	[0.025, 0.975]
Const	−0.6472	0.357	−1.813	0.070	[−1.347, 0.053]
Age	0.0869	0.057	1.520	0.128	[−0.025, 0.199]
Job	0.1129	0.059	1.923	0.054	[−0.002, 0.228]
Marital	−0.0306	0.053	−0.580	0.562	[−0.134, 0.073]
Education	−0.0651	0.058	−1.128	0.259	[−0.178, 0.048]
Default	−0.1122	0.058	−1.928	0.054	[−0.226, 0.002]
Balance	0.0210	0.058	0.365	0.715	[−0.092, 0.134]
Housing	0.0456	0.057	0.793	0.428	[−0.067, 0.158]
Loan	0.3471	0.053	6.499	0.000	[0.242, 0.452]
Contact	0.6468	0.056	11.552	0.000	[0.537, 0.757]
Day	0.0194	0.056	0.344	0.731	[−0.091, 0.130]
Month	−0.0757	0.059	−1.293	0.196	[−0.190, 0.039]
Duration	0.3059	0.054	5.619	0.000	[0.199, 0.413]
Campaign	0.0951	0.055	1.723	0.085	[−0.013, 0.203]
pdays	0.2128	0.057	3.751	0.000	[0.102, 0.324]
Previous	0.1379	0.055	2.497	0.013	[0.030, 0.246]
poutcome	−0.0350	0.056	−0.628	0.530	[−0.144, 0.074]

Additional features, including *job*, *default*, and *campaign*, show marginal significance with p-values between 0.05 and 0.10, implying moderate explanatory power within the analyzed neighborhood. In contrast, variables such as *age*, *marital*, *education*, *balance*, *housing*, *month*, and *poutcome* are not statistically significant, with confidence intervals covering zero, suggesting limited local influence. The regression results provide a robust quantification of local feature contributions.

5.3.5.2 Marketing Predictive Performance of the Local Linear Model

To evaluate the performance of the local linear surrogate model based on Tukey robust regression, several metrics were computed, including the mean squared error (MSE), mean absolute error (MAE), and pseudo R^2. The obtained MSE of 0.0258 and MAE of 0.1117 indicate that the linear approximation closely matches the predicted probabilities of the original model within the perturbed neighborhood, with low levels of approximation error. While MSE is sensitive to larger deviations, MAE provides a more interpretable measure of average absolute deviation, and both confirm a reasonably accurate fit.

The pseudo R^2 value is 0.3216, suggesting that the surrogate model explains approximately 32.16% of the variance in the original model's outputs over the local sample set. This level of fit is considered moderate in the context of local interpretability for nonlinear models XGBoost, providing meaningful support for explanation and uncertainty quantification while leaving room for further enhancement via nonlinear modeling.

5.3.5.3 Marketing Confidence Ellipsoids for Linear Parameters

To quantify the uncertainty associated with parameter estimation, a confidence ellipsoid was constructed based on the covariance matrix derived from the Tukey robust regression model. The model-provided covariance matrix was obtained via the *cov_params()* method, and its generalized inverse was computed using the Moore–Penrose pseudoinverse to ensure numerical stability. The confidence ellipsoid is defined by the quadratic form:

$$(\boldsymbol{\theta} - \hat{\boldsymbol{\theta}})^T \mathbf{C}^{-1} (\boldsymbol{\theta} - \hat{\boldsymbol{\theta}}) \leq \frac{\chi^2_{1-\alpha}(m)}{N} \tag{5.3}$$

where $\hat{\boldsymbol{\theta}}$ denotes the estimated coefficients, $\mathbf{C}$ the covariance matrix, and $m = 17$ the number of parameters. At a 95% confidence level, the chi-square critical value is 27.5871. Therefore, the resulting ellipsoid is given by:

$$\left\{\boldsymbol{\theta} \in \mathbb{R}^{17} \parallel Q(\boldsymbol{\theta}) \leq 27.5871\right\} \tag{5.4}$$

To visualize the uncertainty structure of parameter estimates in the local linear model, six 3D parameter groups were selected, each composed of three variables from the full set. For each group, the corresponding 3×3 submatrix was extracted from the overall covariance matrix and decomposed into eigenvalues and eigenvectors. The semi-axis lengths of each ellipsoid were determined using the 95% chi-square quantile for three degrees of freedom ($\chi^2_{0.95}(3) = 7.8147$), allowing for the construction of confidence ellipsoids that represent the joint variability of parameters in each subspace.

Each ellipsoid was rendered by transforming a unit sphere by scaling and rotation according to the covariance structure. The estimated parameter vector $\hat{\boldsymbol{\theta}}$ was plotted at the center of each ellipsoid to visualize its relative position within the uncertainty region, as shown in Fig. 5.10.

Directional elongation reflects variable-wise estimation uncertainty. All six ellipsoids exhibit prominent elongation along one or more axes, suggesting variable-specific differences in local estimation uncertainty. For example, in the parameter triplet *(contact, day, month)*, the ellipsoid is strongly stretched along the *contact*

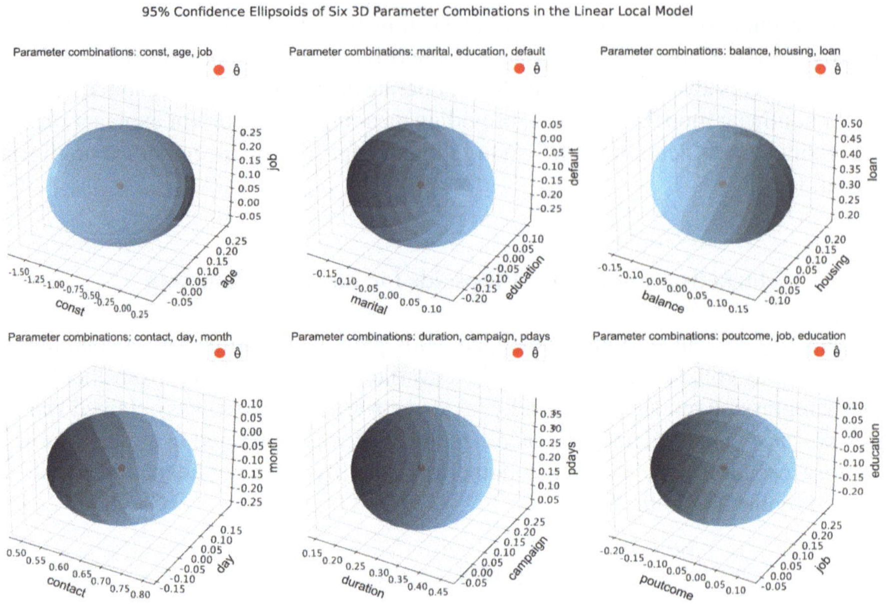

Fig. 5.10 95% confidence ellipsoids of six 3D parameter combinations in the linear local model

axis, indicating that the estimated coefficient of this feature varies more under local perturbations. This result implies that while *contact* plays a key role in prediction, its estimation may be less stable. In contrast, the dimensions corresponding to *day* and *month* are more compact, suggesting relatively robust estimation. The degree of axis elongation provides an intuitive indication of the model's local sensitivity to different features.

Ellipsoid skewness reveals interaction and collinearity between parameters. Beyond axis length, the overall orientation and skewness of the ellipsoids are shaped by the covariance structure among parameters. If two or more features are strongly correlated, the resulting ellipsoid tends to rotate or tilt along the directions of high covariance. For instance, the ellipsoid for the *(duration, campaign, pdays)* combination exhibits a diagonally stretched shape, reflecting the presence of multicollinearity or joint variability. Such interactions indicate that changes in one variable can implicitly affect others. This type of interdependence offers insights into the coupling structure of local interpretations and highlights directions where feature redundancy may occur.

Principal axes may align with prediction-sensitive directions. From an interpretability standpoint, the principal axes of the ellipsoids may align with directions that are most sensitive to prediction outcomes. If a feature exerts a strong influence on the predicted probability, and its associated uncertainty aligns with that direction, the interpretation may be both impactful and unstable. For example, the ellipsoid of *(loan, housing, balance)* is relatively compact and symmetric, indicating that although these features contribute to prediction, their estimates remain stable under local perturbations. In contrast, features with elongated and prediction-aligned uncertainty directions may require cautious interpretation. These geometric characteristics serve as an indirect indicator of predictive sensitivity embedded in the parameter space.

5.3.5.4 Marketing Functional Confidence Band Visualization

To assess the uncertainty associated with model predictions, a function confidence band was constructed for the local linear model, as shown in Fig. 5.11. The procedure is based on evaluating the quadratic form $x^\top \hat{C} x$, where x denotes each neighborhood sample and $\hat{C}$ is the estimated covariance matrix of the regression parameters. The 95% confidence interval for each prediction $f(x) = x^\top \hat{\theta}$ was then determined by scaling this variance by the chi-square critical value with 17 degrees of freedom ($\chi^2_{0.95}(17) \approx 27.5871$). The predicted curve and its associated confidence region were then visualized. The function confidence band captures how prediction uncertainty varies across the local neighborhood. Key observations include:

Fluctuating band widths. The confidence band remains narrow for most sample indices, suggesting stable local predictions. However, certain regions show wider intervals, indicating increased uncertainty and potential model fragility in those areas.

Smooth predicted function. The predicted values are distributed smoothly without abrupt changes, suggesting that the robust regression (using Tukey's biweight)

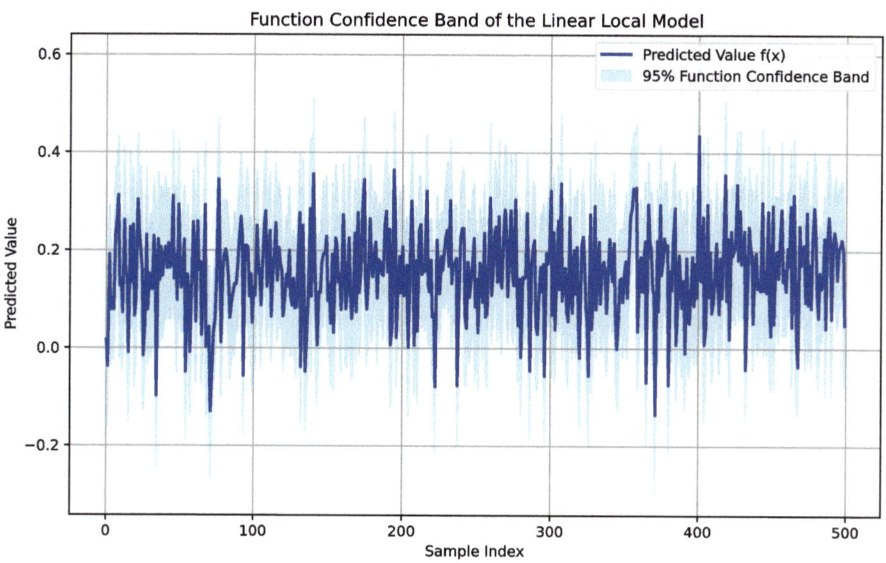

Fig. 5.11 Function confidence band of the linear local model

has effectively mitigated the influence of outliers. This aligns with prior evaluation metrics (e.g., $MSE = 0.0258$, $R^2 \approx 0.322$), which indicate moderate fit quality.

Interpretation near decision thresholds. The band plays a critical role when predictions approach classification thresholds (e.g., 0.497). If the confidence band includes this threshold, the decision becomes less reliable; otherwise, it may be considered stable. This helps provide actionable uncertainty-aware insights for marketing decision-makers.

5.3.5.5 Marketing CICE Calculation Analysis

To address the challenge of the XGBoost model being "predictive but not interpretable" in bank marketing, this section selects representative customer samples and generates perturbed neighborhoods around each to fit local linear regression models, approximating the decision boundary of the original model. By using parameter covariance matrices to construct confidence ellipsoids and bands, we effectively quantify and visualize the prediction uncertainty, enhancing model transparency. The computed results show that the local linear surrogate model achieves high interpretability scores, with minimal deviation between the confidence interval center and the predicted values, indicating strong consistency and stability in approximating the black-box model in local regions. We set the weight of the center deviation and confidence width to be 0.3 and 0.7, respectively, and set the weights of interpretability index and accuracy index to 0.5 respectively. With a CICE value of **0.8029**, this method offers a structurally simple and clearly interpretable local surrogate, capturing the key behavior of the black-box model, making it a useful tool for informed marketing strategy decisions under resource constraints.

Although the local linear model simplifies the representation of the complex decision boundary, it still demonstrates reasonable predictive performance in this experiment. The composite prediction accuracy (PA $\approx$ 0.729) indicates that the local model fits the perturbed neighborhood samples reasonably well, with errors kept within acceptable limits (MSE = 0.0258). Combining this with the interpretability score, the resulting CICE value of **0.8029** suggests that the model achieves an effective balance between interpretability and prediction accuracy. This makes it a promising approach for applications in the financial sector, where both business transparency and risk control are crucial, such as in precise customer identification and resource allocation.

5.3.6 *Marketing Local Nonlinear Approximation and Prediction*

Although the Tukey-based robust linear surrogate provided a moderately acceptable fit within the local neighborhood, its relatively low pseudo-R^2 and limited explanatory power suggested the potential presence of nonlinear structures—particularly near classification thresholds where decision boundaries may exhibit complex curvature. To improve the local approximation of XGBoost behavior, a nonlinear surrogate model was developed based on polynomial feature expansion.

The process begins by identifying the test instance whose predicted probability is closest to the classification threshold. A Gaussian perturbation is applied to its standardized feature vector to generate a set of 500 synthetic neighbors. Each neighbor is then passed through the XGBoost model to obtain the corresponding predicted probabilities.

The features of these neighbors are then expanded into second-degree polynomial terms, focusing on interaction features and truncating the total number of resulting terms to a maximum of 30. A weighted least squares (WLS) regression model is fitted to this nonlinear feature space, capturing higher-order effects and interactions. Finally, the fitted model is used to predict the probability for the original focal instance, forming the basis for subsequent explanation and uncertainty estimation. To ensure a fair and consistent comparison, the nonlinear surrogate model was constructed using the same focal instance and the same set of perturbed neighborhood samples as in the linear approximation. This design enables both models to operate on identical local input data, facilitating a reliable comparison of their fitting accuracy and interpretability.

5.3.6.1 Marketing Estimated Coefficients of Local Nonlinear Model

The nonlinear local model based on polynomial regression exhibits improved explanatory capacity over the linear case. Key observations from Table 5.13 include:

Significant Main Effects. Variables such as *education*, *loan*, *contact*, and *pdays* show statistical significance at the 5% level. The positive signs of *contact* and *pdays* indicate that frequent prior interactions are associated with a higher likelihood of subscription. In contrast, the negative coefficient for *education* suggests more educated clients may be less inclined to accept the offer.

Interaction Terms. Among interaction terms, *age* × *education* and *age* × *campaign* are statistically significant, revealing non-additive relationships. This reinforces the rationale for employing a nonlinear model that can capture feature interactions effectively.

5.3.6.2 Marketing Predictive Performance of Local Nonlinear Model

The weighted least squares (WLS) model constructed using a second-degree polynomial feature expansion demonstrates an improved local approximation of the XGBoost prediction function. The coefficient of determination (R^2) reaches 0.457, with an adjusted R^2 of 0.422, representing a clear improvement over the previous

Table 5.13 Coefficient estimates of the nonlinear local model

Feature	Coef.	Std. Err	t	$P > \|t\|$	[0.025, 0.975]
Const	0.7932	3.177	0.250	0.803	[−5.450, 7.037]
Age	2.2820	4.029	0.566	0.571	[−5.635, 10.199]
Job	0.3082	0.591	0.521	0.602	[−0.854, 1.470]
Marital	−0.2105	0.566	−0.372	0.710	[−1.323, 0.902]
Education	−1.2837	0.531	−2.417	0.016	[−2.327, −0.240]
Default	−0.9785	0.554	−1.767	0.078	[−2.067, 0.110]
Balance	0.5960	0.529	1.128	0.260	[−0.443, 1.635]
Housing	0.3230	0.582	0.555	0.579	[−0.820, 1.466]
Loan	1.2630	0.513	2.462	0.014	[0.255, 2.271]
Contact	1.1291	0.581	1.942	0.053	[−0.013, 2.272]
Day	−0.1067	0.567	−0.188	0.851	[−1.221, 1.008]
Month	−0.8192	0.543	−1.509	0.132	[−1.886, 0.248]
Duration	0.6273	0.541	1.160	0.246	[−0.435, 1.690]
Campaign	−1.0073	0.558	−1.806	0.072	[−2.104, 0.089]
pdays	1.2694	0.558	2.273	0.023	[0.172, 2.367]
Previous	0.2772	0.571	0.486	0.628	[−0.845, 1.399]
poutcome	0.0450	0.069	0.657	0.511	[−0.090, 0.180]
Age job	0.2544	0.768	0.331	0.741	[−1.255, 1.763]
Age marital	−0.3079	0.729	−0.422	0.673	[−1.741, 1.125]
Age education	−1.3835	0.684	−2.022	0.044	[−2.728, −0.039]
Age default	−1.2504	0.720	−1.736	0.083	[−2.665, 0.165]
Age balance	0.6447	0.690	0.934	0.351	[−0.712, 2.001]
Age housing	0.4793	0.761	0.630	0.529	[−1.017, 1.975]
Age loan	0.8917	0.661	1.348	0.178	[−0.408, 2.191]
Age contact	0.4402	0.759	0.446	0.656	[−1.153, 2.034]
Age day	−0.3159	0.725	−0.436	0.663	[−1.741, 1.109]
Age month	−0.9658	0.700	−1.380	0.168	[−2.341, 0.409]
Age duration	0.2446	0.708	0.345	0.730	[−1.145, 1.635]
Age campaign	−1.5889	0.719	−2.209	0.028	[−3.000, −0.178]
Age pdays	1.0630	0.701	1.516	0.130	[−0.313, 2.439]
Age previous	−0.1914	0.716	−0.267	0.790	[−1.596, 1.213]

linear model ($R^2 \approx 0.322$). This suggests that nonlinear terms and interactions effectively capture local structural patterns in the perturbed neighborhood.

The F-statistic is 13.13 with a corresponding p-value of 3.98×10^{-45}, indicating that the regression model is statistically significant overall. At least one of the polynomial features contributes meaningfully to the prediction of the XGBoost outputs within the local region. The Durbin–Watson statistic is 1.924, indicating no strong evidence of residual autocorrelation. However, the Jarque–Bera test yields a p-value far below 0.05, suggesting some deviation from normality in the residual distribution—likely due to asymmetric distribution in perturbations or the influence of higher-order terms. Nonetheless, the nonlinear WLS model offers a more expressive and accurate local approximation, improving interpretability of the XGBoost decision surface.

5.3.6.3 Marketing Confidence Ellipsoids for Nonlinear Model

To quantify parameter uncertainty in the nonlinear local approximation, six three-dimensional parameter combinations were selected to construct confidence ellipsoids, as shown in Fig. 5.12. Each combination includes both main effects and interaction terms derived from polynomial regression. For each triplet, the corresponding 3×3 submatrix was extracted from the covariance matrix of the estimated parameters. The ellipsoid shape was determined by eigenvalue decomposition of the submatrix, where the axes correspond to the square roots of the scaled eigenvalues using the chi-square threshold $\chi^2_{0.95}(3) = 7.815$. Each ellipsoid was rendered in 3D space by linearly transforming a unit sphere, and the estimated coefficient vector $\hat{\theta}$ was plotted at its center. This procedure mirrors the process used in the linear case, but incorporates the richer parameter space and structure induced by nonlinear feature interactions.

The confidence ellipsoids constructed in the nonlinear setting reveal several important patterns regarding the stability and interdependence of parameter estimates.

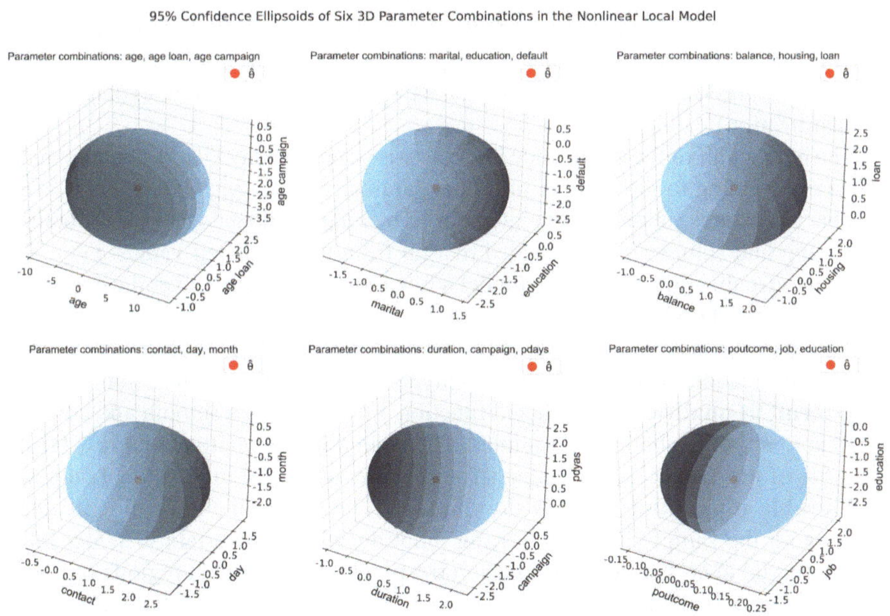

Fig. 5.12 95% confidence ellipsoids of six 3D parameter combinations in the nonlinear local model

Quantification of Joint Uncertainty. Ellipsoid volumes vary substantially across combinations. For instance, the ellipsoid corresponding to *(age, age loan, age campaign)* exhibits a relatively large volume, indicating high joint uncertainty associated with age-related interaction terms. In contrast, the ellipsoid of *(poutcome, job, education)* is more compact, suggesting stable estimates with minimal variance.

Covariance Structure among Marginal Variables. Ellipsoid orientation captures parameter correlations. Combinations such as *(duration, campaign, pdays)* display elongated and skewed shapes, reflecting multicollinearity or joint sensitivity. This suggests that these variables jointly drive variation in the prediction output and are not independently identifiable in the local space.

Sensitivity of Interaction Terms. Combinations that include interaction terms tend to exhibit greater uncertainty and directional distortion. For example, *age, campaign* and *age, education* show wide uncertainty regions, reinforcing the idea that while interaction terms enhance model flexibility, they may also introduce estimation instability.

Overall, the ellipsoids provide a geometric interpretation of local model uncertainty, which complements coefficient-based interpretation and supports robustness-aware explanation in high-stakes decision environments.

5.3.6.4 Marketing Functional Confidence Band for Nonlinear Model

To visualize the prediction uncertainty of the nonlinear local model, a function confidence band was constructed based on the fitted polynomial regression, as shown in Fig. 5.13. Using the estimated parameter vector $\hat{\boldsymbol{\theta}}$ and its covariance matrix $\hat{C}$, the standard error of each prediction was computed via the quadratic form $x^\top \hat{C}\, x$, where x is the feature vector of a neighborhood sample (after polynomial expansion and standardization).

The half-width of the confidence interval was scaled by the 95% chi-square quantile corresponding to the number of model parameters, i.e., $\chi^2_{0.95}(d)$, where d is the number of features (including the intercept). The final confidence band was obtained by adding and subtracting this scaled margin from the predicted values $f(x)$. The resulting visualization reflects the model's functional uncertainty over the entire perturbed neighborhood.

The confidence band constructed for the nonlinear local model provides an intuitive quantification of predictive uncertainty. Several key observations can be made:

Overall fit quality. The predicted function $f(x)$ exhibits a smooth and coherent trend across the neighborhood samples. The band remains relatively narrow for the majority of the instances, indicating strong local consistency in the model output.

Variable uncertainty across the domain. The width of the confidence band varies with the sample index, reflecting input-dependent uncertainty. Wider intervals observed at certain locations may correspond to sparser data regions or highly nonlinear local structures, where the model's predictions are less reliable.

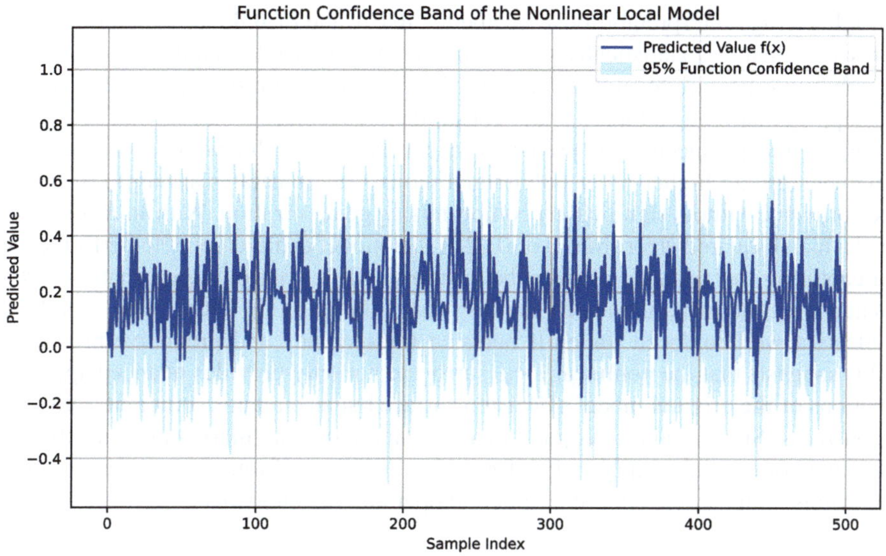

Fig. 5.13 Function confidence band of the nonlinear local model

Stability around decision thresholds. The confidence band offers valuable support for uncertainty-aware interpretation. In classification tasks, if the predicted value lies close to the decision threshold (e.g., 0.497) and the band includes this threshold, the prediction should be considered less stable. Conversely, when the band does not cross the threshold, the decision is more robust.

Improved resolution over the linear model. Compared to the confidence band from the linear approximation, the nonlinear band better adapts to local curvature in the prediction space, offering enhanced fidelity for interpretability and decision support.

5.3.6.5 Marketing CICE Calculation Analysis

To mitigate the practical limitations posed by the low interpretability of XGBoost in bank marketing applications, this section adopts a nonlinear local surrogate modeling approach. Perturbed neighborhoods are generated around a representative customer samples, and local polynomial regressions are fitted to approximate the decision surface of the original black-box model. By leveraging the covariance matrix of the model parameters, function-level confidence bands are constructed to quantify prediction uncertainty in a statistically principled manner. A local interpretability index is introduced, incorporating both the width of the confidence band and the deviation between its center and the predicted value, to evaluate the stability and fidelity of the surrogate explanations. In order to ensure the consistency with the linear local fitting results, we do not adjust the weights. In this case, the nonlinear local model achieves a high average interpretability score and a CICE value of **0.8878**, indicating that it provides clear and statistically grounded explanations while maintaining sufficient flexibility to capture localized decision behaviors, thus offering actionable insights to support marketing strategy under constrained resources.

From the perspective of predictive accuracy, the nonlinear local surrogate—augmented with higher-order polynomial terms—demonstrates strong local fitting performance without significantly compromising generalization. With an R^2 of 0.457 and a mean squared error (MSE) of 0.0211, the model exhibits reliable performance in capturing the variance of perturbed samples while controlling estimation error. When integrated with the interpretability score, the resulting CICE value of **0.8878** reflects a well-balanced trade-off between transparency and predictive precision. This suggests that the nonlinear local surrogate not only provides nuanced approximations of the original model's decision surface but also retains sufficient statistical robustness. Such a balance is particularly valuable in bank marketing contexts that demand both explainability and performance, enabling informed customer targeting decisions with greater scientific rigor and operational confidence.

5.3.7 Marketing Integrated Interpretation and Comparative Analysis

To provide a comprehensive interpretation of the selected instance near the decision boundary, both linear and nonlinear local surrogate models were developed and evaluated within a common perturbed neighborhood. This comparative framework enables a robust assessment of model behavior, stability, and feature influence at the individual level.

Model Comparison. The linear surrogate model based on Tukey's robust regression achieved a pseudo R^2 of 0.322, while the nonlinear model using second-degree polynomial regression yielded a higher adjusted R^2 of 0.422, as shown in Table 5.14. The nonlinear model also exhibited a more expressive feature structure, with several significant interaction terms (e.g., $age \times education$, $age \times campaign$) contributing to predictive power. Confidence bands constructed from each model further demonstrated that the nonlinear surrogate captured more nuanced local curvature and variation, especially in regions where the linear model's predictions appeared less stable.

Feature Influence and Stability. In both models, key features such as *contact*, *pdays*, and *loan* consistently appeared as influential. However, the nonlinear model revealed that their impact varies depending on interactions with other features, which was not observable in the linear setting. Confidence ellipsoids also showed higher joint uncertainty for interaction terms, suggesting greater sensitivity to local perturbations.

Predictive Certainty Assessment. The confidence bands provided visual and quantitative tools to assess the stability of the predicted probability for this instance.

Table 5.14 Comparison of linear and nonlinear local surrogate models

Metric	Linear model (Tukey RLM)	Nonlinear model (Polynomial WLS)
Pseudo R^2	0.322	0.457
Adjusted R^2	–	0.422
MSE	0.0258	0.0211
CICE	0.8029	0.8878
Confidence band width	Medium (mostly uniform)	Variable and adaptive
Significant interactions	None	*age* × *education*, *age* × *campaign*
Visual stability	Moderate	Higher (captures curvature)
Interpretability	High (linear coefficients)	Rich but complex (nonlinear terms)

In the nonlinear model, the predicted probability was found to be close to the classification threshold (0.497), with the confidence interval including the threshold. This suggests that although the model assigns a prediction, the result should be interpreted with caution due to estimation uncertainty.

CICE Index Assessment. The local nonlinear proxy model (polynomial regression based on weighted least squares) has a CICE value of **0.8878**, which is significantly higher than that of the linear model **(0.8029)**, indicating that it performs better in both model interpretation and prediction accuracy. In terms of accuracy, the nonlinear model was superior to the linear model (R^2 = 0.322, MSE = 0.0258) in Pseudo R^2 (0.457) and MSE (0.0211), indicating that it fitted the local disturbance samples more effectively and captured the nonlinear decision boundary of the black-box model. In terms of interpretation, although the structure of the nonlinear model is more complex, including higher-order terms and interaction terms (such as age × education and age × campaign), its explanatory structure is still relatively stable and intuitive by constructing visual confidence belts and evaluating function uncertainty. In particular, it is more expressive than linear models in presenting curvature changes. To sum up, the improvement of CICE shows that the nonlinear agent model can improve the forecasting ability while still maintaining a certain degree of interpretability, which is suitable for bank customer identification and strategy formulation tasks that require high accuracy and transparency in practical applications.

Conclusion. This case study demonstrates the practical utility of combining local surrogate modeling, uncertainty quantification, visual diagnostics and CICE index to interpret complex model predictions. The nonlinear local model provided a more faithful approximation of the original XGBoost decision surface, while the linear model offered simplicity and robustness. Together, they offer complementary views that enhance transparency and support trust in model-informed decision-making.

5.4 Stock Market Time Series Case

The stock market, as a typical complex dynamic system, exhibits significant non-linearity and temporal dependencies due to the influence of various factors such as macroeconomic conditions, industry developments, and market sentiment. This inherent complexity renders accurate stock price prediction a highly challenging yet valuable task. Consequently, modeling and forecasting stock market time series data not only contribute to a deeper understanding of market evolution but also play a crucial role in risk management and strategic decision-making.

Considering the characteristics of stock data, we construct an LSTM model for prediction. While the LSTM model exhibits strong performance in forecasting key stock price indicators such as closing, high, and low prices, its black-box nature poses challenges for practical interpretability in financial applications. To address this limitation, this section introduces a local interpretability analysis by selecting representative time points from historical stock sequences and constructing perturbed temporal neighborhoods around each. Local surrogate models—either linear or nonlinear—are trained to approximate the LSTM's decision function within these neighborhoods, and the corresponding parameter covariance structures are used to derive confidence bands and temporal ellipsoids. Finally, we use a new evaluation index—CICE. These visualization and calculation tools quantify predictive uncertainty and provide intuitive insights into the model's temporal behavior, thereby enhancing transparency and supporting more informed financial decision-making under uncertainty [5].

5.4.1 Stock Key Problem Statement

Stock market behavior can be regarded as a manifestation of complex, interdependent economic dynamics driven by a wide range of endogenous and exogenous factors. These include macroeconomic indicators, industry-specific developments, corporate performance, investor sentiment, and policy interventions. As supported by extensive empirical studies, stock price movements display characteristics such as strong temporal dependence, non-stationarity, and nonlinear interactions, making short- and medium-term forecasting inherently difficult.

High Volatility and Noise. Stock prices exhibit rapid fluctuations over short time scales, often in response to unpredictable market news or behavioral reactions. This introduces significant noise into the data and complicates signal extraction for reliable forecasting.

Multivariate Influences and Feature Interactions. A single price value at a given time is influenced by a constellation of trading-related features, such as opening price, trading volume, market capitalization, and limit events. Capturing the dynamic relationships among these features is crucial for accurate modeling.

Temporal Dependence and Structural Shifts. Historical price sequences often contain patterns that are informative for future trends, yet these patterns may shift abruptly due to structural breaks, policy changes, or macro shocks. Standard linear models struggle to adapt to such changing regimes.

Given the practical need for timely and accurate stock price prediction—particularly for closing price, high price, and low price—developing robust models that can effectively learn from multivariate time series data is of great importance for both academic research and real-world financial decision-making.

We use a multivariate time series forecasting model based on Long Short-Term Memory (LSTM) networks, using the top ten constituent stocks of the SSE 50 Index by index weight as our study objects. By leveraging twelve market trading features from 2020 to 2025, the model aims to predict three key stock price indicators: closing price, highest price, and lowest price. Building on this LSTM-based stock forecasting framework, we place particular emphasis on the interpretability of the model. In financial forecasting, the practical deployment of predictive models depends not only on the accuracy of predictions but also critically on their transparency and interpretability—especially in high-stakes investment contexts where a mere prediction is often insufficient for decision-making. The central challenge, therefore, lies in explaining why the model forecasts a specific price at a given time and assessing the stability and credibility of such explanations under local perturbations.

To address this, we introduce a local surrogate modeling approach atop the LSTM architecture. Representative time points are selected from the historical sequences, and perturbed temporal neighborhoods are constructed around each. Local linear or nonlinear models are then fitted to approximate the LSTM's predictive behavior within these neighborhoods. Moreover, parameter covariance structures are employed to construct confidence ellipsoids, bands, and evaluation index—CICE, allowing for visualization of local prediction stability. This framework not only enhances the interpretability of the model but also improves its trustworthiness in real-world financial applications.

The primary contribution of our work lies in providing time-point-specific explanations of model predictions by integrating local surrogate modeling with uncertainty visualization techniques. Unlike existing approaches that primarily focus on predictive accuracy, our method places greater emphasis on the stability and credibility of explanations—factors that are essential for building managerial confidence in model outputs during real-time decision-making. By fitting local surrogate models and visualizing confidence structures, this approach offers a concrete and actionable pathway for enhancing the transparency of deep learning models in financial forecasting, thereby supporting more interpretable investment analysis and strategy formulation.

5.4.2 Stock Data Description

The dataset used in this study is sourced from the CSMAR (China Stock Market & Accounting Research [1]) database and covers the top ten constituent stocks of the SSE 50 Index by index weight. Specifically, these stocks include Kweichow Moutai (600519), Ping An Insurance (601318), China Merchants Bank (600036), China Yangtze Power (600900), Industrial Bank (601166), Zijin Mining (601899), CITIC Securities (600030), Industrial and Commercial Bank of China (601398), Jiangsu Hengrui Pharmaceuticals (600276), and Yili Group (600887). The time span of the dataset ranges from April 4, 2020, to April 3, 2025, covering a five-year period with a total of 12,093 daily trading records.

Based on industry classification, the selected ten stocks span a diverse set of key sectors in the Chinese A-share market. The **financial sector** includes China Merchants Bank, Industrial Bank, and ICBC in the banking category, Ping An Insurance in insurance, and CITIC Securities in brokerage. The **consumer sector** is represented by Kweichow Moutai and Yili Group in food and beverage, as well as Jiangsu Hengrui Pharmaceuticals in healthcare. The **energy and utilities sector** includes China Yangtze Power, a leading player in electric power generation. Lastly, the **basic materials sector** is represented by Zijin Mining, a major company in the metals and mining industry.

These stocks hold significant weight in the SSE 50 Index and exhibit strong market influence and representativeness. As such, they provide a comprehensive reflection of the core dynamics of China's A-share market, making them highly suitable for subsequent time series forecasting and modeling efforts.

In the dataset employed in this study, each record corresponds to a specific constituent stock on a particular trading day, forming a two-dimensional index structure defined by "time-stock". This structure effectively captures the temporal evolution of individual stocks while also preserving cross-sectional differences among them.

Each record comprises 12 feature variables that span multiple dimensions, including price levels, trading activity, market capitalization, and price limit status. Specifically, price-related variables consist of the opening price (Opnprc), highest price (Hiprc), lowest price (Loprc), and closing price (Clsprc). Trading activity is represented by the number of shares traded (Dnshrtrd) and the trading value (Dnvaltrd). Market capitalization features include free-float market value (Dsmvosd) and total market value (Dsmvtll), capturing the scale of circulating shares and the company's overall valuation, respectively. Price fluctuation and limit status are described by the daily return ratio (ChangeRatio), the down-limit price (LimitDown), the up-limit price (LimitUp), and the limit status indicator (LimitStatus). The dataset is described in Table 5.15.

This dataset exhibits several notable characteristics: (1) Since the sample covers a five-year trading period and the selected stocks hold significant weights in the market, the data is fairly representative. However, this five-year span not only encompasses various market conditions and economic cycles but may also include structural shifts and non-stationary phenomena. (2) The features differ in units, volatility ranges,

Table 5.15 Variable description of stock market time series dataset

Variable name	Role	Type	Missing values
Opnprc	Feature	Continuous	No
Hiprc	Feature/Target	Continuous	No
Loprc	Feature/Target	Continuous	No
Clsprc	Feature/Target	Continuous	No
Dnshrtrd	Feature	Continuous	No
Dnvaltrd	Feature	Continuous	No
Dsmvosd	Feature	Continuous	No
Dsmvtll	Feature	Continuous	No
ChangeRatio	Feature	Continuous	No
LimitDown	Feature	Continuous	No
LimitUp	Feature	Continuous	No
LimitStatus	Feature	Categorical	No

and distribution patterns. For example, the number of shares traded and trading value may display considerable variability across different stocks and time periods, whereas market capitalization indicators generally show relatively stable growth or fluctuation trends. (3) With the exception of "LimitStatus", which is a categorical variable, all other features are continuous. This diversity in data types necessitates appropriate preprocessing—such as standardization or normalization—for features with different scales in the subsequent modeling process.

Table 5.16 presents a part of data for stock code 600519 over the period from June 2020 to April 2025, capturing key metrics on stock prices, trading volumes, and other related financial indicators at specific intervals (almost every four months). The dataset includes 12 variables, which are grouped into two sections: stock price indicators and trading data. Opnprc (Opening Price), Hiprc (Highest Price), Loprc (Lowest Price), and Clsprc (Closing Price) provide insights into the price fluctuations of the stock over the selected period. The highest recorded price (Hiprc) peaked at 2,248.95 on June 1, 2021, reflecting a significant uptrend in stock value. The lowest price during the observed period was 1,359.00 on November 1, 2022, indicating a notable dip during the market slowdown in 2022. The Opnprc and Clsprc values show trends aligned with the highest and lowest prices, but they vary slightly depending on the day-to-day market movements. The trading volume, represented by Dnshrtrd (number of shares traded) and Dnvaltrd (total trade value), demonstrates considerable variation over the period, reflecting market sentiment and stock liquidity. The Dnshrtrd has consistently fluctuated, with a sharp increase on September 1, 2021, at 7,731,610, indicating heightened trading activity during that period. In contrast, the Dnvaltrd—which measures the monetary value of trades—has seen an increase in line with share prices, peaking on November 1, 2022, at approximately 16.86 billion. It should be noted that this stock is also the primary subject of our forthcoming in-depth analysis.

Table 5.16 Stock data for stock code 600519: price and trading information from 2020 to 2025

Trddt	Opnprc	Hiprc	Loprc	Clsprc	Dnshrtrd	Dnvaltrd
2020/6/1	1,381.00	1,420.10	1,381.00	1,419.50	3,547,976.00	4,986,178,912.00
2020/9/1	1,792.00	1,802.00	1,775.01	1,801.98	3,163,992.00	5,661,594,315.00
2021/2/1	2,130.00	2,160.00	2,095.00	2,109.32	2,934,059.00	6,219,639,111.00
2021/6/1	2,222.00	2,248.95	2,206.00	2,240.95	3,061,634.00	6,832,594,466.00
2021/9/1	1,559.00	1,634.99	1,531.10	1,622.01	7,731,610.00	12,370,243,405.00
2022/3/1	1,802.00	1,863.57	1,802.00	1,858.48	4,737,930.00	8,721,229,267.00
2022/7/1	2,055.00	2,067.77	2,015.21	2,029.05	2,159,645.00	4,385,555,299.00
2022/11/1	1,359.00	1,469.40	1,359.00	1,462.01	11,876,648.00	16,859,775,304.00
2023/3/1	1,813.00	1,848.00	1,803.23	1,838.53	2,445,810.00	4,475,592,195.00
2023/6/1	1,618.00	1,674.99	1,618.00	1,635.92	2,864,525.00	4,708,722,933.00
2023/11/1	1,850.00	1,850.00	1,773.79	1,780.99	9,314,308.00	16,747,292,689.00
2024/3/1	1,690.00	1,694.97	1,676.68	1,685.06	2,686,796.00	4,526,845,398.00
2024/7/1	1,468.88	1,472.99	1,428.00	1,440.38	3,216,138.00	4,636,689,118.00
2024/11/1	1,521.00	1,545.13	1,520.80	1,533.75	3,228,392.00	4,954,302,212.00
2025/4/1	1,564.00	1,569.89	1,552.21	1,556.02	1,855,644.00	2,893,739,955.00
Trddt	Dsmvosd	Dsmvtll	ChangeRatio	LimitDown	LimitUp	LimitStatus
2020/6/1	1,783,172,777.10	1,783,172,777.10	0.04	1,229.94	1,503.26	0
2020/9/1	2,263,643,311.64	2,263,643,311.64	0.01	1,607.85	1,965.15	0
2021/2/1	2,649,723,143.50	2,649,723,143.50	0.00	1,904.56	2,327.80	0
2021/6/1	2,815,076,459.91	2,815,076,459.91	0.01	1,996.20	2,439.80	0
2021/9/1	2,037,565,393.58	2,037,565,393.58	0.04	1,402.20	1,713.80	0
2022/3/1	2,334,618,487.34	2,334,618,487.34	0.04	1,611.36	1,969.44	0
2022/7/1	2,548,888,146.09	2,548,888,146.09	-0.01	1,840.50	2,249.50	0
2022/11/1	1,836,573,745.58	1,836,573,745.58	0.08	1,215.00	1,485.00	0
2023/3/1	2,309,557,341.23	2,309,557,341.23	0.01	1,632.37	1,995.11	0
2023/6/1	2,055,039,104.98	2,055,039,104.98	0.00	1,466.01	1,791.79	0
2023/11/1	2,237,275,719.82	2,237,275,719.82	0.06	1,516.12	1,853.04	0
2024/3/1	2,116,768,664.87	2,116,768,664.87	-0.01	1,525.50	1,864.50	0
2024/7/1	1,809,402,187.16	1,809,402,187.16	-0.02	1,320.65	1,614.13	0
2024/11/1	1,926,693,375.75	1,926,693,375.75	0.00	1,375.01	1,680.57	0
2025/4/1	1,954,668,900.76	1,954,668,900.76	0.00	1,404.90	1,717.10	0

Note All amounts are denominated in CNY

Figures 5.14, 5.15, and 5.16 show the trends of Clsprc (Closing Price), Hiprc (Highest Price), and Loprc (Lowest Price) from 2020 to 2025 for nine selected stocks (excluding stock code 600519). Across the three price indicators, the majority of the stocks exhibit moderate to high volatility over the five-year period. While some stocks show a sustained upward or downward trend, others display periodic fluctuations, reflecting sensitivity to market cycles or sector-specific dynamics.

Stocks such as 600030 (CITIC Securities) and 601166 (Industrial Bank) demonstrate relatively stable trends across Hiprc, Loprc, and Clsprc, with gradual upward or horizontal movement, indicating stable market expectations and robust fundamentals. In contrast, stocks like 600276 (Jiangsu Hengrui Medicine) and 600887 (Yili Group) exhibit noticeable mid-period peaks (2022–2023), followed by slight corrections, potentially due to post-pandemic valuation normalization or sector rotation.

Fig. 5.14 Clsprc trend for stocks over the past 5 years excluding stock 600519

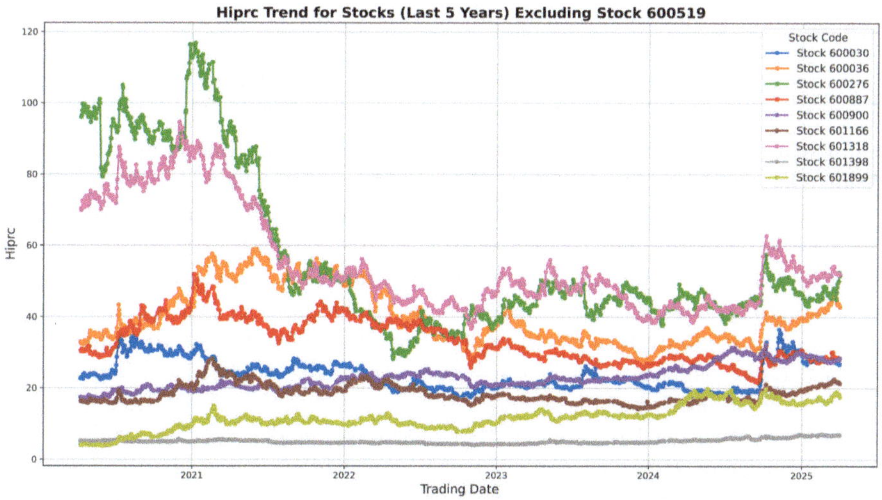

Fig. 5.15 Hiprc trend for stocks over the past 5 years excluding stock 600519

As the central reference for market consensus at the end of each trading cycle, the trend in Clsprc reveals that no stock experienced uninterrupted growth or decline throughout the observed period (As shown in Fig. 5.14). Notably, 600276 and 601318 show post-peak declines, suggesting either overvaluation followed by correction or deteriorating fundamentals. Meanwhile, 600036 and 601166 maintain relatively stable Clsprc trends, reflecting strong balance sheets or investor confidence.

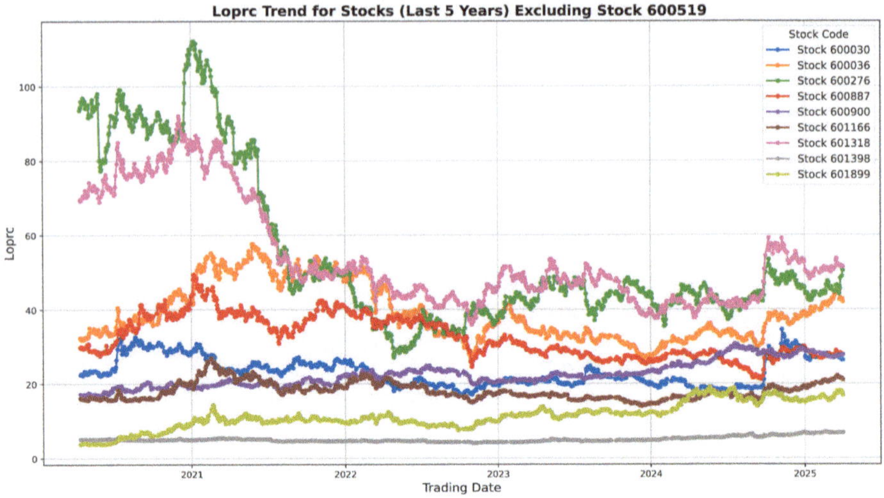

Fig. 5.16 Loprc trend for stocks over the past 5 years excluding stock 600519

As shown in Fig. 5.15, 601899 (Zijin Mining) and 600900 (China Yangtze Power) display upward trends in Hiprc, with 601899 in particular experiencing sharp increases, likely driven by strong commodity pricing. Several stocks show converging or diverging patterns in their Hiprc curves at different time points, indicating sector-wide alignment (e.g., 2021–2022) and differentiated valuation thereafter.

The trend in Loprc (As shown in Fig. 5.16) closely mirrors that of Hiprc, confirming price structure stability and consistent intraday volatility characteristics. For stocks like 600900, where the gap between Hiprc and Loprc is narrower, this suggests lower volatility, possibly due to a more conservative investor base or stable sector performance. In contrast, stocks like 600887 and 601398 exhibit a larger divergence between Hiprc and Loprc, indicating more speculative trading or fluctuations in earnings sentiment.

For most stocks, Clsprc remains between Loprc and Hiprc, reflecting a logical price structure. However, flattening of Hiprc and Clsprc lines, particularly from 2023 onward, for stocks like 601398 (ICBC) may suggest valuation ceilings, indicating limited market expectations for further price appreciation. The above analysis underscores the complexity and dynamic nature of the stock market, highlighting the necessity of constructing well-founded and interpretable models for accurate prediction.

5.4.3 *Stock Forecasting Models and Methodological Frameworks*

To address the complexities and practical challenges inherent in forecasting stock prices within the Chinese A-share market, this section introduces the forecasting models and methodological frameworks implemented in this research. Guided by the data characteristics and analytical motivation previously discussed, the proposed methodology aims not only to deliver robust forecasting performance but also to enhance the interpretability and reliability of model predictions, ensuring their suitability for informed financial decision-making. The remainder of this section is structured as follows: it begins with the presentation of the modeling context and formal problem formulation; subsequently, the rationale behind the selection of Long Short-Term Memory (LSTM) models and their theoretical foundations are detailed. Following this, the complete methodological approach is outlined comprehensively, culminating in a flowchart illustrating the integrated forecasting framework. Collectively, these components represent the methodological core underpinning the predictive and explanatory capabilities of the proposed time series forecasting system.

5.4.3.1 Stock Modeling Background

Accurate forecasting of stock market prices remains a central yet challenging topic in financial modeling, due to the inherent complexity, volatility, and nonlinear dynamics of stock markets. As previously discussed, stock prices in the Chinese A-share market are influenced by numerous interacting factors, such as macroeconomic indicators, industry-specific developments, investor sentiment, and regulatory policies, resulting in strong temporal dependencies and structural shifts. Conventional linear methods typically struggle to adequately capture these complex relationships and dynamic patterns.

To address these issues, this study adopts the Long Short-Term Memory (LSTM) network, a deep learning method well suited for modeling complex temporal dependencies in multivariate sequential data. By employing specialized gating mechanisms, the LSTM effectively captures and preserves relevant historical information, overcoming common limitations associated with traditional recurrent neural networks. Therefore, the use of LSTM models is expected to enhance predictive accuracy and interpretability for key stock price indicators—including closing price, highest price, and lowest price—among the selected top ten weighted stocks from the SSE 50 Index.

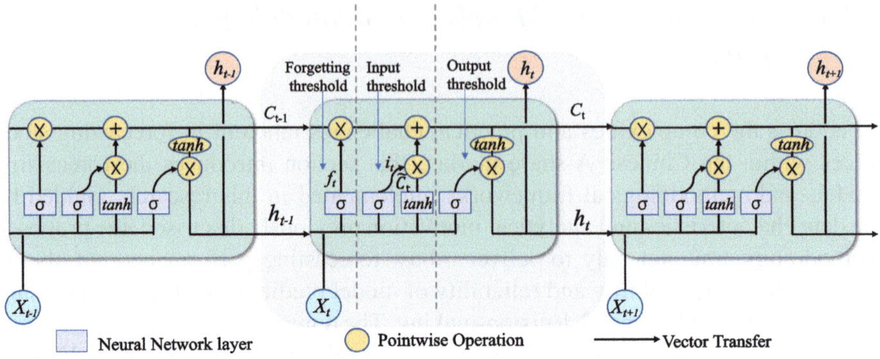

Fig. 5.17 LSTM network architecture

5.4.3.2 Stock Introduction to Long Short-Term Memory Neural Network

LSTM (Long Short-Term Memory) is a special type of recurrent neural network (RNN) designed to address the vanishing gradient problem encountered in traditional RNNs when processing long sequences. Its structure includes a cell state and three gating mechanisms: the input gate, forget gate, and output gate. The functions (As shown in Fig. 5.17) of each gate are as follows:

(1) Forget Gate: Decides which information from the previous cell state C_{t-1} should be retained. The forget gate processes the current input X_t and the previous hidden state h_{t-1} through a sigmoid function, outputting a value between 0 and 1 that represents the retention proportion for each memory unit.

(2) Input Gate: Determines how the current input X_t should update the cell state. It uses a sigmoid function to decide which values to update, and the tanh function generates candidate values to be added to the cell state.

(3) Cell State Update: Based on the outputs from the forget gate and the input gate, the previous cell state C_{t-1} and the new candidate values are combined and updated, resulting in the current cell state C_t.

(4) Output Gate: Decides how much information from the cell state C_t should be output. The output gate processes the current input and previous hidden state through a sigmoid function to determine which parts of the cell state influence the output. Additionally, the tanh function scales the cell state output to values between -1 and 1, producing the current hidden state h_t.

5.4.3.3 Stock Model Selection and Fundamental Principles

In the model construction phase, this study adopts an enhanced multi-task learning framework that leverages a shared bottom architecture with branched outputs to simultaneously forecast key stock price indicators. The implementation is structured around several core components:

(1) To begin with, the model's input layer receives fixed-length historical time series data, where each time step consists of 12 market-related features. This multi-dimensional input captures comprehensive historical trading information and market dynamics, providing sufficient context for temporal pattern recognition.

(2) To effectively extract the complex dynamic dependencies embedded in the sequential data, the model incorporates stacked Long Short-Term Memory (LSTM) layers. By increasing the depth and number of LSTM units, the network gains improved capacity to learn long-term dependencies and nonlinear interactions. The first LSTM layer processes the sequence across all time steps, passing forward temporal representations to the subsequent LSTM layer, which summarizes the sequence into a compact representation for downstream tasks.

(3) Following the LSTM layers, batch normalization is applied to the intermediate feature representations. This operation stabilizes the internal feature distributions, accelerates convergence, and enhances generalization by mitigating the covariate shift across layers, thereby ensuring a smoother learning process.

(4) To further improve the model's ability to capture critical information, a simple attention mechanism is introduced. This mechanism computes attention weights via a fully connected layer to identify and emphasize the most relevant temporal features. The model then applies these weights to reweight the feature representation, allowing it to dynamically focus on time steps that are most influential to the prediction outcomes.

(5) After attention-based weighting, the model employs a shared dense layer to extract low-dimensional representations that are common across all prediction tasks. These shared features serve as the foundation for three separate output branches, each dedicated to predicting one of the target variables: closing price, highest price, and lowest price. The use of branched outputs allows the model to benefit from shared information while enabling task-specific optimization and error evaluation.

(6) The mean squared error (MSE) is used as the primary loss function to evaluate prediction accuracy, capturing the deviations between predicted and actual values. The model is trained using the Adam optimizer with a carefully selected learning rate, ensuring both stability and efficiency in convergence. To further improve robustness, early stopping and learning rate scheduling strategies are employed.

(7) During the training process, to prevent overfitting and avoid the model getting stuck in local minima, early stopping and ReduceLROnPlateau callbacks were employed. The early stopping function monitors the validation loss and automatically halts training if no improvement is observed over a specified number of consecutive epochs. The ReduceLROnPlateau callback, on the other hand, dynamically reduces the learning rate when the validation performance plateaus, thereby further enhancing the model's robustness.

(8) Finally, to improve the stability and accuracy of predictions, this study also explored a model ensemble strategy. By training multiple independent models with different random seeds and averaging their predictions, the strategy aims to reduce

the random errors of individual models and produce smoother, more reliable output results.

5.4.3.4 Stock Methodological Workflow Overview

The constructed time series data is used to train an improved LSTM-based neural network.

(1) Model Training with LSTM

The model employs a sliding window to capture temporal dependencies inherent in financial data. Its architecture consists of two stacked LSTM layers, followed by batch normalization and an attention mechanism to enhance representational capacity. Three output branches are used to predict the closing price, high price, and low price, respectively. The model is optimized using the mean squared error (MSE) loss function and the Adam optimizer, with EarlyStopping and learning rate scheduling techniques applied to ensure convergence and robustness.

(2) Selection of Representative Instances for Interpretation

To analyze the model's behavior at the individual level, a set of representative instances is selected from the test set. These instances are chosen to capture different types of price fluctuations and feature profiles, with particular attention given to samples where the model exhibits larger prediction errors. These samples serve as focal cases for interpretability analysis.

(3) Local Neighborhood Generation and Surrogate Modeling

For each selected instance, a local neighborhood is constructed by introducing controlled perturbations to the input time series. These perturbed samples maintain the structural integrity of the original sequences while simulating local variations. A simple regression model is then fitted within each neighborhood to approximate the local behavior of the original LSTM model.

(4) Confidence Region Construction and Visualization

To assess the stability and reliability of the local surrogate models, confidence ellipsoids and confidence bands, and CICE index are constructed based on the estimated regression parameters and their covariance structure. These visual representations provide an intuitive understanding of local uncertainty and help users evaluate the credibility of the model explanations.

(5) Integration of Predictive and Interpretive Insights

The outputs of the predictive model and the interpretability framework are integrated to support decision-making in financial forecasting contexts. Accurate numerical predictions are complemented by locally interpretable explanations, enabling a transparent and interpretable time series forecasting system.

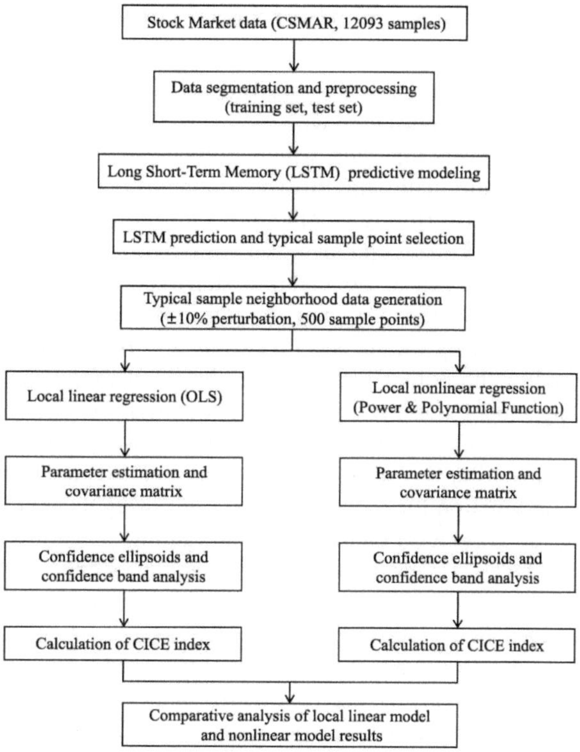

Fig. 5.18 Methodological framework of LSTM prediction model

5.4.3.5 Stock Forecasting Process Flowchart

Figure 5.18 illustrates a complete methodological pipeline that integrates prediction and interpretability for stock market modeling and local explanation. The process begins with stock market data (CSMAR, 12,093 samples), followed by data splitting and preprocessing to construct the training and test sets. A Long Short-Term Memory (LSTM) model is then employed to perform time series forecasting, and representative sample points are selected for analysis. Around each selected point, a local neighborhood is generated by introducing controlled perturbations ($\pm 10\%$, 500 samples), which are then fitted using both local linear models (e.g., ordinary least squares regression) and local nonlinear models (e.g., power functions and polynomial functions).

For each fitted model, the covariance matrix of the parameter estimates is computed and used to construct confidence ellipsoids and confidence bands, providing uncertainty-aware visualizations. Finally, by comparing linear and nonlinear local explanations, the framework enables quantitative interpretation and stability assessment of LSTM predictions at the individual level.

5.4.4 Stock LSTM for Local Interpretation

The model employs two stacked LSTM layers, each consisting of 128 units. The first layer is configured with `return_sequences=True` so that it outputs hidden states at every time step, which is crucial for capturing long-term temporal dependencies. Additionally, `dropout=0.2` and `recurrent_dropout=0.2` are applied to reduce overfitting by randomly omitting portions of the inputs and recurrent connections. The second layer, set with `return_sequences=False`, outputs only the hidden state at the final time step to obtain a compact representation of the entire sequence, while also using the same dropout settings to improve generalization.

Following the second LSTM layer, a Batch Normalization layer is used to normalize the output features, thereby accelerating the training process and enhancing model stability by reducing sensitivity to weight initialization. Subsequently, a fully connected layer (`Dense(128, activation='softmax')`) computes an attention vector that generates normalized attention weights via a softmax activation function for each time step. These weights are then applied through a Multiply layer to perform element-wise weighting on the LSTM outputs, thus highlighting the features corresponding to the key time steps.

The attention-weighted features are further processed by a shared fully connected layer (`Dense(32, activation='relu')`) that extracts a compact, low-dimensional representation. This shared representation is then fed into three separate output layers, each consisting of a single `Dense(1)` unit, which are dedicated to predicting the closing price, high price, and low price, respectively. This architecture effectively integrates feature processing with multi-task prediction objectives.

In this training, the learning rate was dynamically reduced from the initial value of 0.005 to 0.0025, 0.00125, and 0.000625 using the ReduceLROnPlateau callback, thereby refining the model optimization. Meanwhile, Early_Stopping allowed for early termination of training and restoration of the best weights when the validation loss showed no improvement over several epochs, effectively preventing overfitting. The combined use of these two strategies significantly enhanced the stability and generalization performance of the model.

Figure 5.19 shows the prediction results of the closing price (Clsprc), high price (Hiprc), and low price (Loprc), When combining 10 stocks. The output values are not a weighted average of the predictions for the 10 stocks, but rather the simple average of the predictions from multiple models (LSTM models trained with different random seeds) for each target variable (closing price, high price, and low price). This approach aims to reduce random errors from individual models, thereby improving the stability and robustness of the prediction results. The predicted values are in close agreement with the actual values of the three price indicators (closing price, high price, and low price), indicating that the model performs well in capturing the overall trends in the data. The prediction results are smooth and follow the same general pattern as the actual stock prices, demonstrating the model's strong performance in modeling the temporal dynamics of stock prices. A significant surge in prediction error occurs near sample point 180, which may be due to the model's inability to predict accurately

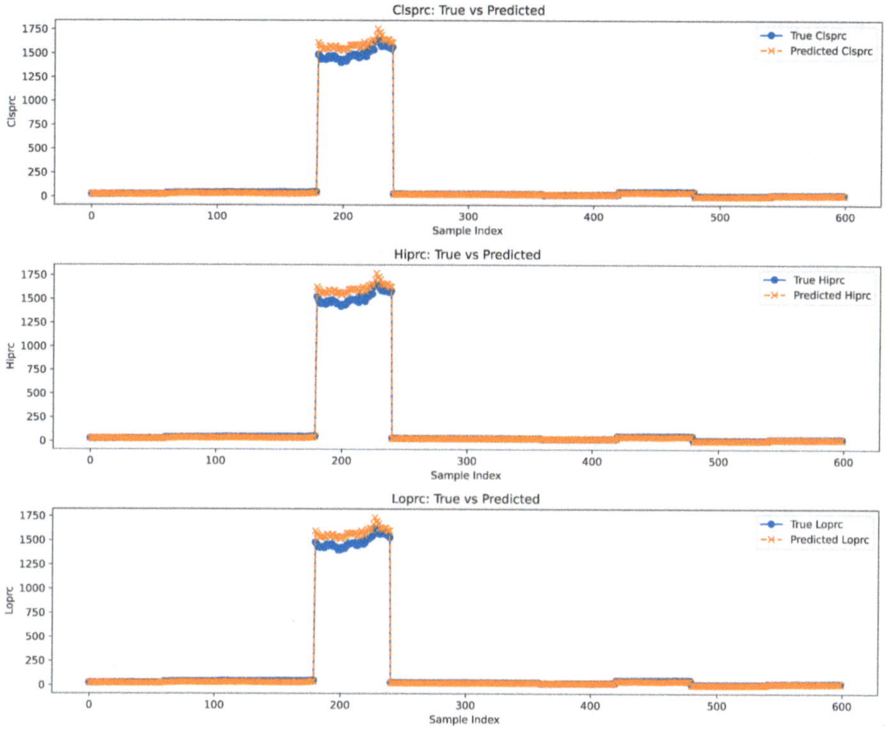

Fig. 5.19 Actual versus predicted stock prices across 10 combined stocks

when handling an anomalous event or market disruption. This could be attributed to a sudden market fluctuation or outliers in the dataset.

The model performs exceptionally well in predicting the closing price, with an R^2 value of 0.9946, indicating that it can accurately explain the fluctuations in the closing price. The MSE is 1049.7, showing small prediction errors, and the MAPE of 17.84% suggests a moderate relative error. The model also demonstrates high accuracy in predicting the high price, with an R^2 value of 0.9947, indicating a good fit. The MSE shows 1041.993, and the MAPE of 17.36%, similar to the closing price, suggests small prediction errors. The model performs well in predicting the low price, with an R^2 value of 0.9948, indicating that it fits the fluctuations of the low price well. The MSE is 997.6548, which is still small, but the MAPE of 20.38% is slightly higher than for the closing and high prices, suggesting that the model is somewhat slower to respond to price movements in the low price.

Next, we also performed predictions for the stock with the highest weight, 600519. The model's learning rate was gradually adjusted using the Reduce_LR-On_Plateau callback, optimizing the training process. Initially, the learning rate started at 0.005 and was reduced several times during training: first to 0.0025 (at epoch 30), then to 0.00125 (at epoch 42), and finally to 0.000625 (at epoch 47). When the validation

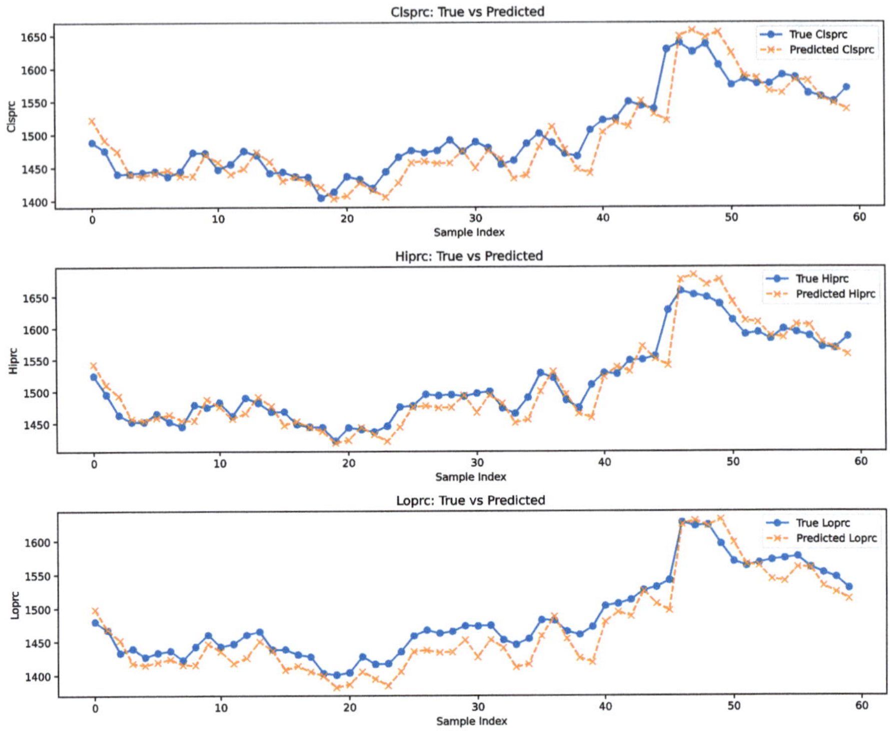

Fig. 5.20 Actual versus predicted stock prices for 600519

loss stopped improving, early stopping was triggered, and the best model weights from epoch 37 were restored, demonstrating the effectiveness of the model training. In subsequent training, the learning rate was adjusted again: to 0.0025 at epoch 50, to 0.00125 at epoch 55, and the best weights were restored at epoch 45. Similarly, early stopping was triggered again at epoch 43 due to the lack of improvement in validation loss.

Figure 5.20 shows the closing price (Clsprc), high price (Hiprc), and low price (Loprc) of the stock 600519. The predicted closing prices follow the true values very closely, indicating that the model captures the general trend and fluctuations of the stock price well. However, some minor discrepancies are visible in the predictions during periods of higher volatility, where the model slightly underperforms in tracking the actual price movements. Similar to the closing price, the predicted high prices closely match the true high prices, demonstrating strong performance in predicting the peaks of stock prices. The predictions are consistent with the general upward and downward trends, although, as with the closing price, some small deviations are present during more volatile periods. The predicted low prices also align well with the true values, but there are some noticeable deviations, especially during rapid price drops. The model may struggle to predict sudden sharp decreases in price, indicating that while it captures general patterns, extreme market movements may introduce error in prediction. Overall, the model demonstrates solid performance in predicting stock prices for 600519, accurately capturing the trends in closing, high, and low prices. However, minor discrepancies may occur, especially during periods of higher volatility or rapid market changes.

The model demonstrates a reasonable performance in predicting the closing price with an R^2 value of 0.8088, indicating that approximately 81% of the variance in the closing price is explained by the model. The MSE of 734.4567 and RMSE of 27.1009 suggest that the model's predictions are relatively accurate, with moderate errors. The MAPE value of 1.30% indicates that the model's prediction error is small in relation to the actual values, reflecting the model's high precision in predicting the closing price. For the high price prediction, the model performs even better with an R^2 value of 0.8710, indicating that over 87% of the variance in the high price is captured by the model. The MSE value of 510.8156 and RMSE of 22.6012 are low, further suggesting accurate predictions. The MAPE value of 1.13% also demonstrates the model's effectiveness in predicting the high price with minimal relative error. The model performs excellently in predicting the low price, with the highest R^2 value of 0.8925, meaning that nearly 89% of the variance in the low price is explained by the model. The MSE of 386.9012 and RMSE of 19.6698 indicate that the model makes small prediction errors. The MAPE of 1.11% is the lowest among the three outputs, reflecting a high level of accuracy in predicting the low price. Overall, the model's performance is robust and highly reliable for stock price forecasting.

By comparing the prediction results for 600519 and the 10 stocks, it is evident that 600519 significantly outperforms many of the other stocks in terms of prediction accuracy and stability, particularly in the MAPE metric. The model for 600519 exhibits much lower relative errors in predicting the closing price, high price, and low price compared to the 10-stock group, demonstrating higher model robustness. As a result, subsequent interpretability analysis can be further explored based on the prediction results for 600519 to investigate the model's performance.

To enable local interpretability, the test instance with index 10 was selected based on its relevance for the model's interpretability analysis. As shown in the table above, this sample, indexed as 10, corresponds to stock 600519 with the trading date of 2025-01-16. The features for this instance include the opening price (Opnprc) of 1474, the highest price (Hiprc) of 1482.66, lowest price (Loprc) of 1442, closing price (Clsprc) of 1446.38, and trading volume (Dnshrtrd) of 2,634,213. Additionally, the market value (Dsmvosd) is 3,840,411,437, and the transaction value (Dnvaltrd) is 1,816,939,374. The sample also shows a change ratio (ChangeRatio) of −0.016917, with the limit down price (LimitDown) set at 1324.14, and the limit up price (LimitUp) set at 1618.4, while the LimitStatus is 0, indicating no limit status. After predicting with the LSTM model, the three stock prices are 1,454.4194, 1,472.509, and 1,433.6167, respectively. This sample is chosen as it provides a balanced view of both price and trading volume, positioned in the area of highest uncertainty within the model's decision-making boundary, making it an ideal candidate for interpretability analysis.

Table 5.17 The mean and standard deviation of the feature values for the instance to be explained

	Mean	Standard deviation
Opnprc	1,508.94	157.42
Hiprc	1,519.07	159.46
Loprc	1,497.58	154.21
Clsprc	1,506.22	157.02
Dnshrtrd	3,089,216.58	1,215,388.40
Dnvaltrd	4,672,353,875.82	1,878,117,759.15
Dsmvosd	1,891,045,519.89	196,676,048.74
Dsmvtll	1,894,517,201.90	196,397,225.82
ChangeRatio	0.00	0.01
LimitDown	1,358.46	140.94
LimitUp	1,660.75	171.45
LimitStatus	0	0

In order to investigate the model's response around the chosen instance, the feature vector was standardized in the same manner as the training data. A set of 500 neighboring samples was created by adding Gaussian noise, sampled from $N(0, 0.1^2)$, to each feature of the original data. This noise injection introduces controlled variations while maintaining the underlying structure of the data, which is ideal for building an interpretable neighborhood suitable for surrogate modeling. Table 5.17 shows the mean and standard deviation of the feature values for the instance to be explained.

5.4.5 *Stock Local Linear Surrogate with OLS Regression*

We used Ordinary Least Squares (OLS) regression to construct a local linear surrogate model. First, feature data was extracted from the neighborhood samples, and then a constant term was added to the feature data for each target variable. A single-target OLS regression model was then fitted. The model was trained using the "statsmodels" library, and the regression results for each target, including coefficients and statistical significance analysis, were output.

5.4.5.1 Stock Estimated Coefficients of the Local Linear Model

Table 5.18 shows the local linear regression results for the clsprc target variable, modeled using the Ordinary Least Squares (OLS) method.

According to the OLS regression results in Table 5.18, the primary features influencing the prediction of Clsprc are Opnprc, Hiprc, Loprc, and Clsprc (the previous closing price). The coefficient for Opnprc is 0.046, indicating a positive relationship

Table 5.18 Coefficient estimates of the OLS linear model for Clsprc

Feature	Coef.	Std. Err	z	$P > \|z\|$	[0.025, 0.975]
Const	649.608	15.901	40.854	0.000	[618.366, 680.851]
Opnprc	0.046	0.003	14.345	0.000	[0.039 , 0.052]
Hiprc	0.076	0.003	23.851	0.000	[0.070 , 0.083]
Loprc	0.067	0.003	21.521	0.000	[0.061 , 0.073]
Clsprc	0.062	0.003	20.139	0.000	[0.056 , 0.068]
Dnshrtrd	0.000	0.000	1.399	0.163	[0.000 , 0.000]
Dnvaltrd	0.000	0.000	2.868	0.004	[0.000 , 0.000]
Dsmvosd	0.000	0.000	21.035	0.000	[0.000 , 0.000]
Dsmvtll	0.000	0.000	26.944	0.000	[0.000 , 0.000]
ChangeRatio	7411.519	5464.621	1.356	0.176	[−3325.572 , 18100]
LimitDown	0.087	0.004	24.604	0.000	[0.080 , 0.093]
LimitUp	0.054	0.003	18.227	0.000	[0.048 , 0.059]
LimitStatus	0	0	nan	nan	[0.000 , 0.000]

with Clsprc, which is statistically significant (p-value far below 0.05). Similarly, Hiprc and Loprc have coefficients of 0.076 and 0.067, respectively, indicating that both have a positive effect on Clsprc, and these features are significant (p-values less than 0.05). The regression coefficient of Clsprc is 0.062, which also shows a significant positive impact, further confirming the dependence of the current closing price on the historical closing price.

The features Dnshrtrd and ChangeRatio show no significant impact on the closing price. The p-value for Dnshrtrd is 0.163, suggesting no significant linear relationship with Clsprc. ChangeRatio has a large p-value of 0.176 and a large coefficient (7411.519), but its high standard error makes it difficult to precisely estimate its influence.

Features such as Dsmvosd, Dsmvtll, and LimitDown are statistically significant (p-values below 0.05), but their coefficients are very small, indicating weak influence on Clsprc. These features may require further consideration regarding their weight and actual impact in practical models. LimitUp and LimitStatushave a more substantial impact on the closing price, with LimitUp showing a positive effect with a coefficient of 0.054. LimitStatus has a coefficient of 0, suggesting no direct influence, but in practical applications, especially during market volatility, it might still need to be considered alongside other factors.

Tables 5.19 and 5.20 present the OLS regression results for Hiprc and Loprc. Similar to the regression results for Clsprc, the features Opnprc, Hiprc, and Loprc show a significant positive influence on the highest and lowest prices. The features Dnshrtrd and ChangeRatiodo not significantly affect the target variables, with p-values greater than 0.05, showing their minor contribution to the regression models. Features such

Table 5.19 Coefficient estimates of the OLS linear model for Hiprc

Feature	Coef.	Std. Err	z	$P > \|z\|$	[0.025, 0.975]
Const	650.666	16.231	40.087	0.000	[618.775 , 682.558]
Opnprc	0.047	0.003	14.343	0.000	[0.040 , 0.053]
Hiprc	0.078	0.003	23.862	0.000	[0.071 , 0.084]
Loprc	0.069	0.003	21.529	0.000	[0.062 , 0.075]
Clsprc	0.064	0.003	20.142	0.000	[0.057 , 0.070]
Dnshrtrd	0.000	0.000	1.400	0.162	[0.000 , 0.000]
Dnvaltrd	0.000	0.000	2.870	0.004	[0.000 , 0.000]
Dsmvosd	0.000	0.000	21.038	0.000	[0.000 , 0.000]
Dsmvtll	0.000	0.000	26.961	0.000	[0.000 , 0.000]
ChangeRatio	7538.066	5578.198	1.351	0.177	[−3422.184 , 18500]
LimitDown	0.088	0.004	24.609	0.000	[0.081 , 0.095]
LimitUp	0.055	0.003	18.239	0.000	[0.049 , 0.061]
LimitStatus	0.000	0.000	nan	nan	[0.000 , 0.000]

Table 5.20 Coefficient estimates of the OLS linear model for Loprc

Feature	Coef.	Std. Err	z	$P > \|z\|$	[0.025, 0.975]
Const	637.067	15.723	40.518	0.000	[606.174 , 667.961]
Opnprc	0.045	0.003	14.350	0.000	[0.039 , 0.051]
Hiprc	0.076	0.003	23.875	0.000	[0.069 , 0.082]
Loprc	0.067	0.003	21.543	0.000	[0.061 , 0.073]
Clsprc	0.062	0.003	20.150	0.000	[0.056 , 0.068]
Dnshrtrd	0.000	0.000	1.400	0.162	[0.000 , 0.000]
Dnvaltrd	0.000	0.000	2.872	0.004	[0.000 , 0.000]
Dsmvosd	0.000	0.000	21.049	0.000	[0.000 , 0.000]
Dsmvtll	0.000	0.000	26.977	0.000	[0.000 , 0.000]
ChangeRatio	7304.351	5403.629	1.352	0.177	[−3312.9 , 17900.0]
LimitDown	0.086	0.003	24.620	0.000	[0.079 , 0.092]
LimitUp	0.053	0.003	18.250	0.000	[0.047 , 0.059]
LimitStatus	0	0	nan	nan	[0 , 0]

as LimitDown and LimitUp (price limits) also exhibit significant effects, confirming the impact of market price restrictions on stock prices. These results are consistent with those found for Clsprc, indicating that these features are crucial for stock price prediction.

The local linear surrogate models for the three target features exhibit high similarity. Therefore, the subsequent analysis will focus solely on the model performance based on Clsprc, further exploring the stability of each feature.

5.4.5.2 Stock Predictive Performance of Local Linear Model

The Clsprc linear OLS regression model's R-squared value is 0.884, indicating that approximately 88.4% of the variance in the closing price can be explained by the selected features, demonstrating strong fitting capability. The F-statistic is 338.7, with a p-value significantly less than 0.05, indicating the overall significance of the model.

The adjusted R-squared value of 0.882 slightly adjusts the R-squared to account for the number of predictors, further confirming the model's appropriateness. The Durbin–Watson statistic of 2.172 indicates that there is no significant autocorrelation in the residuals, which is ideal for ensuring unbiased regression results. In summary, the model demonstrates high explanatory power, statistical significance, and a lack of autocorrelation, indicating that it is a reliable model for predicting the closing price (Clsprc). The results suggest that even a "simple" local linear least squares fitting model exhibits strong interpretability.

5.4.5.3 Stock Confidence Ellipsoids for Linear Model Parameters

To quantify the uncertainty associated with parameter estimation, a confidence ellipsoid was constructed based on the covariance matrix derived from the OLS regression model. The model-provided covariance matrix was obtained via the "cov_params()" method, and its generalized inverse was computed using the Moore–Penrose pseudoinverse to ensure numerical stability. The confidence ellipsoid is defined by the quadratic form:

$$(\boldsymbol{\theta} - \hat{\boldsymbol{\theta}})^T \mathbf{C}^{-1} (\boldsymbol{\theta} - \hat{\boldsymbol{\theta}}) \leq \frac{\chi^2_{1-\alpha}(m)}{N} \tag{5.5}$$

where $\hat{\boldsymbol{\theta}}$ denotes the estimated coefficients, $\mathbf{C}$ the covariance matrix, and $m = 13$ the number of parameters. At a 95% confidence level, the chi-square critical value is 22.362. Therefore, the resulting ellipsoid is given by:

$$\left\{\boldsymbol{\theta} \in \mathbb{R}^{12} \parallel Q(\boldsymbol{\theta}) \leq 22.362.\right\} \tag{5.6}$$

To illustrate the uncertainty in the parameter estimates of the local linear model, six distinct 3D parameter combinations were chosen, each containing three variables from the full parameter set. For each combination, the corresponding 3×3 submatrix was extracted from the overall covariance matrix, and then eigenvalue decomposition was performed. The semi-axis lengths for the confidence ellipsoids were calculated using the 95% chi-square critical value for three degrees of freedom ($\chi^2_{0.95}(3) = 7.8147$), which allowed the construction of ellipsoids representing the combined variability of the parameters in each parameter space. Each ellipsoid was generated by transforming a unit sphere using scaling and rotation operations based on the covariance structure. The estimated parameter vector $\hat{\boldsymbol{\theta}}$ was positioned at the center of each ellipsoid, providing a visual representation of its location within the region of uncertainty.

5.4.5.4 Stock Functional Confidence Band Visualization

Figure 5.21 illustrates six 95% confidence ellipsoids for parameter triplets in the local linear OLS model for Clsprc. In the combinations involving the core price Variables—(const, Opnprc, Hiprc) and (const, Loprc, Clsprc)—the ellipsoids are compact and directionally aligned, indicating strong and stable positive contributions from Hiprc and Loprc to the predicted closing price. In contrast, (const, Dnshrtrd, Dnvaltrd) exhibits weaker and more diffuse influence, consistent with the lower significance of trading-related features in the OLS estimates. The ellipsoid for (Opnprc, ChangeRatio, LimitDown) suggests moderate interaction, with ChangeRatio showing greater influence than the regulatory-bound feature *LimitDown*. (Hiprc, LimitUp, Dsmvtll) confirms the prominent role of Hiprc and LimitUp, while Dsmvtll introduces additional variance. Finally, (Clsprc, Dsmvosd, LimitStatus) reflects minimal contribution from LimitStatus and a relatively indirect role for the market capitalization features, reinforcing the dominance of price-based predictors in closing price estimation.

(1) Directional elongation reflects estimation uncertainty

The elongation of each ellipsoid along specific axes reflects the level of uncertainty associated with the corresponding parameter estimates. For instance, in the combination involving Opnprc and Hiprc, the ellipsoid is noticeably stretched along the price feature directions, indicating higher estimation variance and greater model sensitivity to changes in these variables.

(2) Ellipsoid skewness reveals interaction and collinearity between parameters

The skewed or tilted shape of an ellipsoid suggests potential interaction or collinearity among the selected features. In the case of Opnprc, ChangeRatio, and LimitDown, the observed skewness implies that these features may not act independently, but rather interactively influence the model's predictions of the closing price.

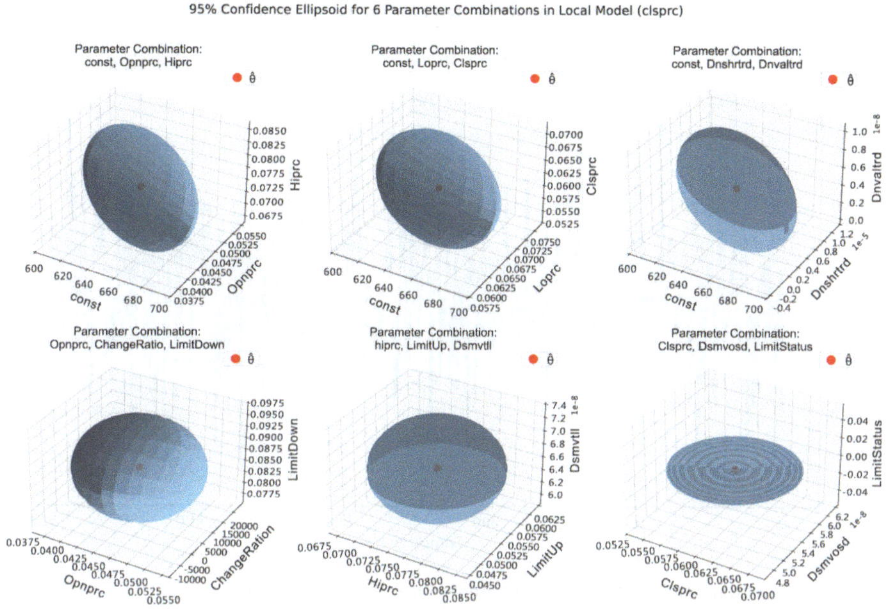

Fig. 5.21 95% confidence ellipsoids of six 3D parameter combinations in the linear local model for Clsprc

(3) Principal axes may align with prediction-sensitive directions

The principal axis of each ellipsoid corresponds to the direction of maximum parameter variation, aligned with the eigenvector of the largest eigenvalue. For example, in the combination involving Hiprc and Loprc, the alignment of the principal axis with these price-related features underscores their dominant role in the model's predictive behavior.

5.4.5.5 Stock Repeated Confidence Band Visualization

To assess the uncertainty associated with model predictions, a function confidence band was constructed for the local linear model, as shown in Fig. 5.22. The procedure is based on evaluating the quadratic form $x^T \hat{C} x$, where x denotes each neighborhood sample and $\hat{C}$ is the estimated covariance matrix of the regression parameters. The 95% confidence interval for each prediction $f(x) = x^T \hat{\theta}$ was then determined by scaling this variance by the chi-square critical value with 13 degrees of freedom ($\chi^2_{0.95}(13) \approx 22.362$).

In practice, the estimated parameter vector $\hat{\theta}$ and the corresponding covariance matrix $\hat{C}$ were extracted from the local OLS regression model. For each row of the design matrix X, the variance component $x^T \hat{C} x$ was computed and then scaled to obtain the confidence half-width. The predicted function $f(x)$ and the associated 95%

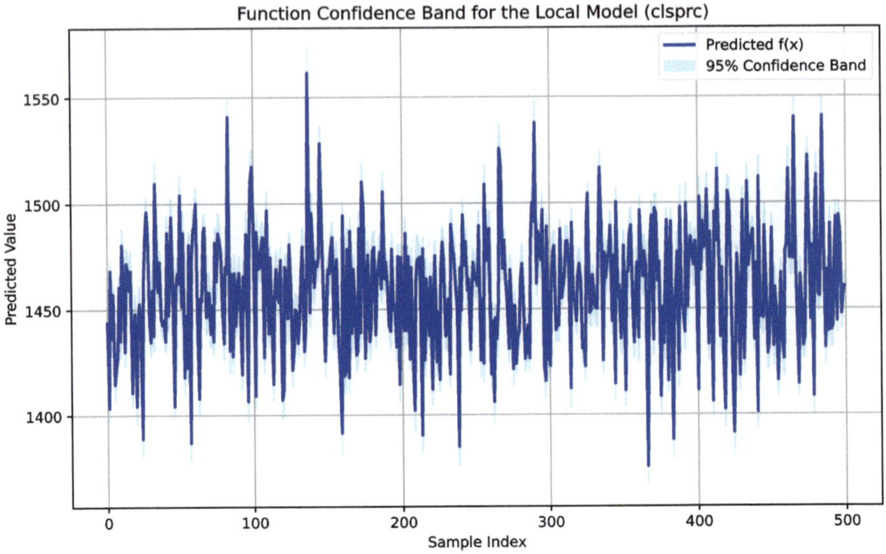

Fig. 5.22 Function confidence band of the linear local model for Clsprc

confidence region were plotted, enabling the visualization of prediction uncertainty across the local neighborhood. The resulting confidence band provides a rigorous and intuitive view of the prediction reliability. It highlights which regions are more stable in terms of prediction and which are associated with greater variability, thus offering valuable insight into the model's local predictive behavior and interpretability.

(1) Stability of the fitted curve

The blue curve represents the predicted function $f(x)$, which exhibits a smooth and continuous pattern across the 500 neighborhood samples. This indicates that the model effectively captures local sample behavior and maintains strong fitting consistency within the local region.

(2) Variation in confidence band width

The light blue shaded area represents the 95% confidence band, reflecting the uncertainty in model predictions. In most regions, the band remains narrow, suggesting stable parameter estimates and high confidence in predictions. However, slight expansions are observed near the boundaries (e.g., early and late sample indices), suggesting reduced certainty at the edges of the local domain.

(3) Localized uncertainty structure

Since the band width is derived from the quadratic form $x^T \hat{C} x$, it captures each sample's sensitivity to the parameter covariance structure. The figure reveals heterogeneity in prediction uncertainty across neighborhood samples, indicating that some inputs exert a stronger influence on model stability.

(4) Model interpretability and credibility

The visualization reinforces the transparency of the model. The confidence band not only demonstrates the prediction range but also conveys the reliability of the estimation. Consistent with the high R^2 from the earlier regression, the plot confirms the model's local validity and interpretability, making it suitable for subsequent sensitivity analysis and decision support.

5.4.5.6 Stock CICE Calculation Analysis

CICE is a composite evaluation metric designed to jointly assess two key aspects of a model: performance assessment (PA) and interpretability. An interpretability score is computed for each sample, and the average across all 500 samples is used as the overall interpretability indicator. The linear model explains 88.4% of the variance, as indicated by a Pseudo R^2 of 0.884. The MSE value of 214.367 leads to $\frac{1}{1+214.367} \approx 0.00464$, which severely diminishes the PA value when averaged with the high R^2.

Wider confidence intervals (i.e., high uncertainty) yield lower values of $\frac{1}{1+\text{width}}$, reducing the interpretability score. Ideally, for symmetric confidence intervals, the center deviation is zero. In this implementation, its contribution is less significant (weighted at 20%) compared to the width (weighted at 80%). The sample-wise interpretability scores are averaged across 500 samples to produce the overall interpretability score, which appears to be low and thus pulls the overall CICE down.

We set the center deviation and confidence width weights to 0.2 and 0.8, respectively. The interpretability index and accuracy index weights are 0.5 and 0.5, respectively. The **CICE** score of **0.3407** reflects a balance between model performance and interpretability. Although the model has a high R^2, the large MSE and wide confidence intervals lower the overall score. Despite the model's strong ability to explain variance, its predictions are accompanied by considerable uncertainty, which may limit its reliability in practical applications. This result highlights the fact that strong explanatory power does not necessarily imply reliable prediction. In real-world applications, model evaluation should consider both accuracy and the credibility of uncertainty quantification.

5.4.6 Stock Local Nonlinear Approximation and Predictive Performance

We will delve into the predictive performance and related analysis based on local nonlinear models. Two power function models are employed: one is a 12-feature, 24-parameter power function model, and the other is a 5-feature, 10-parameter power function model. A 12-feature, 36-parameter poly function model is also used; these models will be used to analyze the impact of feature quantity on model prediction performance.

5.4.6.1 Stock Estimated Coefficients of Local Nonlinear Model

Let the dataset be

$$\mathcal{D} = \{(\mathbf{x}^{(n)}, y^{(n)}) \mid n = 1, 2, \ldots, N\},$$

where each input vector

$$\mathbf{x}^{(n)} = (x_1^{(n)}, x_2^{(n)}, \ldots, x_d^{(n)}) \in \mathbb{R}^d$$

represents d features, and the corresponding target value is $y^{(n)} \in \mathbb{R}$. In the present context, the number of features is $d = 12$.

(1) Model 1

The model assumes a power-law form for each feature x_i (with $i = 1, 2, \ldots, 12$) as: $a_i \cdot x_i^{k_i}$, where a_i and k_i are parameters for each feature. The model prediction is given by the sum of these power-law terms:

$$f(\mathbf{x}) = \sum_{i=1}^{12} a_i \cdot x_i^{k_i}$$

Therefore, the model contains a total of 24 parameters: 12 values for a_i and 12 values for k_i.

(2) Model 2

For each feature x_i (with $i = 1, 2, \ldots, 12$), a quadratic polynomial $c_{i,0} + c_{i,1}x_i + c_{i,2}x_i^2$ is assigned. The model prediction is defined as the sum of these individual polynomials:

$$f(\mathbf{x}) = \sum_{i=1}^{12} \left(c_{i,0} + c_{i,1}x_i + c_{i,2}x_i^2\right)$$

Hence, the model contains a total of 36 parameters.

Taking polynomial function as an example, for each sample $(\mathbf{x}^{(n)}, y^{(n)})$, the residual is defined as

$$r_n(c) = f(\mathbf{x}^{(n)}) - y^{(n)} = \sum_{i=1}^{12} \left(c_{i,0} + c_{i,1}x_i^{(n)} + c_{i,2}\left(x_i^{(n)}\right)^2\right) - y^{(n)}$$

To mitigate the risk of overfitting, an L_2 regularization term is incorporated. Given a regularization parameter $\lambda > 0$, the regularization term is

$$R(c) = \lambda \sum_{i=1}^{12} \sum_{j=0}^{2} c_{i,j}^2$$

Consequently, the overall objective function that is minimized is the sum of the squared residuals along with the regularization term:

$$J(c) = \sum_{n=1}^{N} \left(\sum_{i=1}^{12} \left(c_{i,0} + c_{i,1} x_i^{(n)} + c_{i,2} \left(x_i^{(n)} \right)^2 \right) - y^{(n)} \right)^2 + \lambda \sum_{i=1}^{12} \sum_{j=0}^{2} c_{i,j}^2$$

Figure 5.23 displays the fitting results of the power function and polynomial function models. The power function model fits the data smoothly and consistently, with small deviations from the true values. This indicates that the power function model effectively captures the nonlinear relationships in the data, providing stable and accurate predictions. While the polynomial model also fits the data well, its deviations from the true values are slightly larger compared to the power function model.

Table 5.21 shows the coefficient of the OLS nonlinear model when using all 12 features. Model 1 employs scaling factors a_i and exponents k_i for each feature, capturing the nonlinear relationships between each feature and the target variable. The values of a_i and k_i differ significantly across features, indicating that the influence and nonlinearity of each feature on the target variable vary. Model 2 uses constant $c_{0,i}$, linear $c_{1,i}$, and quadratic $c_{2,i}$ terms to model the relationships between the features and the target variable as quadratic functions. The variations in $c_{0,i}$, $c_{1,i}$, and $c_{2,i}$ indicate that the features have nonlinear influences on the target variable as well. However, the fit of this model is slightly worse compared to the power function model. Model 1 has a cost of 25535.79, an R^2 value of 0.8878, MSE of 102.13, and RMSE of 10.11. These metrics indicate that the power function model explains approximately 88.78% of the variance in the target variable and exhibits low error, suggesting a high level of predictive accuracy. Model 2 has a cost of 30877.25, an R^2 value of 0.8643, MSE of 123.51, and RMSE of 11.11. Compared to the power function model, the polynomial model has a higher cost, lower R^2, and larger error values (MSE), indicating that it provides a poorer fit to the data. Given the superior performance of the power function model, especially its higher R^2 and lower error metrics, it is the more appropriate choice for this dataset. While the polynomial model is more flexible, it does not provide a better fit than the power function model for this particular problem.

However, after carefully observing the above two fitting models, we found many problems. In linear fitting, the feature LimitStatus is considered meaningless, and then in these two models, the coefficient of this feature is very high. Secondly, when we calculate the covariance matrix, it is found that the covariance of these two models tends to 0, which may be due to complete multicollinearity or over-parameterization. Although the fitting effect of the model is good from the image point of view, R^2 is also very high, the generalization ability of the model is not necessarily strong, and the function of the model still needs to be re-evaluated. For example, consider using

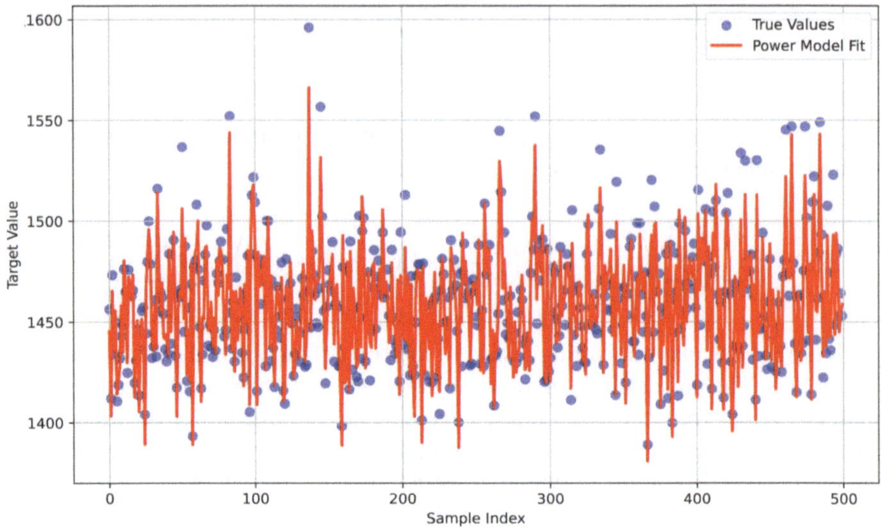

(a) Power function (Model 1)

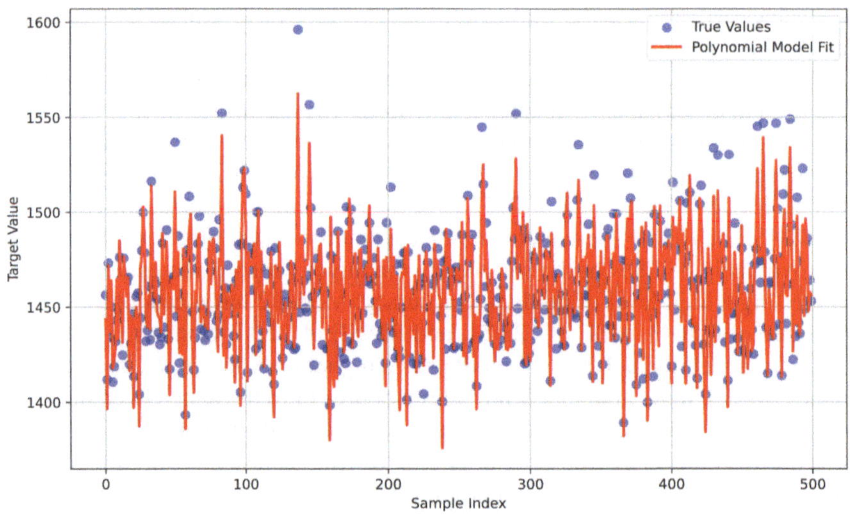

(b) Polynomial function (Model 2)

Fig. 5.23 Local fitting results based on two models

Table 5.21 Coefficient estimates of the OLS nonlinear model of 12 features for Ciprc

	Model 1		Model 2		
	a_i	k_i	$c_{0,i}$	$c_{1,i}$	$c_{2,i}$
Opnprc	0.140035	0.865310	−0.653136	0.202018	−0.000053
Hiprc	0.002195	1.437626	−0.502290	0.191015	−0.000038
Loprc	0.000001	2.380405	0.001707	−0.051861	0.000041
Clsprc	0.000003	2.271746	−0.803654	0.000039	0.000021
Dnshrtrd	3.730406	0.170144	0.286457	0.000000	0.000000
Dnvaltrd	0.000000	2.100371	0.465753	0.000000	0.000000
Dsmvosd	0.000013	0.757819	−0.094268	0.150557	−0.885348
Dsmvtll	0.000000	2.010567	0.633510	0.000209	0.000000
ChangeRatio	2.775638	−0.415535	0.974500	0.000000	0.000000
LimitDown	0.000001	2.524179	−0.706716	0.063786	0.000010
LimitUp	0.004193	1.307577	−0.053843	0.294144	−0.000073
LimitStatus	1.333596	2.051627	0.364972	0.509518	0.754820

simpler models (such as the previously used linear models) or trying to use different power function models that parameterize or reduce the number of parameters. Based on the above analysis, we give the following model.

(3) Model 3

The Model 3 assumes a power-law form for each feature x_i (with $i = 1, 2, \dots, 5$) as: $a_i \cdot x_i^{k_i}$, where a_i and k_i are parameters for each feature. It only consists of the following five features: Opnprc, Hiprc, Loprc, Clsprc, and Dnshrtrd. The model prediction is given by the sum of these power-law terms:

$$f(\mathbf{x}) = \sum_{i=1}^{5} a_i \cdot x_i^{k_i}$$

Therefore, the model contains a total of 10 parameters: 5 values for a_i and 5 values for k_i.

Figure 5.24 shows the fitted results of the nonlinear local surrogate power function model constructed using 10 parameters (the first 5 being $a_1, \dots, a_5$ and the next 5 being $k_1, \dots, k_5$), based solely on 5 features: Opnprc, Hiprc, Loprc, Clsprc, and Dnshrtrd. From the figure, it is evident that the predicted curve (displayed in red) generally follows the overall trend of the true target values (displayed in blue); however, significant deviations are still observed.

The estimated model parameters are as follows. The scaling coefficients a_i are 35.27, 44.21, 52.37, 42.27, and 29.04, while the corresponding power exponents k_i are 0.3103, 0.2656, 0.2720, 0.2853, and 0.0781, respectively. These parameters reflect the degree of nonlinear influence that each feature exerts on the target variable. In particular, larger values of a_i indicate that the corresponding feature contributes

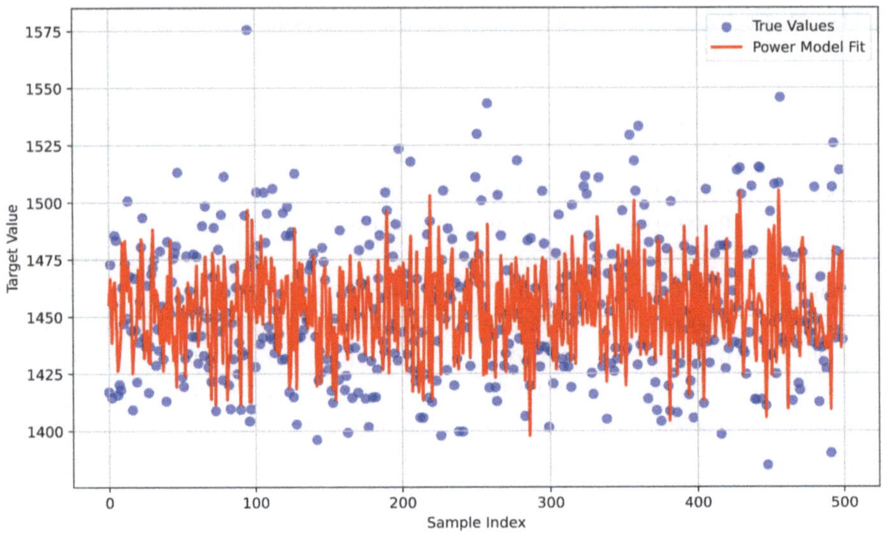

Fig. 5.24 Local fitting results based on 10-parameter power function models

more substantially to the model, while the power exponents determine the extent of curvature in the nonlinear relationship.

As far as the model effect is concerned, the model effect is worse than that of Model 1 and Model 2. The fitting cost is 129976.17, representing the overall objective value that combines the least squares residuals with the regularization term. $R^2 = 0.4136$, which indicates that the model explains approximately 41.36% of the variance in the target variable, suggesting that the current model's explanatory power is relatively limited. The modest R^2 also implies that there may be additional variables or structural aspects not captured by the model. MSE is 518.19 and RMSE is 22.76, the high values of the Mean Squared Error and Root Mean Squared Error indicate that, on average, there is a considerable deviation between the predicted and true values.

As far as the significance level is concerned, the overall significance level of k_i is also higher than that of a_i. However, since the covariance matrix of the fitting model does not have ill-conditioned or singular problems, we will still use the model for further analysis.

5.4.6.2 Stock Confidence Ellipsoids for Nonlinear Model

For each given parameter triple, a 95% confidence ellipsoid is constructed and visualized as follows. First, the critical value of the chi-square distribution with 3 degrees of freedom at the 95% confidence level is computed as approximately 7.8147. This value is used to determine the scale of the ellipsoid. Let $\boldsymbol{\mu}$ denote the estimated

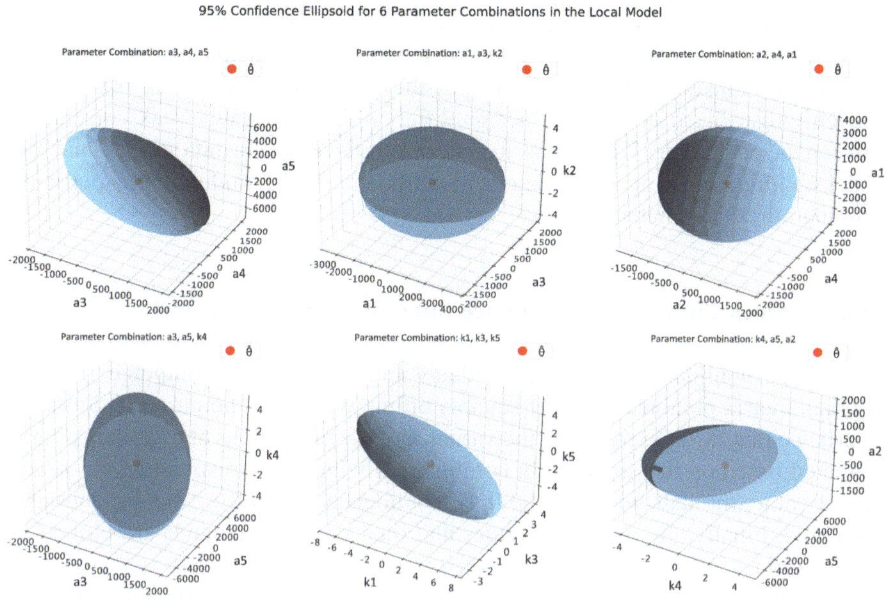

Fig. 5.25 95% confidence ellipsoids of six 3D parameter combinations in the linear local model for Clsprc

mean vector and $\boldsymbol{\Sigma} \in \mathbb{R}^{3\times 3}$ the corresponding covariance submatrix. An eigenvalue decomposition of $\boldsymbol{\Sigma}$ yields eigenvalues $\lambda_1, \lambda_2, \lambda_3$ (sorted in descending order), and an orthogonal eigenvector matrix Q. The semi-axis lengths of the ellipsoid are given by $a_i = \sqrt{\lambda_i \times 7.8147}, \quad i = 1, 2, 3.$

A set of points $\mathbf{p} = (p_1, p_2, p_3)$ is generated on the surface of the unit sphere. These points are mapped to the ellipsoid surface by the transformation

$$\boldsymbol{\theta} = \boldsymbol{\mu} + Q\left(a_1 p_1 \ a_2 p_2 \ a_3 p_3\right)^T$$

This linear transformation scales and rotates the unit sphere to match the orientation and size of the confidence ellipsoid. The resulting ellipsoid surface, along with the center point $\boldsymbol{\mu}$, is plotted in a 3D coordinate system to visually represent the uncertainty of the parameter estimates.

The 95% confidence ellipsoids for various parameter combinations (as shown in Fig. 5.25) visually represent the joint uncertainty of parameter estimates. The length and orientation of each ellipsoid's axes reflect parameter correlations and individual estimation precision. Specifically, longer axes indicate stronger correlations between parameters and higher uncertainty along that direction, while shorter axes denote greater estimation stability. The overall size of the ellipsoid represents the uncertainty level of the parameter estimates, where larger ellipsoids imply greater uncertainty, suggesting a potential need for additional data or improved modeling techniques to enhance accuracy.

For example, for the parameter combination "k1, k3, k5", if the confidence ellipsoid in the figure appears symmetric and the lengths of its principal axes are relatively balanced, it indicates that the estimates of these three parameters are robust, with an evenly distributed uncertainty across different directions and weak correlations between the parameters. This scenario generally reflects strong data support and high model stability, implying a high degree of confidence in the model for this particular combination of parameters.

5.4.6.3 Stock Functional Confidence Band for Nonlinear Model

For a given nonlinear model, the residuals and variance are computed from the Jacobian matrix obtained via least squares fitting, which are then used to derive the covariance matrix. Subsequently, the variance of the predictions at each data point is estimated based on the gradient. In combination with the corresponding chi-square critical value, the half-width of the 95% confidence interval for the predictions is calculated. Finally, the predicted values and their confidence bands are plotted for visualization.

Figure 5.26 presents the predicted values from the nonlinear model along with their 95% confidence bands. The horizontal axis represents the sample index, while the vertical axis shows the predicted function values, $f(x)$. The predicted values change smoothly within the range of approximately 1380 to 1540, indicating that the model's fit is stable and continuous over the sample interval, thereby reflecting overall good performance. The confidence band is displayed as a filled region. Its width is

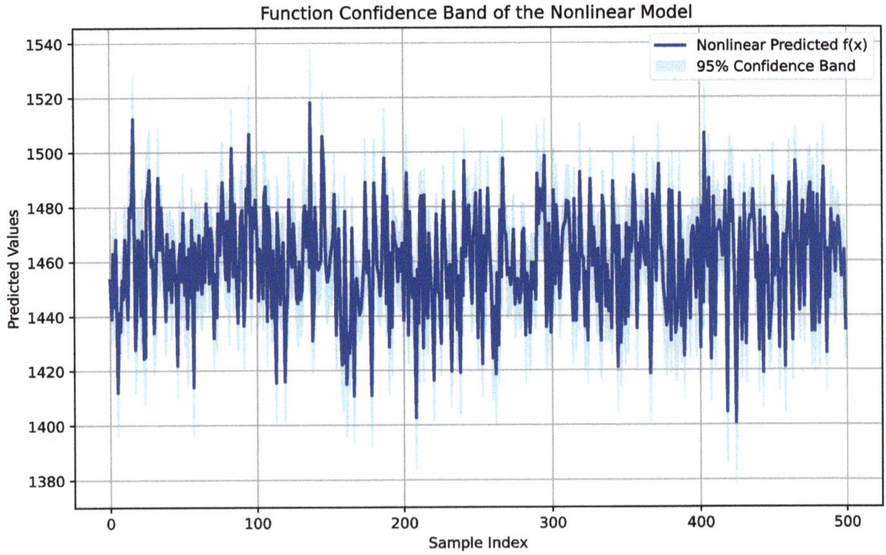

Fig. 5.26 Function confidence band of the nonlinear local model for Clsprc

computed from the prediction variance at each data point, which is estimated using the gradient and the parameter covariance matrix, and adjusted by the corresponding chi-square critical value. The symmetry of the band around the prediction curve suggests that the uncertainty estimation is reasonable and unbiased. And the moderate width of the confidence band implies that the prediction uncertainty within the data-covered region is within an acceptable range. If the confidence band were to widen significantly near certain sample points, it could indicate insufficient data support or higher sensitivity of the model in those areas; however, no such anomalies are evident in the current plot. Thus, it can be inferred that the nonlinear model's fit and predictions are robust, and the 95% confidence intervals provide a reliable measure of confidence for subsequent interpretation and decision-making.

5.4.6.4 Stock CICE Calculation Analysis

In order to ensure the consistency with the linear local fitting results, we do not adjust the weights. Similarly, we also computed the **CICE value** for the power function model, which is only **0.220**, significantly lower than that of the linear model. The corresponding PA value is approximately **0.2078**, indicating that poor performance considerably dragged down the overall CICE level. From the perspective of interpretability, the nonlinear model exhibits stronger sensitivity to input variations. In particular, the power function structure tends to amplify output instability when inputs are small. Since the power exponents $k_i \in (-1, 3)$, the model may respond more aggressively to small input values, resulting in unstable predictions and wider confidence intervals.

Moreover, the power function is highly nonlinear and asymmetric. If the confidence interval center deviates unevenly from the predicted value across samples, interpretability may be further reduced—even when the interval widths are moderate—due to substantial center deviation. It is also worth noting that although the model has fewer parameters, the function space it spans is complex. This increases the likelihood of local oscillations and instability in the fitted output.

In summary, the low CICE value is primarily attributed to **poor interpretability**, rather than merely weak performance. While the nonlinear power function model offers theoretical flexibility, it fails to provide stable and trustworthy prediction intervals in the local surrogate setting, ultimately resulting in **low confidence and poor consistency**.

5.4.7 Stock Integrated Interpretation and Comparative Analysis

To offer an in-depth explanation of the chosen instance near the decision boundary, local surrogate models—both linear and nonlinear—were constructed and analyzed

Table 5.22 Comparison of linear and nonlinear local surrogate models

Metric	Linear model (First-order polynomial with OLS)	Nonlinear model (Power function with OLS)
Pseudo R^2	0.884	0.4136
Adjusted R^2	0.882	0.402
MSE	214.367	518.19
CICE	0.341	0.220

within the same perturbed vicinity. This comparative approach facilitates a rigorous evaluation of model behavior, stability, and the influence of features at the individual level.ped and evaluated within a common perturbed neighborhood. This comparative framework enables a robust assessment of model behavior, stability, and feature influence at the individual level.

The linear surrogate model based on OLS regression achieved a pseudo R^2 of 0.884, and the nonlinear model using power function regression has a lower R^2 of 0.4136, the linear model has higher explanatory power as shown in Table 5.22. And with an MSE of only 214.367, the prediction error for the linear model is substantially lower than that of the nonlinear model, which exhibits an MSE of 518.19—more than twice as high. The CICE indicator also demonstrates the superiority of the linear model, suggesting better robustness and consistency.

The linear model has 13 parameters, while the nonlinear model (power function model) has 10 parameters. Although the linear model has slightly more parameters, its Adjusted R^2 does not change significantly, indicating that the increase in parameters did not lead to overfitting; rather, it better captured the structure of the data. The nonlinear model might suffer from issues such as local optima and sensitivity to the initial starting point, resulting in a less satisfactory fit and an inability to fully reveal the relationship between the input variables and the target variable.

Interpretability and stability are the core components of the CICE metric. Although the nonlinear model theoretically offers stronger expressive capacity, it fails to outperform the linear model in practical local surrogate modeling, particularly in terms of interpretability and stability.

The **advantage of the linear model** lies in its low complexity, strong robustness, and effective control of uncertainty. These characteristics make it more practically usable in local surrogate tasks.

The **shortcomings of the nonlinear model** are not limited to performance (PA); more importantly, it fails to provide stable and reliable prediction intervals, which significantly lowers its overall CICE value.

For the current data and task, the results suggest that a linear model should be adopted as the local surrogate model.

References

1. China Stock Market & Accounting Research (CSMAR), 2025. https://data.csmar.com/.
2. Pope D. Brooks, Thomas and Michael Marcolini. Airfoil Self-Noise. UCI Machine Learning Repository, 1989. https://doi.org/10.24432/C5VW2C.
3. Thomas F Brooks, D Stuart Pope, and Michael A Marcolini. Airfoil self-noise and prediction. Technical report, 1989.
4. Sérgio Moro, Paulo Cortez, and Paulo Rita. A data-driven approach to predict the success of bank telemarketing. *Decision Support Systems*, 62:22–31, 2014.
5. Jiuping Xu, Yanbing Hou, and Asaf H. Hajiyev. Enhancing stock market prediction with LSTM-Attention models: A local interpretability approach for financial decision-making. Reasearch Report, School of Business, Sichuan University, 2025.
6. Jiuping Xu, Yuhan Li, and Asaf H. Hajiyev. Confidence evaluation and local interpretability analysis for airfoil self-noise prediction models. Research Report, School of Business, Sichuan University, 2025.
7. Jiuping Xu, Jianglin Ma, and Asaf H. Hajiyev. Enhancing transparency in bank marketing predictions: Local surrogate modeling and confidence evaluation for XGboost classifiers. Research Report, School of Business, Sichuan University, 2025.

Appendix 1
Python Code for Chap. 3 Some Arithmetic Examples

Example 3.1

```
#Table of Parameter Bias Convergence Under Large-Sample Conditions

np.random.seed(42)

theta_true = np.array([1.5, 0.8, 2.1, -1.2, 0.7, 3.0])
sigma = 0.5

N_values = np.linspace(50, 1000000, 21, dtype=int)

replications = 50

mean_estimates = np.zeros((len(N_values), 6))
bias_estimates = np.zeros((len(N_values), 6))

for idx, N in enumerate(N_values):
    rep_estimates = np.zeros((replications, 6))
    x = np.linspace(0, 1, N)
    X = np.vander(x, N=6, increasing=True)

    for r in range(replications):
        epsilon = np.random.normal(0, sigma, N)
        y = X \@ theta_true + epsilon
        theta_hat = np.linalg.lstsq(X, y, rcond=None)[0]
        rep_estimates[r, :] = theta_hat

    mean_estimates[idx, :] = rep_estimates.mean(axis=0)
    bias_estimates[idx, :] = mean_estimates[idx, :] - theta_true

df_mean = pd.DataFrame(mean_estimates, columns=[f'$\theta${i} Mean' for i in range(6)])
df_mean.insert(0, 'Sample Size', N_values)

df_bias = pd.DataFrame(bias_estimates, columns=[f'$\theta${i} Bias' for i in range(6)])
df_bias.insert(0, 'Sample Size', N_values)


#Figure of Estimator Consistency and Bias Convergence

import numpy as np
```


A. Hajiyev and J. Xu, *High-Dimensional Regression Analysis and Artificial Intelligence*, https://doi.org/10.1007/978-981-95-2514-0

```
import matplotlib.pyplot as plt

theta_true = np.array([1.5, 0.8, 2.1, -1.2, 0.7, 3.0])
sigma = 0.5

N_values = np.linspace(50, 1000000, 20, dtype=int)

replications = 50

mean_estimates = np.zeros((len(N_values), 6))
bias_estimates = np.zeros((len(N_values), 6))

for idx, N in enumerate(N_values):
    rep_estimates = np.zeros((replications, 6))
    x = np.linspace(0, 1, N)
    X = np.vander(x, N=6, increasing=True)

    for r in range(replications):
        epsilon = np.random.normal(0, sigma, N)
        y = X \@ theta_true + epsilon
        theta_hat = np.linalg.lstsq(X, y, rcond=None)[0]
        rep_estimates[r, :] = theta_hat

    mean_estimates[idx, :] = rep_estimates.mean(axis=0)
    bias_estimates[idx, :] = mean_estimates[idx, :] - theta_true


#Quantile Visualization of the Convergence Trajectories of Parameter Biases

theta_true = np.array([1.5, 0.8, 2.1, -1.2, 0.7, 3.0])
sigma = 0.5

N_values = np.arange(50, 1000050, 2500)

theta_hats = []
det_values = []

for N in N_values:
    x = np.linspace(0, 1, N)
    X = np.vander(x, N=6, increasing=True)

    epsilon = np.random.normal(0, sigma, N)
    y = X \@ theta_true + epsilon

    theta_hat = np.linalg.lstsq(X, y, rcond=None)[0]
    theta_hats.append(theta_hat)

    XTX = X.T \@ X
    det_ratio = det(XTX / N)
    det_values.append(N * det_ratio)

theta_hats = np.array(theta_hats)

replications = 50
quantiles = [0.05, 0.25, 0.5, 0.75, 0.95]

theta_estimates = {i: {q: [] for q in quantiles} for i in range(6)}
median_estimates = {i: [] for i in range(6)}

for N in N_values:
    rep_theta = np.zeros((replications, 6))
    x = np.linspace(0, 1, N)
    X = np.vander(x, N=6, increasing=True)
    for r in range(replications):
```

```
            epsilon = np.random.normal(0, sigma, N)
            y = X \@ theta_true + epsilon
            theta_hat = np.linalg.lstsq(X, y, rcond=None)[0]
            rep_theta[r, :] = theta_hat
        for i in range(6):
            for q in quantiles:
                theta_estimates[i][q].append(np.quantile(rep_theta[:, i], q))
            median_estimates[i].append(np.quantile(rep_theta[:, i], 0.5))

X_variances = {i: [] for i in range(6)}
for N in N_values:
    x = np.linspace(0, 1, N)
    X = np.vander(x, N=6, increasing=True)
    var_cols = np.var(X, axis=0)
    for i in range(6):
        X_variances[i].append(var_cols[i])


#Graphical Test for Heteroscedasticity

theta_true = np.array([1.5, 0.8, 2.1, -1.2, 0.7, 3.0])
sigma = 0.5

sample_sizes = [10, 100, 1000, 10000, 100000, 1000000]

plt.figure(figsize=(10, 20))

for idx, N in enumerate(sample_sizes):
    x = np.linspace(0, 1, N)
    X = np.vander(x, N=6, increasing=True)

    epsilon = np.random.normal(0, sigma, N)
    y = X \@ theta_true + epsilon

    theta_hat = np.linalg.lstsq(X, y, rcond=None)[0]
    y_hat = X \@ theta_hat
    residuals = y - y_hat


#Heatmap of Marginal and Joint Parameter Density Distributions

np.random.seed(42)

theta_true = np.array([1.5, 0.8, 2.1, -1.2, 0.7, 3.0])
sigma = 0.5

sample_sizes = [10, 100, 1000, 10000, 100000,1000000]
replications = 200

for N in sample_sizes:
    estimates = np.zeros((replications, 6))

    x = np.linspace(0, 1, N)
    X = np.vander(x, N=6, increasing=True)

    for r in range(replications):
        epsilon = np.random.normal(0, sigma, N)
        y = X \@ theta_true + epsilon
        theta_hat = np.linalg.lstsq(X, y, rcond=None)[0]
        estimates[r, :] = theta_hat


#Confidence Interval

np.random.seed(42)
```

```
theta_true = np.array([1.5, 0.8, 2.1, -1.2, 0.7, 3.0])
sigma = 0.5
sample_sizes = [10, 100, 1000, 10000, 100000, 1000000]
replications = 200

theta_means = []
theta_ci_lower = []
theta_ci_upper = []

for N in sample_sizes:
    estimates = np.zeros((replications, 6))
    x = np.linspace(0, 1, N)
    X = np.vander(x, N=6, increasing=True)
    for r in range(replications):
        epsilon = np.random.normal(0, sigma, N)
        y = X @ theta_true + epsilon
        theta_hat = np.linalg.lstsq(X, y, rcond=None)[0]
        estimates[r, :] = theta_hat
    theta_mean = np.nanmean(estimates, axis=0)
    theta_std = np.nanstd(estimates, axis=0)
    ci_lower = theta_mean - 1.96 * (theta_std / np.sqrt(
        replications))
    ci_upper = theta_mean + 1.96 * (theta_std / np.sqrt(
        replications))
    theta_means.append(theta_mean)
    theta_ci_lower.append(ci_lower)
    theta_ci_upper.append(ci_upper)

theta_means = np.array(theta_means)
theta_ci_lower = np.array(theta_ci_lower)
theta_ci_upper = np.array(theta_ci_upper)

num_params = theta_means.shape[1]
num_cols = 3
num_rows = math.ceil(num_params / num_cols)

fig, axes = plt.subplots(num_rows, num_cols, figsize=(16, 9))
axes = axes.flatten()

for i in range(num_params):
    ax = axes[i]
    means = theta_means[:, i]
    lower = theta_ci_lower[:, i]
    upper = theta_ci_upper[:, i]
    yerr_lower = means - lower
    yerr_upper = upper - means
    yerr = np.vstack([yerr_lower, yerr_upper])
    ax.errorbar(sample_sizes, means, yerr=yerr, fmt='-o',
        capsize=4,
                    ecolor='red', color='blue', markersize=5,
                        label=f'\theta{i} Estimate')
    ax.set_xscale('log')
    ax.set_xlabel('Sample Size (log scale)')
    ax.set_ylabel('Parameter Estimate')
    ax.set_title(f'Parameter \theta{i}')
    ax.grid(True, linestyle=':', alpha=0.7)
    ax.legend()

for j in range(num_params, len(axes)):
    axes[j].axis('off')


#Covariance calculation

np.random.seed(42)

```

```
theta_true = np.array([1.5, 0.8, 2.1, -1.2, 0.7, 3.0])
sigma = 0.5
sample_sizes = [
    100, 200, 300, 400, 500, 600, 700, 800, 900, 1000,
    2000, 3000, 4000, 5000, 10000, 20000, 30000, 50000, 100000, 1000000
]
replications = 200

errors_noN = []
errors_withN = []

for N in sample_sizes:
    x = np.linspace(0, 1, N)
    X = np.vander(x, N=6, increasing=True)

    XtX = X.T @ X
    XtX_inv = np.linalg.inv(XtX)
    C_true_noN = sigma**2 * XtX_inv

    C_true_withN = sigma**2 * N * XtX_inv

    C_tilde_accum = np.zeros((6, 6))

    for r in range(replications):
        eps = np.random.normal(0, sigma, N)
        y = X @ theta_true + eps

        theta_hat = np.linalg.lstsq(X, y, rcond=None)[0]
        residuals = y - X @ theta_hat

        C_tilde = np.zeros((6, 6))
        for k in range(6):
            for l in range(6):
                C_tilde[k, l] = np.sum(residuals**2 * x**(k+l))

        C_tilde /= N

        C_tilde_accum += C_tilde

    C_tilde_avg = C_tilde_accum / replications

    M = (XtX) / N
    M_inv = np.linalg.inv(M)

    C_hat_withN = M_inv @ C_tilde_avg @ M_inv

    C_hat_noN = (1.0 / N) * C_hat_withN

    err_noN = np.linalg.norm(C_hat_noN - C_true_noN, ord='fro')
    err_withN = np.linalg.norm(C_hat_withN - C_true_withN, ord='fro')

    errors_noN.append(err_noN)
    errors_withN.append(err_withN)

    if N == sample_sizes[-1]:
        print(f"\n=== Sample size N = {N} ===")

        print("\n True Cov (noN) = sigma^2 (X^T X)^(-1):")
        print(np.round(C_true_noN, 4))
        print("[reckon] C_hat_noN:")
        print(np.round(C_hat_noN, 4))
        print("Frobenius norm error (noN):", err_noN)

```

```
    print("\n True Cov (withN) = sigma^2 * N * (X^T X)
        ^(-1):")
    print(np.round(C_true_withN, 4))
    print(" C_hat_withN:")
    print(np.round(C_hat_withN, 4))
    print("Frobenius norm error (withN):", err_withN)
```

Example 3.3

```
#Table of Parameter Bias Convergence Under Large-Sample
    Conditions

np.random.seed(42)

theta_true = np.array([1.5, 0.8, 2.1, -1.2, 0.7, 3.0])
sigma = 0.5

sample_sizes = np.linspace(50, 10000050, 20, dtype=int)
replications = 50

def generate_ar1_errors(N, rho, sigma):
    epsilon = np.zeros(N)
    epsilon[0] = np.random.normal(0, sigma)
    for i in range(1, N):
        epsilon[i] = rho * epsilon[i-1] + np.random.normal(0,
            sigma)
    return epsilon

mean_estimates = np.zeros((len(sample_sizes), 6))
bias_estimates = np.zeros((len(sample_sizes), 6))

for idx, N in enumerate(sample_sizes):
    rep_estimates = np.zeros((replications, 6))
    for r in range(replications):
        x = np.linspace(0, 1, N)
        X = np.vander(x, N=6, increasing=True)
        epsilon = generate_ar1_errors(N, rho=0.5, sigma=sigma)
        y = X \@ theta_true + epsilon
        theta_hat = np.linalg.lstsq(X, y, rcond=None)[0]
        rep_estimates[r, :] = theta_hat
    mean_estimates[idx, :] = rep_estimates.mean(axis=0)
    bias_estimates[idx, :] = mean_estimates[idx, :] -
        theta_true

df_bias = pd.DataFrame(bias_estimates, columns=[f'$\theta${i}
    Bias' for i in range(6)])
df_bias.insert(0, 'Sample Size', sample_sizes)

print("Bias Estimates:")
print(df_bias)

#Figure of Estimator Consistency and Bias Convergence

theta_true = np.array([1.5, 0.8, 2.1, -1.2, 0.7, 3.0])
sigma = 0.5

N_values = np.linspace(50, 10000000, 21, dtype=int)

replications = 50
```

```
mean_estimates = np.zeros((len(N_values), 6))
bias_estimates = np.zeros((len(N_values), 6))

def generate_ar1_errors(N, rho, sigma):
    epsilon = np.zeros(N)
    epsilon[0] = np.random.normal(0, sigma)
    for i in range(1, N):
        epsilon[i] = rho * epsilon[i-1] + np.random.normal(0,
            sigma)
    return epsilon

for idx, N in enumerate(N_values):
    rep_estimates = np.zeros((replications, 6))

    x = np.linspace(0, 1, N)
    X = np.vander(x, N=6, increasing=True)

    for r in range(replications):
        epsilon = generate_ar1_errors(N, rho=0.5, sigma=sigma)
        y = X \@ theta_true + epsilon
        theta_hat = np.linalg.lstsq(X, y, rcond=None)[0]
        rep_estimates[r, :] = theta_hat

    mean_estimates[idx, :] = rep_estimates.mean(axis=0)
    bias_estimates[idx, :] = mean_estimates[idx, :] -
        theta_true


#Quantile Visualization of the Convergence Trajectories of
    Parameter Biases

theta_true = np.array([1.5, 0.8, 2.1, -1.2, 0.7, 3.0])
sigma = 0.5

N_values_1 = np.linspace(50, 10000, 60, dtype=int)

N_values_2 = np.linspace(10000, 10000000, 40, dtype=int)

N_values = np.unique(np.concatenate((N_values_1, N_values_2)))

theta_hats = []
det_values = []

def generate_ar1_errors(N, rho, sigma):
    epsilon = np.zeros(N)
    epsilon[0] = np.random.normal(0, sigma)
    for i in range(1, N):
        epsilon[i] = rho * epsilon[i-1] + np.random.normal(0,
            sigma)
    return epsilon

for N in N_values:
    x = np.linspace(0, 1, N)
    X = np.vander(x, N=6, increasing=True)

    epsilon = generate_ar1_errors(N, rho=0.5, sigma=sigma)
    y = X \@ theta_true + epsilon

    theta_hat = np.linalg.lstsq(X, y, rcond=None)[0]
    theta_hats.append(theta_hat)

    XTX = X.T \@ X
    det_ratio = det(XTX / N)
    det_values.append(N * det_ratio)

theta_hats = np.array(theta_hats)
```

```python

replications = 50
quantiles = [0.05, 0.25, 0.5, 0.75, 0.95]

theta_estimates = {i: {q: [] for q in quantiles} for i in range(6)}
median_estimates = {i: [] for i in range(6)}

for N in N_values:
    rep_theta = np.zeros((replications, 6))
    x = np.linspace(0, 1, N)
    X = np.vander(x, N=6, increasing=True)
    for r in range(replications):
        epsilon = np.random.normal(0, sigma, N)
        y = X \@ theta_true + epsilon
        theta_hat = np.linalg.lstsq(X, y, rcond=None)[0]
        rep_theta[r, :] = theta_hat
    for i in range(6):
        for q in quantiles:
            theta_estimates[i][q].append(np.quantile(rep_theta[:, i], q))
        median_estimates[i].append(np.quantile(rep_theta[:, i], 0.5))

fig, axes = plt.subplots(2, 3, figsize=(18, 10))
axes = axes.flatten()
for i in range(6):
    ax = axes[i]
    for q in quantiles:
        ax.plot(N_values, theta_estimates[i][q], label=f'{int(q*100)}\% Quantile', linestyle='--')
    ax.plot(N_values, median_estimates[i], label='Median', color='black', linewidth=2)
    ax.axhline(theta_true[i], color='red', linestyle='-', label='True Value')
    ax.set_xscale('log')
    ax.set_xlabel('Sample Size (N)')
    ax.set_ylabel(f'Estimate of $\theta${i}')
    ax.set_title(f'Convergence Path for \theta{i}')
    ax.legend()


#Graphical Test for Heteroscedasticity

theta_true = np.array([1.5, 0.8, 2.1, -1.2, 0.7, 3.0])
sigma = 0.5

sample_sizes = [10, 100, 1000, 10000, 100000, 1000000, 10000000]

def generate_ar1_errors(N, rho, sigma):
    epsilon = np.zeros(N)
    epsilon[0] = np.random.normal(0, sigma)
    for i in range(1, N):
        epsilon[i] = rho * epsilon[i - 1] + np.random.normal(0, sigma)
    return epsilon

def experiment(N):
    x = np.linspace(0, 1, N)
    X = np.vander(x, N=6, increasing=True)

    epsilon = generate_ar1_errors(N, rho=0.5, sigma=sigma)

    y = X \@ theta_true + epsilon

    theta_hat = np.linalg.lstsq(X, y, rcond=None)[0]
```

```
    return theta_hat , X, y


#Heatmap of Marginal and Joint Parameter Density Distributions

np.random.seed(42)

theta_true = np.array([1.5, 0.8, 2.1, -1.2, 0.7, 3.0])
sigma = 0.5

N_values = [10, 100, 100000, 1000000]

replications = 200

def generate_ar1_errors(N, rho, sigma):
    epsilon = np.zeros(N)
    epsilon[0] = np.random.normal(0, sigma)
    for i in range(1, N):
        epsilon[i] = rho * epsilon[i-1] + np.random.normal(0,
            sigma)
    return epsilon

for N in N_values:
    estimates = np.zeros((replications, 6))

    x = np.linspace(0, 1, N)
    X = np.vander(x, N=6, increasing=True)

    for r in range(replications):
        epsilon = generate_ar1_errors(N, rho=0.5, sigma=sigma)
        y = X \@ theta_true + epsilon
        theta_hat = np.linalg.lstsq(X, y, rcond=None)[0]
        estimates[r, :] = theta_hat

#Confidence Interval

np.random.seed(42)

theta_true = np.array([1.5, 0.8, 2.1, -1.2, 0.7, 3.0])
sigma = 0.5
sample_sizes = [10, 100, 100000, 1000000]
replications = 200

def generate_ar1_errors(N, rho, sigma):
    epsilon = np.zeros(N)
    epsilon[0] = np.random.normal(0, sigma)
    for i in range(1, N):
        epsilon[i] = rho * epsilon[i - 1] + np.random.normal
            (0, sigma)
    return epsilon

theta_means = []
theta_ci_lower = []
theta_ci_upper = []

for N in sample_sizes:
    estimates = np.zeros((replications, 6))
    x = np.linspace(0, 1, N)
    X = np.vander(x, N=6, increasing=True)
    for r in range(replications):
        epsilon = generate_ar1_errors(N, rho=0.5, sigma=sigma)
        y = X @ theta_true + epsilon
        theta_hat = np.linalg.lstsq(X, y, rcond=None)[0]
        estimates[r, :] = theta_hat
    theta_mean = np.nanmean(estimates, axis=0)
    theta_std = np.nanstd(estimates, axis=0)
    ci_halfwidth = 1.96 * (theta_std / np.sqrt(replications))
```

```
    ci_lower = theta_mean - ci_halfwidth
    ci_upper = theta_mean + ci_halfwidth
    theta_means.append(theta_mean)
    theta_ci_lower.append(ci_lower)
    theta_ci_upper.append(ci_upper)

theta_means = np.array(theta_means)
theta_ci_lower = np.array(theta_ci_lower)
theta_ci_upper = np.array(theta_ci_upper)

num_params = theta_means.shape[1]
num_cols = 3
num_rows = math.ceil(num_params / num_cols)

fig, axes = plt.subplots(num_rows, num_cols, figsize=(16, 9))
axes = axes.flatten()

for i in range(num_params):
    ax = axes[i]
    means = theta_means[:, i]
    lower = theta_ci_lower[:, i]
    upper = theta_ci_upper[:, i]
    yerr_lower = means - lower
    yerr_upper = upper - means
    yerr = np.vstack([yerr_lower, yerr_upper])
    ax.errorbar(sample_sizes, means, yerr=yerr, fmt='-o',
        capsize=4,
                ecolor='red', color='blue', markersize=5,
                    label=f'\theta{i} Estimate')
    ax.set_xscale('log')
    ax.set_xlabel('Sample Size (log scale)')
    ax.set_ylabel('Parameter Estimate')
    ax.set_title(f'Parameter \theta{i}')
    ax.grid(True, linestyle=':', alpha=0.7)
    ax.legend()

for j in range(num_params, len(axes)):
    axes[j].axis('off')

#Covariance calculation

np.random.seed(42)

theta_true = np.array([1.5, 0.8, 2.1, -1.2, 0.7, 3.0])
sigma = 0.5
sample_sizes = np.linspace(50, 10000000, 50, dtype=int)
replications = 200
rho = 0.5

def generate_ar1_errors_vectorized(N, rho, sigma, n_reps):
    noise = np.random.normal(0, sigma, size=(n_reps, N))
    epsilons = lfilter([1], [1, -rho], noise, axis=1)
    return epsilons

def process_sample_size(N, replications, chunk_size=50):
    x = np.linspace(0, 1, N)
    X = np.vander(x, N=6, increasing=True)
    XtX = X.T @ X
    XtX_inv = np.linalg.inv(XtX)
    C_true_noN = sigma**2 * XtX_inv
    C_true_withN = sigma**2 * N * XtX_inv
    max_exponent = 10
    x_power = np.empty((max_exponent + 1, N))
    for m in range(max_exponent + 1):
        x_power[m, :] = x ** m
    exp_matrix = np.add.outer(np.arange(6), np.arange(6))
    y_mean = X @ theta_true
```

```
    n_chunks = replications // chunk_size
    remainder = replications % chunk_size
    C_tilde_accum = np.zeros((6, 6))
    def process_chunk(n_reps):
        eps = generate_ar1_errors_vectorized(N, rho, sigma,
            n_reps)
        y_all = y_mean[None, :] + eps
        b_all = y_all @ X
        theta_hats = b_all @ XtX_inv
        fitted_all = X @ theta_hats.T
        residuals_all = y_all.T - fitted_all
        r2_all = residuals_all ** 2
        sums_all = x_power @ r2_all
        sums_avg = np.mean(sums_all, axis=1)
        C_tilde = sums_avg[exp_matrix] / N
        return C_tilde
    for _ in range(n_chunks):
        C_tilde_accum += process_chunk(chunk_size)
    if remainder > 0:
        C_tilde_accum += process_chunk(remainder)
        total_reps = n_chunks * chunk_size + remainder
    else:
        total_reps = n_chunks * chunk_size
    C_tilde_avg = C_tilde_accum / total_reps
    M = (X.T @ X) / N
    M_inv = np.linalg.inv(M)
    C_hat_withN = M_inv @ C_tilde_avg @ M_inv
    C_hat_noN = (1.0 / N) * C_hat_withN
    err_noN = np.linalg.norm(C_hat_noN - C_true_noN, ord='fro'
        )
    err_withN = np.linalg.norm(C_hat_withN - C_true_withN, ord
        ='fro')
    return N, err_noN, err_withN
```

Appendix 2
Python Code for Chap. 4 Some Arithmetic Examples

Example 4.2

```
#Table of Parameter Bias Convergence Under Large-Sample
    Conditions

np.random.seed(42)

alpha_true = np.array([2.0, 0.3, 1.2, 0.5, 0.8, 1.2])
sigma_error = 0.5
rho = 0.5

def mixed_model(x, a1, a2, a3, a4, a5, a6):
    return a1 * np.exp(a2 * x) + a3 * (x ** a4) + a5 * np.sin(
        a6 * x)

def mixed_model_jac(x, a1, a2, a3, a4, a5, a6):
    d_da1 = np.exp(a2 * x)
    d_da2 = a1 * x * np.exp(a2 * x)
    d_da3 = x ** a4
    d_da4 = a3 * (x ** a4) * np.log(x)
    d_da5 = np.sin(a6 * x)
    d_da6 = a5 * x * np.cos(a6 * x)
    return np.vstack((d_da1, d_da2, d_da3, d_da4, d_da5, d_da6
        )).T

def generate_correlated_errors(N, rho, sigma_error):
    noise = np.random.normal(0, sigma_error, size=N)
    epsilon = lfilter([1], [1, -rho], noise)
    return epsilon

sample_sizes = np.linspace(50, 10000000, 20, dtype=int)
replications = 50
num_params = len(alpha_true)

mean_estimates = np.zeros((len(sample_sizes), num_params))
bias_estimates = np.zeros_like(mean_estimates)

def run_single_fit(N):
    x = np.linspace(1, 6, N)
    epsilon = generate_correlated_errors(N, rho, sigma_error)
    y = mixed_model(x, *alpha_true) + epsilon
    try:
```


A. Hajiyev and J. Xu, *High-Dimensional Regression Analysis and Artificial Intelligence*,
https://doi.org/10.1007/978-981-95-2514-0

```
        popt, _ = curve_fit(mixed_model, x, y, p0=alpha_true, jac=mixed_model_jac, maxfev=10000)
    except RuntimeError:
        popt = np.full(num_params, np.nan)
    return popt

if __name__ == "__main__":
    for idx, N in enumerate(sample_sizes):
        rep_estimates = np.zeros((replications, num_params))
        with concurrent.futures.ProcessPoolExecutor() as executor:
            futures = [executor.submit(run_single_fit, N) for _ in range(replications)]
            for i, future in enumerate(concurrent.futures.as_completed(futures)):
                rep_estimates[i, :] = future.result()
        mean_estimates[idx, :] = np.nanmean(rep_estimates, axis=0)
        bias_estimates[idx, :] = mean_estimates[idx, :] - alpha_true

    column_names = [f'\alpha{i+1} Bias' for i in range(num_params)]
    df_bias = pd.DataFrame(bias_estimates, columns=column_names)
    df_bias.insert(0, 'Sample Size', sample_sizes)

    print("Bias Estimates:")
    print(df_bias)

#Figure of Estimator Consistency and Bias Convergence

np.random.seed(42)

alpha1_true, alpha2_true, alpha3_true = 2.0, 0.3, 1.2
alpha4_true, alpha5_true, alpha6_true = 0.5, 0.8, 1.2
sigma_error = 0.5
rho = 0.5
true_params_array = np.array([alpha1_true, alpha2_true, alpha3_true, alpha4_true, alpha5_true, alpha6_true])

N_values_1 = np.linspace(50, 1000000, 60, dtype=int)
N_values_2 = np.linspace(1000000, 10000000, 40, dtype=int)
N_values = np.unique(np.concatenate((N_values_1, N_values_2)))
replications = 20

mean_estimates = np.zeros((len(N_values), 6))
bias_estimates = np.zeros((len(N_values), 6))

def mixed_model(x, a1, a2, a3, a4, a5, a6):
    return a1 * np.exp(a2 * x) + a3 * (x ** a4) + a5 * np.sin(a6 * x)

def residual_and_jac(params, x, y):
    a1, a2, a3, a4, a5, a6 = params
    y_pred = mixed_model(x, a1, a2, a3, a4, a5, a6)
    residuals = y_pred - y
    J = np.zeros((len(x), 6))
    J[:, 0] = np.exp(a2 * x)
    J[:, 1] = a1 * x * np.exp(a2 * x)
    J[:, 2] = x ** a4
    J[:, 3] = a3 * (x ** a4) * np.log(x)
    J[:, 4] = np.sin(a6 * x)
    J[:, 5] = a5 * x * np.cos(a6 * x)
    return residuals, J

def fun_for_ls(params, x, y):
```

```
    res, _ = residual_and_jac(params, x, y)
    return res

def jac_for_ls(params, x, y):
    _, J = residual_and_jac(params, x, y)
    return J

def generate_correlated_errors(N, rho, sigma_error):
    epsilon = np.zeros(N)
    epsilon[0] = np.random.normal(0, sigma_error)
    for i in range(1, N):
        epsilon[i] = rho * epsilon[i-1] + np.random.normal(0, sigma_error)
    return epsilon

def subsample_if_needed(x, y, max_points=100000):
    N = len(x)
    if N <= max_points:
        return x, y
    else:
        idx = np.random.choice(N, size=max_points, replace=False)
        return x[idx], y[idx]

def multi_start_least_squares(x_sub, y_sub, base_guess, n_starts=3):
    best_params = None
    best_cost = np.inf
    param_bounds = ([0, -1, 0, 0, -5, -5], [5, 2, 5, 2, 5, 5])
    for i in range(n_starts):
        perturb = np.random.uniform(-0.2, 0.2, size=len(base_guess))
        init_guess = np.array(base_guess) + perturb
        try:
            result = least_squares(fun_for_ls, init_guess, jac=jac_for_ls, args=(x_sub, y_sub),
                                   method='trf', bounds=param_bounds, max_nfev=5000)
            cost = np.sum(result.fun**2)
            if cost < best_cost:
                best_cost = cost
                best_params = result.x
        except Exception:
            pass
    if best_params is None:
        best_params = [np.nan]*6
    return best_params

def single_replication(N, rho, sigma_error, base_guess):
    x_full = np.linspace(1, 6, N)
    epsilon = generate_correlated_errors(N, rho, sigma_error)
    y_full = (alpha1_true * np.exp(alpha2_true * x_full) +
              alpha3_true * (x_full ** alpha4_true) +
              alpha5_true * np.sin(alpha6_true * x_full) +
              epsilon)
    x_sub, y_sub = subsample_if_needed(x_full, y_full, max_points=100000)
    params = multi_start_least_squares(x_sub, y_sub, base_guess, n_starts=3)
    return params

initial_guess = [1, 0.1, 1, 0.1, 1, 1]

for idx, N in enumerate(N_values):
    results = Parallel(n_jobs=-1)(
```

```
        delayed(single_replication)(N, rho, sigma_error, initial_guess)
        for _ in range(replications)
    )
    rep_estimates = np.array(results)
    mask = np.all(np.abs(rep_estimates) <= 5, axis=1)
    filtered_estimates = rep_estimates[mask, :]
    if len(filtered_estimates) == 0:
        mean_estimates[idx, :] = np.array([np.nan]*6)
        bias_estimates[idx, :] = np.array([np.nan]*6)
        continue
    mean_estimates[idx, :] = np.nanmean(filtered_estimates, axis=0)
    bias_estimates[idx, :] = mean_estimates[idx, :] - true_params_array

#Quantile Visualization of the Convergence Trajectories of Parameter Biases

np.random.seed(42)

alpha1_true = 2.0
alpha2_true = 0.3
alpha3_true = 1.2
alpha4_true = 0.5
alpha5_true = 0.8
alpha6_true = 1.2

sigma_error = 0.5
rho = 0.5
true_params_array = np.array([alpha1_true, alpha2_true, alpha3_true, alpha4_true, alpha5_true, alpha6_true])

N_values_1 = np.linspace(50, 1000000, 60, dtype=int)
N_values_2 = np.linspace(1000000, 10000000, 40, dtype=int)
N_values = np.unique(np.concatenate((N_values_1, N_values_2)))
replications = 20

mean_estimates = np.zeros((len(N_values), 6))
bias_estimates = np.zeros((len(N_values), 6))

def mixed_model(x, a1, a2, a3, a4, a5, a6):
    return a1 * np.exp(a2 * x) + a3 * (x ** a4) + a5 * np.sin(a6 * x)

def residual_and_jac(params, x, y):
    a1, a2, a3, a4, a5, a6 = params
    y_pred = mixed_model(x, a1, a2, a3, a4, a5, a6)
    residuals = y_pred - y
    J = np.zeros((len(x), 6))
    J[:, 0] = np.exp(a2 * x)
    J[:, 1] = a1 * x * np.exp(a2 * x)
    J[:, 2] = x ** a4
    J[:, 3] = a3 * (x ** a4) * np.log(x)
    J[:, 4] = np.sin(a6 * x)
    J[:, 5] = a5 * x * np.cos(a6 * x)
    return residuals, J

def fun_for_ls(params, x, y):
    res, _ = residual_and_jac(params, x, y)
    return res

def jac_for_ls(params, x, y):
    _, J = residual_and_jac(params, x, y)
    return J

def generate_correlated_errors(N, rho, sigma_error):
```

```
    epsilon = np.zeros(N)
    epsilon[0] = np.random.normal(0, sigma_error)
    for i in range(1, N):
        epsilon[i] = rho * epsilon[i-1] + np.random.normal(0,
            sigma_error)
    return epsilon

def subsample_if_needed(x, y, max_points=100000):
    N = len(x)
    if N <= max_points:
        return x, y
    else:
        idx = np.random.choice(N, size=max_points, replace=
            False)
        return x[idx], y[idx]

def multi_start_least_squares(x_sub, y_sub, base_guess,
    n_starts=3):
    best_params = None
    best_cost = np.inf
    param_bounds = ([0, -1, 0, 0, -5, -5], [5, 2, 5, 2, 5, 5])
    for i in range(n_starts):
        perturb = np.random.uniform(-0.2, 0.2, size=len(
            base_guess))
        init_guess = np.array(base_guess) + perturb
        try:
            result = least_squares(fun_for_ls, init_guess, jac
                =jac_for_ls, args=(x_sub, y_sub),
                                   method='trf', bounds=
                                       param_bounds, max_nfev
                                       =5000)
            cost = np.sum(result.fun**2)
            if cost < best_cost:
                best_cost = cost
                best_params = result.x
        except Exception:
            pass
    if best_params is None:
        best_params = [np.nan]*6
    return best_params

def single_replication(N, rho, sigma_error, base_guess):
    x_full = np.linspace(1, 6, N)
    epsilon = generate_correlated_errors(N, rho, sigma_error)
    y_full = (alpha1_true * np.exp(alpha2_true * x_full) +
              alpha3_true * (x_full ** alpha4_true) +
              alpha5_true * np.sin(alpha6_true * x_full) +
              epsilon)
    x_sub, y_sub = subsample_if_needed(x_full, y_full,
        max_points=100000)
    params = multi_start_least_squares(x_sub, y_sub,
        base_guess, n_starts=3)
    return params

initial_guess = [1, 0.1, 1, 0.1, 1, 1]

for idx, N in enumerate(N_values):
    results = Parallel(n_jobs=-1)(
        delayed(single_replication)(N, rho, sigma_error,
            initial_guess)
        for _ in range(replications)
    )
    rep_estimates = np.array(results)
    mask = np.all(np.abs(rep_estimates) <= 5, axis=1)
    filtered_estimates = rep_estimates[mask, :]
    if len(filtered_estimates) == 0:
        mean_estimates[idx, :] = np.array([np.nan]*6)
```

```
        bias_estimates[idx, :] = np.array([np.nan]*6)
        continue
    mean_estimates[idx, :] = np.nanmean(filtered_estimates, axis=0)
    bias_estimates[idx, :] = mean_estimates[idx, :] - true_params_array
    initial_guess = mean_estimates[idx, :].copy()

quantiles = [0.05, 0.25, 0.5, 0.75, 0.95]
theta_estimates = {i: {q: [] for q in quantiles} for i in range(6)}
median_estimates = {i: [] for i in range(6)}
initial_guess = np.array([1, 0.1, 1, 0.1, 1, 1], dtype=float)

for N in N_values:
    rep = 10
    with Parallel(n_jobs=-1) as parallel:
        rep_theta = parallel(
            delayed(single_replication)(get_cached_x_y0(N)[0], get_cached_x_y0(N)[1], N, rho, sigma_error, initial_guess)
            for _ in range(rep)
        )
    rep_theta = np.array(rep_theta)
    initial_guess = np.nanmean(rep_theta, axis=0)
    for i in range(6):
        for q in quantiles:
            theta_estimates[i][q].append(np.nanquantile(rep_theta[:, i], q))
        median_estimates[i].append(np.nanquantile(rep_theta[:, i], 0.5))

fig, axes = plt.subplots(2, 3, figsize=(18, 10))
axes = axes.flatten()

for i in range(6):
    ax = axes[i]
    for q in quantiles:
        ax.plot(N_values, theta_estimates[i][q], linestyle='--', label=f'{int(q*100)}\% Quantile')
    ax.plot(N_values, median_estimates[i], color='black', linewidth=2, label='Median')
    ax.axhline(true_params_array[i], color='red', linestyle='-', label='True Value')
    ax.set_xscale('log')
    ax.set_xlabel('Sample Size (N)', fontsize=12)
    ax.set_ylabel(rf'Estimate of $\alpha_{{{i+1}}}$', fontsize=12)
    ax.set_title(rf'Convergence Path for $\alpha_{{{i+1}}}$', fontsize=14)
    ax.legend(fontsize=10)
    ax.grid(True, linestyle='--', alpha=0.5)

for j in range(6, len(axes)):
    axes[j].set_visible(False)

#Graphical Test for Heteroscedasticity

np.random.seed(42)

alpha1_true = 2.0
alpha2_true = 0.3
alpha3_true = 1.2
alpha4_true = 0.5
alpha5_true = 0.8
alpha6_true = 1.2

```

```
sigma_error = 0.5
rho = 0.5

def mixed_model(x, a1, a2, a3, a4, a5, a6):
    return a1 * np.exp(a2 * x) + a3 * (x ** a4) + a5 * np.sin(a6 * x)

def generate_correlated_errors(N, rho, sigma_error):
    epsilon = np.zeros(N)
    epsilon[0] = np.random.normal(0, sigma_error)
    for i in range(1, N):
        epsilon[i] = rho * epsilon[i-1] + np.random.normal(0, sigma_error)
    return epsilon

np.random.seed(42)

sample_sizes = [10, 100, 1000, 10000, 100000, 1000000, 10000000]

for idx, N in enumerate(sample_sizes):
    x = np.linspace(1, 10, N)
    epsilon = generate_correlated_errors(N, rho, sigma_error)
    y = mixed_model(x, alpha1_true, alpha2_true, alpha3_true, alpha4_true, alpha5_true, alpha6_true) + epsilon
    try:
        popt, _ = curve_fit(mixed_model, x, y, p0=[1, 1, 1, 1, 1, 1])
    except RuntimeError:
        popt = [np.nan] * 6
    y_fitted = mixed_model(x, *popt)
    residuals = y - y_fitted
    plt.subplot(len(sample_sizes), 1, idx + 1)
    plt.scatter(y_fitted, residuals, alpha=0.6)
    plt.axhline(0, color='red', linestyle='--')

#Heatmap of Marginal and Joint Parameter Density Distributions

np.random.seed(42)

alpha1_true = 2.0
alpha2_true = 0.3
alpha3_true = 1.2
alpha4_true = 0.5
alpha5_true = 0.8
alpha6_true = 1.2

sigma_error = 0.5
rho = 0.5

N_values = [10, 100, 10000, 100000, 1000000, 5000000, 8000000, 10000000]
replications = 50

mean_estimates = np.zeros((len(N_values), 6))
bias_estimates = np.zeros((len(N_values), 6))

def mixed_model(x, a1, a2, a3, a4, a5, a6):
    return a1 * np.exp(a2 * x) + a3 * (x ** a4) + a5 * np.sin(a6 * x)

def generate_correlated_errors(N, rho, sigma_error):
    epsilon = np.zeros(N)
    epsilon[0] = np.random.normal(0, sigma_error)
    for i in range(1, N):
        epsilon[i] = rho * epsilon[i-1] + np.random.normal(0, sigma_error)
```

```
    return epsilon

for idx, N in enumerate(N_values):
    rep_estimates = np.zeros((replications, 6))
    x = np.linspace(1, 6, N)
    for r in range(replications):
        epsilon = generate_correlated_errors(N, rho, sigma_error)
        y = (alpha1_true * np.exp(alpha2_true * x) +
             alpha3_true * (x ** alpha4_true) +
             alpha5_true * np.sin(alpha6_true * x) +
             epsilon)
        try:
            params, covariance = curve_fit(mixed_model, x, y, p0=[0.7, 0.1, 1, 0.5, 1, 1])
        except RuntimeError:
            params = [np.nan] * 6
        rep_estimates[r, :] = params
    mean_estimates[idx, :] = np.nanmean(rep_estimates, axis=0)
    true_params_array = np.array([alpha1_true, alpha2_true, alpha3_true,
                                  alpha4_true, alpha5_true, alpha6_true])
    bias_estimates[idx, :] = mean_estimates[idx, :] - true_params_array
    param_cols = [f'\theta{i}' for i in range(6)]

#Confidence Interval

np.random.seed(42)

alpha1_true = 2.0
alpha2_true = 0.3
alpha3_true = 1.2
alpha4_true = 0.5
alpha5_true = 0.8
alpha6_true = 1.2

sigma_error = 0.5
rho = 0.5

N_values = [10, 100, 10000, 100000, 1000000, 10000000]
replications = 50

def mixed_model(x, a1, a2, a3, a4, a5, a6):
    return a1 * np.exp(a2 * x) + a3 * (x ** a4) + a5 * np.sin(a6 * x)

def generate_correlated_errors(N, rho, sigma_error):
    epsilon = np.zeros(N)
    epsilon[0] = np.random.normal(0, sigma_error)
    for i in range(1, N):
        epsilon[i] = rho * epsilon[i-1] + np.random.normal(0, sigma_error)
    return epsilon

theta_means = np.zeros((len(N_values), 6))
theta_cis_lower = np.zeros((len(N_values), 6))
theta_cis_upper = np.zeros((len(N_values), 6))

for idx, N in enumerate(N_values):
    rep_estimates = np.zeros((replications, 6))
    x = np.linspace(1, 6, N)
    for r in range(replications):
        epsilon = generate_correlated_errors(N, rho, sigma_error=sigma_error)
        y = (alpha1_true * np.exp(alpha2_true * x) +
```

```
             alpha3_true * (x ** alpha4_true) +
             alpha5_true * np.sin(alpha6_true * x) +
             epsilon)
        try:
            params, covariance = curve_fit(mixed_model, x, y, p0=[0.7, 0.1, 1, 0.5, 1, 1])
        except RuntimeError:
            params = [np.nan] * 6
        rep_estimates[r, :] = params
    theta_mean = np.nanmean(rep_estimates, axis=0)
    theta_std = np.nanstd(rep_estimates, axis=0)
    ci_halfwidth = 1.96 * (theta_std / np.sqrt(replications))
    theta_means[idx, :] = theta_mean
    theta_cis_lower[idx, :] = theta_mean - ci_halfwidth
    theta_cis_upper[idx, :] = theta_mean + ci_halfwidth
    print(f"Sample Size N = {N}")
    for i in range(6):
        print(f"\theta{i}: mean = {theta_mean[i]:.3f}, 95\% CI = [{theta_cis_lower[idx, i]:.3f}, {theta_cis_upper[idx, i]:.3f}]")
    print("-" * 50)

#Covariance calculation

n_cpu = multiprocessing.cpu_count()
blas_threads = max(1, n_cpu // 4)
os.environ["OMP_NUM_THREADS"] = str(blas_threads)
os.environ["OPENBLAS_NUM_THREADS"] = str(blas_threads)
os.environ["MKL_NUM_THREADS"] = str(blas_threads)
os.environ["NUMEXPR_NUM_THREADS"] = str(blas_threads)
print(f"[INFO] CPU={n_cpu}, BLAS Threads per process={blas_threads}")

c_value = 2.34

@njit
def tukey_loss_numba(theta, X, y, c=c_value):
    r = y - X @ theta
    loss = np.where(np.abs(r) <= c, (c**2 / 6) * (1 - (1 - (r/c)**2)**3), (c**2 / 6))
    return np.sum(loss)

@njit
def tukey_grad_numba(theta, X, y, c=c_value):
    r = y - X @ theta
    grad_factor = np.where(np.abs(r) <= c, r * (1 - (r/c)**2)**2, 0.0)
    return -X.T @ grad_factor

def safe_inverse(H, threshold=1e8):
    cond = np.linalg.cond(H)
    if cond < threshold:
        return np.linalg.solve(H, np.eye(H.shape[0]))
    else:
        print(f"Warning: Hessian ill-conditioned (cond={cond:.2e}), using pinv()")
        return np.linalg.pinv(H)

def tukey_regression_with_cov(X, y, c=c_value):
    theta0, _, _, _ = np.linalg.lstsq(X, y, rcond=None)
    r0 = y - X @ theta0
    w0 = np.where(np.abs(r0) <= c, (1 - (r0/c)**2)**2, 1e-6)
    theta0_new, _, _, _ = np.linalg.lstsq(X * np.sqrt(w0[:, None]), y * np.sqrt(w0), rcond=None)
    res = minimize(fun=lambda th: tukey_loss_numba(th, X, y, c),
                   x0=theta0_new,
```

```
                      jac=lambda th: tukey_grad_numba(th, X, y, c),
                      method='L-BFGS-B')
    theta_hat = res.x
    r = y - X @ theta_hat
    factor = np.where(np.abs(r) <= c, (1 - (r/c)**2) * (1 - 5*(r/c)**2), 0.0)
    H = X.T @ (factor[:, None] * X)
    cov_est = sigma**2 * safe_inverse(H)
    return theta_hat, cov_est

np.random.seed(42)
theta_true = np.array([1.5, 0.8, 2.1, -1.2, 0.7, 3.0, 0.5, -0.6, 1.1, -0.9, 2.2])
sigma = 0.5
sample_sizes = np.linspace(50, 10000000, 50, dtype=int)
replications = 50
errors_noN = []

def run_replication(seed):
    np.random.seed(seed)
    eps = np.random.normal(0, sigma, size=X_global.shape[0])
    y = X_global @ theta_true + eps
    _, cov_est = tukey_regression_with_cov(X_global, y, c=c_value)
    return cov_est

for idx, N in enumerate(sample_sizes):
    print(f"Processing sample size: {N} (step {idx+1}/{len(sample_sizes)})")
    x = np.linspace(0, 1, N)
    X = legvander(2 * x - 1, deg=10)
    XtX = X.T @ X
    C_true = sigma**2 * safe_inverse(XtX)
    global X_global
    X_global = X
    tasks = [42 + r for r in range(replications)]
    with concurrent.futures.ProcessPoolExecutor() as executor:
        results = executor.map(run_replication, tasks)
        cov_est_accum = sum(results)
    avg_cov_est = cov_est_accum / replications
    err = np.linalg.norm(avg_cov_est - C_true, ord='fro')
    errors_noN.append(err)
    if N == sample_sizes[-1]:
        print(f"\n=== Sample size N = {N} ===")
        print("\n[True Covariance]:")
        print(np.round(C_true, 4))
        print("\n[Estimated Covariance]:")
        print(np.round(avg_cov_est, 4))
        print("\nFrobenius norm error:", err)
```

Example 4.4

```
#Table of Parameter Bias Convergence Under Large-Sample Conditions

c_value = 2.34

def tukey_loss(theta, X, y, c=c_value):
```

```python
    r = y - X.dot(theta)
    loss = np.where(np.abs(r) <= c,
                    (c**2 / 6) * (1 - (1 - (r/c)**2)**3),
                    (c**2 / 6))
    return np.sum(loss)

def tukey_grad(theta, X, y, c=c_value):
    r = y - X.dot(theta)
    grad_factor = np.where(np.abs(r) <= c, r * (1 - (r/c)**2)**2, 0.0)
    return -X.T.dot(grad_factor)

def tukey_regression(X, y, c=c_value):
    theta0 = np.zeros(X.shape[1])
    res = minimize(tukey_loss, theta0, args=(X, y, c),
                   jac=tukey_grad, method='L-BFGS-B',
                   options={'maxiter': 50, 'ftol': 1e-5})
    return res.x

theta_true = np.array([1.5, 0.8, 2.1, -1.2, 0.7, 3.0, 0.5, -0.6, 1.1, -0.9, 2.2])
sigma = 0.5
replications = 50
np.random.seed(42)

sample_sizes = np.linspace(50, 10000050, 20, dtype=int)
mean_estimates = np.zeros((len(sample_sizes), len(theta_true)))
bias_estimates = np.zeros_like(mean_estimates)

for idx, N in enumerate(sample_sizes):
    print(f"Running N = {N}")

    x = 0.5 * (1 - np.cos((2 * np.arange(1, N + 1) - 1) * np.pi / (2 * N)))
    z = 2 * x - 1
    X = np.empty((N, 11))
    for i in range(11):
        X[:, i] = np.polynomial.legendre.legval(z, np.eye(11)[i])

    y_mean = X @ theta_true

    def run_replication():
        eps = np.random.normal(0, sigma, N)
        y = y_mean + eps
        return tukey_regression(X.copy(), y)

    theta_hats = Parallel(n_jobs=min(12, os.cpu_count() // 2),
                          max_nbytes=1e6)(
        delayed(run_replication)() for _ in range(replications)
    )

    theta_hats = np.array(theta_hats)
    mean_estimates[idx, :] = theta_hats.mean(axis=0)
    bias_estimates[idx, :] = mean_estimates[idx, :] - theta_true

column_names = [f'\theta{i} Bias' for i in range(len(theta_true))]
df_bias = pd.DataFrame(bias_estimates, columns=column_names)
df_bias.insert(0, 'Sample Size', sample_sizes)

print(``Bias Estimates:'')
print(df_bias)
```

```
#Figure of Estimator Consistency and Bias Convergence

c_value = 2.34

def tukey_loss(theta, X, y, c=c_value):
    r = y - X.dot(theta)
    loss = np.where(np.abs(r) <= c,
                    (c**2 / 6) * (1 - (1 - (r/c)**2)**3),
                    (c**2 / 6))
    return np.sum(loss)

def tukey_grad(theta, X, y, c=c_value):
    r = y - X.dot(theta)
    grad_factor = np.where(np.abs(r) <= c, r * (1 - (r/c)**2)
        **2, 0.0)
    return -X.T.dot(grad_factor)

def tukey_regression(X, y, c=c_value, theta0=None):
    if theta0 is None:
        theta0 = np.zeros(X.shape[1])
    res = minimize(tukey_loss, theta0, args=(X, y, c), jac=
        tukey_grad, method='BFGS')
    return res.x

def build_legendre_matrix(x, order=10):
    x_mapped = 2.0 * x - 1.0
    N = len(x)
    M = np.zeros((N, order+1))
    for i in range(order+1):
        coeff = np.zeros(i+1)
        coeff[-1] = 1.0
        M[:, i] = legval(x_mapped, coeff)
    return M

theta_true = np.array([1.5, 0.8, 2.1, -1.2, 0.7, 3.0, 0.5,
    -0.6, 1.1, -0.9, 2.2])
sigma = 0.5
N_values = np.unique(np.logspace(np.log10(50), np.log10
    (10000000), num=20, dtype=int))
theta_hats = []
det_values = []
prev_theta = None
last_N = None
n_rep = 5

def run_replication(N, theta0):
    x = np.linspace(0, 1, N)
    X = build_legendre_matrix(x, order=10)
    epsilon = np.random.normal(0, sigma, N)
    y = X @ theta_true + epsilon
    theta_hat = tukey_regression(X, y, c=c_value, theta0=
        theta0)
    theta_hat = np.clip(theta_hat, -4, 4)
    XTX = X.T @ X
    det_ratio = np.linalg.det(XTX / N)
    return theta_hat, N * det_ratio

for N in N_values:
    if last_N is not None and (N / last_N > 5):
        prev_theta = None
    last_N = N
    results = Parallel(n_jobs=-1)(
        delayed(run_replication)(N, prev_theta) for _ in range
            (n_rep)
    )
    theta_reps = np.array([res[0] for res in results])
    det_vals = [res[1] for res in results]
```

```
    theta_avg = np.mean(theta_reps, axis=0)
    theta_hats.append(theta_avg)
    det_values.append(np.mean(det_vals))
    prev_theta = theta_avg

theta_hats = np.array(theta_hats)
bias_estimates = theta_hats - theta_true

#Quantile Visualization of the Convergence Trajectories of
    Parameter Biases

c_value = 2.34

def tukey_loss(theta, X, y, c=c_value):
    r = y - X.dot(theta)
    loss = np.where(np.abs(r) <= c,
                    (c**2 / 6) * (1 - (1 - (r/c)**2)**3),
                    (c**2 / 6))
    return np.sum(loss)

def tukey_grad(theta, X, y, c=c_value):
    r = y - X.dot(theta)
    grad_factor = np.where(np.abs(r) <= c, r * (1 - (r/c)**2)
        **2, 0.0)
    return -X.T.dot(grad_factor)

def tukey_regression(X, y, c=c_value, p0=None):
    if p0 is None:
        p0 = np.zeros(X.shape[1])
    res = minimize(tukey_loss, p0, args=(X, y, c), jac=
        tukey_grad, method='BFGS')
    return res.x

theta_true = np.array([1.5, 0.8, 2.1, -1.2, 0.7, 3.0, 0.5,
    -0.6, 1.1, -0.9, 2.2])
sigma = 0.5
N_values = np.logspace(np.log10(50), np.log10(10000000), num
    =20, dtype=int)

x_cache = {}
y0_cache = {}

def get_cached_x_y0(N):
    if N not in x_cache:
        x = np.linspace(0, 1, N)
        X = np.vander(x, N=11, increasing=True)
        y0 = X.dot(theta_true)
        x_cache[N] = x
        y0_cache[N] = y0
    return x_cache[N], y0_cache[N]

def get_replications(N):
    if N <= 1e4:
        return 20
    else:
        return 10

def single_rep_fit(x, y0, sigma, p0):
    X = np.vander(x, N=11, increasing=True)
    epsilon = np.random.normal(0, sigma, len(x))
    y = y0 + epsilon
    theta_hat = tukey_regression(X, y, c=c_value, p0=p0)
    return theta_hat

theta_hats = []
p0_global = np.zeros(11)

```

```
for N in N_values:
    rep = get_replications(N)
    x, y0 = get_cached_x_y0(N)
    rep_estimates = Parallel(n_jobs=-1)(
        delayed(single_rep_fit)(x, y0, sigma, p0_global)
        for _ in range(rep)
    )
    rep_estimates = np.array(rep_estimates)
    avg_params = np.nanmean(rep_estimates, axis=0)
    theta_hats.append(avg_params)
    p0_global = avg_params

theta_hats = np.array(theta_hats)

quantiles = [0.05, 0.25, 0.5, 0.75, 0.95]
theta_estimates = {i: {q: [] for q in quantiles} for i in
    range(len(theta_true))}
median_estimates = {i: [] for i in range(len(theta_true))}

p0_global = np.zeros(11)

for N in N_values:
    rep = get_replications(N)
    x, y0 = get_cached_x_y0(N)
    rep_theta = Parallel(n_jobs=-1)(
        delayed(single_rep_fit)(x, y0, sigma, p0_global)
        for _ in range(rep)
    )
    rep_theta = np.array(rep_theta)
    p0_global = np.nanmean(rep_theta, axis=0)
    for i in range(len(theta_true)):
        for q in quantiles:
            theta_estimates[i][q].append(np.nanquantile(
                rep_theta[:, i], q))
        median_estimates[i].append(np.nanquantile(rep_theta[:,
            i], 0.5))

fig, axes = plt.subplots(4, 3, figsize=(18, 12))
axes = axes.flatten()
for i in range(len(theta_true)):
    ax = axes[i]
    for q in quantiles:
        ax.plot(N_values, theta_estimates[i][q], linestyle='--
            ',
                label=f'{int(q*100)}\% Quantile')
    ax.plot(N_values, median_estimates[i], color='black',
        linewidth=2, label='Median')
    ax.axhline(theta_true[i], color='red', linestyle='-',
        label='True Value')
    ax.set_xscale('log')
    ax.set_xlabel('Sample Size (N)')
    ax.set_ylabel(rf'Estimate of $\theta_{{{i}}}$')
    ax.set_title(rf'Convergence Path for $\theta_{{{i}}}$')
    ax.legend()

if len(axes) > 11:
    for ax in axes[11:]:
        ax.axis('off')

#Graphical Test for Heteroscedasticity

c_value = 2.34

def tukey_loss(theta, X, y, c=c_value):
    r = y - X.dot(theta)
    loss = np.where(np.abs(r) <= c,
                    (c**2 / 6) * (1 - (1 - (r/c)**2)**3),
```

```
                                (c**2 / 6))
    return np.sum(loss)

def tukey_grad(theta, X, y, c=c_value):
    r = y - X.dot(theta)
    grad_factor = np.where(np.abs(r) <= c, r * (1 - (r/c)**2)**2, 0.0)
    return -X.T.dot(grad_factor)

def tukey_regression(X, y, c=c_value, p0=None):
    if p0 is None:
        p0 = np.zeros(X.shape[1])
    res = minimize(tukey_loss, p0, args=(X, y, c), jac=tukey_grad, method='BFGS')
    return res.x

np.random.seed(42)

theta_true = np.array([1.5, 0.8, 2.1, -1.2, 0.7, 3.0, 0.5, -0.6, 1.1, -0.9, 2.2])
sigma = 0.5

def generate_normal_errors(N, sigma):
    return np.random.normal(0, sigma, N)

sample_sizes = [20, 100, 1000, 10000, 100000, 1000000, 10000000]

for idx, N in enumerate(sample_sizes):
    x = np.linspace(0, 1, N)
    X = np.vander(x, N=11, increasing=True)
    epsilon = generate_normal_errors(N, sigma)
    y = X @ theta_true + epsilon
    theta_est = tukey_regression(X, y)
    y_fitted = X.dot(theta_est)
    residuals = y - y_fitted
    plt.subplot(len(sample_sizes), 1, idx + 1)
    plt.scatter(y_fitted, residuals, alpha=0.6)
    plt.axhline(0, color='red', linestyle='--')

#Heatmap of Marginal and Joint Parameter Density Distributions

np.random.seed(42)

c_value = 2.34
lambda_reg = 0.001

def tukey_loss(theta, X, y, c=c_value):
    r = y - X.dot(theta)
    loss = np.where(np.abs(r) <= c,
                    (c**2 / 6) * (1 - (1 - (r/c)**2)**3),
                    (c**2 / 6))
    reg = lambda_reg * np.sum(theta**2)
    return np.sum(loss) + reg

def tukey_grad(theta, X, y, c=c_value):
    r = y - X.dot(theta)
    grad_factor = np.where(np.abs(r) <= c, r * (1 - (r/c)**2)**2, 0.0)
    grad = -X.T.dot(grad_factor)
    grad_reg = 2 * lambda_reg * theta
    return grad + grad_reg

def tukey_regression(X, y, c=c_value, theta0=None):
    if theta0 is None:
        theta0 = np.zeros(X.shape[1])
```

```
    res = minimize(tukey_loss, theta0, args=(X, y, c), jac=
        tukey_grad, method='BFGS')
    theta_est = np.clip(res.x, -4, 4)
    return theta_est

def build_legendre_matrix(x, order=10):
    x_mapped = 2.0 * x - 1.0
    N = len(x)
    M = np.zeros((N, order+1))
    for i in range(order+1):
        coeff = np.zeros(i+1)
        coeff[-1] = 1.0
        M[:, i] = np.polynomial.legendre.legval(x_mapped,
            coeff)
    return M

theta_true = np.array([1.5, 0.8, 2.1, -1.2, 0.7, 3.0, 0.5,
    -0.6, 1.1, -0.9, 2.2])
sigma = 0.5
N_values = [10, 100, 10000, 100000, 1000000, 5000000, 6000000,
     7000000, 8000000, 9000000, 10000000]
replications = 20

theta_hats = []
theta_std = []
theta_ci_lower = []
theta_ci_upper = []
det_values = []

prev_theta = None
last_N = None

def run_replication(N, theta_true, sigma, theta0, use_legendre
    =True):
    x = np.linspace(0, 1, N)
    if use_legendre:
        X = build_legendre_matrix(x, order=10)
    else:
        X = np.vander(x, N=11, increasing=True)
    epsilon = np.random.normal(0, sigma, N)
    y = X @ theta_true + epsilon
    theta_hat = tukey_regression(X, y, c=c_value, theta0=
        theta0)
    XTX = X.T @ X
    det_ratio = np.linalg.det(XTX / N)
    return theta_hat, N * det_ratio

for N in N_values:
    if last_N is not None and (N / last_N > 5):
        prev_theta = None
    last_N = N
    results = Parallel(n_jobs=-1)(
        delayed(run_replication)(N, theta_true, sigma,
            prev_theta, True) for _ in range(replications)
    )
    estimates = np.array([res[0] for res in results])
    dets = np.array([res[1] for res in results])
    mean_est = estimates.mean(axis=0)
    std_est = estimates.std(axis=0)
    ci_lower = mean_est - 1.96 * std_est / np.sqrt(
        replications)
    ci_upper = mean_est + 1.96 * std_est / np.sqrt(
        replications)
    theta_hats.append(mean_est)
    theta_std.append(std_est)
    theta_ci_lower.append(ci_lower)
    theta_ci_upper.append(ci_upper)
```

```python
    det_values.append(dets.mean())
    prev_theta = mean_est.copy()
    df = pd.DataFrame(estimates, columns=[f'\theta{i}' for i in range(11)])
    g = sns.pairplot(df, diag_kind="kde", kind="kde", plot_kws={'fill': True, 'cmap': 'viridis'})
    g.fig.suptitle(f'Parameter Space Density (N = {N})', y=1.02)
    plt.savefig(f"4.44Parameter_Space_Density_N_{N}.pdf", format='pdf')
    plt.show()

theta_hats = np.array(theta_hats)
theta_std = np.array(theta_std)
theta_ci_lower = np.array(theta_ci_lower)
theta_ci_upper = np.array(theta_ci_upper)
det_values = np.array(det_values)

for idx, N in enumerate(N_values):
    print(f"Sample Size N={N}")
    for i in range(11):
        print(f"\theta{i}: mean = {theta_hats[idx, i]:.3f}, 95\% CI = [{theta_ci_lower[idx, i]:.3f}, {theta_ci_upper[idx, i]:.3f}]")
    print(f"Design Matrix Condition: {det_values[idx]:.3e}")
    print("----------------------------------------")

#Confidence Interval

num_params = theta_hats.shape[1]
num_samples = len(N_values)
param_indices = range(num_params)

num_cols = 4
num_rows = math.ceil(num_params / num_cols)

fig, axes = plt.subplots(num_rows, num_cols, figsize=(16, 9))
axes = axes.flatten()

for i in param_indices:
    ax = axes[i]
    means = theta_hats[:, i]
    lower = theta_ci_lower[:, i]
    upper = theta_ci_upper[:, i]
    yerr_lower = means - lower
    yerr_upper = upper - means
    yerr = np.vstack([yerr_lower, yerr_upper])
    ax.errorbar(N_values, means, yerr=yerr, fmt='-o', capsize=4,
                ecolor='red', color='blue', markersize=5,
                label=f'\theta{i} Estimate')
    ax.set_xscale('log')
    ax.set_xlabel('Sample Size (log scale)')
    ax.set_ylabel('Parameter Estimate')
    ax.set_title(f'Parameter \theta{i}')
    ax.grid(True, linestyle=':', alpha=0.7)
    ax.legend()

for j in range(num_params, len(axes)):
    axes[j].axis('off')

#Covariance calculation

n_cpu = multiprocessing.cpu_count()
blas_threads = max(1, n_cpu // 4)
os.environ["OMP_NUM_THREADS"] = str(blas_threads)
os.environ["OPENBLAS_NUM_THREADS"] = str(blas_threads)
```

```
os.environ["MKL_NUM_THREADS"] = str(blas_threads)
os.environ["NUMEXPR_NUM_THREADS"] = str(blas_threads)
print(f"[INFO] CPU={n_cpu}, BLAS Threads per process={
    blas_threads}")

c_value = 2.34

@njit
def tukey_loss_numba(theta, X, y, c=c_value):
    r = y - X @ theta
    loss = np.where(np.abs(r) <= c, (c**2 / 6) * (1 - (1 - (r/
        c)**2)**3), (c**2 / 6))
    return np.sum(loss)

@njit
def tukey_grad_numba(theta, X, y, c=c_value):
    r = y - X @ theta
    grad_factor = np.where(np.abs(r) <= c, r * (1 - (r/c)**2)
        **2, 0.0)
    return -X.T @ grad_factor

def safe_inverse(H, threshold=1e8):
    cond = np.linalg.cond(H)
    if cond < threshold:
        return np.linalg.solve(H, np.eye(H.shape[0]))
    else:
        print(f"Warning: Hessian ill-conditioned (cond={cond
            :.2e}), using pinv()")
        return np.linalg.pinv(H)

def tukey_regression_with_cov(X, y, c=c_value):
    theta0, _, _, _ = np.linalg.lstsq(X, y, rcond=None)
    r0 = y - X @ theta0
    w0 = np.where(np.abs(r0) <= c, (1 - (r0/c)**2)**2, 1e-6)
    theta0_new, _, _, _ = np.linalg.lstsq(X * np.sqrt(w0[:,
        None]), y * np.sqrt(w0), rcond=None)
    res = minimize(fun=lambda th: tukey_loss_numba(th, X, y, c
        ),
                   x0=theta0_new,
                   jac=lambda th: tukey_grad_numba(th, X, y, c
                       ),
                   method='L-BFGS-B')
    theta_hat = res.x
    r = y - X @ theta_hat
    factor = np.where(np.abs(r) <= c, (1 - (r/c)**2) * (1 -
        5*(r/c)**2), 0.0)
    H = X.T @ (factor[:, None] * X)
    cov_est = sigma**2 * safe_inverse(H)
    return theta_hat, cov_est

np.random.seed(42)
theta_true = np.array([1.5, 0.8, 2.1, -1.2, 0.7, 3.0, 0.5,
    -0.6, 1.1, -0.9, 2.2])
sigma = 0.5
sample_sizes = np.linspace(50, 10000000, 50, dtype=int)
replications = 50
errors_noN = []

def run_replication(seed):
    np.random.seed(seed)
    eps = np.random.normal(0, sigma, size=X_global.shape[0])
    y = X_global @ theta_true + eps
    _, cov_est = tukey_regression_with_cov(X_global, y, c=
        c_value)
    return cov_est

for idx, N in enumerate(sample_sizes):
```

```
    print(f"Processing sample size: {N} (step {idx+1}/{len(sample_sizes)})")
    x = np.linspace(0, 1, N)
    X = legvander(2 * x - 1, deg=10)
    XtX = X.T @ X
    C_true = sigma**2 * safe_inverse(XtX)
    global X_global
    X_global = X
    tasks = [42 + r for r in range(replications)]
    with concurrent.futures.ProcessPoolExecutor() as executor:
        results = executor.map(run_replication, tasks)
        cov_est_accum = sum(results)
    avg_cov_est = cov_est_accum / replications
    err = np.linalg.norm(avg_cov_est - C_true, ord='fro')
    errors_noN.append(err)
    if N == sample_sizes[-1]:
        print(f"\n=== Sample size N = {N} ===")
        print("\n[True Covariance]:")
        print(np.round(C_true, 4))
        print("\n[Estimated Covariance]:")
        print(np.round(avg_cov_est, 4))
        print("\nFrobenius norm error:", err)
```

Appendix 3
Python Code for Realistic Arithmetic in Chap. 5

Section 5.2.1 Airfoil Self-Noise Prediction

```python
import pandas as pd
import numpy as np
import matplotlib.pyplot as plt
from sklearn.svm import SVR
from sklearn.preprocessing import StandardScaler
from sklearn.pipeline import Pipeline
from sklearn.model_selection import train_test_split, GridSearchCV
from sklearn.metrics import mean_squared_error, r2_score

df = pd.read_csv("airfoil_self_noise.csv")

X = df[['Frequency (Hz)',
        'Angle of Attack (deg)',
        'Chord Length (m)',
        'Free-stream Velocity (m/s)',
        'Suction Side Displacement Thickness (m)']]

y = df['Scaled Sound Pressure (dB)']

X_train, X_test, y_train, y_test = train_test_split(X, y, test_size=0.2, random_state=42)

pipe = Pipeline([
    ('scaler', StandardScaler()),
    ('svr', SVR())
])

param_grid = {
    'svr__C': [1, 10, 100],
    'svr__gamma': ['scale', 0.1, 0.01],
    'svr__epsilon': [0.1, 0.5, 1]
}

grid_search = GridSearchCV(pipe, param_grid, cv=5, scoring='r2', verbose=2, n_jobs=-1)
grid_search.fit(X_train, y_train)

print("Best Parameters:", grid_search.best_params_)
best_model = grid_search.best_estimator_

```


A. Hajiyev and J. Xu, *High-Dimensional Regression Analysis and Artificial Intelligence*, https://doi.org/10.1007/978-981-95-2514-0

```
y_pred = best_model.predict(X_test)

import numpy as np

sample_idx = 10
x_instance = X_test.iloc[sample_idx]
y_true_instance = y_test.iloc[sample_idx]

y_pred_instance = best_model.predict([x_instance])[0]

import numpy as np
import pandas as pd

x_array = x_instance.values if hasattr(x_instance, 'values') else x_instance

num_samples = 500
noise = np.random.randn(num_samples, X.shape[1]) * 0.1

neighbors = x_array + x_array * noise

neighbors_df = pd.DataFrame(neighbors, columns=X.columns)

neighbors_preds = best_model.predict(neighbors_df)

import statsmodels.api as sm

X_local = neighbors_df.copy()
y_local = neighbors_preds

X_local_const = sm.add_constant(X_local)

local_model = sm.OLS(y_local, X_local_const).fit()

import pandas as pd
import numpy as np
import statsmodels.api as sm

theta_hat = local_model.params

X_local_const = sm.add_constant(X_local)

cov_matrix = local_model.cov_params()

cov_matrix_df = pd.DataFrame(cov_matrix, columns=local_model.params.index, index=local_model.params.index)

import numpy as np
import matplotlib.pyplot as plt
from mpl_toolkits.mplot3d import Axes3D
import scipy.stats as stats
import pandas as pd

triples = [
    ("Frequency (Hz)", "Angle of Attack (deg)", "Chord Length (m)"),
    ("Frequency (Hz)", "Chord Length (m)", "Suction Side Displacement Thickness (m)"),
    ("Frequency (Hz)", "Chord Length (m)", "Free-stream Velocity (m/s)"),
    ("Angle of Attack (deg)", "Chord Length (m)", "Suction Side Displacement Thickness (m)"),
    ("Angle of Attack (deg)", "Free-stream Velocity (m/s)", "Suction Side Displacement Thickness (m)"),
    ("Chord Length (m)", "Free-stream Velocity (m/s)", "Suction Side Displacement Thickness (m)")
]
```

```
chi2_val_3d = stats.chi2.ppf(0.95, df=3)

for i, triple in enumerate(triples):

    mean_3d = np.array([theta_hat[triple[0]], theta_hat[triple[1]], theta_hat[triple[2]]])

    cov_3d = cov_matrix_df.loc[list(triple), list(triple)].values

    eigvals, eigvecs = np.linalg.eigh(cov_3d)

    order = eigvals.argsort()[::-1]
    eigvals = eigvals[order]
    eigvecs = eigvecs[:, order]

    radii = np.sqrt(eigvals * chi2_val_3d)

    u = np.linspace(0, 2 * np.pi, 50)
    v = np.linspace(0, np.pi, 50)
    x = np.outer(np.cos(u), np.sin(v))
    y = np.outer(np.sin(u), np.sin(v))
    z = np.outer(np.ones_like(u), np.cos(v))

    ellipsoid = np.zeros((x.shape[0], x.shape[1], 3))
    for j in range(x.shape[0]):
        for k in range(x.shape[1]):
            point = np.array([x[j, k], y[j, k], z[j, k]])
            ellipsoid[j, k, :] = mean_3d + eigvecs @ (radii * point)

import numpy as np
import matplotlib.pyplot as plt

chi2_val_full = 11.0705
cov_mat = cov_matrix.values
var_factor = np.array([np.dot(x, np.dot(cov_mat, x)) for x in X_local_const.values])

half_width = np.sqrt(chi2_val_full * var_factor)

f_pred = local_model.predict(X_local_const)

import numpy as np
import pandas as pd
from sklearn.metrics import mean_squared_error, r2_score
import statsmodels.api as sm

def calculate_width(L, U):
    return U - L

def calculate_center_deviation(L, U, f_true):
    center = (L + U) / 2
    return abs(center - f_true)

def calculate_PA(y_true, y_pred):
    r2 = r2_score(y_true, y_pred)
    mse = mean_squared_error(y_true, y_pred)
    return 0.5 * r2 + 0.5 * (1 / (1 + mse))

def compute_CICE(X, y_true, y_pred, L, U):

    n_samples = X.shape[0]

    interpretability_scores = []
    for i in range(n_samples):
```

```
        width = calculate_width(L[i], U[i])
        center_dev = calculate_center_deviation(L[i], U[i], y_true[i])

        interpretability_index = 0.8 * (1 / (1 + width)) + 0.2 * (1 / (1 + center_dev))
        interpretability_scores.append(interpretability_index)

    PA = calculate_PA(y_true, y_pred)

    avg_interpretability = np.mean(interpretability_scores)

    CICE = 0.5 * avg_interpretability + 0.5 * PA
    return CICE, avg_interpretability, PA

import numpy as np
import pandas as pd
import matplotlib.pyplot as plt
from scipy.optimize import least_squares
from mpl_toolkits.mplot3d import Axes3D
import scipy.stats as stats
from sklearn.metrics import r2_score, mean_squared_error

def multi_power_model(X, *params):
    a1, a2, a3, a4, a5, k1, k2, k3, k4, k5 = params
    x1, x2, x3, x4, x5 = X[:, 0], X[:, 1], X[:, 2], X[:, 3], X[:, 4]
    return (a1 * x1**k1 + a2 * x2**k2 + a3 * x3**k3 +
            a4 * x4**k4 + a5 * x5**k5)

def regularized_resid(params, X, y, reg_lambda):
    r = multi_power_model(X, *params) - y
    penalty = np.sqrt(reg_lambda) * params
    return np.concatenate([r, penalty])

def fit_power_model_reg(X_data, y_data, init_guess, reg_lambda=0.01, max_nfev=8000):
    res = least_squares(regularized_resid, init_guess, args=(X_data, y_data, reg_lambda), max_nfev=max_nfev)
    return res
X_nl = neighbors_df.copy()
y_nl = neighbors_preds

init_guess_nl = np.ones(10)
res_nl = fit_power_model_reg(X_nl.values, y_nl, init_guess_nl)
params_nl = res_nl.x
cost_nl = res_nl.cost
y_pred_nl = multi_power_model(X_nl.values, *params_nl)

r2_nl = r2_score(y_nl, y_pred_nl)
mse_nl = mean_squared_error(y_nl, y_pred_nl)
rmse_nl = np.sqrt(mse_nl)

theta_hat_nl = params_nl

def gradient_multi_power_model(x, params):
    a1, a2, a3, a4, a5, k1, k2, k3, k4, k5 = params
    grad = np.zeros(10)
    grad[0] = x[0] ** k1
    grad[1] = x[1] ** k2
    grad[2] = x[2] ** k3
    grad[3] = x[3] ** k4
    grad[4] = x[4] ** k5
    grad[5] = a1 * x[0] ** k1 * np.log(x[0] + 1e-8)
    grad[6] = a2 * x[1] ** k2 * np.log(x[1] + 1e-8)
    grad[7] = a3 * x[2] ** k3 * np.log(x[2] + 1e-8)
```

```
    grad[8] = a4 * x[3] ** k4 * np.log(x[3] + 1e-8)
    grad[9] = a5 * x[4] ** k5 * np.log(x[4] + 1e-8)
    return grad

J_local = np.array([gradient_multi_power_model(x, theta_hat_nl
    ) for x in X_nl.values])

J_local_df = pd.DataFrame(J_local, columns=["d(a1)", "d(a2)",
    "d(a3)", "d(a4)", "d(a5)",
                                        "d(k1)", "d(k2)",
                                            "d(k3)", "d(
                                           k4)", "d(k5)"
                                           ])

param_names = ["a1", "a2", "a3", "a4", "a5", "k1", "k2", "k3",
    "k4", "k5"]

N_nl = X_nl.shape[0]
p_nl = len(params_nl)
r_data = res_nl.fun[:N_nl]
dof_nl = max(N_nl - p_nl, 1)
sigma_sq_nl = np.sum(r_data**2) / dof_nl
J_data = res_nl.jac[:N_nl, :]
cov_nl = sigma_sq_nl * np.linalg.pinv(J_data.T @ J_data)

cov_nl_df = pd.DataFrame(cov_nl, index=param_names, columns=
    param_names)


triples_nl = [
    ("a3", "a4", "a5"),
    ("a1", "a3", "k1"),
    ("a2", "a4", "k2"),
    ("a3", "k2", "k3"),
    ("k1", "k3", "k4"),
    ("k3", "k4", "k5")
]

chi2_val_3d_nl = stats.chi2.ppf(0.95, df=3)
fig_nl = plt.figure(figsize=(18, 12))

for i, triple in enumerate(triples_nl):
    mean_3d = np.array([params_nl[param_names.index(t)] for t
        in triple])
    cov_3d = cov_nl_df.loc[list(triple), list(triple)].values
    eigvals, eigvecs = np.linalg.eigh(cov_3d)
    order = eigvals.argsort()[::-1]
    eigvals, eigvecs = eigvals[order], eigvecs[:, order]
    radii = np.sqrt(eigvals * chi2_val_3d_nl)

    u = np.linspace(0, 2 * np.pi, 50)
    v = np.linspace(0, np.pi, 50)
    x = np.outer(np.cos(u), np.sin(v))
    y = np.outer(np.sin(u), np.sin(v))
    z = np.outer(np.ones_like(u), np.cos(v))

    ellipsoid = np.zeros((x.shape[0], x.shape[1], 3))
    for j in range(x.shape[0]):
        for k in range(x.shape[1]):
            point = np.array([x[j, k], y[j, k], z[j, k]])
            ellipsoid[j, k, :] = mean_3d + eigvecs @ (radii *
                point)

def gradient_multi_power_model(x, params):
    a1, a2, a3, a4, a5, k1, k2, k3, k4, k5 = params
    grad = np.zeros(10)
    grad[0] = x[0] ** k1
```

```
    grad[1] = x[1] ** k2
    grad[2] = x[2] ** k3
    grad[3] = x[3] ** k4
    grad[4] = x[4] ** k5
    grad[5] = a1 * x[0] ** k1 * np.log(x[0] + 1e-8)
    grad[6] = a2 * x[1] ** k2 * np.log(x[1] + 1e-8)
    grad[7] = a3 * x[2] ** k3 * np.log(x[2] + 1e-8)
    grad[8] = a4 * x[3] ** k4 * np.log(x[3] + 1e-8)
    grad[9] = a5 * x[4] ** k5 * np.log(x[4] + 1e-8)
    return grad

f_pred_nl = np.array([multi_power_model(x.reshape(1, -1), *
    params_nl)[0] for x in X_nl.values])
gradients = np.array([gradient_multi_power_model(x, params_nl)
     for x in X_nl.values])
var_factors_nl = np.array([grad.T @ cov_nl @ grad for grad in
    gradients])
chi2_val_full_nl = 27.5871
half_width_nl = np.sqrt(chi2_val_full_nl * var_factors_nl)

import numpy as np
import pandas as pd
from sklearn.metrics import mean_squared_error, r2_score
import statsmodels.api as sm

def calculate_width(L_nl, U_nl):
    return U_nl - L_nl
def calculate_center_deviation(L_nl, U_nl, f_true_nl):
    center_nl = (L_nl + U_nl) / 2
    return abs(center_nl - f_true_nl)

def calculate_PA(y_true_nl, y_pred_nl):
    r2_nl = r2_score(y_true_nl, y_pred_nl)
    mse_nl = mean_squared_error(y_true_nl, y_pred_nl)
    return 0.5 * r2_nl + 0.5 * (1 / (1 + mse_nl))

def compute_CICE(X_nl, y_true_nl, y_pred_nl, L_nl, U_nl):

    n_samples_nl = X_nl.shape[0]

    interpretability_scores_nl = []
    for i_nl in range(n_samples_nl):
        width_nl = calculate_width(L_nl[i_nl], U_nl[i_nl])
        center_dev_nl = calculate_center_deviation(L_nl[i_nl],
            U_nl[i_nl], y_true_nl[i_nl])

        interpretability_index_nl = 0.8 * (1 / (1 + width_nl))
            + 0.2 * (1 / (1 + center_dev_nl))
        interpretability_scores_nl.append(
            interpretability_index_nl)

    PA_nl = calculate_PA(y_true_nl, y_pred_nl)

    avg_interpretability_nl = np.mean(
        interpretability_scores_nl)

    CICE_nl = 0.5 * avg_interpretability_nl + 0.5 * PA_nl
    return CICE_nl, avg_interpretability_nl, PA_nl
```

5.2.2 Bank Marketing Decisions

```
import pandas as pd
import numpy as np
from xgboost import XGBClassifier, plot_importance
from sklearn.model_selection import train_test_split,
    RandomizedSearchCV
from sklearn.metrics import roc_auc_score, precision_score,
    recall_score, f1_score, confusion_matrix, roc_curve,
    precision_recall_curve, accuracy_score
import matplotlib.pyplot as plt

df = pd.read_csv(r"D:\.csv")

df['y'] = df['y'].map({'yes':1, 'no':0})

X = df.drop('y', axis=1)
y = df['y']

X_train, X_test, y_train, y_test = train_test_split(X, y,
    test_size=0.3, random_state=42)

cat_cols = ['job', 'marital', 'education', 'default', 'housing
    ', 'loan', 'contact', 'month', 'poutcome']
X.loc[X_train.index, :] = X_train
X.loc[X_test.index, :] = X_test

m = 20

global_mean = y_train.mean()
for col in cat_cols:
    tmp = X_train.copy()
    tmp['y'] = y_train.values

    stats = tmp.groupby(col)['y'].agg(['mean', 'count'])
    smoothing = (stats['count'] * stats['mean'] + m *
        global_mean) / (stats['count'] + m)

    X_train[col] = X_train[col].map(smoothing)
    X_test[col] = X_test[col].map(smoothing)
    X_test[col].fillna(global_mean, inplace=True)

scale = (y_train == 0).sum() / (y_train == 1).sum()

param_grid = {
    'max_depth': [4,5,6,7],
    'min_child_weight': [1,5,10],
    'gamma': [0,0.5,1],
    'subsample': [0.7,0.8,0.9],
    'colsample_bytree': [0.7,0.8,0.9],
    'scale_pos_weight': [scale, scale*0.8, scale*1.2]
}

model = XGBClassifier(use_label_encoder=False, eval_metric='
    logloss', random_state=42)

search = RandomizedSearchCV(model, param_distributions=
    param_grid, n_iter=30, scoring='roc_auc', cv=5, verbose=2,
     n_jobs=1)
search.fit(X_train, y_train)

best_model = search.best_estimator_

best_model.fit(
    X_train, y_train,
```

```
    eval_set=[(X_test, y_test)],
    early_stopping_rounds=20,
    verbose=True
)

y_pred_prob = best_model.predict_proba(X_test)[:,1]

precision, recall, thresholds = precision_recall_curve(y_test,
    y_pred_prob)
f1_scores = 2 * precision * recall / (precision + recall + 1e
    -6)

best_threshold = thresholds[np.argmax(f1_scores)]

y_pred_adjusted = (y_pred_prob >= best_threshold).astype(int)

import numpy as np

diffs = np.abs(y_pred_prob - best_threshold)
sample_idx = diffs.argmin()
x_instance = X_test.iloc[sample_idx]
y_true_instance = y_test.iloc[sample_idx]

y_pred_instance = best_model.predict([x_instance])[0]
y_pred_prob_instance = best_model.predict_proba([x_instance])
    [:, 1][0]

from sklearn.preprocessing import StandardScaler
import pandas as pd
import numpy as np

diffs = np.abs(y_pred_prob - best_threshold)
sample_idx = diffs.argmin()
x_instance = X_test.iloc[sample_idx]
y_true_instance = y_test.iloc[sample_idx]

x_instance_encoded = x_instance.copy()

for col in cat_cols:
    tmp = X_train.copy()
    tmp['y'] = y_train.values
    stats = tmp.groupby(col)['y'].agg(['mean', 'count'])
    smoothing = (stats['count'] * stats['mean'] + m *
        global_mean) / (stats['count'] + m)
    value = x_instance[col]
    x_instance_encoded[col] = smoothing.get(value, global_mean
        )

X_train_encoded = X_train.copy()
for col in cat_cols:
    tmp = X_train.copy()
    tmp['y'] = y_train.values
    stats = tmp.groupby(col)['y'].agg(['mean', 'count'])
    smoothing = (stats['count'] * stats['mean'] + m *
        global_mean) / (stats['count'] + m)
    X_train_encoded[col] = X_train[col].map(smoothing)

scaler = StandardScaler()
scaler.fit(X_train_encoded)

x_scaled = scaler.transform([x_instance_encoded])[0]

num_samples = 500
noise = np.random.randn(num_samples, X_train_encoded.shape[1])
    * 0.1
neighbors = x_scaled + noise
```

```
neighbors_df = pd.DataFrame(neighbors, columns=X_train_encoded
    .columns)

neighbors_unscaled = pd.DataFrame(
    scaler.inverse_transform(neighbors_df),
    columns=neighbors_df.columns
)

neighbors_pred_prob = best_model.predict_proba(
    neighbors_unscaled)[:, 1]

import statsmodels.api as sm
import statsmodels.robust.norms as norms

X_local_const = sm.add_constant(neighbors_df)
local_model_tukey = sm.RLM(neighbors_pred_prob, X_local_const,
     M=norms.TukeyBiweight()).fit()

import numpy as np
from sklearn.metrics import mean_squared_error,
    mean_absolute_error

y_pred_lime = local_model_tukey.predict(X_local_const)

mse_lime = mean_squared_error(neighbors_pred_prob, y_pred_lime
    )
mae_lime = mean_absolute_error(neighbors_pred_prob,
    y_pred_lime)

ss_res = np.sum((neighbors_pred_prob - y_pred_lime) ** 2)
ss_tot = np.sum((neighbors_pred_prob - np.mean(
    neighbors_pred_prob)) ** 2)
pseudo_r2_lime = 1 - ss_res / ss_tot

import numpy as np

cov_params = local_model_tukey.cov_params()

import numpy as np
import scipy.stats as stats

theta_hat_np = theta_hat.values if hasattr(theta_hat, 'values'
    ) else np.array(theta_hat)
m = len(theta_hat_np)

alpha = 0.05
chi2_val = stats.chi2.ppf(1 - alpha, m)

cov_inv = np.linalg.pinv(cov_params)

def quadratic_form(theta):
    diff = theta - theta_hat_np
    return diff.T @ cov_inv @ diff

import numpy as np
import matplotlib.pyplot as plt
from mpl_toolkits.mplot3d import Axes3D
import scipy.stats as stats

triples = [
    ("const", "age", "job"),
    ("marital", "education", "default"),
    ("balance", "housing", "loan"),
    ("contact", "day", "month"),
    ("duration", "campaign", "pdays"),
    ("poutcome", "job", "education")
]
```

```
chi2_val_3d = stats.chi2.ppf(0.95, df=3)

for i, triple in enumerate(triples):
    mean_3d = np.array([theta_hat[triple[0]], theta_hat[triple[1]], theta_hat[triple[2]]])
    cov_3d = cov_params.loc[list(triple), list(triple)].values

    eigvals, eigvecs = np.linalg.eigh(cov_3d)
    order = eigvals.argsort()[::-1]
    eigvals = eigvals[order]
    eigvecs = eigvecs[:, order]

    radii = np.sqrt(eigvals * chi2_val_3d)

    u = np.linspace(0, 2 * np.pi, 50)
    v = np.linspace(0, np.pi, 50)
    x = np.outer(np.cos(u), np.sin(v))
    y = np.outer(np.sin(u), np.sin(v))
    z = np.outer(np.ones_like(u), np.cos(v))

    ellipsoid = np.zeros((x.shape[0], x.shape[1], 3))
    for j in range(x.shape[0]):
        for k in range(x.shape[1]):
            point = np.array([x[j, k], y[j, k], z[j, k]])
            ellipsoid[j, k, :] = mean_3d + eigvecs @ (radii * point)

import numpy as np
import matplotlib.pyplot as plt

chi2_val_full = 27.5871

cov_mat = cov_params.values

var_factor = np.array([np.dot(x, np.dot(cov_mat, x)) for x in X_local_const.values])
half_width = np.sqrt(chi2_val_full * var_factor)

f_pred = local_model_tukey.predict(X_local_const)

for i in range(len(f_pred)):
    lower = f_pred[i] - half_width[i]
    upper = f_pred[i] + half_width[i]

def explain_instance_with_poly(x_instance, X, model, num_samples=500, poly_degree=2, max_features=30):

    import numpy as np
    import pandas as pd
    from sklearn.preprocessing import StandardScaler, PolynomialFeatures
    import statsmodels.api as sm

    if isinstance(x_instance, pd.Series):
        x_array = x_instance.values
    elif isinstance(x_instance, np.ndarray):
        x_array = x_instance
    else:
        raise ValueError

    if len(x_array) != X.shape[1]:

    scaler = StandardScaler()
    X_scaled = pd.DataFrame(scaler.fit_transform(X), columns=X.columns)
    x_scaled = scaler.transform([x_array])[0]
```

```

    noise = np.random.randn(num_samples, X.shape[1]) * 0.1
    neighbors = x_scaled + noise
    neighbors_df = pd.DataFrame(neighbors, columns=X.columns)

    neighbors_unscaled = pd.DataFrame(scaler.inverse_transform(neighbors_df), columns=X.columns)
    neighbors_pred_prob = model.predict_proba(neighbors_unscaled)[:, 1]

    poly = PolynomialFeatures(degree=poly_degree, interaction_only=True, include_bias=False)
    X_poly = poly.fit_transform(neighbors_df)
    poly_feature_names = poly.get_feature_names_out(X.columns)
    X_poly_df = pd.DataFrame(X_poly[:, :max_features], columns=poly_feature_names[:max_features])
    X_poly_df = sm.add_constant(X_poly_df)

    model_local = sm.WLS(neighbors_pred_prob, X_poly_df)
    results = model_local.fit()

    x_poly = poly.transform([x_scaled])[:, :max_features]
    x_poly_df = sm.add_constant(
    pd.DataFrame(x_poly, columns=poly_feature_names[:max_features]),
    has_constant='add'
    )

    pred = results.predict(x_poly_df)[0]
    pred_var = (x_poly_df.values @ results.cov_params().values @ x_poly_df.values.T).item()
    pred_std = np.sqrt(pred_var)
    z = 1.96
    lower = pred - z * pred_std
    upper = pred + z * pred_std

diffs = np.abs(y_pred_prob - best_threshold)
sample_idx = diffs.argmin()

x_instance = X_test.iloc[sample_idx]

x_instance_encoded = X_test.iloc[sample_idx]

pred, ci, results = explain_instance_with_poly(x_instance_encoded, X_train, best_model)

from sklearn.metrics import mean_squared_error

y_true = results.model.endog
y_pred = results.fittedvalues

mse = mean_squared_error(y_true, y_pred)

import matplotlib.pyplot as plt
from mpl_toolkits.mplot3d import Axes3D
import numpy as np
import scipy.stats as stats

theta_hat = results.params
cov_params = results.cov_params()

triples = [
    ("age", "age loan", "age campaign"),
    ("marital", "education", "default"),
    ("balance", "housing", "loan"),
    ("contact", "day", "month"),
    ("duration", "campaign", "pdays"),
```

```
    ("poutcome", "job", "education")
]

chi2_val_3d = stats.chi2.ppf(0.95, df=3)

fig = plt.figure(figsize=(18, 12))

for i, triple in enumerate(triples):
    mean_3d = np.array([theta_hat[t] for t in triple])
    cov_3d = cov_params.loc[list(triple), list(triple)].values

    eigvals, eigvecs = np.linalg.eigh(cov_3d)
    order = eigvals.argsort()[::-1]
    eigvals = eigvals[order]
    eigvecs = eigvecs[:, order]
    radii = np.sqrt(eigvals * chi2_val_3d)

    u = np.linspace(0, 2 * np.pi, 50)
    v = np.linspace(0, np.pi, 50)
    x = np.outer(np.cos(u), np.sin(v))
    y = np.outer(np.sin(u), np.sin(v))
    z = np.outer(np.ones_like(u), np.cos(v))

    ellipsoid = np.zeros((x.shape[0], x.shape[1], 3))
    for j in range(x.shape[0]):
        for k in range(x.shape[1]):
            point = np.array([x[j, k], y[j, k], z[j, k]])
            ellipsoid[j, k, :] = mean_3d + eigvecs @ (radii *
                point)

from scipy.stats import chi2

X_poly_df = pd.DataFrame(results.model.exog, columns=results.
    model.exog_names)

X_local_const = X_poly_df

f_pred = results.predict(X_local_const)

df = X_local_const.shape[1]
chi2_val = chi2.ppf(0.95, df)

cov_mat = results.cov_params().values
half_width = np.sqrt(np.array([
    x @ cov_mat @ x.T for x in X_local_const.values
]) * chi2_val)
```

5.2.3 Multivariate Time Series Prediction for Stock Market

```

import pandas as pd

file_path = 'TRD_Dalyr.xlsx'

dfs = pd.read_excel(file_path, sheet_name=None)

df_all = pd.DataFrame()
for sheet_name, df in dfs.items():
    df_all = pd.concat([df_all, df], ignore_index=True)
```

```python

selected_stocks = ['600519', '601318', '600036', '600900', '601166', '601899', '600030', '601398', '600276', '600887']

df_selected = df_all[df_all['Stkcd'].astype(str).isin(selected_stocks)]

df_selected['Trddt'] = pd.to_datetime(df_selected['Trddt'])

df_selected.sort_values(by=['Stkcd', 'Trddt'], inplace=True)

import numpy as np

feature_cols = ['Opnprc', 'Hiprc', 'Loprc', 'Clsprc', 'Dnshrtrd', 'Dnvaltrd', 'Dsmvosd', 'Dsmvtll', 'ChangeRatio', 'LimitDown', 'LimitUp', 'LimitStatus']
target_cols = ['Clsprc', 'Hiprc', 'Loprc']

def create_sequences(df, window_size, feature_cols, target_cols):
    X, y = [], []
    for stock in df['Stkcd'].unique():
        df_stock = df[df['Stkcd'] == stock].sort_values(by='Trddt')
        if len(df_stock) <= window_size:
            continue
        for i in range(window_size, len(df_stock)):
            X.append(df_stock.iloc[i-window_size:i][feature_cols].values)
            y.append(df_stock.iloc[i][target_cols].values)
    return np.array(X), np.array(y)

X_15, y_15 = create_sequences(df_selected, window_size=15, feature_cols=feature_cols, target_cols=target_cols)
X_30, y_30 = create_sequences(df_selected, window_size=30, feature_cols=feature_cols, target_cols=target_cols)
X_60, y_60 = create_sequences(df_selected, window_size=60, feature_cols=feature_cols, target_cols=target_cols)

import numpy as np
import pandas as pd
import math
import matplotlib.pyplot as plt
from keras.models import Model
from keras.layers import Input, LSTM, Dense, Dropout, BatchNormalization, Multiply
from keras.optimizers import Adam
from keras.callbacks import EarlyStopping, ReduceLROnPlateau
from sklearn.metrics import mean_squared_error, mean_absolute_error, r2_score
from sklearn.preprocessing import StandardScaler

feature_cols = ['Opnprc', 'Hiprc', 'Loprc', 'Clsprc', 'Dnshrtrd', 'Dnvaltrd', 'Dsmvosd', 'Dsmvtll', 'ChangeRatio', 'LimitDown', 'LimitUp', 'LimitStatus']
target_cols = ['Clsprc', 'Hiprc', 'Loprc']

train_end_date = pd.to_datetime('2025-01-01')
test_end_date = pd.to_datetime('2025-04-03')

def create_train_test_sequences_with_ids(df, window_size, feature_cols, target_cols, train_end_date, test_end_date):
    X_train, y_train, stock_ids_train = [], [], []
    X_test, y_test, stock_ids_test = [], [], []
    for stock in df['Stkcd'].unique():
        df_stock = df[df['Stkcd'] == stock].sort_values(by='Trddt')
```

```
        for i in range(window_size, len(df_stock)):
            target_date = df_stock.iloc[i]['Trddt']
            sample_X = df_stock.iloc[i-window_size:i][
                feature_cols].values
            sample_y = df_stock.iloc[i][target_cols].values
            if target_date < train_end_date:
                X_train.append(sample_X)
                y_train.append(sample_y)
                stock_ids_train.append(stock)
            elif train_end_date <= target_date <=
                test_end_date:
                X_test.append(sample_X)
                y_test.append(sample_y)
                stock_ids_test.append(stock)
    return (np.array(X_train), np.array(y_train),
        stock_ids_train,
            np.array(X_test), np.array(y_test), stock_ids_test
                )

window_size = 30
X_train, y_train, stock_ids_train, X_test, y_test,
    stock_ids_test = create_train_test_sequences_with_ids(
    df_selected, window_size, feature_cols, target_cols,
        train_end_date, test_end_date)

scaler_X = StandardScaler()
X_train_2d = X_train.reshape(-1, X_train.shape[-1])
X_train_scaled_2d = scaler_X.fit_transform(X_train_2d)
X_train_scaled = X_train_scaled_2d.reshape(X_train.shape)

X_test_2d = X_test.reshape(-1, X_test.shape[-1])
X_test_scaled_2d = scaler_X.transform(X_test_2d)
X_test_scaled = X_test_scaled_2d.reshape(X_test.shape)

scaler_y = StandardScaler()
y_train_scaled = scaler_y.fit_transform(y_train)
y_test_scaled = scaler_y.transform(y_test)

def build_multi_output_model(window_size, num_features):
    inputs = Input(shape=(window_size, num_features))
    x = LSTM(128, return_sequences=True, dropout=0.2,
        recurrent_dropout=0.2)(inputs)
    x = LSTM(128, return_sequences=False, dropout=0.2,
        recurrent_dropout=0.2)(x)

    x = BatchNormalization()(x)

    attention_probs = Dense(128, activation='softmax', name='
        attention_vec')(x)
    attention_probs = Multiply()([attention_probs, x])

    shared = Dense(32, activation='relu')(attention_probs)
    output_clsprc = Dense(1, name='clsprc')(shared)
    output_hiprc = Dense(1, name='hiprc')(shared)
    output_loprc = Dense(1, name='loprc')(shared)

    model = Model(inputs=inputs, outputs=[output_clsprc,
        output_hiprc, output_loprc])
    optimizer = Adam(learning_rate=0.005)
    model.compile(optimizer=optimizer,
                  loss='mse',
                  metrics=['mae','mae','mae'])
    return model
model = build_multi_output_model(window_size, len(feature_cols
    ))
model.summary()
```

```
early_stopping = EarlyStopping(monitor='val_loss', patience
    =10, verbose=1, restore_best_weights=True)
reduce_lr = ReduceLROnPlateau(monitor='val_loss', factor=0.5,
    patience=5, verbose=1)

y_train_scaled_dict = {
    'clsprc': y_train_scaled[:, 0],
    'hiprc': y_train_scaled[:, 1],
    'loprc': y_train_scaled[:, 2]
}
y_test_scaled_dict = {
    'clsprc': y_test_scaled[:, 0],
    'hiprc': y_test_scaled[:, 1],
    'loprc': y_test_scaled[:, 2]
}

history = model.fit(X_train_scaled, y_train_scaled_dict,
    epochs=100, batch_size=64,
                    validation_split=0.1, verbose=1,
                    callbacks=[early_stopping, reduce_lr])

y_pred_scaled = model.predict(X_test_scaled)
y_pred_scaled = np.concatenate(y_pred_scaled, axis=1)
y_pred = scaler_y.inverse_transform(y_pred_scaled)

def build_and_train_model(seed):
    import tensorflow as tf
    tf.random.set_seed(seed)
    model_tmp = build_multi_output_model(window_size, len(
        feature_cols))
    model_tmp.fit(X_train_scaled, y_train_scaled_dict, epochs
        =100, batch_size=64,
                  validation_split=0.1, verbose=0,
                  callbacks=[early_stopping, reduce_lr])
    return model_tmp

seeds = [42, 7, 2020]
ensemble_models = [build_and_train_model(s) for s in seeds]
predictions = [model_tmp.predict(X_test_scaled) for model_tmp
    in ensemble_models]

predictions_avg = []
for i in range(3):
    preds = np.mean([pred[i] for pred in predictions], axis=0)
    predictions_avg.append(preds)
predictions_avg = np.concatenate(predictions_avg, axis=1)
y_pred_ensemble = scaler_y.inverse_transform(predictions_avg)

plt.figure(figsize=(12, 10))
titles = ['Clsprc', 'Hiprc', 'Loprc']
for i in range(3):
    plt.subplot(3,1,i+1)
    plt.plot(y_test[:, i], label=f"True {titles[i]}", marker='
        o', linestyle='-')
    plt.plot(y_pred[:, i], label=f"Predicted {titles[i]}",
        marker='x', linestyle='--')
    plt.title(f'{titles[i]}: True vs Predicted')
    plt.xlabel("Sample Index")
    plt.ylabel(titles[i])
    plt.legend()
plt.tight_layout()
plt.savefig("LSTM_prediction_600519.pdf",format = "pdf")
plt.show()

def mape_func(y_true, y_pred):
    return np.mean(np.abs((y_true - y_pred) / (y_true + 1e-8))
        ) * 100
```

```
def smape_func(y_true, y_pred):
    return 100 * np.mean(2 * np.abs(y_true - y_pred) / (np.abs
        (y_true) + np.abs(y_pred) + 1e-8))

titles = ['Clsprc', 'Hiprc', 'Loprc']
metrics_results = {}

for i, name in enumerate(titles):
    mse_val = mean_squared_error(y_test[:, i], y_pred[:, i])
    rmse_val = math.sqrt(mse_val)
    mae_val = mean_absolute_error(y_test[:, i], y_pred[:, i])
    r2_val = r2_score(y_test[:, i], y_pred[:, i])
    mape_val = mape_func(y_test[:, i], y_pred[:, i])
    smape_val = smape_func(y_test[:, i], y_pred[:, i])

    metrics_results[name] = {
        "MSE": mse_val,
        "RMSE": rmse_val,
        "MAE": mae_val,
        "R2": r2_val,
        "MAPE": mape_val,
        "SMAPE": smape_val
    }

sample_idx = 10

x_instance_orig = X_test[sample_idx]
y_true_instance = y_test[sample_idx]
stock_id_instance = stock_ids_test[sample_idx]

x_instance_scaled = X_test_scaled[sample_idx]
x_instance_scaled = np.expand_dims(x_instance_scaled, axis=0)

y_pred_instance_scaled = model.predict(x_instance_scaled)
y_pred_instance_scaled = np.concatenate(y_pred_instance_scaled
    , axis=1)
y_pred_instance = scaler_y.inverse_transform(
    y_pred_instance_scaled)[0]

import numpy as np
import pandas as pd

x_instance_window = x_instance_orig

x_array = x_instance_window.values if hasattr(
    x_instance_window, 'values') else x_instance_window

num_samples = 500
noise = np.random.randn(num_samples, *x_array.shape) * 0.1

neighbors = x_array + x_array * noise
neighbors_df = pd.DataFrame(neighbors.reshape(-1, len(
    feature_cols)), columns=feature_cols)

neighbors_scaled = scaler_X.transform(neighbors_df)

neighbors_scaled_3d = neighbors_scaled.reshape(num_samples,
    30, 12)

neighbors_preds_scaled = model.predict(neighbors_scaled_3d)

neighbors_preds_scaled = np.concatenate(neighbors_preds_scaled
    , axis=1)

neighbors_preds = scaler_y.inverse_transform(
    neighbors_preds_scaled)
```

```
X_local_final = neighbors_df.iloc[neighbors_df.index \% 30 ==
    (30 - 1)].reset_index(drop=True)

y_local_df = pd.DataFrame(y_local, columns=['clsprc', 'hiprc',
    'loprc'])

X_local_final = X_local_final.apply(pd.to_numeric, errors='
    coerce').dropna()
y_local_df = y_local_df.apply(pd.to_numeric, errors='coerce')

models = {}
for col in y_local_df.columns:
    X_local_const = sm.add_constant(X_local_final)
    model = sm.OLS(y_local_df[col], X_local_const).fit()
    models[col] = model
    print(model.summary())

import numpy as np
import pandas as pd
import statsmodels.api as sm
import matplotlib.pyplot as plt
from matplotlib.patches import Ellipse
import matplotlib.transforms as transforms

for target in ['clsprc', 'hiprc', 'loprc']:
    model = models[target]
    conf_int = model.conf_int(alpha=0.05)
    coeff_summary = pd.concat([model.params, conf_int], axis
        =1)
    coeff_summary.columns = ['Coefficient', 'CI_lower', '
        CI_upper']

model_cls = models["clsprc"]
theta_hat = model_cls.params
cov_matrix = model_cls.cov_params()
cov_matrix_df = pd.DataFrame(cov_matrix, columns=model_cls.
    params.index, index=model_cls.params.index)

triples = [
    ("const", "Opnprc", "Hiprc"),
    ("const", "Loprc", "Clsprc"),
    ("const", "Dnshrtrd", "Dnvaltrd"),
    ("Opnprc", "ChangeRatio", "LimitDown"),
    ("Hiprc", "LimitUp", "Dsmvtll"),
    ("Clsprc", "Dsmvosd", "LimitStatus")
]

chi2_val_3d = stats.chi2.ppf(0.95, df=3)

fig = plt.figure(figsize=(18, 12))
for i, triple in enumerate(triples):
    mean_3d = np.array([theta_hat[triple[0]], theta_hat[triple
        [1]], theta_hat[triple[2]]])

    cov_3d = cov_matrix_df.loc[list(triple), list(triple)].
        values

    eigvals, eigvecs = np.linalg.eigh(cov_3d)

    order = eigvals.argsort()[::-1]
    eigvals = eigvals[order]
    eigvecs = eigvecs[:, order]

    radii = np.sqrt(eigvals * chi2_val_3d)

    u = np.linspace(0, 2 * np.pi, 50)
```

```
    v = np.linspace(0, np.pi, 50)
    x = np.outer(np.cos(u), np.sin(v))
    y = np.outer(np.sin(u), np.sin(v))
    z = np.outer(np.ones_like(u), np.cos(v))

    ellipsoid = np.zeros((x.shape[0], x.shape[1], 3))
    for j in range(x.shape[0]):
        for k in range(x.shape[1]):
            point = np.array([x[j, k], y[j, k], z[j, k]])
            ellipsoid[j, k, :] = mean_3d + eigvecs @ (radii *
                point)

local_model = models["clsprc"]

theta_hat = local_model.params

X_local_const = sm.add_constant(X_local_final)

cov_matrix = local_model.cov_params()
cov_mat_np = cov_matrix.values

df_params = len(theta_hat)
chi2_val_full = stats.chi2.ppf(0.95, df=df_params)

var_factor = np.array([np.dot(x, np.dot(cov_mat_np, x)) for x
    in X_local_const.values])

half_width = np.sqrt(chi2_val_full * var_factor)

f_pred = local_model.predict(X_local_const)

eps = 1e-6

def multi_power_model_all(X, *params):
    a = params[:12]
    k = params[12:]
    f_val = np.zeros(X.shape[0])
    for i in range(12):
        base = np.maximum(X[:, i], eps)
        f_val += a[i] * np.power(base, k[i])
    return f_val

def regularized_resid_all(params, X, y, reg_lambda):
    r = multi_power_model_all(X, *params) - y
    penalty = np.sqrt(reg_lambda) * params
    return np.concatenate([r, penalty])

def fit_power_model_multi_start(X_data, y_data, num_starts=10,
    reg_lambda=0.1, max_nfev=8000):
    best_cost = np.inf
    best_res = None
    for i in range(num_starts):
        init_guess = np.concatenate([np.random.uniform(0.5,
            1.5, 12), np.random.uniform(0.5, 1.5, 12)])
        res = least_squares(regularized_resid_all, init_guess,
            args=(X_data, y_data, reg_lambda), max_nfev=
            max_nfev)
        if res.cost < best_cost:
            best_cost = res.cost
            best_res = res
    return best_res

X_local_final = neighbors_df.iloc[neighbors_df.index \% 30 ==
    (30 - 1)].copy()
X_nl = X_local_final.iloc[:, :12].copy()

y_nl = neighbors_preds
```

```
if y_nl.ndim > 1 and y_nl.shape[1] > 1:
    y_nl = y_nl[:, 0]


X_nl = X_nl.apply(pd.to_numeric, errors='coerce').astype(np.
    float64)
X_nl_arr = np.asarray(X_nl.values, dtype=np.float64)
y_nl_arr = np.asarray(y_nl, dtype=np.float64)

num_starts = 20
reg_lambda = 0.1
res_nl = fit_power_model_multi_start(X_nl_arr, y_nl_arr,
    num_starts=num_starts, reg_lambda=reg_lambda)
params_nl = res_nl.x
cost_nl = res_nl.cost
y_pred_nl = multi_power_model_all(X_nl_arr, *params_nl)


r2_nl = r2_score(y_nl_arr, y_pred_nl)
mse_nl = mean_squared_error(y_nl_arr, y_pred_nl)
rmse_nl = np.sqrt(mse_nl)

eps = 1e-6

def multi_power_model_6d(X, *params):
    a = params[:5]
    k = params[5:]
    f_val = np.zeros(X.shape[0])
    for i in range(5):
        base = np.maximum(X[:, i], eps)
        f_val += a[i] * np.power(base, k[i])
    return f_val

def regularized_resid_6d(params, X, y, reg_lambda):
    r = multi_power_model_6d(X, *params) - y
    penalty = np.sqrt(reg_lambda) * params
    return np.concatenate([r, penalty])

def fit_power_model_6d_multi_start(X_data, y_data,
                                   num_starts=10, reg_lambda
                                       =0.1, max_nfev=8000):
    best_cost = np.inf
    best_res = None
    param_dim = 10

    for i in range(num_starts):
        init_guess = np.concatenate([
            np.random.uniform(0.5, 1.5, 5),
            np.random.uniform(0.5, 1.5, 5)
        ])
        res = least_squares(
            regularized_resid_6d,
            init_guess,
            args=(X_data, y_data, reg_lambda),
            max_nfev=max_nfev
        )

        if res.cost < best_cost:
            best_cost = res.cost
            best_res = res

    return best_res

X_local_final = neighbors_df.iloc[neighbors_df.index \% 30 ==
    (30 - 1)].copy()

```

```
selected_cols = ['Opnprc', 'Hiprc', 'Loprc', 'Clsprc', 'Dnshrtrd']
X_nl = X_local_final[selected_cols].copy()

y_nl = neighbors_preds
if y_nl.ndim > 1 and y_nl.shape[1] > 1:
    y_nl = y_nl[:, 0]

X_nl = X_nl.apply(pd.to_numeric, errors='coerce').astype(np.float64)
X_nl_arr = np.asarray(X_nl.values, dtype=np.float64)
y_nl_arr = np.asarray(y_nl, dtype=np.float64)

num_starts = 20
reg_lambda = 0.1
res_nl = fit_power_model_6d_multi_start(
    X_nl_arr, y_nl_arr,
    num_starts=num_starts,
    reg_lambda=reg_lambda
)

params_nl = res_nl.x
cost_nl = res_nl.cost
y_pred_nl = multi_power_model_6d(X_nl_arr, *params_nl)

r2_nl = r2_score(y_nl_arr, y_pred_nl)
mse_nl = mean_squared_error(y_nl_arr, y_pred_nl)
rmse_nl = np.sqrt(mse_nl)

triples_nl = [
    ("a3", "a4", "a5"),
    ("a1", "a3", "k2"),
    ("a2", "a4", "a1"),
    ("a3", "a5", "k4"),
    ("k1", "k3", "k5"),
    ("k4", "a5", "a2")
]

chi2_val_3d_nl = stats.chi2.ppf(0.95, df=3)
fig_nl = plt.figure(figsize=(18, 12))

for i, triple in enumerate(triples_nl):
    mean_3d = np.array([params_nl[param_names.index(t)] for t in triple])
    cov_3d = cov_nl_df.loc[list(triple), list(triple)].values
    eigvals, eigvecs = np.linalg.eigh(cov_3d)
    order = eigvals.argsort()[::-1]
    eigvals, eigvecs = eigvals[order], eigvecs[:, order]
    radii = np.sqrt(eigvals * chi2_val_3d_nl)
    u = np.linspace(0, 2 * np.pi, 50)
    v = np.linspace(0, np.pi, 50)
    x = np.outer(np.cos(u), np.sin(v))
    y = np.outer(np.sin(u), np.sin(v))
    z = np.outer(np.ones_like(u), np.cos(v))

    ellipsoid = np.zeros((x.shape[0], x.shape[1], 3))
    for j in range(x.shape[0]):
        for k in range(x.shape[1]):
            point = np.array([x[j, k], y[j, k], z[j, k]])
            ellipsoid[j, k, :] = mean_3d + eigvecs @ (radii * point)

J = res_nl.jac
n = X_nl_arr.shape[0]
p = len(params_nl)
sigma2 = (2 * res_nl.cost) / (n - p)
cov_nl = sigma2 * np.linalg.inv(J.T @ J)
```

```
print("Estimated covariance matrix shape:", cov_nl.shape)

def gradient_multi_power_model_6d(x, params):
    a1, a2, a3, a4, a5, k1, k2, k3, k4, k5 = params
    grad = np.zeros(10)
    grad[0] = x[0] ** k1
    grad[1] = x[1] ** k2
    grad[2] = x[2] ** k3
    grad[3] = x[3] ** k4
    grad[4] = x[4] ** k5
    grad[5] = a1 * (x[0] ** k1) * np.log(x[0] + 1e-8)
    grad[6] = a2 * (x[1] ** k2) * np.log(x[1] + 1e-8)
    grad[7] = a3 * (x[2] ** k3) * np.log(x[2] + 1e-8)
    grad[8] = a4 * (x[3] ** k4) * np.log(x[3] + 1e-8)
    grad[9] = a5 * (x[4] ** k5) * np.log(x[4] + 1e-8)
    return grad

f_pred_nl = np.array([multi_power_model_6d(x.reshape(1, -1), *
    params_nl)[0] for x in X_nl_arr])

gradients = np.array([gradient_multi_power_model_6d(x,
    params_nl) for x in X_nl_arr])

var_factors_nl = np.array([grad.T @ cov_nl @ grad for grad in
    gradients])

chi2_val_full_nl = 22.362
half_width_nl = np.sqrt(chi2_val_full_nl * var_factors_nl)

eps = 1e-6
def multi_poly_model_5d(X, *params):
    n_samples = X.shape[0]
    f_val = np.zeros(n_samples)
    for i in range(12):
        c0 = params[3*i]
        c1 = params[3*i + 1]
        c2 = params[3*i + 2]
        f_val += c0 + c1 * X[:, i] + c2 * (X[:, i]**2)
    return f_val

def regularized_resid_poly(params, X, y, reg_lambda):
    r = multi_poly_model_5d(X, *params) - y
    penalty = np.sqrt(reg_lambda) * params
    return np.concatenate([r, penalty])

def fit_poly_model_multi_start(X_data, y_data, num_starts=10,
    reg_lambda=0.1, max_nfev=8000):
    best_cost = np.inf
    best_res = None
    param_dim = 36
    for i in range(num_starts):
        init_guess = np.random.uniform(-1, 1, param_dim)
        res = least_squares(
            regularized_resid_poly,
            init_guess,
            args=(X_data, y_data, reg_lambda),
            max_nfev=max_nfev
        )
        if res.cost < best_cost:
            best_cost = res.cost
            best_res = res
    return best_res

X_local_final = neighbors_df.iloc[neighbors_df.index \% 30 ==
    (30 - 1)].copy()

```

```
selected_cols = ['Opnprc', 'Hiprc', 'Loprc', 'Clsprc', '
    Dnvaltrd', 'Dsmvosd', 'ChangeRatio', 'Dnshrtrd', 'Dsmvtll'
    , 'LimitDown', 'LimitUp', 'LimitStatus']
X_nl = X_local_final[selected_cols].copy()

y_nl = neighbors_preds
if y_nl.ndim > 1 and y_nl.shape[1] > 1:
    y_nl = y_nl[:, 0]

X_nl = X_nl.apply(pd.to_numeric, errors='coerce').astype(np.
    float64)
X_nl_arr = np.asarray(X_nl.values, dtype=np.float64)
y_nl_arr = np.asarray(y_nl, dtype=np.float64)

num_starts = 20
reg_lambda = 0.1
res_poly = fit_poly_model_multi_start(X_nl_arr, y_nl_arr,
                                      num_starts=num_starts,
                                      reg_lambda=reg_lambda)
params_poly = res_poly.x
cost_poly = res_poly.cost
y_pred_poly = multi_poly_model_5d(X_nl_arr, *params_poly)

r2_poly = r2_score(y_nl_arr, y_pred_poly)
mse_poly = mean_squared_error(y_nl_arr, y_pred_poly)
rmse_poly = np.sqrt(mse_poly)
```

Index

A. Hajiyev and J. Xu, *High-Dimensional Regression Analysis and Artificial Intelligence*,
https://doi.org/10.1007/978-981-95-2514-0

MIX
Papier aus verantwortungsvollen Quellen
Paper from responsible sources
FSC® C105338

If you have any concerns about our products,
you can contact us on
ProductSafety@springernature.com

In case Publisher is established outside the EU,
the EU authorized representative is:
Springer Nature Customer Service Center GmbH
Europaplatz 3, 69115 Heidelberg, Germany

Printed by Libri Plureos GmbH
in Hamburg, Germany